Water Supply and Sanitation

Water Supply and Sanitation

Athar Hussain
Professor & Head
Civil Engineering and Geoinformatics Department
Netaji Subhas University of Technology (West Campus)
(Formerly CBPGEC Jaffarpur)
New Delhi

Alpana Gupta
Formerly as a BIRAC Innovator
IIT, Delhi

I.K. International Pvt. Ltd.
NEW DELHI

Published by
I.K. International Pvt. Ltd.
4435-36/7, Ansari Road, Daryaganj
New Delhi-110 002 (India)
E-mail: info@ikinternational.com
Website: www.ikbooks.com

ISBN 978-93-90620-77-7

Published by Krishan Makhijani for I.K. International Pvt. Ltd., 4435-36/7, Ansari Road, Daryaganj, New Delhi-110 002.

Preface

Water scarcity and sanitation are two major challenges that affect over 40% of the global population and it is predicted that this percentage could increase due, in part, to global warming and desertification. Human beings are the main cause of water pollution. More than 80% of the wastewater resulting from human activities is dumped in rivers or the sea without any kind of water treatment, producing contamination.

To respond to this crucial need, appropriate technologies pertaining to water treatment and supply are discussed in this book. A unique feature of the book is the way in which chapters interact. Cross-references between theory and application foster overall integration of the subject matter. This volume aims to assist process engineers, water treatment plant operators and managers, environmental consultants, water treatment plant operators, UG and PG students of civil, environmental, and chemical engineering.

Chapter 1 provides an overview on the assessment of various types of water demands and population forecasting methods. Chapters 2 provides the detail design of water intake structures for surface and groundwater sources. Chapter 3 addresses the various parameters that pertain to the quality and characteristics of water used for public supplies along with standard limits. Chapters 4 and 5 provide detailed design of various methods used in water treatment systems and is intended for the scientists and engineers to learn about the design aspects of each water treatment unit. Chapter 6 deals with water disinfection processes used after the treatment and before its supply to the city. Chapter 7 provides the detail design of various types of water supply and distribution system intended for a city. Chapters 8 and 9 include water transmission pipes, pumps and pumping stations and appurtenances.

The book will be of interest to a broad audience of analytical chemists, environmental chemists, water management operators and technologists working in the field of wastewater treatment, and newcomers who want to learn more about the topic.

However, the recommendations of various Indian Standards pertaining to the subject including Water Manual by Central Public Health and Environmental

Engineering under Ministry of Urban Development has been incorporated and followed. All the chapters have been very well organised with clear definitions, design principles and theories. Further suggestions for the improvement of this book is highly encouraged.

Athar Hussain
Alpana Gupta

Contents

1

CHAPTER

Water Demands

1.1 INTRODUCTION

There are many aspects in which humans affect the environment. Popular effects include reduced water quality, increased pollution and greenhouse gas emissions, natural resource degradation and contribution to global climate change. Some of these are the direct product of human behaviour, while others are side effects which are part of a sequence of actions and reactions. This chapter deals with requirements of water for domestic use, industrial use, institutional use and the factors controlling the per capita demand. A comprehensive study is also done on population forecasting methods and design periods.

1.2 WATER POLLUTION

Probably the most vital example of adverse impact on environment due to humans is water contamination. We need water to live obviously but a few people know how much we need and how much is available. Consider these statistics from the United Nations Environment Programme, just 2.5% of which is freshwater and 2.5%, less than 1% of which is available to everyone. Citizens need up to 50 L (13 gallons) of fresh water a day for drinking, cooking and cleaning. This does not take into account the endless gallons of water needed to raise animals for food or treatment, and 70% of all freshwater is used for irrigation. About 2% of farms are organic, according to the Organic Farming Research Foundation. This is equal to about 69% of our freshwater supply being polluted with industrial pesticides, herbicides, and artificial fertilizers, but also combined with fossil fuels and heavy agricultural machinery pollution. The acid rain adds to these chemical compounds.

Since very little can survive in an acidic environment, acid rain has adverse effects on plants, animals and marine life, as well as on humans and even on houses, sculptures or other objects. Acid rain also contaminates our dwindling supply of freshwater, and so the water contamination process begins. According to the U.S. Environmental Protection Agency, 45% of assessed stream miles, 47% of assessed lake acres, and 32%

of assessed bay and estuarine square miles were not clean enough to support uses such as swimming or fishing. The following reasons and possible sources for this include: sediments, pathogens and habitat alterations from agricultural activity and hydrologic modifications, Excessive nutrients, metals and organic enrichment from agricultural activity and atmospheric deposition that leads the movement of pollutants from one environment to another, such as from water to air. The unaccounted industrial and municipal discharges lead to increase of heavy metals concentration in surface and groundwater bodies.

1.3 LAND POLLUTION

Land pollution, resulted in the degradation of the Earth's surfaces and soil, is caused by human activity and a misuse of natural resources. Causes of land pollution and degradation include:

- **Urban sprawl:** Natural ecosystems are destroyed, usually through inefficient or reckless planning, to make space for development. Urban sprawl usually contributes to unexploited land waste (such as unnecessary highways, decorative and unused areas, etc.).
- **Poor agricultural practices:** CAFO (Confined Animal Feeding Operations) manure runoff, the use of chemical fertilizers, herbicides and pesticides, the production of monocultures (only one crop season after season) and the deforestation needed to increase farmland all lead to degradation and pollution.
- **Personal consumption:** The need of our consumer society to have more, greater and better "things," as well as our connection to and use of waste, has resulted in land stripping, unsustainable mining and pollution from industrial activities.
- **Industrial activities:** Production of chemically treated products, low product quality, corrupt practices (such as illegal dumping), and significant pollution impact both local and far-reaching regions.

None of this takes into account industrial dumping, reducing landfill space, garbage, over-production of manufactured products, radioactive waste and more. All land contamination should thus be avoided by our actions. Two of the necessary actions are proper planning and proper usage of natural resources. For example, livestock may be taken out of CAFO and permitted to graze on mountainous or wooded areas that are unsuitable for buildings or crops. Using organic and sustainable farming techniques can eliminate our need for chemical uses. Clean energy, such as wind or solar power, will gradually start replacing coal or nuclear power stations. And as consumers, by seeking higher quality goods, environmental ethical practices from companies and a move toward renewable energy, we will lessen our human effect on the environment.

1.4 AIR POLLUTION

As per the U.S. Environmental Protection Agency, from 1990 to 2008: 14% decrease in ozone, 78% decrease in mercury, 35% decrease in nitrogen dioxide, 68% decrease in carbon monoxide and 59% decrease in sulfur dioxide. And some of the quality standards, including ozone, are far from ideal. The question is: what amount of air pollution is really okay? Shouldn't we be striving for technologies or practices that virtually eliminate major polluters, such as: better public transit, electric cars, rail, and city planning, etc., to replace heavy commutes, fossil fuels, air travel and un-walkable communities, retrofitting or otherwise upgrading the existing factories with greener technologies such as solar panels or air filtration systems. The largest human effect that we can have on the world is voting with our pocket book and our actions. Through encouraging companies using greener technology, we will begin a movement towards sustainable solutions. We can have a global influence at a local level by participating in urban planning and zoning. And by buying less "stuffs," consuming less animal products and sourcing our food from sustainable farms, we can spread the word that there are better options and they don't have to cater for mega-corporations and lobby pressure.

Some important impacts of human activities on environment are outlined below:

- **Deforestation:** To provide an growing population with timber and farmland, a significant number of forest trees are cut down, and the forest area is converted to farmland. The rate of deforestation is so rapid that India alone loses about 1.5 million hectares of forest cover each year. The deforestation cycle results in decreasing rainfall, rising global temperature, depletion of top soil, altering climatic conditions, and so on.
- **Industrialization:** While man's industrial activities provide a basic need for society, the same releases a lot of contaminants to the atmosphere at the same time. Environmental contaminants cause loss of raw materials, health hazards, increase in death rates, crop damage, render the environment unfit for living organisms, etc.
- **Loss of ecological balance**: The inappropriate use, misuse and mismanagement of biosphere resources contribute to ecosystem disruptions or environmental imbalances.
- **Air pollution**: The anthropogenic release of various air contaminants into the atmosphere is causing a variety of feared phenomena such as green house effect, depletion of ozone layers, acid rain, smog formation, etc.
- **Water pollution**: Human actions related to hazardous waste collection, solid waste, urban waste, agricultural and industrial wastes are causing the environment to be unfit for day-to-day use. In addition, the contaminated water spreads or contributes to various diseases.

- **Increased consumption of natural resources***:* Since the beginning of the industrial period, natural resources have been continuously used to produce one or more goods for day-to-day societal use.
- **Production of waste***:* Rapid industrialization and unplanned urbanization release a lot of solid or liquid or gaseous toxic waste materials which cause a number of serious environmental hazards.
- **Extinction of wildlife***:* Because forests are natural habitats of wildlife (both plants and animals), deforestation leads to important wildlife extinction and biodiversity loss.
- **Habitual destruction***:* Commercial and industrial practices associated with mining, dam building, fishing, farming, etc., cause habitat loss that is a gateway to pollution.
- **Noise pollution***:* The man-made noise produced by mechanized vehicles, factories, ships, airplanes, social activities, etc., creates noise pollution that affects both biotic and biotic environmental components.
- **Radiation pollution***:* The radiation from radioactive substances used in nuclear reactors and nuclear weapons will significantly affect body genetic materials (DMA, RHA).
- **Soil erosion***:* Anthropogenic processes such as deforestation and overgrazing cause soil erosion causing reduction of soil moisture, reduction of productivity, decrease in soil fertility, etc.

1.5 NEED FOR PROTECTED WATER SUPPLY

Protected water source means water source that is treated to remove impurities and keeps public health safe. Both physical and bacterial agents can pollute water. Water is a healthy carrier of germ-causing diseases too. The causes of disease outbreak are linked to polluting water and inadequate sanitation hospitals continue to be overwhelmed with the sick due to lack of awareness about health.

However, improvement in public health safety over the past few decades has been targeted by ensuring clean water and sanitation to all residents in developing countries. In 1977, a campaign known as "HEALTH FOR ALL BY THE YEAR 2000 A.D." was announced by the united nations. The working group appointed by the planning commission when proposing methods for achieving the above-mentioned target stressed that potable water should be made available to the entire population from safe water supply. Supply of uncontaminated water to the community alone will reduce the morbidity levels. The complete outline of water supply scheme is as shown in Fig. 1.1.

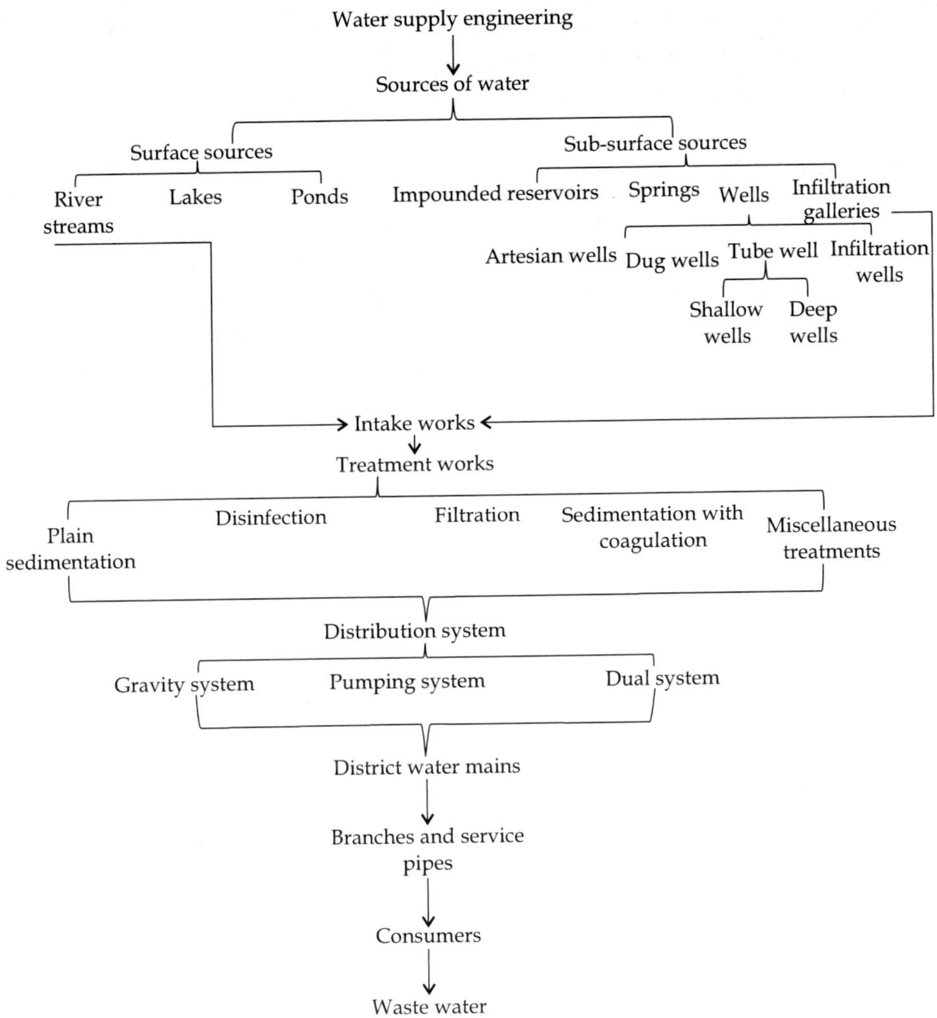

Fig. 1.1: Water supply engineering (Rao, 2005)

The objectives of the community water supply system are:

- To provide better quality and wholesome water to the customers for drinking purposes.
- To provide sufficient quantity of water to satisfy at least the individuals' minimum needs and to provide appropriate arrangements for emergencies such as fire fighting, festivals, etc.
- To compensate for potential demands due to population growth, living standards, storage and transportation.
- Collection and transportation of good quality water to avoid water contamination at source.

- Maintaining the treatment units and delivery system in good condition with sufficient staff and materials.
- To build and operate an efficient and effective operating system.

1.6 WATER DEMANDS

When planning a water supply scheme for a town or area, it is important to decide the total quantity of water provided by the area for various purposes. As a matter of fact the engineer's first task is to assess the town's water demand and then find appropriate sources of water to satisfy the demand. But as there are so many factors involved in demand of water, it is not possible to accurately determine the actual demand. In assessing the water requirement, which is very close to actual requirement, some analytical formulae and thumb rules are employed. The various types of water demands of a city or town follow:

- Domestic water demand
- Industrial demand
- Institution and commercial demand
- Demand for public use
- Fire demand
- Loses and waste

1.6.1 Domestic Water Demand

The quantity of water required for drinking, bathing, cooking, washing, etc., in the houses is called domestic water demand and depends mainly on people's habits, social status, climatic conditions and customs. As per IS: 1172-1963, the domestic water consumption in India is about 135 L/capita/day under normal conditions.

But in developing countries, because of the use of air coolers, air conditioners, lawn care, automatic household appliances, this figure could be 350 L/capita/day.

Table 1.1: Domestic water consumption in India (L/c/d)

S. No.	Purposes	Water consumption (L/c/d)
1.	Drinking	5
2.	Cooking	5
3.	Bathing	55
4.	Clothes washing	20
5.	Utensils washing	10
6.	House washing	10
7.	Flushing & W.C.	30
	Total	**135**

1.6.2 Industrial Demand

The water needed in the industries depends primarily on the type of industries that exist in the town. Industrial use is made of the water provided by factories, paper mills, cloth mills, cotton mills, breweries, sugar refineries, etc. The quantity of industrial water demand is around 20 to 25% of the city's overall demand.

1.6.3 Institution and Commercial Demand

This category includes colleges, universities, commercial buildings and shopping centres including office buildings, factories, supermarkets, hotels, shopping centre's, health centre's, hospitals, churches, cinema houses, railway and bus stations etc. As per IS: 1172-1963, the criteria for water supply to public buildings other than dwellings is shown in Table 1.2.

Table 1.2: Water demand for public buildings

S. No.	Building type	Water requirement (L/c/d)
1.	Factories where bathrooms are required to be provided.	45
2.	Factories where no bathrooms are required to be provided	30
	Hospitals per bed	
3.	No. of beds not exceeding 100	340
4.	No. of beds exceeding 100. Nurses homes	450
5.	Hostels	135
6.	Offices	45
7.	Restaurants (per seat)	70
8.	Hotel (per bed)	180
9.	Cinema concert halls and theatres(per seat)	15
	Schools	
10.	Day schools	45
11.	Boarding schools	135
12.	Garden, sports grounds	35 per sqm.
13.	Animal/vehicles	45

Demand for public use: Amount of water needed for public service purposes, such as washing and sprinkling on roads, sewer cleaning, watering of public parks, gardens, public fountains, etc., is publicly pursued. To meet the water demand for public use, the construction of the water works for a city is made up of 5% of total use. The water specifications for public use are taken as shown in Table 1.3.

Table 1.3: Water demand for public use

S. No.	Purpose	Water requirements
1.	Public parks	1.4 L/m^2/d
2.	Street washing	1.0-1.5 L/m^2/d
3.	Sewer cleaning	4.5 L/head/d

1.6.4 Fire Demand

Fire may occur due to defective electrical wires by short circuiting, fire catching objects, fires, criminal ill-intention or some other unexpected mishaps. If fires are not adequately managed and extinguished in the least possible time, they cause significant damage and can result in burns. All major cities have fire-fighting squads. As large quantities of water are needed to extinguish the fire. Provision is made in the water work to supply ample quantity of water or to hold it as a reserve in the water supply for this reason. In the cities there are fire hydrants on the water supply at a distance of 100 to 150 m. The quantity of water required for fire-fighting is generally calculated by using different empirical formulae (Eq. 1.1-1.4). For Indian conditions Kuichling formula gives satisfactory results.

$$Q = 3182\sqrt{P} \tag{1.1}$$

where,

Q = Quantity of water required in L/min and

P = Population of town or city in thousands

Buston's formula:

$$Q = 5663\sqrt{p}(1.2) \tag{1.2}$$

Freeman's formula:

$$Q = 1136\left(\frac{P}{5}+10\right) \tag{1.3}$$

and

$$F = 2.8\sqrt{P} \tag{1.4}$$

where,

F = number of simultaneous fire streams

$$\text{Q} = 4637\sqrt{p(1-0.01\sqrt{p})} \tag{1.5}$$

Since the overall fire extinguishing water demand is typically very low, the consumption rate is very high. It depends on buildings with size, congestion and fire resistance. The minimum fire demand cap is the volume and supply rate that will extinguish the city's largest propitious fire. In all street corners and at appropriate intermediate points, fire hydrants of 15 to 20 cm diameter are usually given. These are normally connected to mains water supply. Pumps mounted on fire brigade vehicles are rushed to the site when fire happens and connected to fire hydrants from where they throw water jet under very high pressure. Pressure varies from 1 to 2 kg/cm^2. For one fire three streams each of 1100 L/min are required for a moderate fire. The fire flow provided by National Board of Fire Underwriters is 40500 L/min for a city of one lakh population. Assuming an average total flow rate of 150 L/day. Then the flow rate needed for the fire demand is much higher than the total flow rate necessary for the purposes. Nevertheless, the fire demand requirement is made only for fire flows of 3 to 5 hours. The overall amount of water measured on an annual basis is typically very limited, since fires only break out a few times a year.

1.7 LOSSES AND WASTES

The entire water that goes into the pipes delivery does not hit the consumers. For which following are the reasons:

- Losses due to defective joints in the pump, fractured and damaged tubing, defective valves and fittings.
- Because of losses, customers keep their taps or public taps open even though they don't use the water and allow water to be wasted continuously.
- Damages attributable to illegal and illegitimate connections while calculating a city's total water quantity; allocation of 15% of the total water quantity is made to account for damages, robberies and water wastage.

1.8 PER CAPITA DEMAND

If "Q" is the total quantity of water required for various purposes by a town per year and "P" is population of town, then per capita demand will be as shown in Eq. 1.5;

$$\text{Per capita demand} = Q/(P \times 365) \quad \text{L/d} \tag{1.6}$$

The city's per capita demand depends on various factors such as living standard, number of persons, form of commercial sites in a city, etc. For an average Indian town, the requirement of water in various uses is listed in Table 1.4.

Table 1.4: Per capita demand of water for an average Indian town

S. No.	Purposes of water	Per capita demand (L/c/d)
1.	Domestic purpose	135
2.	Industrial use	40
3.	Public use	25
4.	Fire demand	15
5.	Losses, wastage and thefts	55
	Total	**270**

The total quantity of water required by a town per day shall be 270 multiplied with the total population in L/day.

1.8.1 Factors Affecting Per Capita Demand

The following are the main factors that affect per capita demand of a city or town.

- **Climatic conditions:** The amount of water available in hot and dry areas is more than cold countries due to the use of air coolers, air conditioners, water sprinkling in lawns, gardens, courtyards, room washing, more clothes washing and bathing etc. But often the quantity of water required in very cold countries might be more due to waste, since in these places people often keep their taps open and water flows continuously for fear of water freezing in the taps and use hot water to keep the rooms warm.
- **Size of community:** Water demand is more likely to increase with the size of the city as more water is needed for street washing, sewage running, park and gardens maintenance.
- **Living standard of the people:** The town's per capita demand rises with people's living conditions due to the use of air conditioners, room coolers, lawn care, flush, latrines and automatic home appliances, etc.
- **Industrial and commercial activities:** Since the quantity of water needed in some industries is much greater than domestic demand, the city's per capita demand would significantly increase its presence in the city. As a matter of fact the water that the factories provide has no direct link with the town's population.
- **Pressure in the distribution system:** The rate of water consumption raises the building's pressure and even with the necessary pressure at the farthest point the water consumption will increase automatically. This rise in quantity is largely attributed to people's free use of water compared to barely having it and more water loss attributed to leakage, wastage, theft, etc.
- **System of sanitation:** Cities with water carriage network per capita demand would be more than the city where this network is not being used.
- **Cost of water:** The water cost directly affects its demand. If the cost of water is rising, the citizens can use less water than when the cost is small.

1.9 POPULATION FORECASTING METHODS AND PROBLEMS

When the design duration is set the next step is to assess a city's population based on factors such as births, deaths, relocation, and annexation. The city's future development depends mainly on trade expansion, construction industries, and surrounding area, mining discoveries, railway station building, etc. will generate rapid rise, sluggish growth, stagnant conditions, or even population decline. It is best to research the development of other similar towns, which have formed under the same circumstances, for population prediction, since the growth of the expected town would be more or less on the same lines. The following are the standard methods by which the forecasting population is done.

- Arithmetical increase method
- Geometrical increase method
- Incremental increase method
- Decreasing rate method
- Simple graphical method
- Comparative graphical method
- Master plan method
- The logistic curve method
- The apportionment method

1.9.1 Arithmetical Increase Method

This method is based on the assumption that the population is increasing at a constant rate. The rate of change of population with time is constant. The population after n decades can be determined by the Eq. 1.7;

$$P_n = P_o + nx \tag{1.7}$$

where,

P_o = Population at present;

n = No. of decades

x = Average increase in population

1.9.2 Geometrical Increase Method

The average percentage of last few decades' growth is estimated in this method; the population forecast is made on the basis that percentage increase per decade will be the same. The population at the end of an n decade is calculated using Eq. 1.8 followed by the rate of growth of population by arithmetic mean method (Eq. 1.9) and geometric mean method (Eq. 1.10).

$$P_n = P_o + (1 + r)^n \tag{1.8}$$

where,

P_o = Population at present

n = Number of decades

r = Rate of growth

Rate of Growth of Population (r)

1. **Arithmetic mean method**

$$r = r_1 + r_2 + r_3 + \ldots / n \tag{1.9}$$

2. **Geometric mean method**

$$r = \sqrt[n]{r_1, r_2, r_3 \ldots\ldots r_n} \tag{1.10}$$

1.9.3 Incremental Increase Method

This approach is an improvement over the two methods set out above. The annual population growth is calculated by the mathematical process, to which is applied once for each future decade the sum of the net incremental increase as shown in Eq. 1.11.

$$P_n = P_o + nx + \frac{n(n+1)x}{2} \tag{1.11}$$

where,

x = Average increase in population

y = Average incremental increase

Note: Arithmetic increase method gives the lowest value of population forecasted in comparison to geometric and incremental increase methods. Hence, it is generally adopted for old cities. Geometric increase method gives the highest value of population forecasted in comparison to arithmetic and incremental increase method. Hence, it is generally adopted for young cities.

Example 1.1: The population of five decades has been given below:

Year	Population
1940	7000
1950	11,000
1960	16,000
1970	21,500

Calculate the probable population in the year 1980, 1990 and 2000 by the following methods:

(i) Arithmetical increase method

(ii) Geometrical increase method

(iii) Incremental increase method

Solution:

Year	Population	Increase in population	Increment increase	Percentage increase in population
1940	7000	–	–	–
1950	11,000	4000	–	$\frac{4000}{7000} \times 100 = 57.1\%$
1960	16,000	5000	1000	$\frac{5000}{11,000} \times 100 = 45.4\%$
1970	21,500	5500	500	$\frac{5500}{16,000} \times 100 = 34.3\%$
Total		**14,500**	**1500**	**136.8**
Average		**4833**	**750**	**45.60**

(i) Arithmetical increase method

Year	Population
1980	21,500 + 1 × 4833 = 26,333
1990	26,333 + 1 × 4833 = 31,166
2000	31,166 + 1 × 4833 = 35,999

(ii) Geometrical increase method

1980	$21{,}500+\frac{45.60\times 21{,}500}{100}$ = 31,304
1990	$31{,}304+\frac{45.60\times 31{,}304}{100}$ = 45,578
2000	$45{,}578+\frac{45.60\times 45{,}578}{100}$ = 66,361

(iii) The population at the end of various decades by incremental increase method shall be as follows:

In the year 1980, P_{1980} = 21,500 + 1 × (4833 + 750)

P_{1980} = 27,083

P_{1990} =27,083 + (4833 + 750) × 1

P_{1990} = 32,666

and P_{2000} = 32,666 + (4833 + 750)

P_{2000} = 38,249

1.9.4 Decrease Rate of Growth Method or Decreasing Rate Method

If a very old city's complete growth is plotted, the curve can be shown to have S-shape, which means that early growth occurs at an growing rate, the latter growth at a decreasing rate, which implies that saturation limit is reached. In this process the average decrease in percentage increase is worked out and subtracted for each successive decade from the current percentage increase.

Example 1.2: The populations of five decades from 1940 to 1980 are given below. Find out the population in decades 1990, 2000 and 2010 by using decrease rate of growth method.

Year	Population
1930	25,000
1940	28,000
1950	32,500
1960	40,000
1970	45,000

Solution:

Year	Population	Increase in population	Percentage increase in population	Decrease in percentage increase
1930	25,000			
1940	28,000	3000	$\frac{3000}{25,000}\times 100 = 12.0\%$	
1950	32,500	4500	$\frac{4500}{28,000}\times 100 = 16.1\%$	(–4.1%)
1960	40,000	7500	$\frac{7500}{32,500}\times 100 = 23.1\%$	(–7.0%)
1970	45,000	5000	$\frac{5000}{40,000}\times 100 = 12.5\%$	(+10.6%)
Total		**20,000**		(–0.5%)
Average		**5000**		(–0.17%)

Now the population at the end of decade shall be as follows:

Year	Net present increase in population	Population
1980	12.50 – (–0.17) = 12.67	$4500+\frac{12.67\times 4500}{100} = 50,702$
1990	12.67 – (–0.17) = 12.84	$50,702+\frac{12.84\times 50,702}{100} = 57,212$
2000	12.84 – (–0.17) = 13.01	$57,212+\frac{13.01\times 57,212}{100} = 64,654$

1.9.5 Simple Graphical Method

In this approach the populations of the last few decades are correctly plotted on the graph with respect to the decade at a suitable rate. The curve is extended smoothly to estimate future population. The graph of the present city is plotted from the outset, and shows the growth curve. The following data have been noted from the census department:

Year	Population
1940	7000
1950	11,000
1960	16,000
1970	21,500

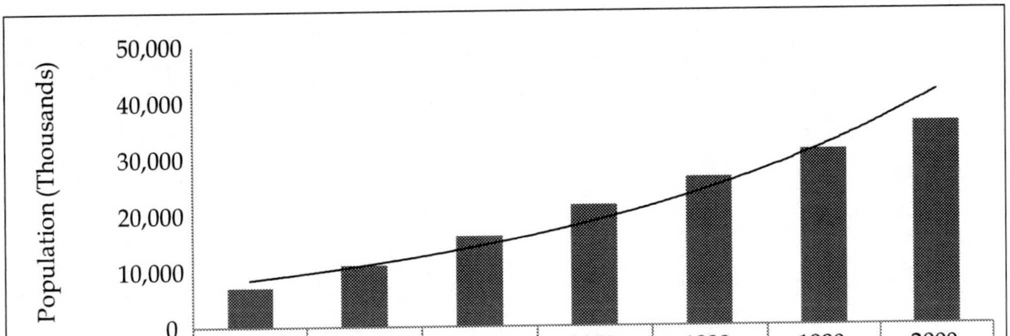

	1940	1950	1960	1970	1980	1990	2000
■ Population	7000	11,000	16,000	21,500	26,333	31,166	35,999

Fig. 1.2: Typical population growth curve of a city

After extending the curve and determining the forecasts of population at various decades, following results are noted on the graph of Fig. 1.2.

Year	Population
1980	26,333
1990	31,166
2000	35,999

1.9.6 Comparative Graphical Method

First of all, the cities having similar conditions and characteristics to the city whose future population is to be projected are selected in this process. The city under consideration is then presumed to grow as the selected related towns have grown in the past. This approach has a rational context, and if development statistics of similar cities are available very accurate and reliable results can be obtained. Consider Fig. 1.2, let the population of a city *A* be given for 4 decades (say 1940, 1950, 1960 and 1970) then the population time curve is plotted. Now, suppose the population of city *A* should be projected at the end of the year 2010. And let the available data show that in the year 1970, this city *A* reached the present population of 42,500. Then the data available for specific cities is analyzed in Band *C*. Let it be noticed that in the year 1940, city *B* exceeded 42,500 then its curve is plotted beyond the year 1940.

This curve for city *B*, however, should start from point *P* which represents the current population of city *A*. Likewise, from the year it exceeded 42,500 onwards the population of city *C* is plotted. Now the curve for city *A* is balanced carefully between the curves of cities *B* and *C* as shown in Fig. 1.3.

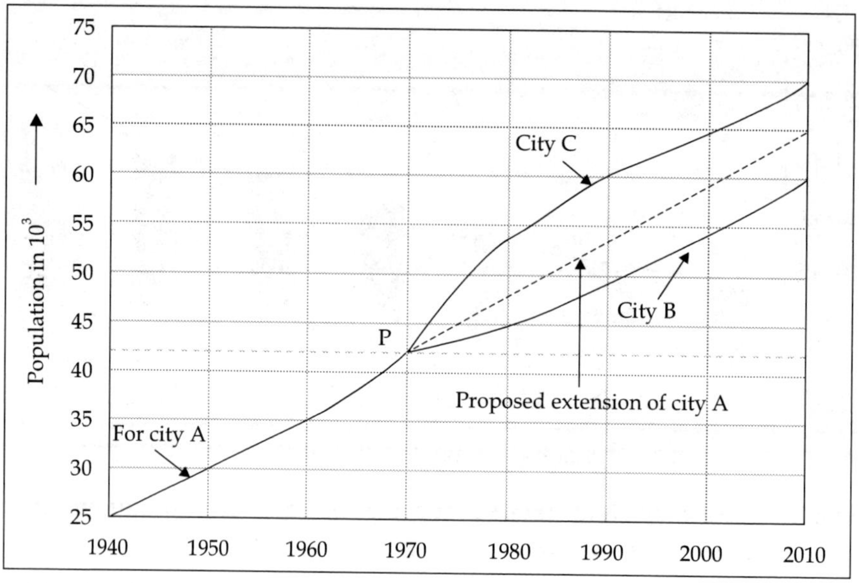

Fig. 1.3: Comparative graphical method for population forecasting of a city

Figure 1.2 shows the development of cities *A*, *B*, *C* and *D*. The development of the city '*X*' under consideration has been shown by dotted line. Future expected population in any decade can be directly determined from this graph.

1.9.7 The Master Plan Method or Zoning Method

Towns and cities are not allowed to evolve in a haphazard way. Their master plans are planned for the growth of the towns and cities. The town is divided into different sections, such as shopping centres, industrial areas, residential areas, hospitals, colleges, parks, etc. According to the master plan, the potential growth of the cities is strictly governed by various corporate and other local laws. The master plans are planned for 25-30 years of town growth. The densities of population for different areas of the towns to be built are also set as shown in Table 1.5.

Table 1.5: Common population densities

S. No.	Type of area	Population density (number of persons per hectare)
1.	Residential areas	
	• Large lots of single family residences	10-40
	• Small lots of single family residences	40-90
	• Multiple family quarters	90-250
	• Tenement houses	250-3500
2.	Commercial areas	40-75
3.	Industrial areas	10-40
4.	Parks, playgrounds, etc.	25-125

1.9.8 Logistic Method

If a town's population is plotted regarding time, the curve thus obtained under normal conditions shall be as shown in Fig. 1.4. Curve *JK* demonstrates the town's early development at a growing rate of $dP/dt \propto P$. The rate of growth between points *K* to *M* follows the dP/dt = constant curve. The transitional curve, *KM*, also passes through the inflection point *L*. Later on the growth from *M* to *N* follows the decreasing trend i.e., $dP/dt \propto (P_s - P)$ where, *p* is town population at point *t* from the origin *j* and P_s is population saturation value. The *JKLMN* curve in S-shaped form is called the logistic curve as indicated in Fig. 1.4.

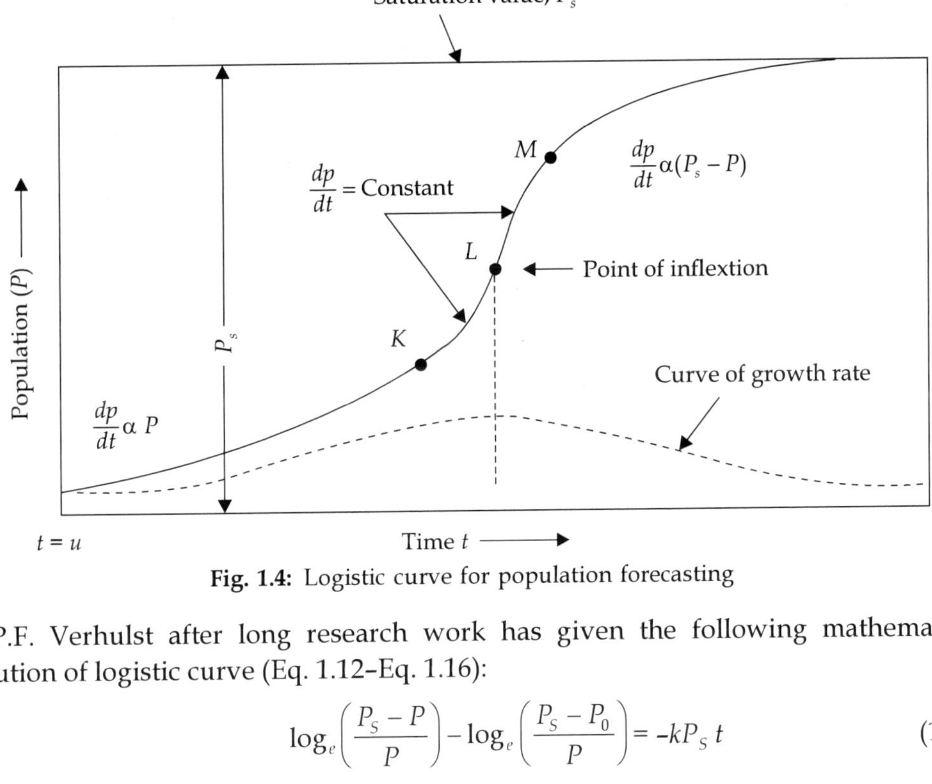

Fig. 1.4: Logistic curve for population forecasting

P.F. Verhulst after long research work has given the following mathematical solution of logistic curve (Eq. 1.12–Eq. 1.16):

$$\log_e\left(\frac{P_S - P}{P}\right) - \log_e\left(\frac{P_S - P_0}{P}\right) = -kP_S\, t \tag{1.12}$$

where;

P_0 = The population of the town at point *j*

P_S = Saturation population

P = Population at time *t* from the origin *j*

K = Constant

Now solving equation for *P*,

$$P = \frac{P_S}{1 + \frac{P_S - P_0}{P_0}\log_e^{-1} - kP_S\, t} \tag{1.13}$$

Substituting $\frac{P_S - P_0}{P_0}$ = m (a constant) and $-kP_S = n$ (another constant), the formula will be:

$$P = \frac{P_S}{1 + m\log_e^{-1}(n.t)} \tag{1.14}$$

This is the required equation of logistic curve.

McLean further suggested that if three pairs of the characteristic values P_0, P_1 and P_2 at times t_0, t_1 and t_2 which are extending over the useful range of the population are so chosen that $t_0 = 0$, t_1, and $t_2 = 2t_1$ the saturation values P_S and constants m and n can be determined from the following equation:

$$P_S = \frac{2P_0\, P_1\, P_2 - P_1^2\,(P_0 + P_2)}{P_0\, P_2 - P_1} \tag{1.15}$$

$$\frac{P_S - P_0}{P_0} = m \text{ (a constant)}$$

$$n = \frac{2.3}{t_1}\log_{10}\left[\frac{P_0(P_S - P_1)}{P_1(P_S - P_0)}\right] \tag{1.16}$$

Example 1.3: Following is the population of a city as noted from the census department

Year	Population
1950	35,000
1960	78,000
1970	1,15,000

Determine (a) the saturation population, and (b) the expected population in 1980 and the equation of logistic curve.

Solution: For logistic curve here,

$P_0 = 35{,}000;\ t_0 = 0$

$P_1 = 78{,}000\ ;\ t_0 = 10$ years

$P_2 = 1{,}15{,}000\ ;\ t_0 = 20$ years

Using equation,

$$P_S = \frac{2\,P_0\, P_1\, P_2 - P_1^2\,(P_0 - P_2)}{P_0\, P_2 - P_1}$$

$$= \frac{2 \times 35{,}000 \times 78{,}000 + 1{,}15{,}000 - 78{,}000^2(35{,}000 + 1{,}15{,}000)}{35{,}000 \times 1{,}15{,}000 - 78{,}000^2}$$

$$= \frac{6{,}27{,}900 \times 10^9 - 6084 \times 10^6(1{,}15{,}000)}{4025 \times 10^6 - 6084 \times 10^6}$$

$$= \frac{-2847 \times 10^5}{-2059}$$

$$= 1.38 \times 10^5$$

Hence, the saturation population = 1,38,000

From equation,

$$\frac{P_S - P_0}{P_0} = m = \frac{1,38,000 - 35,000}{35,000} = 2.95$$

Now using equation,

$$n = \frac{2.3}{t_1} \log_{10} \left[\frac{P_0 (P_S - P_1)}{P_1 (P_S - P_0)} \right]$$

$$n = \frac{2.3}{10} \log_{10} \left[\frac{35,000(1,38,000 - 78,000)}{78,000(1,38,000 - 35,000)} \right]$$

$$n = \frac{2.3}{10} \log_{10} \left[\frac{35 \times 60}{78 \times 103} \right]$$

$$= 0.23 \log_{10} 0.26 = 0.23\ (1.4150)$$

$$= 0.25 \times (-0.585) = -0.146$$

Now substituting the values of P_s m and n in equation;

$$P = \frac{P_S}{1 + m \log_e^{-1} (n.t)}$$

$$P = \frac{1,38,000}{1 + 2.95 \log_e^{-1} (-0.146.30)}$$

Which is the required equation of logistic curve.

or,

$$P = \frac{1,38,000}{1 + 2.95x}$$

where, $x = \log_e^{-1} (0.146 \times 30)$

Now evaluating the value of x;

$$\log_e^{-1} (0.146 \times 30) = x; \log_e x = -4.38$$

or, $$2.3 \log_{10} x = -4.38; \log_{10} x = -1.9043$$

or, $$\log_{10} x = 2.0957$$

Taking antilog;

$$x = 0.01247$$

Now, putting the value of in the above expressions;

$$P = \frac{1,38,000}{1 + 2.95 \times 0.01247}$$

$$P = \frac{1,38,000}{1.03} = 1,33,980$$

Population of the town after 10 years in 1980 will be 1,33,980.

1.9.9 The Apportionment Method

It is also known as the method of estimating future population ratios. In this system the total census population is expressed as the percentage of the entire country's population. Census department gathers the population of the region under consideration, and the population of the world over the last four to five decades. For these decades the town ratio under consideration to the national population is estimated. A graph is now plotted between these and the time ratios. The extension of this graph will provide the proportion corresponding to the future years for which population forecasting is to be made. The ratio thus obtained is multiplied at the end of the planned period by the projected national population to determine the city's estimated national population. This approach is ideal for those towns and cities which are likely to evolve according to national development.

1.10 WATER QUANTITY ESTIMATION

The quantity of water required for municipal uses for which the water supply scheme has to be designed requires following data:

1. Water consumption rate (per capita demand in litres/day/head)
2. Population to be served
3. Quantity = Per capita demand x population

Average daily per capita demand is calculated using Eq.1.17 followed by maximum daily demand using Eq. 1.18 and maximum hourly demand indicated in Eq. 1.19-Eq. 1.23.

Fluctuation in rate of demand

$$\text{Average daily per capita demand} = \frac{\text{Quantity required in 12 months}}{365 \times \text{population})} \qquad (1.17)$$

If this average demand is supplied at all the times, it will not be sufficient to meet the fluctuations.

- Seasonal variation: During summer, fire breakouts are usually higher, so the demand grows.
- Daily variation is activity dependent. During Sundays and festival days, people draw more water, thus rising demand in these days.
- Hourly variations are very significant as they have a wide range of variations. The bulk of the daily requirement is taken during active household working hours, i.e., six to ten in the morning and four to eight in the evening. The necessity

is negligible at certain hours. In fact, if a fire breaks out, a large quantity of water is needed to be distributed in a limited period of time, necessitating the need for a maximum hourly supply limit.

Therefore, there must be sufficient volume of water available to satisfy the peak demand. The supply tanks, service reservoirs, and distribution tanks must be properly proportioned to accommodate all the fluctuations. Water is supplied directly by pumping, and the pumps and distribution system have to be built to meet the peak demand. The effect of monthly variability affects storage reservoir design, and the hourly variations affect pump design and service reservoir design. As the population shrinks, the rate of fluctuation increases.

$$\text{Maximum daily demand} = 1.8 \times \text{average daily demand} \quad (1.18)$$

Maximum hourly demand of maximum day, i.e., peak demand

$$= 1.5 \times \text{average hourly demand} \quad (1.19)$$

$$= \frac{1.5 \times \text{maximum daily demand}}{24} \quad (1.20)$$

$$= \frac{1.5 \times (1.8 \times \text{average daily demand})}{24} \quad (1.21)$$

$$= \frac{2.7 \times \text{average daily demand}}{24} \quad (1.22)$$

$$= 2.7 \times \text{annual average hourly demand} \quad (1.23)$$

1.10.1 Design Periods & Population Forecast

This quantity should be worked out with due provision for the estimated requirements of the future. The future period for which a provision is made in the water supply scheme is known as the design period. Design period is estimated based on the following:

(i) Useful life of the component, considering obsolescence, wear, tears, etc.

(ii) Expandability aspect.

(iii) Anticipated rate of growth of population, including industrial, commercial developments & migration-immigration.

(iv) Available resources.

(v) Performance of the system during initial period.

EXERCISE

1. Define per capita demand and how it is estimated? A town water supply is given 135 liters per capita per day. Suggest a suitable breakup under different heads.

2. Describe National Board of Fire Under Writers Formulae used for different types of populations.
3. The population of 5 decades from 1930 to 1970 are given below. Find out the population after one, two and three decades beyond the last known decade by using arithmetic increase method.

Year	1930	1940	1950	1960	1970
Population	25,000	28,000	34,000	42,000	47,000

4. The population of 5 decades from 1930 to 1970 is given below. Find out the population after one, two and three decades beyond the last known decade by using geometric increase method.

Year	1930	1940	1950	1960	1970
Population	25,000	28,000	34,000	42,000	47,000

5. Name the different methods and empirical formulae used for calculating the fire demand? Also define coincident draft and how it is used for estimating design draft?
6. In two periods each of 20 years a city has grown from 30000 to 170000 and then to 3,00,000. Determine (a) saturation population, (b) equation of logistic curve, and (c) the expected population after 20 years.
7. A town water supply is given at 225 litres per head per day. Suggest a possible breakup of the consumption on different heads. What allowance should be made in India for loss and wastage in water supply?
8. The population of 5 decades from 1930 to 1970 is given below. Find out the population after one, two and three decades beyond the last known decade by using decreasing rate method.

Year	1930	1940	1950	1960	1970
Population	25,000	28,000	34,000	42,000	47,000

9. A water supply has to be designed for a city having population of 1,00,000. Estimate the important kinds of drafts which may be required to be recorded for an average water consumption of 250 Lpcd. Also record the required capacities of the major components of the proposed water works system for a city using a river as a source of supply. Assume suitable data and figures where needed.
10. The population of 5 decades from 1930 to 1970 are given below. Find out the population after one, two and three decades beyond the last known decade by using Incremental increase method.

Year	1930	1940	1950	1960	1970
Population	25,000	28,000	34,000	42,000	47,000

11. Work out the population of year 2000 from a data given below by using simple graphical method.

Year	1880	1890	1900	1910	1920	1930	1940
Population	25,000	27,500	33,000	39,000	45,000	54,500	61,000

12. Sketch the fluctuations in terms of water demand for Indian system.
13. Define the logistic curve method along with the standard equation for determining the future populations of a locality.
14. The average increase of population of a town per decade over a period of 6 decades was 4100 and average % age increase was 12%. If the population at the end of 6 decades was 2,20,000, estimate the population two decades later by geometric increase method.
15. Incremental increase method is better than arithmetic increase and geometric increase methods for population estimation. Why.

REFERENCES

1. Rao, P. V. (2005). Water Supply Engineering (For the Course of Water Supply and Sanitary Engineering), in *Water Supply and Sanitary Engineering* (p. 119). Hyderabad: Telugu Akademi, Hyderabad on behalf of State Institute of Vocational Education Govt. of Andhra Pradesh, Hyderabad.
2. Sengupta. (n.d.). Forecasting of Population: 9 Methods | Water Quantity | Water Engineering. Retrieved from https://www.engineeringenotes.com/water-engineering-2/water-quantity/forecasting-of-population-9-methods-water-quantity-water-engineering/16158

2

CHAPTER

Sources of Water Supply

2.1 INTRODUCTION

Water is the most abundant compound in the nature as it occupies 75 per cent of the earth. Approximately 97.3 per cent of the water is stored throughout the great saline oceans and 2.14 per cent is stored in the poles of ice caps, glaciers, which are not useful either. The remaining 0.56 per cent water on the earth is available as usable fresh water for all living beings. The total quantity of water available on the earth in various forms and regions is shown in Table 2.1.

This chapter revolves around the sources of water such as ponds, lakes, reservoirs and wells. From the intake or pumping of the water to the distribution system is also discussed. Intake structures (dry intake tower, wet intake tower), pumping mains (layout, pipes and maintenance) is highlighted. Besides this, various distribution systems like dead end, ring, and circular are studied for water supply and distribution.

Table 2.1: Water availability on land, atmosphere and in oceans

Location	Volume (m^3)	% of Total
Land areas		
Freshwater lakes	152×10^{12}	0.009
Saline lakes	104×10^{12}	0.008
Inland seas, rivers	1.25×10^{12}	0.001
Soil moisture	67×10^{12}	0.005
Groundwater	8350×10^{12}	0.005
Icecaps and glaciers	52×10^{12}	0.61
Total	$37{,}800 \times 10^{12}$	2.8
Atmosphere		
Water vapour (clouds)	13×10^{12}	0.001
Oceans		
Water in the oceans	$13{,}20{,}000 \times 10^{12}$	97.3
Total on planet	$13{,}60{,}000 \times 10^{12}$	100

2.2 Hydrological Concepts

Hydrology is the science that deals with the flow of water on the earth's surface, under the earth's surface, the evaporation of water as water vapours from the surface of the earth and the transportation of this water from surface through the soil to underground water table known as underground aquifers. This is regarded as the hydrological process of water that returns to the atmosphere by evaporation and transpiration in the form of precipitation under favourable climatic conditions.

2.2.1 Precipitation

It may be defined as the water evaporated from natural water bodies and vegetation present on the earth's surface such as streams, rivers, sea, ponds, wet surfaces, trees and plants through the heat generated in atmosphere by the sun. However, during the condensation process of these water vapours in the atmosphere it returns to the earth as precipitation in the form of rain, hails, dew, sleet, etc. The majority of the precipitation occurs as rain as compared to other sources which are relatively very low. The precipitated water travels or moves on the earth in different ways such as:

- **Run-off:** This may be defined as the portion of water that after precipitation flows over the land in the form of rivers and streams. Also some of the precipitated water flows into the pools, lakes and ponds.
- **Infiltration:** This is a part of precipitation that percolates in the soil and replenishes soil moisture, recharges groundwater aquifers and ultimately supports the stream flows during dry periods.
- **Evaporation:** Much of the precipitation is often evaporated in the form of water from streams, rivers, reservoirs and wet surfaces due to the heat of the sun.
- **Evapotranspiration:** Transpiration is a process by which soil moisture taken up by vegetation is eventually evaporated as it exists in plant pores. Therefore the combined form of evaporation and transpiration is known as evapo-transpiration process. The roots of the trees suck water from the earth, and some of it evaporates through the leaves in the form of transpiration in the atmosphere.

2.3 WATER SOURCES

Water sources can be broadly divided into types as (i) surfaces sources, and (ii) Sub-surface sources

However, the surface sources are further divided into various forms as

(a) Streams
(b) Rivers
(c) Ponds
(d) Lakes
(e) Impounding reservoirs

2.3.1 Streams and Rivers

Rivers and streams are the primary source of water from the soil. In summer season the river water quality is better as compared to monsoon season where due to heavy rainfalls the river run-off water carries clay, sand, silt, etc. This makes the river water turbid and therefore needs special treatments. Some rivers are snow-fed and seasonal and have water throughout the year and therefore no provision of water storage is needed in that region. However, during extreme summer season some rivers dry up as a whole or sometimes in part. In such regions special arrangements of water storage or water availability are made in order to satisfy the water demand. Most of the towns located near the rivers discharge their wastewater into these rivers thus making them polluted and therefore much care should be taken.

2.3.2 Natural Ponds and Lakes

The presence of natural basins in mountains created by natural springs with impermeable beds and streams are known as "lakes". Water quality in the natural ponds and lakes depends on the ability of the basin, catchment area, annual rainfall, ground porosity, etc. Even at higher altitudes lakes and ponds contain almost pure water which can be used without any treatment. Yet some ponds are being created as a result of the construction activities, roads and railways works may consists of vast quantities of impurities and cannot therefore be used for water supply directly.

2.3.3 Impounding Reservoirs

The flow in some rivers is very low and cannot fulfill the demands during hot weathers. Under these circumstances water may be collected by constructing a bank, a weir, or a dam across the river at certain places where minimum area of land is submerged in the water and maximum amount of water can be diverted and extracted. Suspended impurities accumulate in the bottom of lakes and rivers, however, algae, plants, vegetables and organic growth occur in their beds, causing poor smell, taste and colour in water therefore requires thorough treatment before water supplying.

2.3.4 Infiltration Wells

In order to obtain high flow of water, the infiltration wells are constructed in series across the banks of river. In order to collect the river water seeping through their bottoms these wells are closed at top and open at bottom. They are generally constructed using brick masonry with open joints as shown in Fig. 2.1.

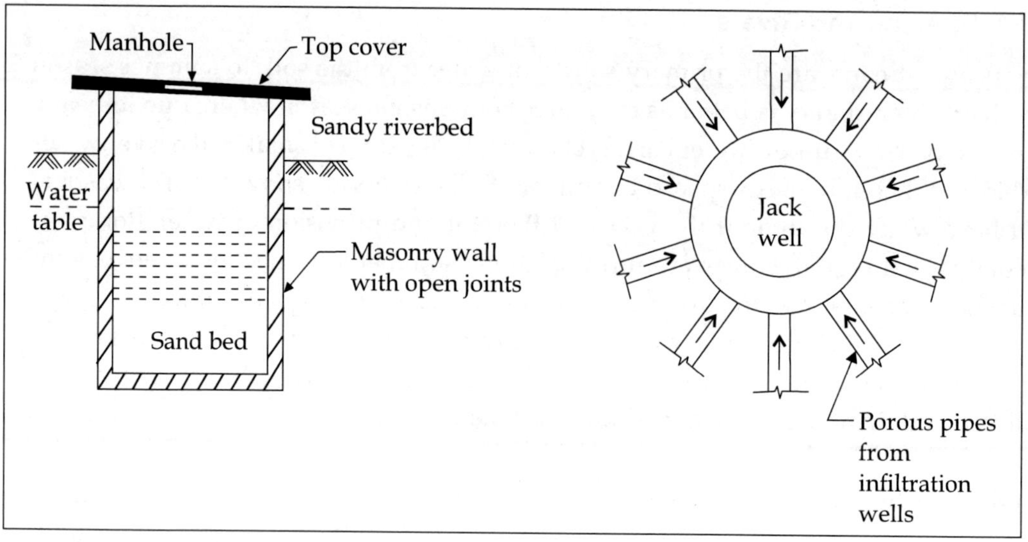

Fig. 2.1: Infiltration and jack well

The manholes are installed at the top cover for well inspection purposes. The water flows through the bottom of these wells and gets filtered to some degree as it pass through different sand and soil layers. The infiltrated water of the well is taken to sump well also known as jack well through perforated pipes. Thereafter, the water from jack well is pumped to the treatment plant for purification process as indicated.

2.3.5 Springs

Often groundwater reappears in the form of springs at the ground surface. Springs usually possess limited quantity of water, and are therefore suitable for hill towns. Some springs discharge hot water due to the presence of sulfur and are sometimes very useful for curing the skin diseases patients. In India such type of springs are found in Sohana in Haryana state. Also on the banks of Sutlej river at Tattapani near Shimla and Manikaran near Manali on river Parvati in Himachal Pradesh such type of springs with sulfur mixed water are available. These springs may be of different types.

2.3.5.1 Types of Springs

(i) **Gravity springs:** As the earth's surface drops sharply, and the ground water table rises high the water carrying stratum is exposed to the atmosphere and gravity springs form (Fig. 2.2). The flow of such springs varies with the rise and fall of water table.

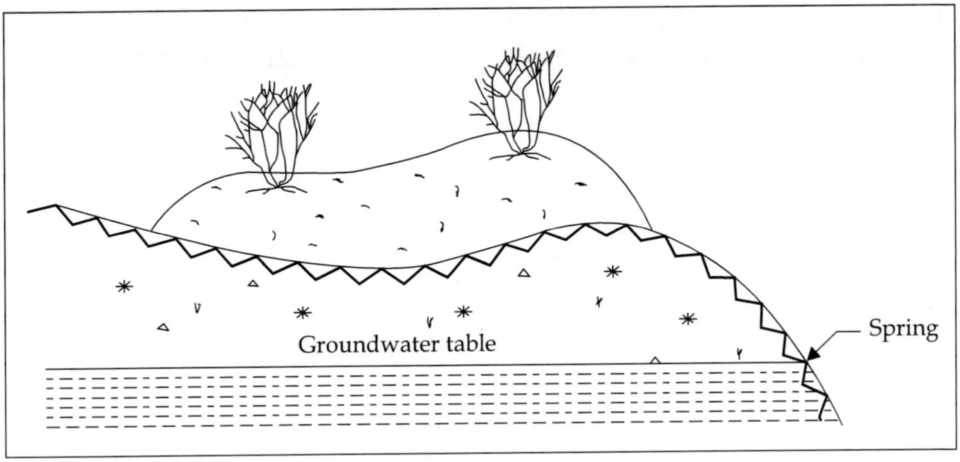

Fig. 2.2: Gravity spring

(ii) Surface springs: This happens when an impermeable layer that protects the reservoir of groundwater becomes inclined and out crops thus causing the ground water table to rise upwards and ultimately exposes to the atmosphere as shown in Fig. 2.3. The water availability from such spring is not very certain and may fluctuate largely.

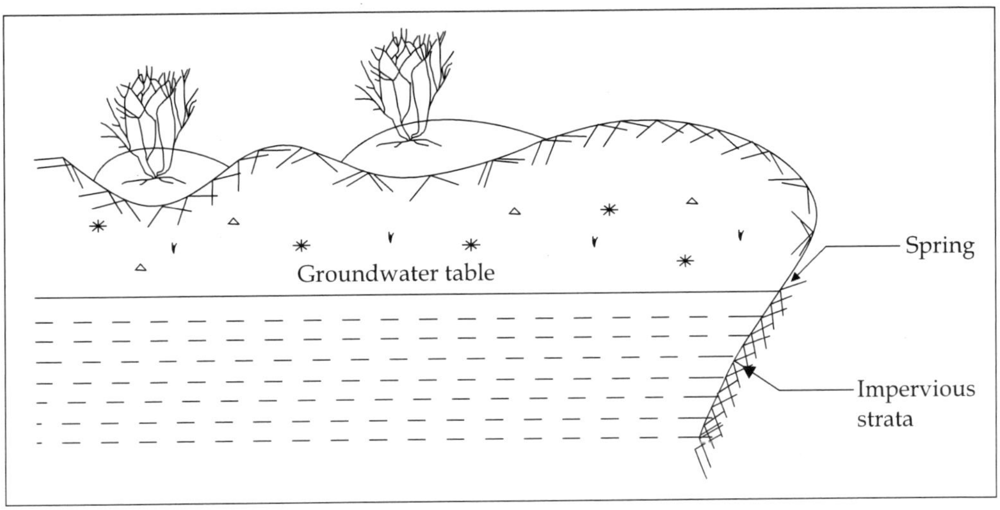

Fig. 2.3: Surface spring

(iii) Artesian springs: When the water storage is under pressure the groundwater rises in the upper impermeable stratum through a fissure or loose layer the springs formed are known as artesian springs (Fig. 2.4). The water carrying stratum has too much hydraulic gradient and is closed between two impervious stratum, the Artesian spring is formed from deep springs. Due to high pressure such springs provide higher yields and almost uniform quantity of water. The water collected from such springs can be directly or after minor treatment be stored and used.

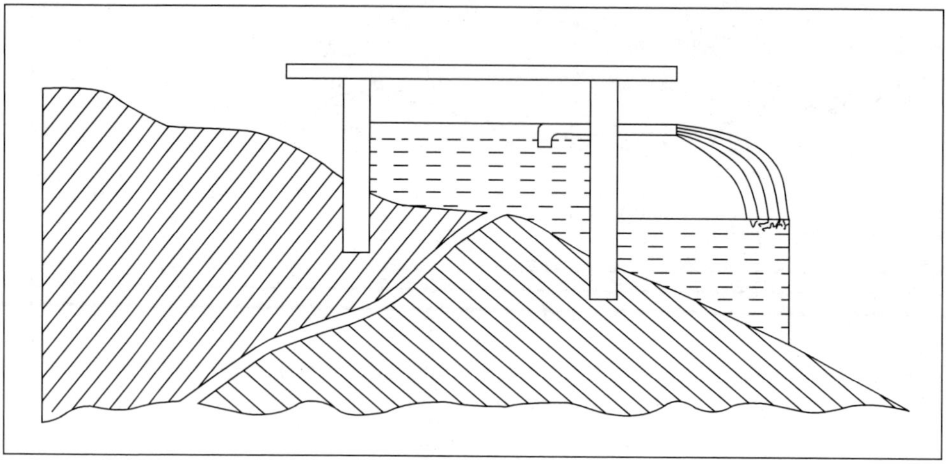

Fig. 2.4: Artesian spring

2.3.6 Wells

A well is described as an artificial hole or a pit created for the purpose for extracting the groundwater to the surface for different usages. In earlier times in India around 75 to 85 per cent of the Indian population depended on such wells. The three factors which form the basis of the theory of wells are: (i) geological conditions of the earth's surface, (ii) porosity of various layers, and (iii) quantity of water, which is absorbed and stored in different layers. The open wells are masonry wells with diameters ranging from 2-9 m with 20 m depth and low discharge rate of 18 m^3/hour. The masonry wall thickness of such well varies from 0.5 to 0.75 m. Due to limited depth of such well the water can only be withdrawn at critical velocity for the soil. However at higher velocities the disturbance in soil grains may result in subsequent subsidence of the well lining.

However, constructing a 8-10 cm diameter borehole in the centre of the well is helpful in improving the yield as additional water available in aquifer or fissures in rocks can be tapped easily. Based on above conditions the open wells may be classified as:

(i) Shallow wells

(ii) Deep wells

2.3.6.1 Shallow Wells

The well resting on previous strata and draws water from surrounding soil and strata is known as shallow well. Shallow wells are constructed in the uppermost layer of the surface of the earth. Well diameter ranges from 2 to 6 m, with an average depth of 7 m. Shallow wells from the inside maybe lined or unlined. A shallow well draws water from topmost water bearing strata and the water may be contaminated during monsoon season. This may be due to the percolation of rainwater through the soil and carrying the dissolved minerals, and decomposed organic matter. Such wells are also referred to as drawing wells or gravity wells or open wells or drag wells or percolation wells as shown in Fig. 2.5. The water can be drawn by means of simple bucket tied to a rope. However, due to possible contamination of external activities on the surface such wells are generally covered in many parts of India and fitted with reciprocating type handpumps for extracting the water safely.

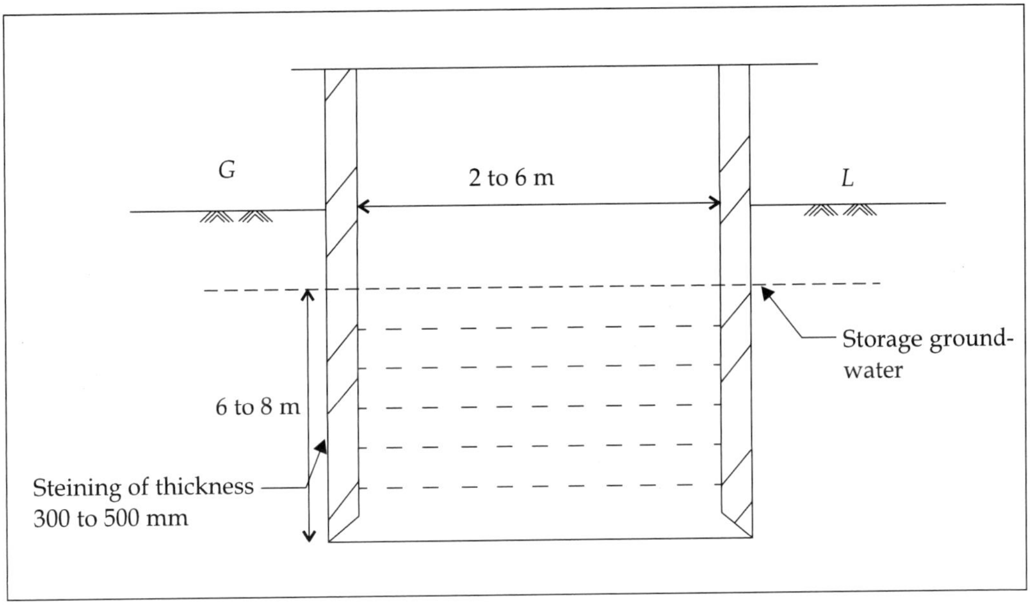

Fig. 2.5: Shallow well

Quantity of water available from shallow wells is limited because their supply source is only the uppermost layer of the earth and can also dry up in summer. They are also not suitable for public water supply schemes. The amount of water collected from shallow wells is better than the water from the rivers but requires purification. Regardless of the pollution of the effluent, the shallow wells should be built away from septic tanks, soak pits, etc. The shallow wells are used as a source of water supply for small villages, undeveloped municipal towns, isolated buildings, etc.

Cavity formation in shallow wells: With continuous withdrawal of water from open well at a constant rate, the water table with respect to time usually lowers down as compared to its nearby or outside level. This difference of water head between the two levels is known as *depression head*. Therefore, the surrounding or nearby water enters the well to fulfil this loss of water. This movement of water from nearby soil towards the well creates a loss of head and a drop of water surface occurs towards the well. This continuous movement of water through the reducing soil areas is observed with gradual increase in flow velocity towards the well. However as per Darcy's law the gradual increase in flow velocity is due to the increase in hydraulic gradient. At the initial level water surface falls slowly but falls very rapidly as it approaches towards the well. Therefore, this water surface level forms a curve from the point of travelling towards the well and the same is known as *cone of depression*. The point at which this curve ends up from the centre line of the well is known as *radius of influence of the well*. If more amount of water is withdrawn from the well will increase in depression head and ultimately the flow velocity will be increased near the stratum of the well. At this condition the flow velocity increases from the critical flow velocity value that can lift the sand particles of the soil and ultimately increases the effective area of the well. This may be unsafe for the shallow wells as this increase in hollowness due to removal of sand particles can ultimately result subsidence of well lining. The maximum withdrawal rate of such wells is restricted.

2.3.6.2 Deep Wells

A well which rests on an impervious layer and draws its supply from previous layer lying below the impervious layer is known as deep well. The deep wells get their water quota from an aquifer below the impermeable layer as shown in Fig. 2.6. This impervious layer comprising of clay, cemented sand, kankar or other hard materials is also known as *mota* layer. This mota layer provides the structural support to the well

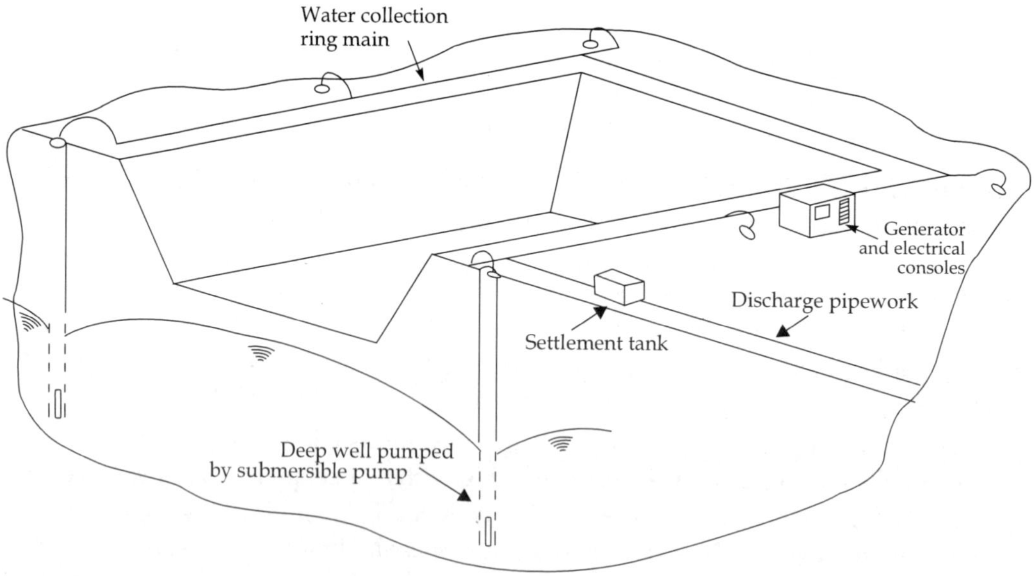

Fig. 2.6: Deep wells

as it proves to be useful even the unlined or partly lined wells. Deep well theory is based on the flow of water from the outcrop to the deep well site. However, the water in deep well is free from any such contamination as compared to shallow wells. The outcrop is the position of access to the atmosphere by aquifers. At the outcrop the rain water enters and is properly filtered until it reaches the deep well site. But it dissolves some salts, and thus becomes hard. In these situations, it would be necessary to take some treatment to eliminate water hardness.

The yield from previous layer of deep wells is higher as compared to shallow wells thus ensuring for high discharge for longer durations. Also the cavity or hollow formation is not unsafe as the lining rests on the hard layer. Therefore, the hollow and much larger area of such deeps well provides higher yields which if compared with sallow wells would prove to be costly.

Depth of deep well should be determined in such a way that out crop position should be not too close to well. The water available at a higher level in the atmosphere is also referred to deep wells as pressure wells.

(i) **Construction of open wells:** Based on the mode of construction required the well may be classified as below:

(a) **Type-I:** The impervious lining wells that are constructed with masonry lining and rest on mota layer that provides most reliable and stable type of wells for safe water supplies. During the construction a pit is dug till the soft soil strata appears and thereafter masonry lining is built on a kerb for a few metres above the ground level. This kerb is a circular ring made of RCC, timber or steel having cut bottom edge and top part flat that is wide enough to support the well thickness of the well lining is known as *steining*. This kerb is lowered in to the pit by imparting loads externally from the top that sinks down as the excavation is carried out further. Reinforcement with vertical steel bars in the well lining (steining) is provided accordingly. The excavation and sinking continues by removing the extra water through pumps. Alternatively, a simple instrument consisting of self-closing bucket known as Jham is used for further deep excavations. This jham when thrown into the well strikes at the bottom of the well thus dislodging some loose material. The pulling out of this jham the soil cutting gets retained with oozing out of water. This sinking continues till it reaches the mota layer. A timber lined small diameter bore hole is made in the centre of well through this mota layer. When the well is lined with impervious linings the flow is not radial as the water enters from the bottom of the well. This flow turns up to be spherical when a cavity is formed at the bottom.

(b) **Type-II:** The wells comprising of previous linings with dry brick or stone as lining without any mortar and water enters through the pores of these linings radially. The wells from the bottom are capped with concrete. If not capped properly the flow combination of radial and spherical may occur. Such wells possess suitable strata of coarse sand or gravels.

(c) **Type-III:** The wells without any lining with shallow depths are generally constructed by farmers for irrigation purposes with water table available at a very shallow depth. The wells are constructed in hard soils that can withstand vertically without any support. Although the wells are not much costly but prove to be dangerous as the soil subsides and collapses after some time.

(ii) **Yield of open well:** The yield of an open well can be determined by using theoretical method such as estimating through velocity of ground water or by practical approach. Also the practical tests such as pumping tests or recuperating tests can be carried out and the observed data can be utilized for designing purpose. The later is useful in assessing the yield of open wells as well as tube wells also through confined aquifers.

(a) **Theoretical method:** In an aquifer the water moves towards the well with velocity (v) through opening surface area (A_a) of aquifer as shown in Eq. 2.1;

$$Q = A_a \times v \tag{2.1}$$

And $v = \eta . v_a$ where, v_a is the actual flow velocity of the ground water and v is the velocity through which the water moves towards the well through soil with porosity η. (Eq. 2.2). A_a is the area of an aquifer joining then well and can be determined by using equation $= \pi \cdot d \cdot H$ where, D is the diameter of well with height, H.

Therefore, $Q = \eta \cdot v_a \cdot A_a$ (2.2)

The v_a values can be computed by using various techniques and methods such as Slichter's method, Hazen's formula, or by chemical and electrical methods of measurement. In the above equation, the velocity of the groundwater flow (v_a) can be estimated by using Slichter's or Hazen's empirical equations. However, in the actual field it can also be measured by using chemical tracers, such as a dye, or by electrical (S) travelled by a tracer or dye in time t between two observation wells, directly indicate the ground flow velocity as $v_a = S/t$. Therefore, K value in the test can be determined as in Eq. 2.3 and Eq. 2.4;

$$v = K \cdot i \quad \text{and} \quad v_a = \frac{S}{t} \tag{2.3}$$

$$\text{or } K = \frac{v}{i} = \frac{\eta \cdot v_a}{i} \quad \text{where,} \quad i = \frac{H_L}{S} \tag{2.4}$$

H_L is difference in water surface elevations of two wells.

Example 2.1: In a field test, a tracer took 8 hours of time to travel through an aquifer from one well to another. The two wells are 50 m apart with a water level difference of 0.45 m. Calculate (i) velocity of flow, and (ii) the coefficient of permeability (K). The porosity of the soil medium is 24%. (iii) Also calculate the coefficient of intrinsic permeability for the aquifer in Darcy's, if viscosity of water $v = 0.015\ \text{cm}^2/\text{sec}$.

Solution: Velocity of groundwater flow as $v_a = \frac{S}{t} = \frac{50}{8} = 6.25\frac{\text{m}}{\text{h}} = \frac{6.25 \times 100}{60 \times 60} = 0.17\frac{\text{cm}}{\text{s}}$

$v = \eta.v_a = 0.24 \times 0.17 = 0.041$ cm/s

Hydraulic gradient between the two wells as, $i = \frac{H_L}{S} = \frac{0.45}{50} = \frac{1}{222}$

The value of $K = \frac{v}{i} = \frac{0.041}{1/222} = 9.11\frac{\text{cm}}{\text{s}}$

Therefore, the coefficient of permeability $K = 9.11$ cm/s

Therefore, the intrinsic permeability coefficient, $K_0 = \frac{K \cdot v}{g} = \frac{9.11 \times 0.015}{981}$

$$= 1.3 \times 10^{-4} \text{ cm}^2$$

$$1 \text{ darcy} = 9.87 \times 10^{-9} \text{ cm}^2$$

Therefore, K_0 (darcy) $= \frac{1.3 \times 10^{-4} \text{cm}^2}{9.87 \times 10^{-9} \text{cm}^2} = 13171$ darcys **Ans.**

(b) Field methods of yield measurement: The yield of open well can be determined by using various methods such as given below.

1. *Pumping test:* Also known as equilibrium pumping test where a pump is being installed and water is withdrawn at a very high rate that causes the visible and measurable drawdown in water level. Thereafter, the water level is made constant by changing the rate of pumping and adjusting it. The developed equilibrium condition thus indicates the rate of pumping equal to the rate of yield. This pumping rate will provide the yield value at that particular drawdown. The yield "Q" at a drawdown of "s" can be estimated by known yield value of Q_1 at a known drawdown s_1 using Darcy's law: as shown below in Eq. 2.5;

$$Q = K.i.A = K\left(\frac{s}{L}\right)A = \frac{K}{L}.A \cdot s = C \cdot A \cdot s \tag{2.5}$$

where, s = Depression head or drawdown in well

i = Hydraulic gradient = s/L

L = Length travelled by the water through pores (cm)

K = Coefficient of permeability (cm/sec)

A = Cross-sectional area of well (cm^2)

However, at known discharge Q_1 and known drawdown s_1, the Eq. 2.6 is as:

$$Q_1 = C \cdot A \cdot s_1 \tag{2.6}$$

Under cavity formation the area of well A is considered to be 4/3 of the actual cross-sectional area at the bottom of the well. Therefore, putting the values of Q_1, s_1 and A in the above equation the value of C can be determined. Also, the increase in depression head(s) results in increase in the velocity. The *s* value at the point of critical velocity is known as critical depression head. However, the depression head, also known as working head, is kept equal to 1/3 of critical depression head. Therefore, the safe yield simulates the working head and critical yield represents the critical depression head.

2. *Recuperating test*: The test is adapted under condition such that the water level has to be maintained constant and the pumping rates are difficult to adjust. In this method the total water at that particular interval of time is drained out and the pumping is stopped. The rise in water level under different time intervals is measured and noted. The discharge can be computed as shown in Fig. 2.7.

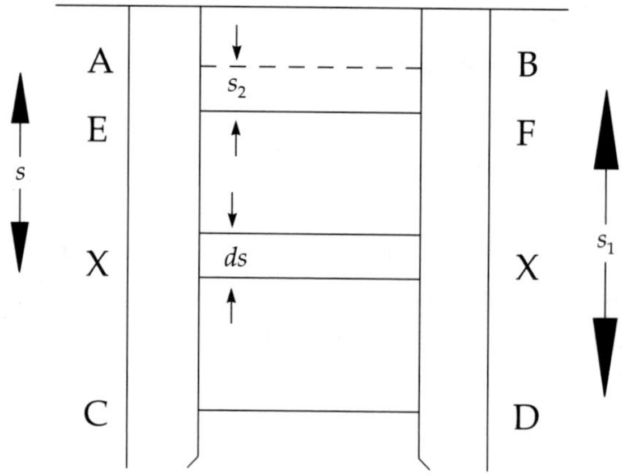

Fig. 2.7: Recuperating test

AB = Water level of the well before pumping

CD = Water level of the well after stoppage of pumping

s_1 = Initial depression level at the start up of pumping

s_2 = Depression level of well after stoppage of pumping at time, T

The water level after stoppage of pumping can be assumed at position of X-X at any time t and the depression head at that time may be assumed to be s.

The rate of depression head can be assumed to be ds at any time dt and therefore the water level will restore by a value of s_1-s at any time t after the pumping is stopped. It will againtry to restore with ds value at time dt.

Therefore, the amount of water entering the well in small interval of time dt can be calculated as in Eq. 2.7:

$$\frac{dV}{ds} = A \tag{2.7}$$

where, A is the bottom cross-sectional area of the well.

In addition to this, if Q is the flow rate of water recharging the well at any time t under depression head s, then volume of water entering the well can be calculated as shown in Eq. 2.8.

$$\frac{dV}{dt} = Q \quad \text{but} \quad Q \propto s$$

And therefore $Q = C' \cdot s$ (2.8)

where C' is a constant that depends on the soil through which water enters the well.

Therefore, $dV = C' \cdot s \cdot dt$ (2.9)

Hence, equating Eq. 2.8 and 2.9, we get

$-A.d_s = C'.s.dt$ the (–ve) sign signifies that decrease in s with respect to increase in t and therefore

$$\frac{C' \cdot dt}{A} = -\left(\frac{ds}{s}\right)$$

Thus, integrating the limits as $s = s_1$ when $t = 0$ and $s = s_2$ when $t = T$

Therefore,

$$\frac{C'}{A}\int_0^t dt = -\int_{s_1}^{s_2} \frac{ds}{s} \text{ or } \frac{C'}{A} \left| t \right|_0^T = -\left| \log_e .s \right|_{s_1}^{s_2}$$

$$\frac{C.'}{A}(T) = -\log_e \frac{s_2}{s_1} = -2.3\log_{10} \frac{s_2}{s_1} = 2.3 \log_{10} \frac{s_1}{s_2}$$

$$\frac{C'}{A} = \frac{2.3}{T} \log_{10} \frac{s_1}{s_2}$$

C'/A is known as *specific yield* or *the specific capacity* of the open well in m^3/m^2 area under unit depression head. Therefore, knowing the value of C'/A, discharge Q under constant depression head H can be computed as below in Eq. 2.10.

$$Q = C'.s = \left(\frac{C'}{A}\right) \cdot A \cdot s = \frac{2.3}{T} \log_{10} \frac{s_1}{s_2} A \cdot s \tag{2.10}$$

When recuperation test is not carried out, the following rough values of C'/A is given by Marriot is used for calculating the other parameters.

Table 2.2: Rough C'/A values by Marriot

S. No.	Soil type	C'/A (m^3/hour/m^2 of area under unit drawdown)
1.	Clay	0.25
2.	Fine sand	0.50
3.	Coarse sand	1.00

Example 2.2: Design an open well in fine sand for a yield of 0.005 m^3/sec with a depression head of 3.5 m.

Solution: Given discharge, Q = 0.005 m^3/sec = 0.005 × 60 x 60 m^3/hour

$$Q = 18 \text{ m}^3/\text{h}$$

From Table 2.2; value of C'/A for fine sand may be taken as 0.5 m^3/h/m^2/m of depression head

Depression head = 3.5 m

Therefore, using equation 5, we have

$$Q = \left(\frac{C'}{A}\right) \cdot A \cdot s$$

Putting the values in equation, we get

18 = 0.5 × A × 3.5 and

therefore, $A = \dfrac{18}{0.5 \times 3.5} = 10.28 \text{ m}^2$

Considering d to be diameter of the well, therefore $\dfrac{\pi d^2}{4} = 10.28$ and, therefore,

$$d = \sqrt{\frac{10.28 \times 4}{\pi}} = 3.6 \text{ m}.$$

Example 2.3: A recuperation test was being carried out for an open well that depressed by pumping up to 3.0 m with 1.75 m of water being recuperated in 95 minutes.

(a) Determine the yield from a well of 3.5 m diameter with a depression head of 3 m.

(b) Find the diameter of well with yield of 12 L/sec under depression head of 2.75 m.

Solution: From the given equation: $\dfrac{C'}{A} = \dfrac{2.3}{T} \log_{10} \dfrac{s_1}{s_2}$

Given that initial drawdown s_1 = 3.0 m and final drawdown s_2 = 3.0 – 1.75 = 1.25 m

T = 95 minutes = 95/60 = 1.58 hours

Therefore, $\dfrac{C'}{A} = \dfrac{2.3}{1.58} \log_{10} \dfrac{3.0}{1.25} = 0.55$ m^3/h/m^2/m of depression head

Now from part (a) the yield of well with 3.5 m diameter with depression head value of 3 m can be computed by using equation as under:

$$Q = \left(\frac{C'}{A}\right) \cdot A \cdot s = 0.55 \times \left(\frac{\pi}{4} \times 3.5^2\right) \times 3.0 = 15.87 \frac{m^3}{h} = 4.4 \frac{L}{s}$$ **Ans.**

Also in part (b) with Q value of 12 L/sec and depression head of 2.75 m, therefore

$$Q = 12 \frac{L}{s} = \frac{12 \times 60 \times 60}{1000} = 43.2 \frac{m^3}{h}$$

and s = 2.75 m, therefore,

$$Q = \left(\frac{C'}{A}\right) \cdot A \cdot s \quad \text{and}$$

therefore, 43.2 = (0.55) × A × 2.75

$$\text{Therefore, } A = \left(\frac{43.2}{0.55 \times 2.75}\right) = 28.56 \text{ m}^2$$

$$\text{And} \left(\frac{\pi}{2} \times d^2\right) = 28.56 \text{ and therefore } d = 6 \text{ m}$$ **Ans.**

2.3.7 Tube Wells

The mechanical pumping of small discharges from open wells is not to economical as such discharges are limited to 3 to 6 L/sec. Therefore, in order to extract high quantity of water with large discharge values mechanically a long pipe or a tube also known as tube well is bored and drilled deep into the ground thus intercepting one or more water bearing stratums. The discharge of an open well is smaller, because: (i) open wells can tap only the topmost or at the most the next lower water bearing stratum; and (ii) water from open wells can be withdrawn only at the velocity equal to or smaller than the critical velocity for the soil, so as to avoid the danger of well subsidence. But in the tube wells, larger discharges can be obtained by getting a larger velocity as well as a larger cross-sectional area of the water bearing stratum. Since, we have an enormous storage of groundwater in India, the tube wells provide excellent method of providing water supplies, although they are generally used for irrigation.

2.3.7.1 Tube Wells in Alluvial Soils

Most of our land in India from the Himalayas to the Vindhya mountains a part of the Indo-Gangetic plain as well as coastal areas and Narmada valley consists of deep alluvial soils. The subsoil water slowly penetrates and is stored in the porous sand and gravel beds which are extensively found in India, except that in the desert areas. Tube wells can be easily installed in such soils and are very useful for irrigation. It is in this context that the tube wells are assuming greater and greater importance for tapping groundwater resources. Deep tube wells are as deep as 70 to 300 m, and tap more than one aquifer. In India these tube wells are constructed by the state governments. Such wells may yield as high as 200 to 220 L/sec. The general average yield from the standard tube wells is however of the order of 40 to 45 L/sec.

The diameter of the hole of a typical well is 0.6 m upto 60 m depth and then 0.56 m below 60 m. The diameter of the strainer of such a well is 0.25 m and drawdown is 10 m. The typical tube wells are being drilled by heavy duty rotary drilling rigs consisting of direct rotary as well as reverse rotary. However, in hard boulder areas the percussion drilling rigs may be used in. In India around 50,000 of such deep tube wells are available and each year about 1000 such wells are being constructed. Most of the deep tube wells in India have been constructed by using mild steel slotted pipe screens with gravel packing. However, the shallow tube wells with 20 to 70 m depth are constructed by individual farmers privately for the irrigation of their personal fields. The yield value of such wells lies between 15-20 L/sec if constructed on proper aquifer and irrigates about 8 hectares. There are about 50 lakh such shallow tube wells in India and every year 2.5 1akh of such wells are being installed by the farmers.

2.3.7.2 Tube wells in Hard Rocky Soils

The construction of tube well in rocky areas for irrigation is very difficult and troublesome. Therefore, in such areas if no alternative source of water is available then only the tube well are constructed. In such rocky areas using the hole hammer rigs also known as DTH rigs the isolated holes of 10 to 15 cm diameter are drilled with a depth range of 100 – 300 m. The bore hole in such a tube well is able to stand on its own in the bottom portion with only a tube is pushed in the upper weathered zone. These wells usually depend on joints and fissures in the rocks for their water supply. These types of tube wells are generally not able to yield more than 5-10 L/sec even after a heavy drawdown of 20 to 30 m and are mostly constructed in southern parts of India.

2.3.7.3 Types of Tube Wells

The tube wells can be classified based on the entry of the water either through a cavity or through a screen as: (i) cavity type tube wells, and (ii) screen type tube wells.

(i) **Cavity type tube wells:** The cavity type tube well draws the water from the bottom of the well as compared to screen well that draws water from the sides. Similarly, the flow in a cavity well is completely spherical type as compared to radial type flow in screen well. The water in such a gravity well is drawn from the bottom side that can tap only one water bearing stratum. The tube well proves to be very economical, as it requires only plain well pipe. This pipe is lowered into the bore made through the ground strata up to the required depth. This allows the well to tap the requisite aquifer. The principle behind the working of cavity type tube well is similar to that of a deep open well, with the only difference that a cavity tube well may even tap the lower or still lower stratum, as shown in Fig. 2.8.

A cavity-type tube well essentially consists of a pipe bored through the soil and resting on the bottom of a strong clay layer. A cavity is formed at the bottom, and the water from the aquifer enters the well pipe through this cavity (Fig. 2.8). In the initial stage of pumping, fine sand comes out with water and consequently a hollow or a cavity is formed. As the spherical area of the cavity increases outwards, the radial critical velocity decreases for the same discharge, thus reducing the flow velocity and consequently stopping the entry of sand. Hence, the flow in the beginning is sandy but becomes clean with the passage of time. The cavity of such a tube well, however needs to be developed carefully and slowly by using a centrifugal pump rather than a compressor or a turbine pump.

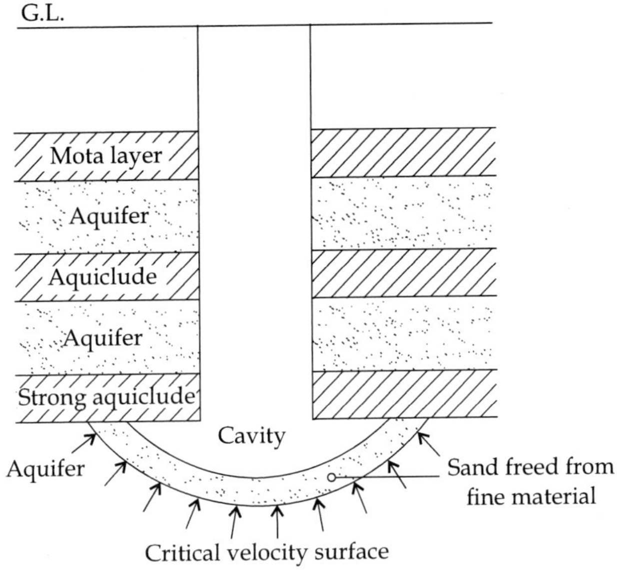

Fig. 2.8: Cavity type tube well

To begin with, the water is pumped from the well at a low discharge rate. When the discharging water becomes clear, the drawdown may be increased slightly which may result in further sand being drawn out. The process is repeated till the normal drawdown and clear discharge is obtained. The pumping is then stopped for an hour or so, and then resumed again. The discharge after restarting may again contain sand. The pumping is continued till the water is clear again. The procedure may be repeated till the tube well is fully developed, which is shown by the sand-free discharge coming out even on the resumption of pumping after a closing interval. Cavity type tube wells can however be used only for small supplies, particularly for domestic supplies.

(ii) Screen type tube wells: Screen type tube wells are most widely adopted and have been extensively constructed in our country particularly for irrigation purpose. So much so that whenever we refer to a tube well, we generally mean a "screen well". All the state tube wells constructed in U. P. from where the technique of tube well construction started in 1931, are exclusively of this type. Such a well can easily tap a number of aquifers, and hence does not depend only on one aquifer, like a cavity well. Screen type tube wells can be further divided into the following two types: (i) strainer tube wells, and (ii) slotted pipe gravel-pack tube wells. A strainer tube well uses strainer lengths lowered into the bore hole and located opposite the water bearing formations, whereas, plain pipe lengths are located opposite the non water bearing formations, as shown in Fig. 2.9. A bail plug is provided at the bottom. Water enters the well through these strainers from the sides, and the flow is radial.

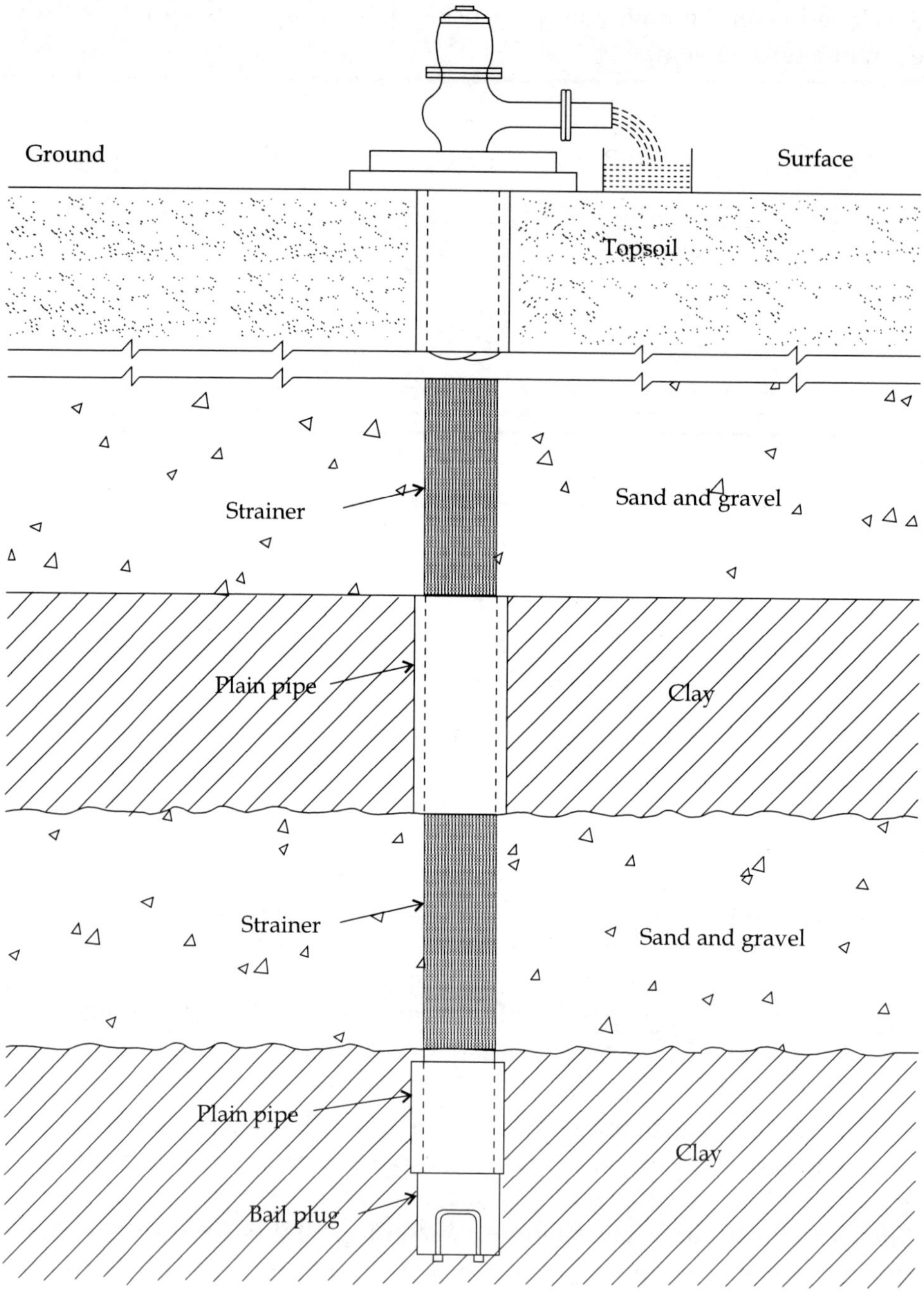

Fig. 2.9: A typical section of strainer tube well

A strainer (Fig. 2.10) essentially consists of a perforated or a slotted pipe with a wire mesh wrapped around the pipe with a small annular space between the two. The wire screen prevents the sand particles from entering the well. The water, therefore, enters the well pipe through the fine mesh and the particles of size larger than the size of the fine mesh are prevented from entering the well. This reduces the danger of sand removal, and hence larger flow velocities can be permitted. The perforated pipe is made to have the cross-sectional area of its openings equal to that in the wire mesh, so that no change of velocity occurs between the two. An annular space between the pipe and the wire mesh is certainly required, otherwise the wires of the mesh will cover a large part of the area or the pipe openings. A strainer type tube well is generally unsuitable for fine sandy strata, because in that case, the size of the mesh openings will have to be considerably reduced, which may result in choking of the strainer; and if the mesh openings are kept bigger, the well will start discharging sand.

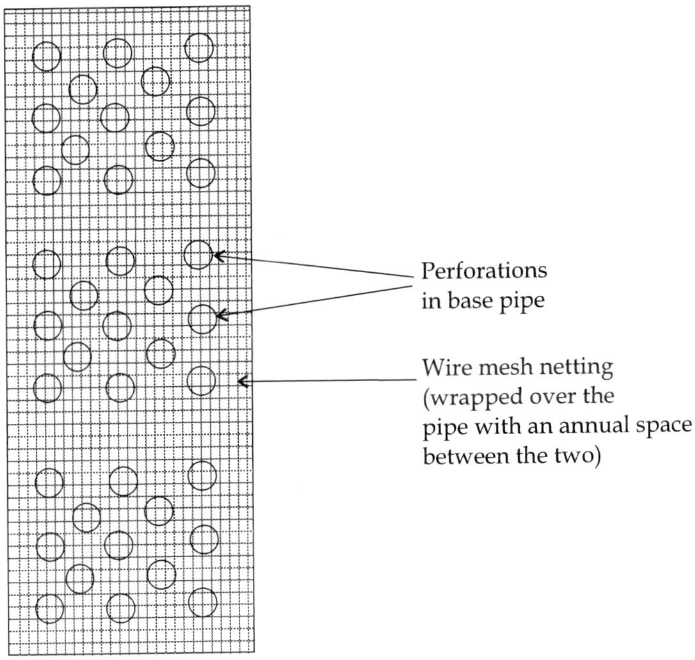

Fig. 2.10: A typical strainer section

A slotted pipe gravel pack tube well, on the other hand, uses a slotted pipe without being covered by any wire mesh. Such slotted pipe lengths are located opposite the water bearing formations, as is done in the strainers in a strainer tube well. These slotted pipes should strictly be referred to as "screen pipes", although, they are often called "slotted pipe strainers" or "pipe strainers", thereby having little scope to differentiate between such slotted pipes (i.e., screen pipes) and the perforated pipes with mesh coverings (i.e., strainers): After placing the assembly of the plain and slotted pipes in the borehole, a mixture of gravel and bajri (called gravel shrouding) is poured into the borehole between the well pipe assembly and the casing pipe, so as to surround the well pipe by optimum thickness of gravel pack. The gravel packing is specifically required around the screen pipes, but since the material is poured from the ground level into the borehole, the pack shall be installed in the entire depth of the well below the top level of the shallowest screen as shown in Fig2.11.Sometimes the gravel pack is provided even above the top level of the shallowest screen, and up to the ground level, so as to obtain a stable and an efficient tube well. Gravel pack wells are generally provided in fine aquifers, where the effective grain size (D10) may be less than 0.25 mm, and uniformity coefficient (Cu) may be 2.0 or less. Gravel packing also enables the use of single size slots in the screen pipes, irrespective of the fact that fine as well as coarser aquifers are tapped simultaneously and in alternating layers.

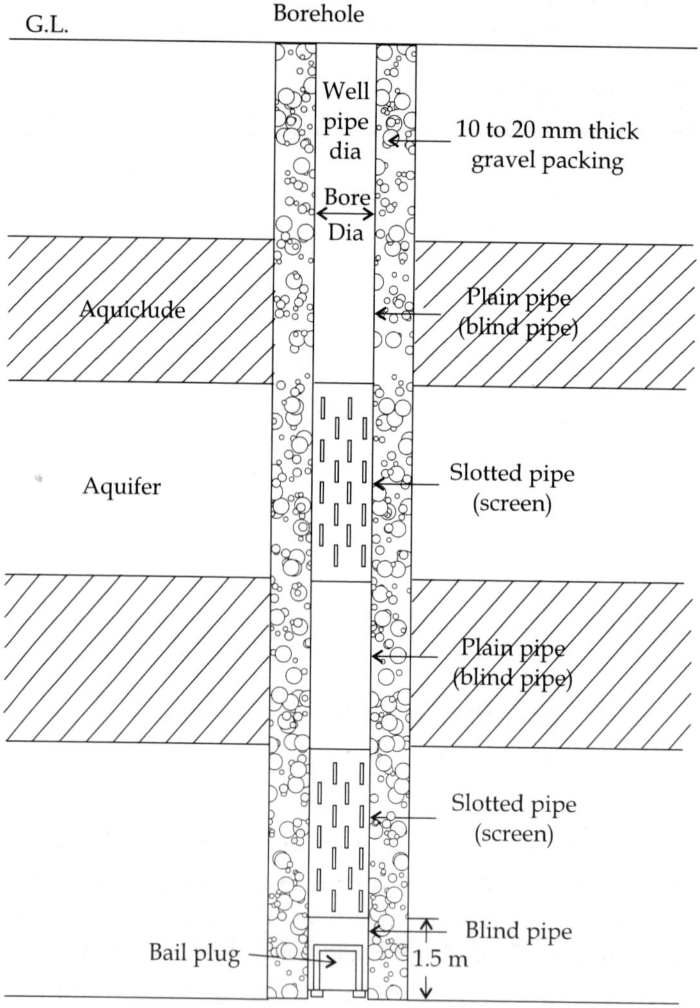

Fig. 2.11: Typical section of slotted pipe gravel well tube well

Gravel pack is therefore highly preferred for deep tube wells, which tap more than one aquifer. Due to this reason, most of the state deep tube wells have been constructed in our .country in fine alluvial soils, by providing cast in situ gravel packs, throughout the full height of the wells. Precast gravel packs have also come into the market, where the slotted pipes are mounted with prepack gravel filters, popularly called Ashim filters. In these filters, graded hard gravel grains are coated and bonded together with water-proof chemicals over the outer surface of the slotted pipe, providing a highly permeable screen. Such prepack gravel filters are found suitable for shallow and medium deep tube wells in aquifers composed of fine sand, coarse sand and gravel. Prepacking is claimed to be provide economy in the quantity of gravel used and protects against dislodging of the gravel envelope. However, such filters are so far not being used in larger projects in our country.

2.3.7.4 Theoretical Approach to Computation of Coefficient of Permeability (K) and Discharge Capacity of an Aquifer (Q)

The theoretical analysis of groundwater flow has been discussed as to compute K value for a given aquifer by performing a field test of observing water levels in one or two observation wells by knowing the discharge being pumped, in the main well. This K value of the given aquifer can be used to compute Q value for a well or a tube well to be driven through that aquifer, at any given drawdown and it will indicate the specific capacity of the well. The equation given by Thiem can be derived as follows.

(a) **Derivation of Thiem's equation for unconfined aquifer:** Consider a non-artesian well in which water is pumped heavily to cause sufficient drawdown. The water level in the neighbourhood of such a well will also fall down as the water level in the main well decreases. This forms an inverted cone of depression all around the well (Fig. 2.12).

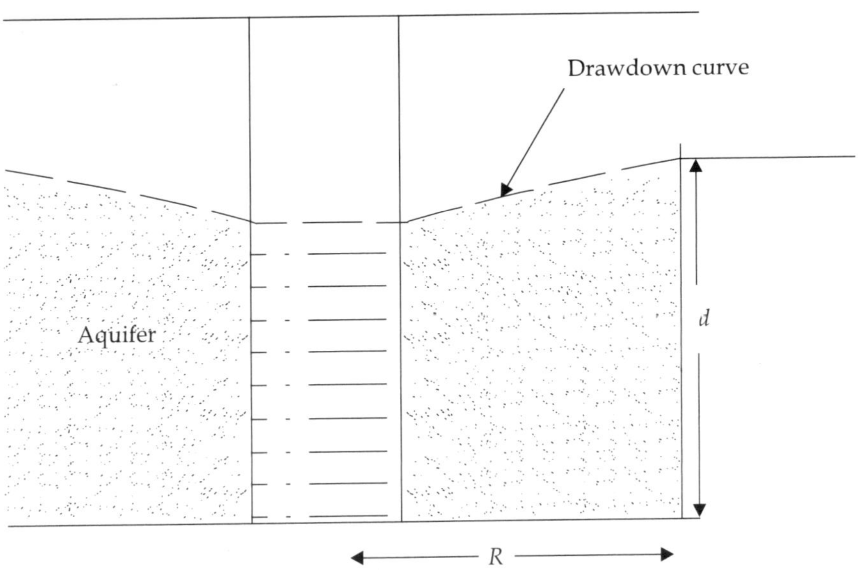

Fig. 2.12: Drawdown curve in an unconfined aquifer

The base of this cone may be considered a circle of radius R and is known as the circle of influence as shown in Fig. 2.12. This inverted cone of depression inclined side is known as the drawdown curve. The two observation wells lying within the circle of influence of the main pumped well are to be driven be numbered as 1 and 2 (Fig. 2.13) and let them be at distances of r_1 and r_2 from the main well (centre-to-centre distance). Let d be the depth of the well or the aquifer, below the static water table.

It is assumed that the pumping is adjusted to equilibrium conditions indicating that the rate of pumping becomes equal to the rate of yield, and thus causing the water level to attain a constant value. The formula is also known as equilibrium formula. At this equilibrium position let s_1 and s_2 be the drawdowns in the two corresponding observation wells.

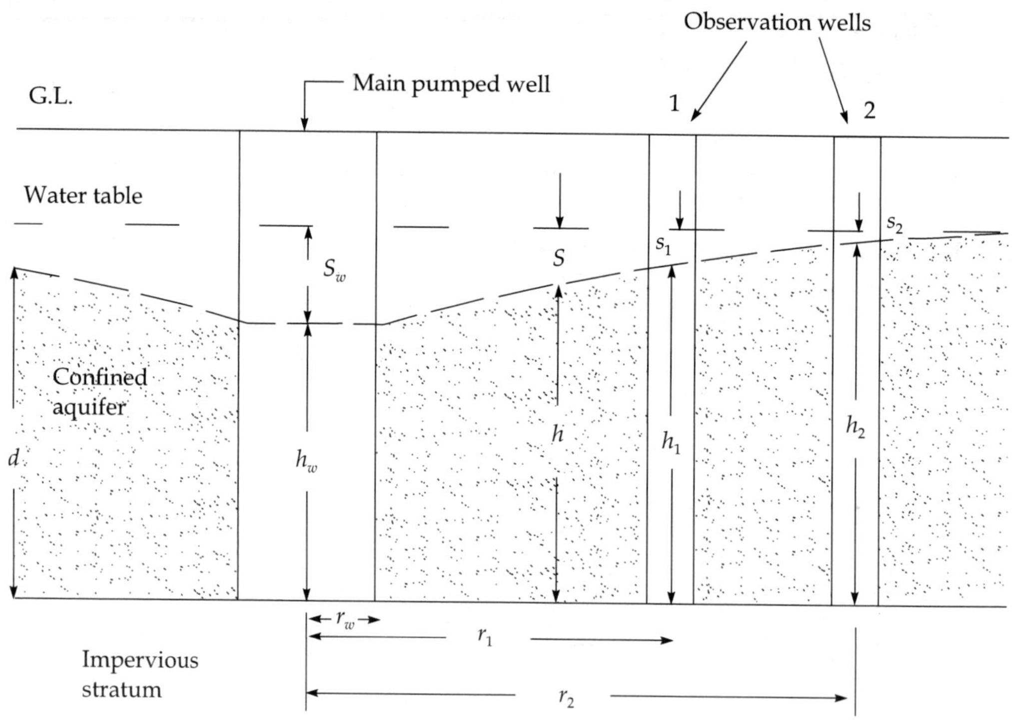

Fig. 2.13: Derivation based on Thiem's formula for unconfined aquifer

The flow through soil in any concentric cylindrical section of the water as specified by Darcy's law can be represented by Eq. (2.11) as:

$$Q = K \cdot i \cdot A \tag{2.11}$$

where,

i = Hydraulic gradient

The cylindrical coordinates with radius r and height of cone of depression h at a distance r from the main well is formulated in Eq. 2.12. However, using Darcy's law and assuming a small water surface inclination to be small and the tangent can be used to calculate the gradient as shown in Fig. 2.14.

$$i = \frac{dh}{dl} \approx \frac{dh}{dr} = \frac{dh}{dr} \tag{2.12}$$

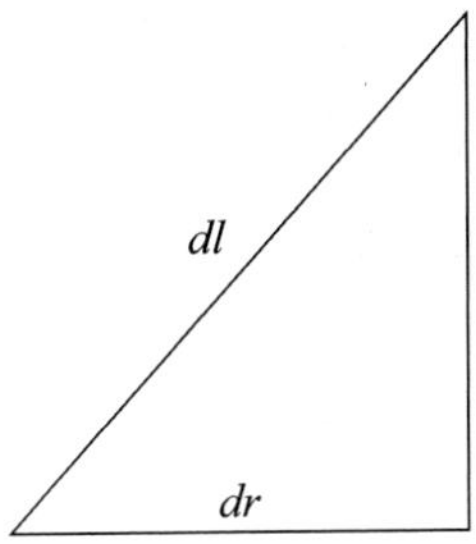

Fig. 2.14: Calculation of gradient using Darcy's law

It is assumed that the water flows through the aquifer up to its full height below the water table. The flow is also assumed to be radial and horizontal means that the velocity distribution is assumed to be uniform as shown Fig. 2.15

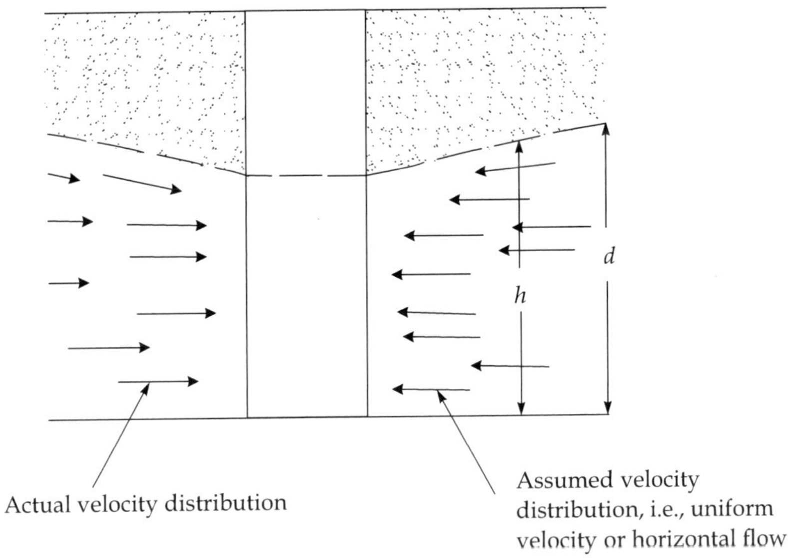

Fig. 2.15: Typical flow pattern in an unconfined aquifer

The flow area, $A = 2\pi rh$

Substituting the values of i and A in Darcy's equation as shown below in Eq. 2.13;

$$Q = K \cdot i \cdot A = K\frac{dh}{dr}2\pi rh = 2\pi Krh\frac{dh}{dr} \tag{2.13}$$

Therefore, $\dfrac{dr}{r} = 2\pi Kh\dfrac{dh}{Q}$

Integrating the above Eq. 2.13 and putting the limits as r_1 to r_2 and h_1 to h_2, we get

$$\int_{r_1}^{r_2} \frac{dr}{r} = \int_{h_1}^{h_2} \frac{2\pi K}{Q} h \cdot dh$$

where, Q = Constant under steady state conditions.

K = Soil permeability (constant at every point as soil is considered homogenoeus and isotropic)

Therefore, $$\int_{r_1}^{r_2} \frac{dr}{r} = \frac{2\pi K}{Q} \int_{h_1}^{h_2} h \cdot dh$$

or $$\int_{r_1}^{r_2} \frac{dr}{r} = \int_{h_1}^{h_2} \frac{2\pi K}{Q} h \cdot dh$$

or $$\left|\log_e r\right|_{r_1}^{r_2} = \frac{2\pi K}{Q} \left|\frac{h^2}{2}\right|_{h_1}^{h_2}$$

or $$\log_e \frac{r_2}{r_1} = \frac{2\pi K}{Q}\left[\frac{h_2^2 - h_1^2}{2}\right] = \frac{\pi K}{Q}\left[h_2^2 - h_1^2\right]$$

or $$K = \frac{Q \log_e \frac{r_2}{r_1}}{\pi\left(h_2^2 - h_1^2\right)}$$

Therefore, $$Q = \frac{\pi K\left(h_2^2 - h_1^2\right)}{2.3 \log_{10} \frac{r_2}{r_1}} \tag{2.14}$$

The Thiem's formula can be more simplified as :

$\left(h_2^2 - h_1^2\right) = (h_2 - h_1)(h_2 + h_1)$ but $h_2 - h_1 = s_1 - s_2$

As compared to the thickness of water bearing material the amount of drawdown is very small, considering the thickness h_1 and h_2 in terms of d, it can be written as shown in Eq. 2.15;

$h_2 + h_1 \approx d + d = 2d$ and therefore

$$\left(h_2^2 - h_1^2\right) = (s_1 - s_2)(2d) = 2\,d(s_1 - s_2) \tag{2.15}$$

Putting the value in Eq. 2.14, we get Eq. 2.16;

We get $$Q = \frac{2\pi K d(s_1 - s_2)}{2.3 \log_{10} \frac{r_2}{r_1}} \quad \text{But } T = K.\,d \tag{2.16}$$

Therefore, we get Eq. 2.17;

$$Q = \frac{2\pi T(s_1 - s_2)}{2.3 \log_{10} \frac{r_2}{r_1}} \tag{2.17}$$

The various assumptions in the above derivation are summarized below as:

1. The aquifer is homogeneous, isotropic and of infinite and areal extent, so that its coefficient of transmissibility or permeability is constant everywhere.
2. The well has been sunk through the full depth of the aquifer and it receives water from the entire thickness of the aquifer.
3. Pumping has continued for a sufficient time at a uniform rate, so that the equilibrium stage or steady flow conditions have reached.
4. Flowlines are radial and horizontal, and flow is laminar.
5. The inclination of the water surface is small so that its tangent can be used in place of sine for the hydraulic gradient in Darcy's equation.

(b) Derivation of Thiem's equation for confined aquifer: For confined aquifers the above derived equation is modified in case of artesian or pressure aquifer. The condition can be described in Fig. 2.16. The flow in actual is radial and horizontal (Fig. 2.17) while other assumptions remain the same in this case also.

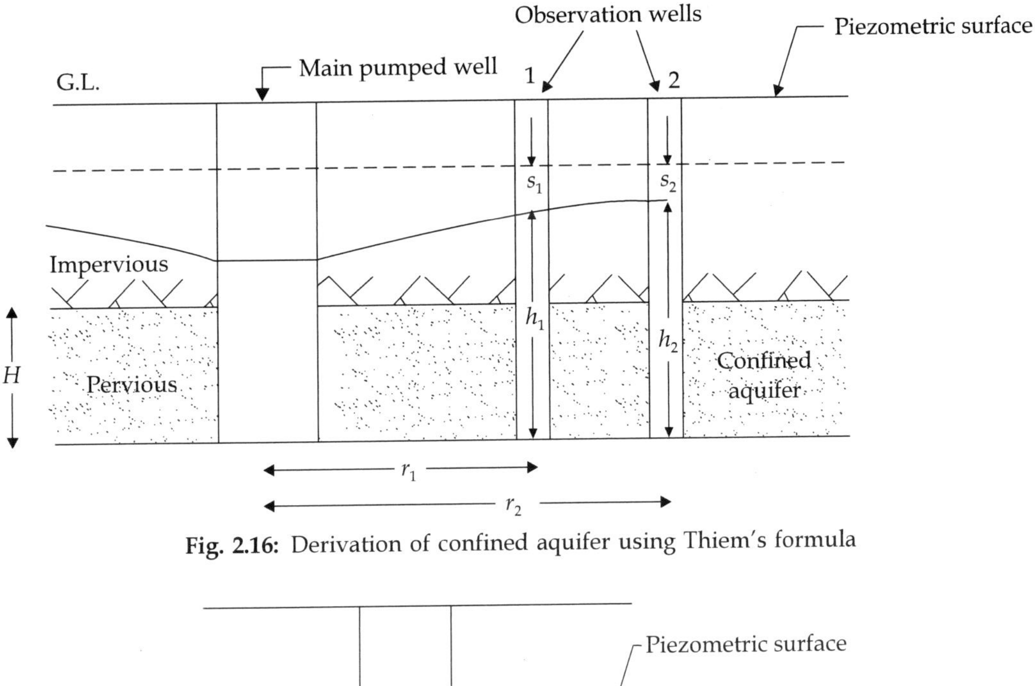

Fig. 2.16: Derivation of confined aquifer using Thiem's formula

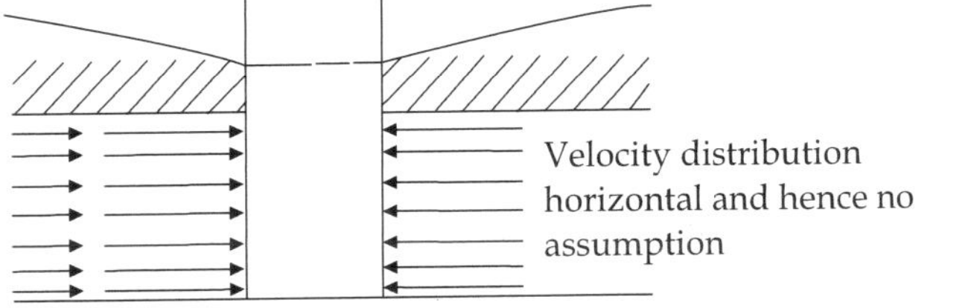

Fig. 2.17: Flow pattern in a confined aquifer

We know that $Q = K.i.A = k\frac{dh}{dr}2\pi rH$

Where the water withdrawal from well of radius r and height H

$Q = 2\pi KrH\frac{dh}{dr}$ and, therefore $\frac{dr}{r} = \frac{2\pi KH}{Q}, dh$

Integrating the above equation and putting the limits as r_1 to r_2 and h_1 to h_2, we get

$$\int_{r_1}^{r_2}\frac{dr}{r} = \int_{h_1}^{h_2}\frac{2\pi KH}{Q}dh$$

where, Q = Constant under steady state conditions

K = Soil permeability (constant at every point as soil is considered homogenous and isotropic)

Therefore, $$\int_{r_1}^{r_2}\frac{dr}{r} = \frac{2\pi KH}{Q}\int_{h_1}^{h_2}dh$$

or $$\int_{r_1}^{r_2}\frac{dr}{r} = \int_{h_1}^{h_2}\frac{2\pi KH}{Q}.dh$$

or $$\left|\log_e r\right|_{r_1}^{r_2} = \frac{2\pi K}{Q}\left|h\right|_{h_1}^{h_2}$$

or $$\log_e\frac{r_2}{r_1} = \frac{2\pi KH}{Q}[h_2 - h_1]$$

Therefore, $$Q = \frac{2\pi KH(h_2 - h_1)}{2.3\log_{10}\frac{r_2}{r_1}}$$

but $h_2 - h_1 = s_1 - s_2$

where, $s_1 - s_2$ is the difference of water levels between the two observation wells under steady state conditions and therefore, we obtain Eq. 2.18;

$$Q = \frac{2\pi KH(s_1 - s_2)}{2.3\log_{10}\frac{r_2}{r_1}} = \frac{2\pi T(s_1 - s_2)}{2.3\log_{10}\frac{r_2}{r_1}} \quad \text{where } T = K.\ H \qquad (2.18)$$

Limitations: Thiem's formula is based on various assumptions however in actual practice all the conditions are always get fulfilled. This may be due to the fact that an aquifer is not fully homogeneous, or the well might have been dug half way through the aquifer, or permeability may not be uniform, or the groundwater table may be inclined sometimes. Also the base of the cone may not be circular or the equilibrium conditions might have not fully reached. However, it is very difficult to assess the effects of these factors, and despite the fact that with such various limiting assumptions, Thiem's formula is widely used and accepted in groundwater problems, as such limitations are adjusted by appropriate calculations.

(i) **Dupuit's equilibrium formulae:** The Thiem's formulas are the refined forms of the original Dupuit's formulas where no such observation wells are constructed. The main well is pumped so as to get sufficient drawdown, and the rate of pumping is so adjusted as to establish equilibrium conditions such that the rate of inflow becomes equal to the rate of outflow, and the water level in the well becomes constant.

The assumptions made in the Thiem's formulas also hold good for the Dupuit's formulas with only difference that the integration done between the limits of r_1 and r_2 in Thiem's formulas changes. The integration carried out between the limits r_w and R, with r_w to be the radius of the main pumped well and R is the radius of influence. The radius of influence is the distance from the centre of the pumped well to the point, where the drawdown is zero or is insignificant.

(a) **Dupuit's formula for gravity well or unconfined aquifer:** The derivation for Dupuit's formula for unconfined aquifer can be derived by using the Fig. 2.18 below as:

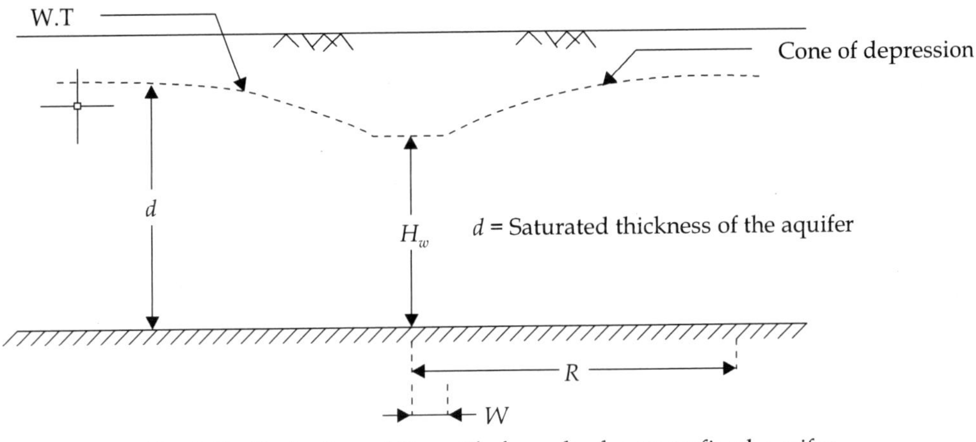

Fig. 2.18: Derivation of Dupuit's formula for unconfined aquifer

We know that $Q = K \cdot i \cdot A = k\dfrac{dh}{dr}2\pi rh$

where, K = Coefficient of permeability

$Q = 2\pi Krh\dfrac{dh}{dr}$ and, therefore $\dfrac{dr}{r} = \dfrac{2\pi K}{Q}h \cdot dh$

Integrating the above equation and putting the limits as r_w to R and h_w to d, we get Eq. 2.19.

$$\int_{r_w}^{R}\frac{dr}{r} = \int_{h_w}^{d}\frac{2\pi K}{Q}h \cdot dh \tag{2.19}$$

where, Q = Constant under steady state conditions

K = Soil permeability (constant at every point as soil is considered homogenous and isotropic)

Therefore, $$\int_{r_w}^{R} \frac{dr}{r} = \int_{h_w}^{d} \frac{2\pi K}{Q} h \cdot dh$$

or $$\left|\log_e r\right|_{r_w}^{R} = \frac{2\pi K}{Q}\left|\frac{h^2}{2}\right|_{h_w}^{d}$$

or $$\log_e \frac{R}{r_w} = \frac{\pi K}{Q}[d^2 - h_w^2]$$

or $$2.3 \log_{10} \frac{R}{r_w} = \frac{\pi K}{Q}[d^2 - h_w^2]$$

or $$K = \frac{2.3\, Q \log_{10} \dfrac{R}{r_w}}{\pi(d^2 - h_w^2)}$$

Therefore, $Q = \dfrac{\pi K(d^2 - h_w^2)}{2.3 \log_{10} \dfrac{R}{r_w}}$ for unconfined aquifer (2.20)

The R values given by various investigators are as (i) by Slitcher 150 m (ii) Tolman 300 m but in actual can be calculated as:

$R = C.Q$ where C is a constant and Q is the discharge.

Putting the value of R in Eq. 2.20 it becomes Eq. 2.21.

$$Q = \frac{\pi K(d^2 - h_w^2)}{2.3 \log_{10} \dfrac{CQ}{r_w}} \qquad (2.21)$$

where, Q is calculated by hit and trial procedure.

(b) Dupuit's formula for pressure well or confined aquifer (artesian wells): The Dupuit's formula for confined aquifer can be derived by using Fig. 2.19 as below:

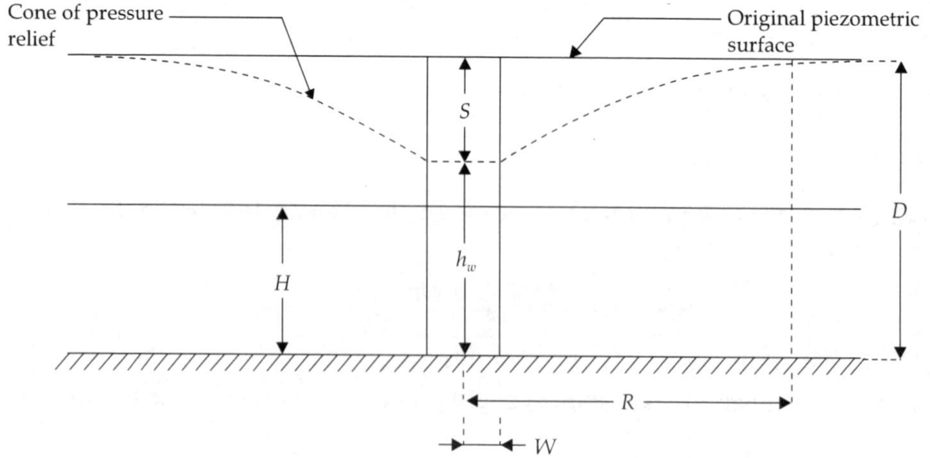

Fig. 2.19: Derivation of Dupuit's formula for confined aquifer

We know that $Q = K \cdot i \cdot A = K\dfrac{dh}{dr}2\pi rH$

Where the water withdrawal from well of radius r and height H

$Q = 2\pi KrH\dfrac{dh}{dr}$ and, therefore $\dfrac{dr}{r} = \dfrac{2\pi KH}{Q}, dh$

Integrating the above equation and putting the limits as r_w to R and h_w to D, we get

$$\int_{r_w}^{R}\frac{dr}{r} = \int_{h_w}^{D}\frac{2\pi KH}{Q}dh$$

where, D = Height of aquifer or well depth below the original level.

K = Soil permeability (constant at every point as soil is considered homogeneous and isotropic)

Therefore, $$\int_{r_w}^{R}\frac{dr}{r} = \frac{2\pi K}{Q}, H\int_{h_w}^{D}dh$$

or $$\left|\log_e\right|_{r_w}^{R} = \frac{2\pi K}{Q}, H\left|h\right|_{h_w}^{D}$$

or $$\log_e\frac{R}{r_w} = \frac{2\pi KH}{Q}[D - h_w]$$ shown below in Eq. 2.22;

Therefore, $$Q = \frac{2\pi KH(D - h_w)}{2.3\log_{10}\dfrac{R}{r_w}} \tag{2.22}$$

But $s = D - h_w$ = drawdown

where, H = Total height of the confined aquifer or well

h_w = Artesian pressure in the well

r_w = Radius of the well

D = Initial artesian pressure at the bottom of the aquifer or the initial height of the piezometric surface from the bottom of the well.

Therefore, we obtain Eq. 2.23;

$$Q = \frac{2\pi KH \cdot s}{2.3\ \log_{10}\left(\dfrac{R}{r_w}\right)} \tag{2.23}$$

(ii) Partial penetration of an aquifer by a well

$$Q = \frac{\pi K(d^2 - h_w^2)}{2.3\ \log_{10}\dfrac{CQ}{r_w}} \text{ (for unconfined wells)} \tag{2.24}$$

$$Q = \frac{2\pi KH \cdot s}{2.3\ \log_{10}\left(\dfrac{R}{r_w}\right)} \text{ (for confined wells)} \tag{2.25}$$

The above equations (Eq. 2.24 - Eq. 2.25) have been derived, respectively for a gravity well and for an artesian well, which penetrate throughout the aquifer. But, if a well does not penetrate upto the bottom of the aquifer, these formulas will not be applicable, as the nature of the flow will become three dimensional. It will not only be radial but will also have an upward component, as shown in Fig. 2.20. The yield of such a well is found to be more than that of a fully penetrating well of the same depth. Kozeny has given a correction factor, and according to him the discharge (Q_p) through such a well is given as follows:

Discharge (Q_p) for a partially penetrating gravity well is shown in Eq. 2.26;

$$Q_p = \frac{\pi K(d^2 - h_w^2)}{2.3 \log_{10} \frac{CQ}{r_w}} \left[1 + 7.\sqrt{\frac{r_w}{2d_1}} . \cos \frac{\pi d_1}{2d} \right] \tag{2.26}$$

where, d_1 = Actual penetration depth below the water table

d = Actual depth of the aquifer below the water table

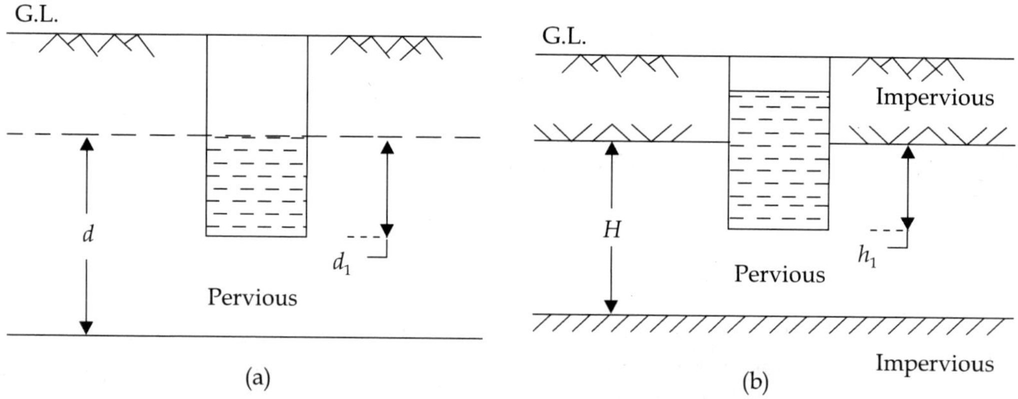

Fig. 2.20: Partially penetrating gravity (a) unconfined well, and (b) confined well

Similarly, the discharge (Q_p) for a partially penetrating artesian well as shown in Eq. 2.27;

$$Q_p = \frac{2\pi K H_1 \cdot (D - h_w)}{2.3 \log_{10} \left(\frac{R}{r_w} \right)} \cdot \left[1 + 7 \cdot \sqrt{\frac{r_w}{2H_1}} \cdot \cos \frac{\pi H_1}{2H} \right] \tag{2.27}$$

where, H = Full depth of the confined aquifer.

H_1 = Depth upto which the well penetrates, Where D, h_w, rand R have the same usual meaning as earlier

(iii) Spherical flow in a well: Figure 2.21 shows a special case of partially penetrating well, where the well just penetrates up to the top surface of the semi-infinite porous medium. In this case, $H_1 = 0$, and the equation is not applicable because

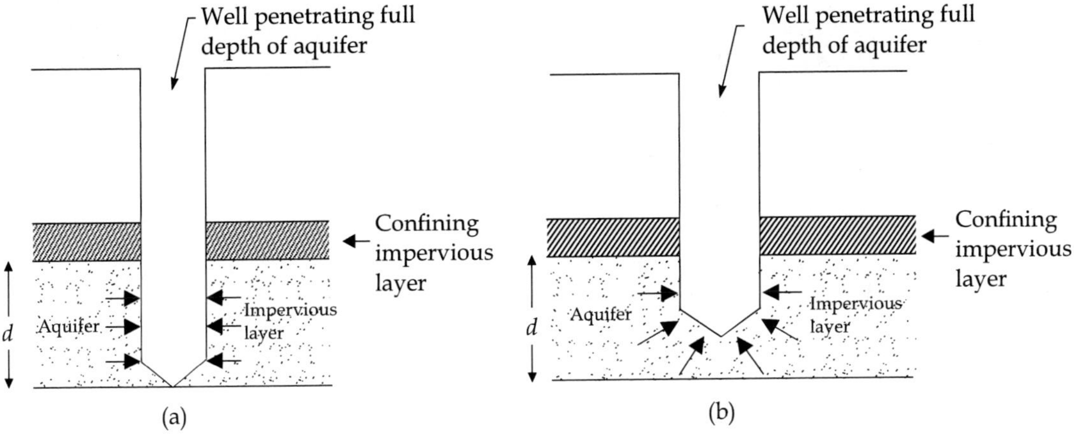

Fig. 2.21: Partial penetration of an aquifer (a) flow lines in a well fully penetrating the aquifer, and (b) flow lines in a well partly penetrating the aquifer

the flow towards the well is purely spherical. The discharge (Q_s) from such a well can be calculated from the Eq. 2.28 below as:

$$Q = 2\,\pi\,Kr_w(D - h_w) \tag{2.28}$$

In case of radial flow in a fully penetrating well discharge represented by Eq. 2.29;

$$Q = \frac{2\pi KH\cdot(D-h_w)}{2.3\log_{10}\left(\dfrac{R}{r_w}\right)}$$

$$\frac{Q_s}{Q} = \frac{2\pi Kr_w.(D-h_w)}{\left[\dfrac{2\pi KH.(D-h_w)}{2.3\log_{10}\left(\dfrac{R}{r_w}\right)}\right]} = \frac{2.3\log_{10} r_w\left(\dfrac{R}{r_w}\right)}{H} \tag{2.29}$$

$$\frac{Q_s}{Q} = 2.3\left(\frac{r_w}{H}\right)\log_{10}\left(\frac{R}{r_w}\right)$$

For example,

$r_w = 12$ cm $= 0.12$ m $\quad R = 125$ m

$\dfrac{R}{r_w} = \dfrac{125}{0.12} = 1042$ and H = Thickness of confined aquifer = 25 m

$$\frac{Q_s}{Q} = 2.3\left(\frac{r_w}{H}\right)\log_{10}\left(\frac{R}{r_w}\right) = 2.3\left(\frac{0.12}{25}\right)\log_{10}(1042) = 0.033 = \frac{1}{30}$$

This indicates the yield in case of spherical flow is much less than that in a radial flow. (Fig. 2.22)

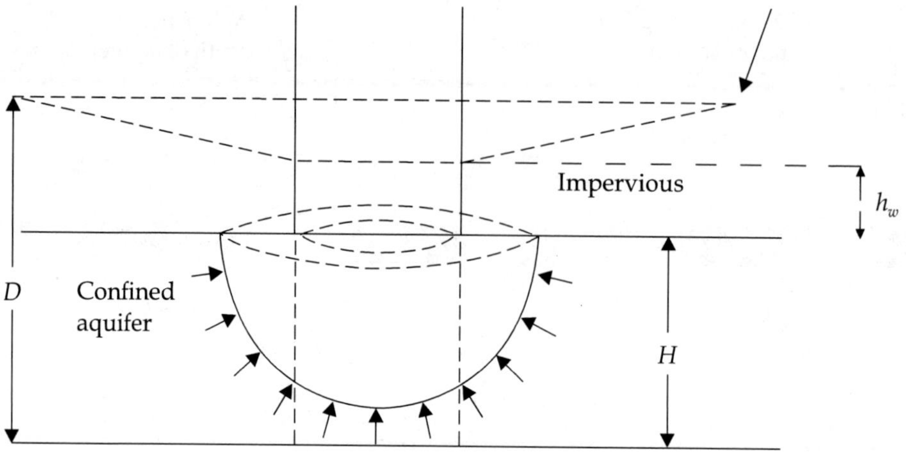

Fig. 2.22: Spherical flow in a well

(iv) Interference among wells: In case of two or more wells are constructed in such a way that they are near to each other and their cones of depressions interact, they are said to interfere. Such interference of wells decreases the discharges of such interfering wells.

Muskat proposed the following formulas for computation of discharges from such interfering wells. These formulas have been found to yield reliable results.

(a) For confined aquifers (artesian wells)

(i) For two artesian identical wells of the same diameter, drawdown, and discharge over the same period of time at a distance B apart, (Eq. 2.30);

$$Q_1 = Q_2 = \frac{2\pi KH(D-h_w)}{2.3\log_{10}\left(\dfrac{R^2}{r_w \cdot B}\right)} \tag{2.30}$$

(ii) For three artesian identical wells at distances B apart, in a pattern of equilateral triangle, (Eq. 2.31);

$$Q_1 = Q_2 = Q_3 = \frac{2\pi KH(D-h_w)}{2.3\log_{10}\left(\dfrac{R^3}{r_w \cdot B^2}\right)} \tag{2.31}$$

(iii) For three artesian identical wells at distances B apart but in a straight line, (Eq. 2.32 - Eq. 2.33);

$$Q_1 = Q_3 = \frac{\dfrac{2\pi KH}{2.3}(D-h_w)\log_{10}\left(\dfrac{B}{r_w}\right)}{\log_{10}\left(\dfrac{R}{B}\right).\log_{10}\left(\dfrac{B}{r_w}\right)+\log_{10}\left(\dfrac{B}{2.r_w}\right).\log_{10}\left(\dfrac{R}{r_w}\right)} \tag{2.32}$$

$$Q_2 = \frac{\frac{2\pi KH}{2.3}(D - h_w)\log_{10}\left(\frac{B}{2.r_w}\right)}{2.\log_{10}\left(\frac{R}{B}\right).\log_{10}\left(\frac{B}{r_w}\right) + \log_{10}\left(\frac{B}{2.r_w}\right).\log_{10}\left(\frac{R}{r_w}\right)} \tag{2.33}$$

where,

Q_1 and Q_3 are the discharges of the outer wells, and Q_2 is the discharge of the middle well.

Total discharge, $Q = Q_1 + Q_2 + Q_3$

(b) For unconfined aquifers (gravity wells): In case of unconfined aquifers the equation of confined aquifers can be replaced by the terms such as HD by $\frac{d^2}{2}$ and $H\,h_w$ by $\frac{h_w^2}{2}$

(i) For two identical gravity wells at distance B apart, formula would therefore, become Eq. 2.34;

$$Q_1 = Q_2 = \frac{2\pi K\left(\frac{d^2}{2} - \frac{h_w^2}{2}\right)}{2.3\log_{10}\left(\frac{R^2}{r_w.B}\right)} = \left[\frac{\pi K\left(d^2 - h_w^2\right)}{2.3\log_{10}\left(\frac{R^2}{r_w.B}\right)}\right] \tag{2.34}$$

(ii) For three identical gravity wells at distances B apart, in a pattern of equilateral triangle, formula is as shown in Eq. 2.35;

$$Q_1 = Q_2 = Q_3 = \frac{2\pi K\left(\frac{d^2}{2} - \frac{h_w^2}{2}\right)}{2.3\log_{10}\left(\frac{R^3}{r_w.B^2}\right)} = \left[\frac{\pi K\left(d^2 - h_w^2\right)}{2.3\log_{10}\left(\frac{R^3}{r_w.B^2}\right)}\right] \tag{2.35}$$

The difference between the actual free surface and the Dupuit's base pressure curve, in a gravity well, arises due to the Dupuit's assumption of horizontal and radial flow. In other words, in an unconfined aquifer, the velocity distribution will not be horizontal near the well but Dupuit assumed it to be so, and this deviation from the actual field conditions gives rise to the difference between the actual water table and the Dupuit's computed water table. The actual water table deviates more and more flow the Dupuit's computed water table in the direction of flow, as shown in Fig. 2.23. The fact that the actual water table lies above the Dupuit's computed water table can be explained by the fact the Dupuit's flows are all assumed horizontal, whereas the actual prevailing velocities of the same magnitude have downward vertical component, so that a greater saturated thickness is required for the same discharge. This discrepancy indicates that the water table does not follow the parabolic path (as given byDupuit).

Nevertheless, for flatter slopes, where the sine and tangents nearly equal, it closely predicts the water table position except near the outflow.

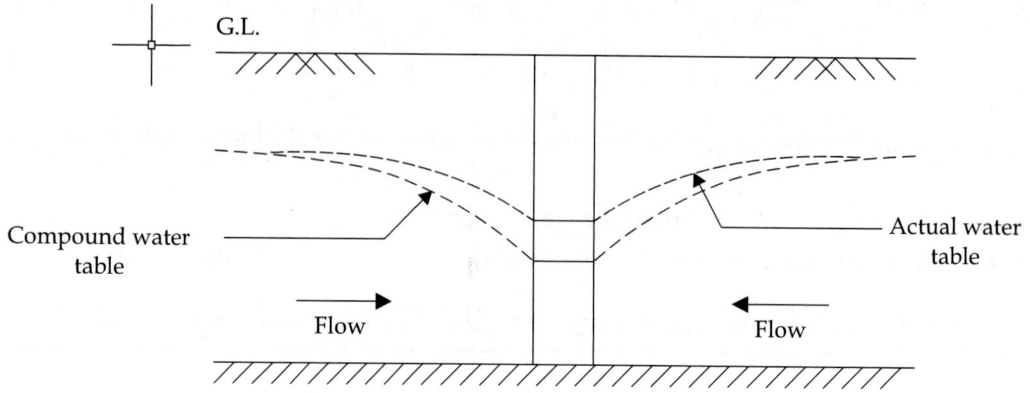

Fig. 2.23: Seepage surface and free surface curve

(c) Equation of free surface curve: The free surface of water in the cone of depression surrounding a pumped well is the surface of water under the atmospheric pressure. And this free surface curve does not coincide with the Dupuit's base pressure curve but lies slightly above the same. The free surface curve is represented by the following Eq. 2.36;

$$Q = \left[\frac{\pi K}{2\,C}\right]\frac{(d-h).d}{\log_{10}\left(\dfrac{R}{0.1.d}\right)} \tag{2.36}$$

where (r, h) are any point on the curve

R = Radius of the influence

K = Coefficient of permeability

C = A constant, the value of which depends on the value of $\frac{r}{R}$, the value can be determine from the curve as Fig. 2.24.

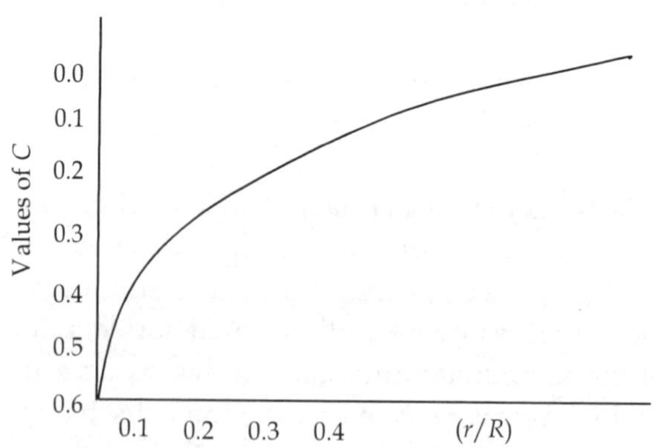

Fig. 2.24: C values to be used in equation

(d) Well loss and specific capacity

Well loss: When water is being pumped out of a well, the drawdown caused includes not only the drawdown which is given by the logarithmic drawdown curve, but also includes a certain drawdown caused by the flow of water through the well screen and axial movement within the well. The drawdown caused by the flow through the screen and the axial movement within the well brings the water level in the well from *AB* to *CD* (Fig. 2.25).

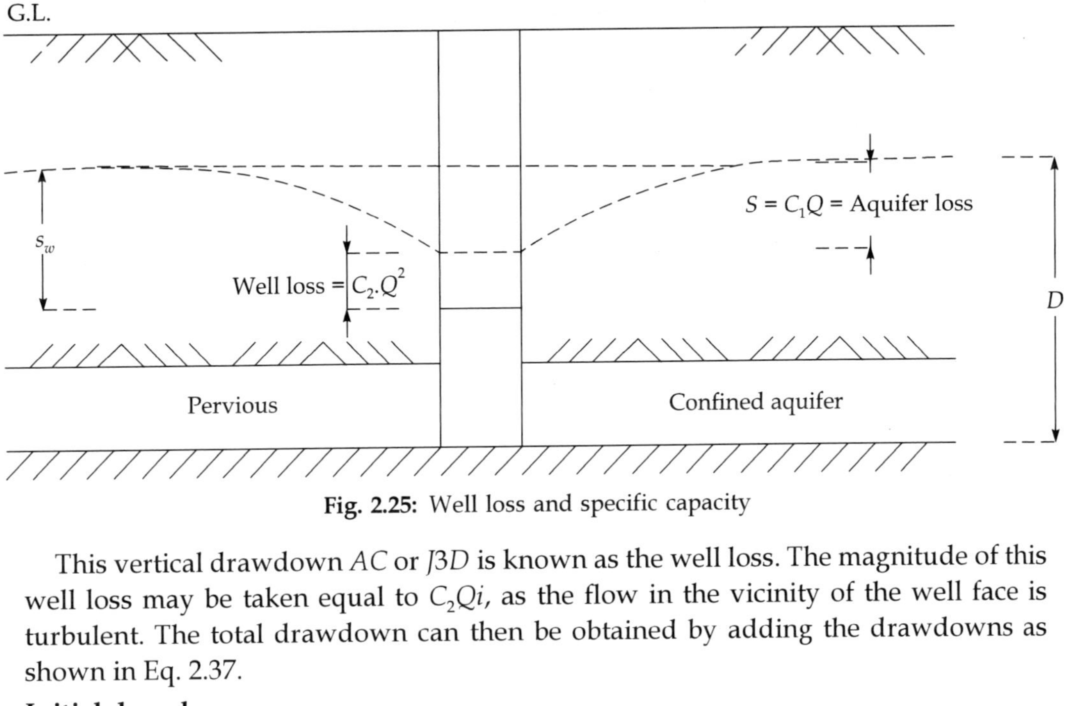

Fig. 2.25: Well loss and specific capacity

This vertical drawdown *AC* or *J3D* is known as the well loss. The magnitude of this well loss may be taken equal to C_2Qi, as the flow in the vicinity of the well face is turbulent. The total drawdown can then be obtained by adding the drawdowns as shown in Eq. 2.37.

Initial drawdown

$$Q = \frac{2\,\pi\,KHs}{2.3\log_{10}\left(\frac{R}{r_w}\right)}$$

or

$$s = \frac{2.3\log_{10}\left(\frac{R}{r_w}\right)}{2\,\pi\,KH} = C_1\,Q \text{ where the term } C_1 \text{ is a constant} = \frac{2.3\log_{10}\left(\frac{R}{r_w}\right)}{2\,\pi\,KH} \tag{2.37}$$

The total drawdown can be calculated as = $A = C_1Q + C_2Q^2$

where C_1Q is the aquifer loss and C_2Q^2 is called well loss

In case of an unclogged or unencrusted well screen, where the screen size is compatible to the surrounding porous media, the well loss is mainly caused by the axial movement inside the well up to the pump intake; because the well loss caused by the water entering the well through the screen is very small. However, when clogging of screen increases, the well loss increases, which adversely affects the pump efficiency.

Specific capacity

The specific capacity of a well is defined as the well yield per unit of drawdown. Therefore, Eq. 2.38 is as below:

$$\text{Specific capacity} = \frac{\text{Discharge of well}}{\text{drawdown}} = \frac{Q}{C_1Q + C_2Q^2} = \frac{1}{C_1 + C_2Q} \quad (2.38)$$

The specific capacity decreases with increase in discharge.

(e) Efficiency of a well: It has been discussed above that discharge from a well is approximately proportional to the drawdown s, (where = C_1Q neglecting well loss).discharge per unit of drawdown was called specific capacity of the well; this specific capacity will be different for different well designs. For determining the best drawdown discharge conditions for a well, the same may be operated under varying drawdown conditions, and then graph may be plotted between discharge and drawdown (called yield drawdown curve) as shown in Fig. 2.26.

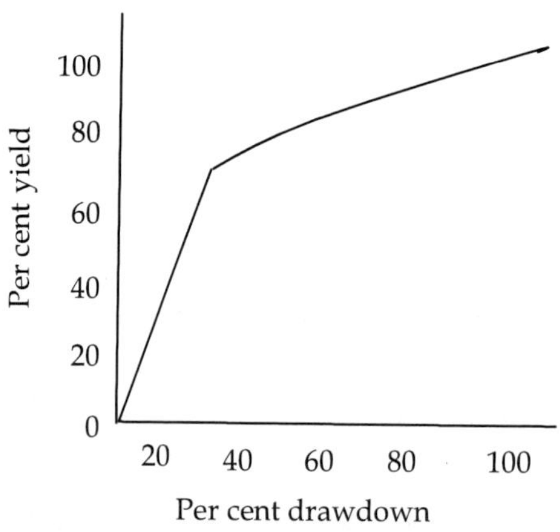

Fig. 2.26: Yield drawdown curve

The curve obtained is a straight line up to a certain stage of drawdown, beyond which the drawdown increases disproportionately to the yield. This places an optimum and efficient limit to the drawdown, which may be allowed to be created in a well. This is generally found to be 70% of the maximum drawdown which can be created in a well

Example 2.4: The pumping has been carried out for a fine sand and gravel upto depth of 18 m with a bed of clay being identified with normal groundwater level at the surface. The holes for observation were made at distances of 3.25 m and 8.0 m from the pumped well with a discharge of 4 L/s at a steady state in about 25 h. The drawdown at 3.25 m was 1.75 m and at 8 m was 0.4 m, compute the coefficient of permeability of the soil?

Solution: $Q = 4\ \text{L/s}$

$r_1 = 3.25$ m $s_1 = 1.75$ m; $h_1 = 18.0 - 1.75 = 16.25$ m

$r_2 = 8.0$ m, $s_2 = 0.40$ m; $h_2 = 18 - 0.40 = 17.60$ m

Putting the values in Thiem's equation, we get

$$Q = \frac{\pi K(h_2^2 - h_1^2)}{2.3\log_{10}\frac{r_2}{r_1}} = \frac{4}{1000} = \frac{\pi K(17.6^2 - 16.25^2)}{2.3\log_{10}\frac{8.0}{3.25}} = \frac{\pi K(309.76 - 264.06)}{2.3 \times 0.391} = \frac{143.57K}{0.899}$$

Therefore, $K = \frac{0.899 \times 4}{1000 \times 143.57} = 0.0025\frac{\text{cm}}{\text{s}}$ **Ans.**

Example 2.5: A tube well of 15 cm diameter penetrates fully an artesian aquifer 27 m thick. Determine permeability of the aquifer if a steady discharge of 30 Lps is obtained from the well under a drawdown of 3.0 m at the well face. Take radius of influence equal to 200 m. (ESE Mains, Civil Engg., 2012)

Solution: Diameter of tube well d_w = 15 cm = 15 cm, Radius of influence (R) = 200 m

$$r_w = \frac{0.15}{2} = 0.075\ \text{m}$$

Discharge, Q = 30 Lps = 0.030 m^3/s

Thickness of aquifer, H = 27 m

Drawdown, s = 3 m

The discharge from a fully penetrating tube well in a confined aquifer is given by,

$$Q = \frac{2\pi KHS}{2.3\log_{10}\left(\frac{R}{r_w}\right)}$$

$$0.03 = \frac{2 \times \pi \times K \times 27 \times 3}{2.3\log_{10}\left(\frac{200}{0.075}\right)}$$

$$2 \times \pi \times K \times 27 \times 3 = 0.2364$$

$$K = 4.645 \times 10^{-4}\ \text{m/sec}$$

2.3.7.5 Design of a Strainer Tube Well

The design of a .strainer type tube well essentially consists of designing the size of the tube, the size of the borehole, the length of the strainer, and type and horse power of the pumping arrangement required to lift the water. The design considerations for these parts of a tubewell1arediscussed below.

Size of the tube: The diameter of the tube (pipe) is decided from the considerations of permissible flow velocity through the tube. Since the strainers are installed at different levels, the velocity of water in a tube of a fixed size will not be constant but will be increasing towards the top. Hence, it is theoretically possible to reduce the size of the tube from the top towards the bottom, such that the velocity is more or less constant throughout the tube length. However, it is generally not economical to install varying size tube, and hence, a single sized or at the most a two sized tube (in case the aquifers are available at too much different depths) may be used in actual practice. The velocity through the tube may be limited between 1.5-to 4.5 m/sec. Knowing the design discharge of the tube well and the velocity, the area of the tube and hence, its diameter can be easily calculated. The nearest available size in the market may then be used.

Size of bore hole: The size of bore hole should be at least 5 cm bigger than the size of the tube, so as to facilitate the lowering of the tube in the hole. If gravel packing is provided then the bore size will be decided by making provision for the gravel thickness also.

Length of the strainer: After deciding the diameter of the tube, the length of the strainer required to obtain the design discharge may be calculated for unconfined or confined aquifer classes, respectively, as given below:

(a) Unconfined aquifer: We have the equation as Eq. 2.39

$$Q = \frac{\pi K(d^2 - h_w^2)}{2.3\log_{10}\left(\frac{R}{r_w}\right)} \tag{2.39}$$

If s is the drawdown or depression head, as in the above equation, therefore

$$d - h_w = s$$

$$Q = \frac{\pi K(d-h_w)(d+h_w)}{2.3\log_{10}\left(\frac{R}{r_w}\right)} = \frac{\pi K.s(\overline{h_w + s} + h_w)}{2.3\log_{10}\left(\frac{R}{r_w}\right)} = \frac{2\pi K.s\left(h_w + \frac{s}{2}\right)}{2.3\log_{10}\left(\frac{R}{r_w}\right)} \tag{2.39 a}$$

$$h_w + \frac{s}{2} = \frac{2.3\ Q\log_{10}\left(\frac{R}{r_w}\right)}{2\pi\ K.s}$$

$$h_w = \frac{2.3\ Q\log_{10}\left(\frac{R}{r_w}\right)}{2\pi\ K.s} - \frac{s}{2} \tag{2.40}$$

h_w is the length of the strainer needed and r_w is the radius of the well (Eq. 2.40).The value of the radius of influence (R) can be assumed between 250 to 500 m, as the variation in R can change the discharge up to a maximum of 12%, due to logarithmic relation between Q and R. The value of permeability coefficient (K) may be assumed suitably. The value of h_w can be calculated if the design discharge and the depression head or drawdown is known.

(b) Confined aquifer: From the given Eq. 2.41

$$Q = \frac{2\ \pi\ KHs}{2.3\ \log_{10}\left(\frac{R}{r_w}\right)} \tag{2.41}$$

$$\text{Therefore, } H = \frac{2.3\, Q \log_{10}\left(\dfrac{R}{r_w}\right)}{2\pi\, K s} \tag{2.42}$$

where H is the length of the strainer and can be calculated by using Eq. 2.42. The strainer length being designed can be provided in one or more aquifers, depending upon the site availabilities.

2.4 INTAKES FOR COLLECTING SURFACE WATER

The specific condition when withdrawal conduit is not a connected part of a dam or any other related arrangement and the water is withdrawn from a surface such as a lake or a river or a reservoir, therefore the access of the withdrawal water required is to be given by constructing an intake structure at the entry of the conduit. The intake structure functions in safely withdrawing water from the source over a determined range of pool levels. This withdrawal water is thereafter discharged into the conduit with the help of gravity flow and is known as intake conduit and is taken to water treatment plant through this conduit. After intake of water through the conduits it may be taken to the treatment plant directly or may be stored in intake well when the gravity flow is not possible in case of direct river supplies. Thereafter, it is taken to treatment plants through conduits after lifting by low lift pumps. The main objective of constructing an intake structure at the entrance of the conduit is to protect the conduit from being damaged or clogged by ice, trash and debris. These intake structure supporting the end of the conduit pipe may be consisting of a simple concrete block up to huge concrete towers housing consisting of intake gates, screens, pumps, etc.

Intake governing factors for site selection

The site required for constructing the intake structures should be selected based on the following:

- The location of intake structure must be selected at the source of purer zone of water in order to cater the best quality of water so that the load of treating water may be reduced on the treatment plant and the selected site should be in close proximity to the treatment plant so that the water transmission cost to the city can be reduced.
- It is advised to avoid the construction of intake at the downstream or in the vicinity of the point of disposal of wastewater. However if it is unavoidable under certain circumstances it is desirable to the construct a weir or a barrage at upstream of the disposal point and install the intake upstream of the barrage. The outbreak of jaundice in 1956 in Delhi caused the contamination of water by Najafgarh drain meeting with Yamuna water at intake point. Thereafter, the Wazirabad barrage in Delhi was being constructed at the upstream of Nazafgarh drain in 1958. Such unseen epidemic needs to be considered while locating the

intake well itself so that people or end users can be saved from water-borne diseases.

- It is advisable to not to construct an intake near the navigation channels so that chances of intake water getting polluted due to the discharge of refuse and waste from ships and boats may be minimized.
- The selection of site should be such that so as to allow maximum withdrawal of water during expansion of treatment of plant due to increase in population with respect to time.
- The intake should have enough provision of drawing water during the driest period of the year and must be located in deep waters, sufficiently away from the shore line. So that during dry periods, at low water level, the intake may not be left dry. However, at river meanders in case of water scarcity during summer seasons, a weir or a barrage must be constructed to divert the water towards the intake point easily. This will ensure adequate water quantity near the intake well or intake tower standing within the under sluice pocket of the river and the water can be directly transmitted to the treatment plant and will ensure constant supply throughout the year. In case of Wazirabad barrage in Delhi, the same was being constructed when the water was not reaching the intake well during summer seasons and creating the trouble. The constructed barrage resolved the problem of water supply during dry periods.
- It is also ensured that the selected intake site should not get flooded and the floodwater should not lodge in the surrounding area of the intake.
- The intake should not be constructed on sharp curves of meandering rivers and it is advisable to construct them on concave banks of the river and avoid the construction on convex banks. This would make the water to remain available on the concave side as on the convex bank side the water may not remain available due to silting and subsequent blockages.

2.4.1 Types of Intake Structures

The various types of intake structures depending upon the source of water can be described as:

(i) Simple submerged intake
(ii) Intake towers
(iii) Medium size river intake
(iv) Canal intake
(v) Intake for dam sluice-ways

2.4.1.1 Simple Submerged Intake

Submerged intakes are generally used to draw water from natural lakes and some rivers and are constructed entirely under water. It consists of an intake conduit laid below the bed of the lake or river with its inlet end placed in the middle of the lake or river projecting above the bed of the lake or river. The withdrawal pipes are generally taken up to the sump well at shore from where the water is lifted by pumps. The inlet end of the intake conduit is protected by concrete or timber crib and it is provided with mouth inlet covered with a screen. In case of lakes where silt tends to settle down, the intake opening is generally kept about 2 to 2.5 m above the bottom of the lake and thus to avoid the entry of silt and sediment. (Fig. 2.27).

The water entering the intake conduit flows under gravity to the bank of the lake or river and collected in a sump well from where it is pumped to the treatment plant. The submerged intake structures are economical and costs less. It also does not cause any hindrance to navigation, causes only little obstruction to the flow of the river and therefore not affected by any floating material. It is also not affected from ice if formed on the surface. The intake structures should be placed in streams or intakes at a place where they may not get buried under sediment under deep water. These are widely used intakes for small water supply projects drawing water from streams and lakes having relatively little change in water surface elevation throughout the year.

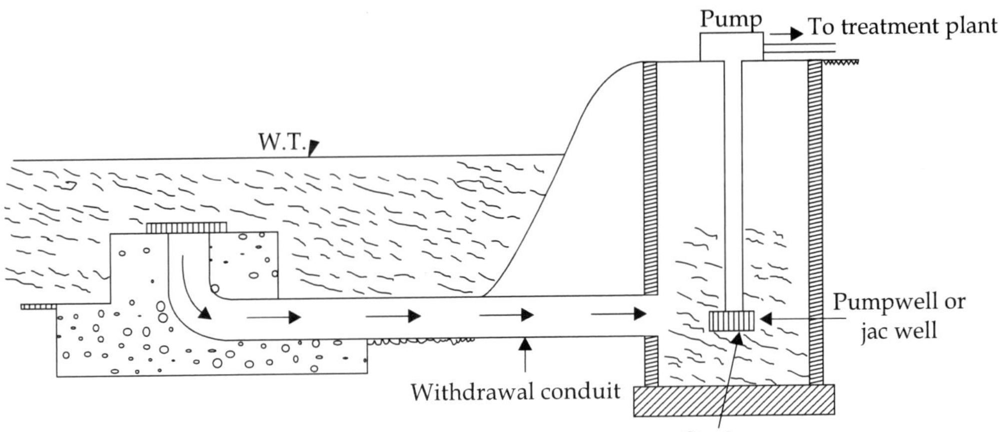

Fig. 2.27: Simple submerged intake made of concrete block

2.4.1.2 Intake Towers and Their Types

Intake towers are used to draw water from reservoirs, lakes or rivers in which there is either a wide fluctuation in water level or the desire to draw water at the most desirable depth, or both. An intake tower is a well type masonry or concrete structure provided with a number of openings at different levels which permit the withdrawal of water from the source of supply. The openings (usually known as intake ports) are provided with intake pipes or penstocks with screens at their inlet ends to prevent the entry of floating debris, and valves to control the entry of water through them. The intake tower houses the valve operating devices, pumps, etc., which are installed in a control room usually located in the top portion of the tower. The intake tower provided in a reservoir created by constructing an earth dam is located at some distance away from the top of the dam. Similarly, in lakes and in some rivers the intake towers are usually located at some distance away from the shore line.

As such for operating the valves, pumps, etc., access to the tower is provided by means of a footbridge connecting the tower and the dam or the shoreline. Further, in order to reach upto the bottom of the intake tower from the floor of the control room a ladder is usually provided. In addition, a ladder is normally installed to reach the bottom of the intake tower from the control room level.

The intake towers may be classified as:

(i) Dry intake towers

(ii) Wet intake towers

(i) Dry intake towers: A dry intake tower is so designed that it has no water inside it when gates are being closed as compared to wet intake towers when gates are close it is full of water. In a dry intake tower as the ports are closed the additional buoyant forces act on it and therefore it requires heavy construction work as compared to wet intake towers. The intake structures possess an advantage that water can be withdrawn at any preferred level by opening the respective ports. The structures are constructed in such a manner in the river so that it may not interact with navigation. While designing the dry intake structures various forces such as hydrostatic forces, wind and earthquake forces, ice debris and other forces caused by waves needs to be considered. The basic difference between dry intake and wet intake is that in a wet intake tower the water enters through the entry ports and then it enters the conduit pipe through separate gate controlled openings. However, in case of dry intake tower the water through withdrawal pipe is withdrawn directly through the gate entry openings as shown through Fig. 2.28. The interior of the tower is thus accessible for inspection and operation.

(ii) Wet intake towers: The wet intake towers are constructed with concrete filled circular shell containing water up to reservoir level. It consists of inside vertical shaft that connects the withdrawal pipe. The water from withdrawal pipe can be directly taken to treatment plant when lifting is not required such as in case of reservoirs. However, in case of river with sump well low lift is required. The withdrawal pipes are constructed in such a manner that either the conduits rest on the riverbed or embedded under the riverbed and acts as tunnels to carry water with openings made into the outer concrete shell as shown in Fig. 2.29. In order to control the flow of water into the shaft and into the withdrawal conduit, the gates are usually provided on the shaft. The water through the withdrawal conduit is taken to treatment plant directly if the treatment plant is located at lower elevation. If the treatment plant is located at higher elevation the water through the withdrawal conduit is taken to the pump house for lifting and transmission.

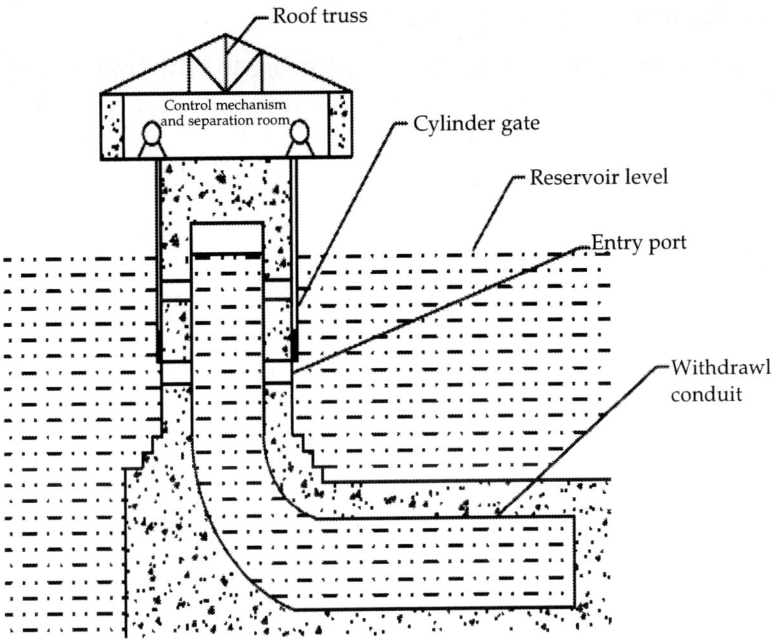

Fig. 2.28: Dry intake towers constructed on a river

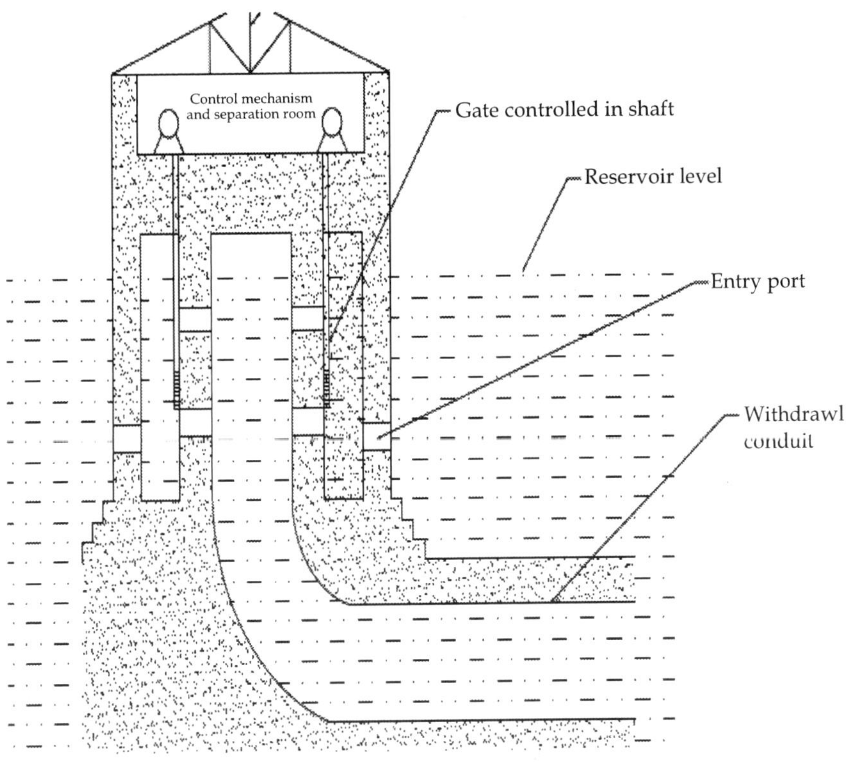

Fig. 2.29: Wet intake tower in constructed in a river or a reservoir

2.4.1.3 Medium Sized River Intake Structures

These types of intake structures are usually constructed to withdraw water from all type of rivers and lakes via between submerged intakes made for small streams and intake towers constructed for reservoirs. These river intake structures can be broadly classified as:

(i) Twin well type

(ii) Single well type

(i) Twin well type of intake structure: It is the most common type of intake structures constructed on all types of river near the bank of it. The details of the structure is shown through Fig. 2.30.A typical example of such an intake structure in India is constructed at Krishna River in Belgaum district of State Karnataka. The typical parts of such an intake structure consist of:

(a) Inlet pipe

(b) Intake well

(c) Jack well (sump)

The circular shape or an oblong well as intake well or collector well is constructed away from the riverbank in the riverbed so water can be withdrawn even during low flow conditions or dry weather flow conditions. The well is generally constructed either with concrete or masonry and raised above the high flood level (HFL) of the river. The well is covered with wooden or concrete sleepers and is made accessible from the river bank through footbridge. This is connected to jack well, which is constructed on the riverbed, by inlet pipe. The ports through the well are made to withdraw water and are fitted with vertical bar screens. The screens are bars of 20 mm diameter placed vertically at a c/c clear spacing of 30–50 mm fixed on iron frame. The flow through velocity through the screens is restricted to 15-20 cm/s in order to prevent the entering of floating material or debris. In order to withdraw water from different levels at different seasons or flow periods the screens in the ports are arranged at two or three levels of the well.

The lower level ports allow water from river during dry weather flow period or low flow conditions. The upper port allows the withdrawal of water during high flow conditions or high flood season. However, the middle level ports helps to withdraw water during fluctuation in river water and under normal river level. Sometimes the screens are provided in the overall vertical length of the well and the withdrawal of water is controlled through gates so that water can be withdrawn from a desired height and level.

The jack well also known sump well constructed on the river bank joins this intake well connected with R.C.C. intake pipe as shown in Fig. 2.8. This section of pipe with depth of 1-3 m is very much helpful for providing space in accumulation of silt and sediment deposits. The intake pipe connecting the jack well and inlet well is a type of non pressure conduit and laid at aslope of around 1 in 200 towards the jack well. The pipe size is designed based on the flow velocity through the pipe should not exceed 1.2 m/s with diameter of the pipe not less than 45 cm and is designed on gravity flow conditions. The water entering the jack well is lifted by pumps and is delivered to the treatment plant units though the rising main. The jack well connecting the intake or inlet pipe is constructed on high ground in close proximity of the river bank. The jack well should be laid on the foundation consisting of hard strata with bearing capacity of not less than 450 kN/m^2. It is advisable to provide lining inside the jack well with and for high firmness should be laid on R.C.C. raft type foundation. This would also be helpful in accessibility in well during operation and maintenance.

For lifting of water the centrifugal pumps or turbine pumps of required horse power may be used. The diameter of the jack well may be kept around 4–5 m so as to accommodate the required pipelines and pumps used for lifting water and should also be accessible for cleaning and maintenance purpose.

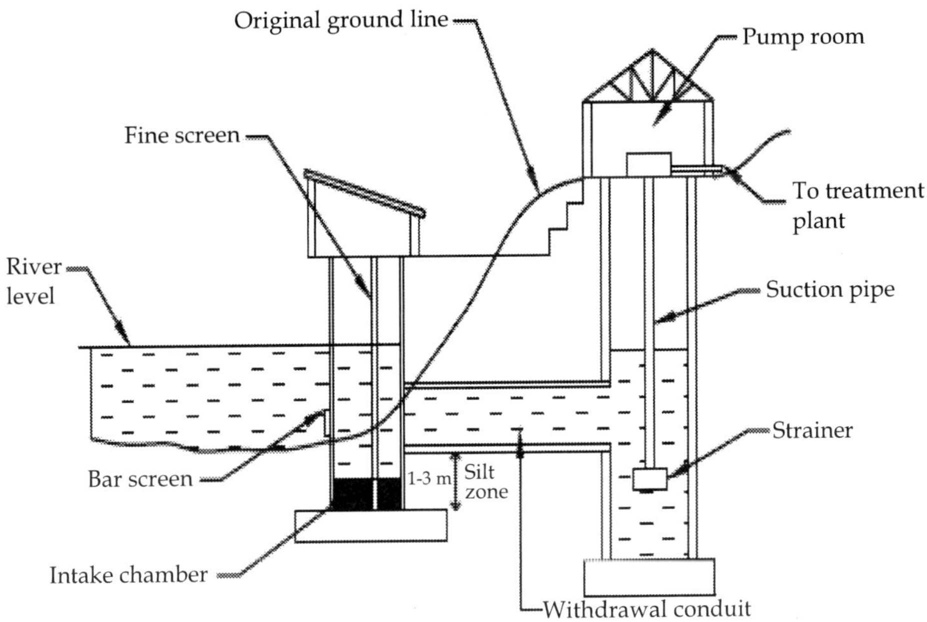

Fig. 2.30: River intake structure with twin well type

In order to avoid the constraints during maintenance generally two pumps are provided in parallel so that if one stops functioning the other one can be started without delay. In case of establishing the turbine pumps the depth of jack well should be kept 3.5 m below the lowest water level of the well. For establishing of centrifugal pumps the depth should be kept around 2 m or so and for high efficiency of the pumps the static head should be limited to 4m and should not exceed to 6 m in any case.

In order to avoid the vortex formation in the jack well or in sump well the intake pipe should be so adjusted and placed near the side walls of the sump with right angels. Also the intake pipe should be extended with long sweep bend and discharged it below the water level. This will prevent the free cascading flow into the water surface. However, in case of such round wells to prevent the vortex formation around the pump suction the vertical baffle bards can be installed opposite the entrance of the pump.

(ii) Single well type on river intake: The weir across the river is generally constructed by impounding the water in alluvial rivers. The channel is sometimes taken off from the upstream side of such a weir the water from the channel is collected and lifted for further supplied to the desired location. Therefore in such a case for collecting water the intake structures the form of a single jack well or sump well with direct entry ports is constructed across this off-take channel. The intake structures of such a type is shown in Fig. 2.31.

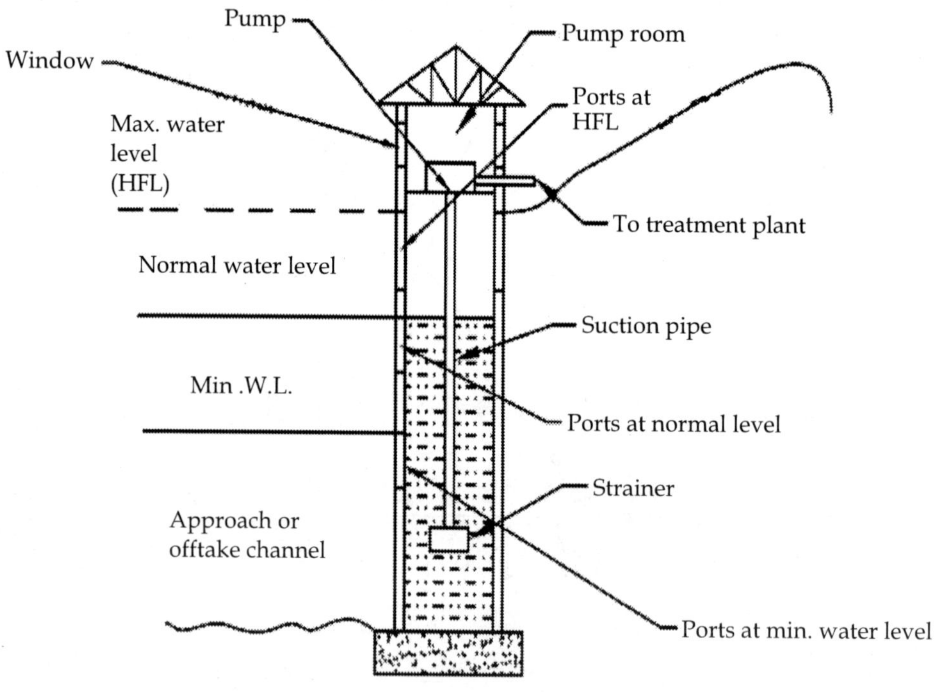

Fig. 2.31: Single well type intake structure constructed across the off-take channel from a river

This single well arrangement eliminates the construction of separate intake well and inlet pipe as required in twin well type. The jack well in itself is provided with openings or ports fitted with bar screens as provided in twin well structures separately. This will help to withdraw the more clear water as the sediment entering the intake will usually be less in the off-take channel. This low sediment either settles down at the bottom of the jack well or may be lifted with water being pumped by using pumps.

The modern-day pumps are enough advanced to easily lift the water containing sediments. The operation and maintenance along with cleansing of the jack well can be carried out easily by stopping the entry of water into the well.

2.4.1.4 Canal Intake

An intake chamber is generally constructed on the bank of canal section. The water through an inlet pipe covered fine screens on its top enters the chamber. The construction of this intake well reduces the flow area in the canal and ultimately results in increase of the flow velocity. Therefore, a pitching is provided near the intake on the downstream and upstream portion of canal. The water from this chamber can be transmitted through conduits to a desired location in city or to the sump house of the water treatment plant. The water entry in the intake well is made through the coarse screens that are placed 0.15 m above the lowest bed level of the canal so that the bed sediment or load may be restricted. The perforations as fine screens fitted with a bell mouth are also provided at the top end of inlet conduit to allow fairly clear water to reach the inlet duct. The entry of water in the outlet pipe is controlled by providing an outlet valve. In order to determine the diameter and size of the withdrawal conduit the velocity through this outlet conduit should not be greater than 1.5 m/s.

The water level in a canal remains more or less constant, inlets at various levels need not be given. Usually, a ladder is installed in the intake chamber in order to reach the bottom of the intake chamber from the manhole atop as shown in Fig. 2.11.The coarse screen with grid of vertical mild steel bars of size 15 to 20 mm diameter with size of opening between each bar for water entry should not be less than 25 to 50 mm. The flow velocity through the coarse screen should not be greater than 0.15 m/s under ideal conditions. The fine screen normally consists of a wire net with openings of 6 mm or less. The flow velocity of the intake conduit through the bell mouth inlet is limited to around 0.3 m/s or so. Additionally, the flow velocity in the intake conduit is typically maintained as 1.0 to 1.5 m/s. Thus, knowing the discharge and putting the value of flow velocity the appropriate diameter of the intake conduit can be assessed accordingly. The total head loss in the intake conduit reaching to the water treatment plant unit can be determined by using Hazen-Williams formula as shown in Eq. 2.43;

$$V = 0.849\, C_H R^{0.63} S^{0.54} \tag{2.43}$$

where,

V = Velocity of the flow (m/s)

R = Hydraulic radius = $A/P = D/4$ (for circular conduits)

D = Diameter of the conduit (m)

A = Area of the conduit (m^2)

P = Wetted parameter of the pipe (m)= πD, when flowing at full

S = Slope = Head loss per unit length of the pipe or conduit = H_L/L (m/m)

C_H = Hazen-Williams smoothness constant of pipe

2.4.1.5 *Intake for Dams Sluice-ways*

The sluiceways of the dams can be utilized for withdrawal of water from its reservoirs. The withdrawal water can be utilized for water supplying to a city from a single purpose water reservoir or multipurpose water reservoir. In earthen dam the intake is constructed near the upstream toe of the dam for withdrawal of water for water supplying. However, in case of masonry dams the intake structures are established inside the body of dam as shown in Fig. 2.32. However, with central impervious cores the earthen dams have intake well or valve towers located within their body as shown in Fig. 2.32.

In all types of such intakes the intake conduits are installed at different levels and connected with common conduits carrying the water to the sluiceway tunnels passing through the body of the dam. The valves are provided in the conduits so as to open and close the water intake as per the requirement and at permissible desirable levels. The arrangement is almost similar to the dry intake towers of a river.

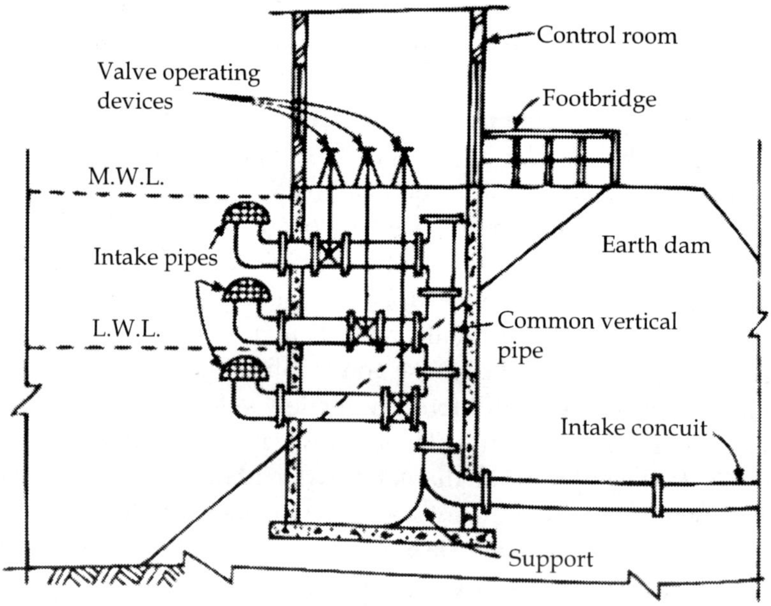

Fig. 2.32: Intake tower constructed on upstream toe of the dam

Portable Intakes or Moving Intakes

Typically, portable intakes or mobile intakes are used to draw water from rivers or streams with sloping banks and having wide variations in water surface levels. Such intakes are often used to avoid the construction of other forms of intakes in the river or stream where moving shoals, lack of foundation in the stream bed or other conditions. In addition, these intakes can also be used in emergencies such as the need to make arrangements for water supply for a mela or festival site, or during wartime, etc. A portable intake consists of a pump mounted on a carriage travelling on a track laid at the intake site as shown in Fig. 2.33. The wagon is pushed and positioned in such a way that the pump suction pipe is kept immersed in the water of the river. A strainer is installed at the end of the suction line, so that the water drained from the river is completely free of suspended solids.

The water drained from the river is distributed via the pumping plant's discharge tubing. When the water level increases or decreases in the channel, the discharge pipe is removed and the carriage along with the pump is placed on the track before the suction pipe is balanced accordingly. Then the discharge pipe is reconnected to the correct level. Intake arrangements are fixed on a movable and floating platform when it is not possible to construct any structure on the riverbed.

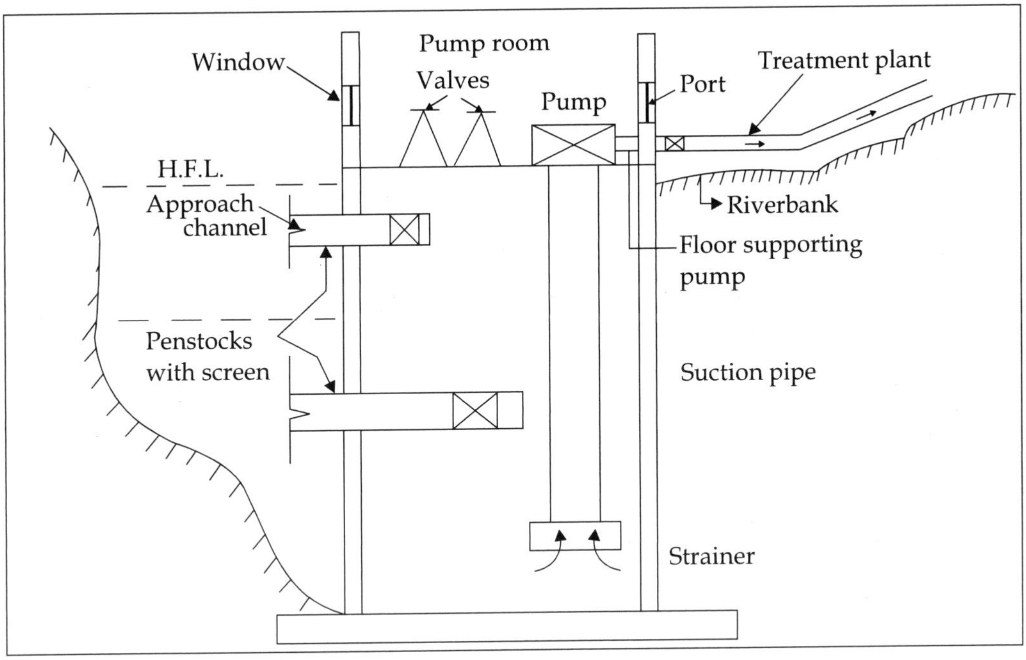

Fig. 2.33: River intake structures

Lake intake

The mainly submersible intakes are used to collect water from the lakes. These intakes are built below water level in the bed of the lake; in order to draw water even in dry season. Such intakes have so many advantages such as no navigational barrier, no floating bodies danger and no ice related trouble. Since these intakes draw small quantities of water, they are not used in large water supply systems, or on rivers or reservoirs as shown in Fig. 2.34.

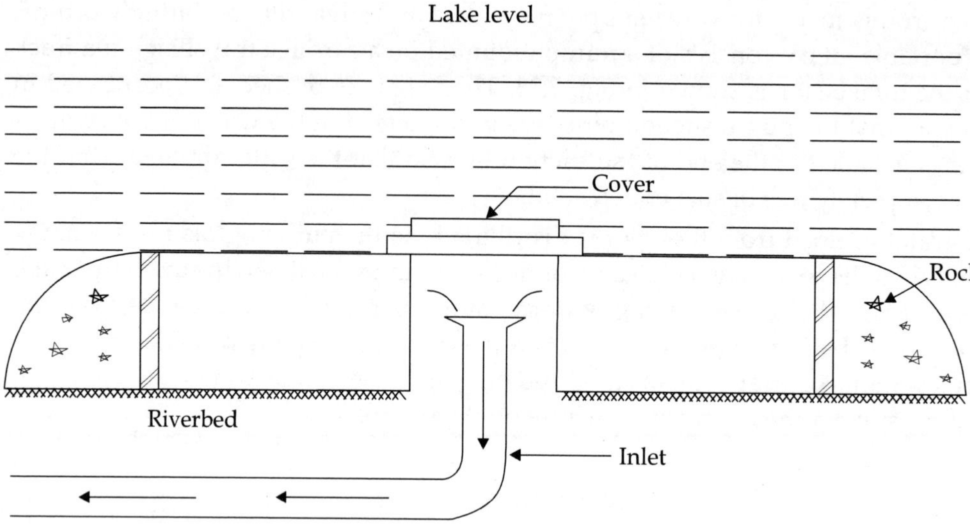

Fig. 2.34: Lake intake

Example 2.6: Design a bell mouth canal intake for a city of 1,00,000 persons drawing water from the canal which runs only 12 hours a day with a depth of 2 m. Also, calculate the head loss in intake conduit if the treatment works are 450 m away. Draw a neat sketch of canal intake. Given average consumption per person to be 200 L/d. Assume the velocity through the screens and bell mouth to be less than 16 cm/s and 32 cm/s respectively.

Solution: Discharge required by the city = $200 \times 1{,}00{,}000 = 20{,}000{,}000\ L/d = 20$ MLD

Since the canal runs only for 12 hours a day, this whole daily flow is required to be drawn in 12 hours.

$$\text{Intake load per hour} = \frac{20\text{ ML}}{12\text{h}} = \frac{20\text{ ML}}{12\text{h}} = 1.67\frac{\text{ML}}{\text{h}}$$

$$= \frac{1.67 \times 10^3\text{m}^3}{\text{h}} = \frac{1.67 \times 10^3\text{m}^3}{60 \times 60\text{ s}} = 0.46\frac{\text{m}^3}{\text{s}}$$

Design of coarse screens: The area of coarse screen can be calculated by assuming the vertical iron bars of 20 mm diameter with clear spacing of 3 to 5 cm from centre-to-centre distance. Assuming the velocity through screens to be 0.16 m/sec

$$\text{Area} = \frac{\text{Discharage entering the well through this coarse screen}}{\text{Velocity through the screen}}$$

Therefore, the minimum area of coarse screens $= \dfrac{0.46}{0.16}\text{m}^2 = 2.875\text{ m}^2$

Height of screen provided can be calculated as = 2.0 – 0.15 – 0.3* = 1.55 m

(*Assuming the minimum water level in the canal to be 0.3 m below the FSL, thereby keeping the top of screen to be height - 0.3 mabove the bed level)

$\therefore$ Minimum length of screen (opening) required $= \dfrac{2.875 \text{ m}^2}{1.55 \text{ m}} = 1.86$ m

Now, assuming the clear opening widths between vertical bars to be 40 mm each, we have

No. of openings $= \dfrac{1.86}{0.04} = 47$ (say n)

$\therefore$ No. of bars $= n - 1 = 47 - 1 = 46$

Length occupied by bars of 20 mm $\phi = 46 \times 0.02$ m $= 0.92$ m

$\therefore$ Total length of screen = 1.86 + 0.92 = 2.78 m; say 2.8 m

Hence, provide 2.8 m length of coarse screen with effective height of 1.55 m in the rectangular intake well.

Design of bell mouth entry (Fig. 2.35)

$$\text{Area of bell mouth entry} = \frac{\text{Discharge}}{\text{Velocity through the bell mouth}} = \frac{0.46 \text{ m}^3/\text{sec}}{0.32 \text{ m/sec (given)}}$$

$$= 1.44 \text{ m}^2$$

If d is the diameter of bell mouth entry of intake, then;

$$\therefore d = \sqrt{\frac{1.44 \times 4}{\pi}} = 1.35 \text{ m; say } 1.3 \text{ m}$$

Hence, use 1.3 m ϕ bell mouth provided with perforations for fine screens.

Design of intake conduit

Assuming the flow velocity through the conduit as 1.5 m/sec, we have

$$\text{The area of conduit} = \frac{\text{Discharge}}{\text{Velocity}} = \frac{0.46}{1.5} \text{ m}^2 = 0.31 \text{ m}^2$$

Therefore, the diameter (D) of the conduit can be calculated as :

$$\therefore D = \sqrt{\frac{0.31 \times 4}{\pi}} = 0.62 \text{ m; say } 0.6 \text{ m}$$

Therefore, provide the conduit with 0.6 m diameter, based on this value the flow velocity should be calculated.

$$\text{Therefore, flow velocity through 0.6 m diameter pipe} = \frac{0.46}{\frac{\pi}{4}(0.6)^2} \text{ m}^? = 1.6 \text{ m/sec}$$

Head loss through the conduit up to treatment plant unit can be calculated by suitable Hazen-Williams equation as:

$$V = 0.85\, C_H\, R^{0.63} S^{0.54}$$

where, C_H = Coefficient of pipe material = 130 for cast iron pipe.

R = Hydraulic mean depth of pipe

$$= \frac{d}{4} \text{ (for pipes running full)} = \frac{0.6}{4} = 0.15 \text{ m}$$

V = Flow velocity through pipe = 1.6 m/sec

S = Slope of energy line = H_L/L = head loss per unit length of pipe

$\therefore 1.6 = 0.85 \times 130 \times (0.15)^{0.63} \times S^{0.54}$

$$\text{or, } S^{0.54} = \frac{1.6}{0.85 \times 130 \times (0.15)^{0.63}} = \frac{1.6}{0.85 \times 130 \times 0.30} = 0.0482$$

$$\therefore S = (0.0482)^{\frac{1}{0.54}} = (0.0482)^{1.85} = 3.67 \times 10^{-3}$$

$$\text{Now, } S = \frac{H_L}{L} = \frac{\text{Head loss}}{\text{Length of pipe}}$$

$$\therefore H_L = L \times S = 450 \times 3.67 \times 10^{-3} = 1.65 \text{ m}$$

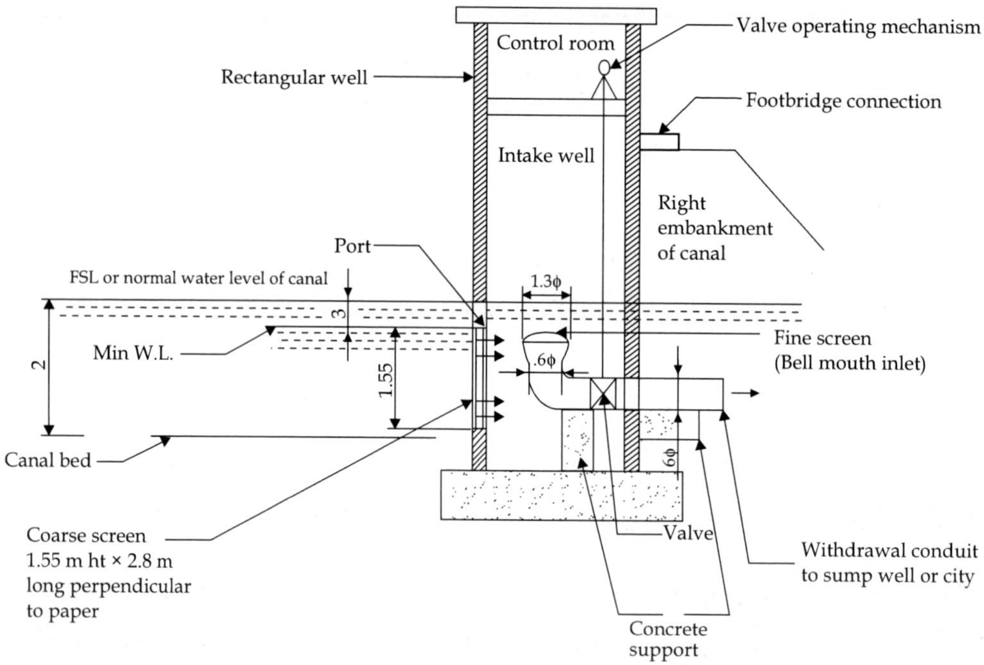

Fig. 2.35: Design of bell mouth entry for canal intake

Example 2.7: Design a river intake structure with the following facilities as (i) number and size of openings in intake well, (ii) size, shape and height of intake well, and (iii) gravity pipe for raw water connecting intake well and jack well. The data available is:

RL of riverbed = 98 m

RL of lowest water level = 101 m

RL of normal water level = 113 m

RL of high water level = 122 m

Population to be served = 75,000

Average water demand = 220 L/head

Assume any other required data suitably

Solution: Average water demand to be pumped = 75,000 × 220 = 16.5 MLD

Maximum water demand = 1.8 × 16.5 MLD = 29.7 MLD

Assuming the pumping to take place for 16 hours a day, discharge to be pumped

$$= \frac{29.6 \times 10^6}{10^3 \times 16 \times 60 \times 60} \text{ m}^3/\text{s} = 0.515 \text{ m}^3/\text{s}$$

Design of inlet well

Let us provide an oblong shaped inlet well, with openings provided at 3 levels; one layer of opening(s) shall be kept below RL 101.0 (lowest river level); and the other may be kept below HFL stage, i.e., below RL 122 m. A middle layer of ports can be provided just below the normal river water level of RL 113 m.

Assume the screens with 50 mm openings with 20 mm diameter of each screen bar. The flow through velocity in the bar screens may be assumed to be 0.15 m/s.

Therefore, area of openings required at each level $= \frac{Q}{v} = \frac{0.515}{0.15} \text{m}^2 = 3.43 \text{ m}^2$

Let us provide the height of screen to be 1m at the openings, therefore the clear length of the openings required $= \frac{3.43 \text{ m}^2}{1.0 \text{ m}} = 3.43 \text{ m}$

Therefore, number of openings required @ 50 mm opening size $= \frac{3.43}{0.05} = 69$ (say n)

Therefore, number of bars = $(n - 1)$ = 69 – 1 = 68

Therefore, total length occupied by 68 bars @ 20 mm dia bars = 68 × 0.02 = 1.36 m

The total length of screens can be computed as = 3.43 + 1.36 = 4.79 m (say 4.8 m)

Let us provide 2 ports at each level with size of each port to be 1 m height x 2.4 m length. Therefore, 6 screened ports are required with 2 ports at each of the three levels.

These 2 screened ports will be provided within the well steining between RL 101 m to 100 m, each having 2.4 m length

The other 2 screened ports shall similarly be provided between RL 113 m to 112 each having 2.4 m length

The remaining 2 screened ports will then be provided at RL between 122 m to RL 121 m with length of screen to be 2.4 m.

These ports can be fitted in an oblong well, consisting of rectangular length of 5 m (sufficient to fix 2 bar screens each of length 2.4 m) and provided with circular ends. The well can have a width of say 3 m.

This inlet well can be sunk into the riverbed by say 3 m below the riverbed, as to provide space for accumulation of sand and silt. So let us keep the bottom of inlet well at RL 95 m.

Also, considering the free board of 2 m over the river HFL to fix the bottom level of roof of the well, therefore provide an invert level of roof at RL 124 m. Therefore, the overall height of the inlet well is = 124 m – 95 m = 29 m.

Design of gravity pipe connecting intake well and jack well as shown in Fig. 2.36.

The intake pipe shall be designed to a gravity flow with maximum velocity of 1.2 m/s.

$$\text{Area of pipe required} = \frac{\pi}{4}d^2 = \frac{Q}{v} = \frac{0.515}{1.2}$$

$$d = \sqrt{\frac{4}{\pi} \times 0.429} = 0.74 \text{ m}$$

Therefore, using 0.75 m diameter intake pipe, the velocity would be

$$= \frac{0.515}{\frac{\pi}{4}(0.75)^2} = 1.16 \text{ m/s}$$

Using Manning's formula;

$$v = \frac{1}{n}R^{2/3}\sqrt{S} \text{ and assuming } n = 0.017$$

$$1.16 = \frac{1}{0.017}\left(\frac{0.75}{4}\right)^{2/3}\sqrt{S}$$

$$S = \frac{1}{235}; \text{ say around 1 in 200}$$

Therefore, the intake pipe with 0.75 m diameter can be laid with slope of 1 in 235.

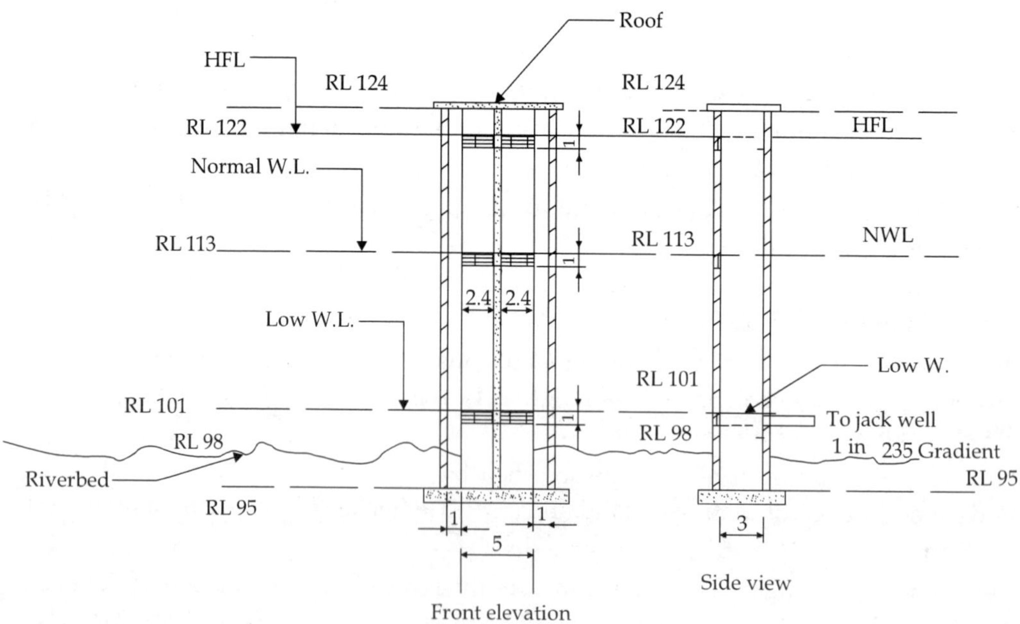

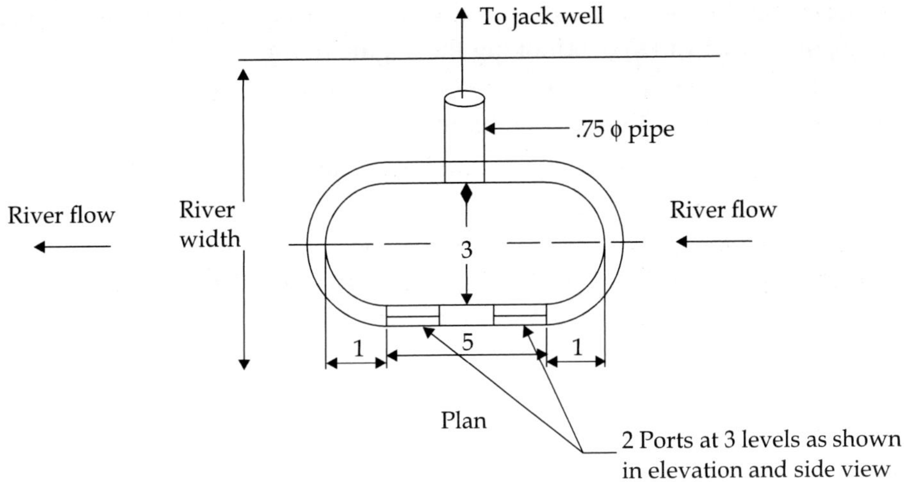

Fig. 2.36: Design of jack well for river intake structures

2.5 PUMPING STATIONS

2.5.1 Lay-out of Pumping Stations

No overall criterion can be developed for the configuration of the pumping stations, as local factors will be of great influence. Generally, the model is the product of a design phase in which all the above design considerations are taken into consideration. Generally speaking, it can be said that the layout of a pumping station is a reasonable fit of all station functions, with ample space to pass between machinery for erection and maintenance purposes, but without excessive empty spaces, either horizontal or vertical. In general, the flow lines should be as short as possible and there should be no unnecessary twists in the piping.

Spaces that may be required are:

- Pump hall/engine room
- Transformer station including high voltage switch gear
- Low voltage switch gear, switch board, control desk, often combined and stores
- Workshop and stores
- Office, toilet (including water supply)
- (Central) heating and/or ventilation
- Generator room

All spaces are to be well lighted. Lighting can also be required outside. Whether a pumping station should have all of these facilities would obviously rely on the nature of the plant and its location. The layout should be constructed in such a way that the free movement of moving cranes and crabs is not hampered by the existence of diesel engine supply pipelines or exhaust pipes. The machinery would need to transfer to the building's workshop floor and/or entrance doors. Proper railings are required along stairs or on platform (catwalks) overlooking the engine room.

Flooring of the station should be such that it will not become slippery by oil leaks from equipment or by grease lost during maintenance work.

Drainage openings for the scrubbing should be provided.

The site configuration of the grounds around the station will ensure sufficient access to the station, for both the erection and possibly removal with electronic devices. They should be able to supply petrol. Proper fencing is necessary, particularly for automated stations. Where necessary, there should be ample space available for potential station extension.

Design procedure

Many steps in sequence have to be taken in a standard design process.

All applicable design requirements should be gathered in the first phase. Preliminary choices would have to be made in the second phase about the number and size of pumps to be used, the types of drives and transmissions, and the auxiliary equipment. Some of these choices would require further elaboration, and plans will be made for a house.

When these first setups are first tested, one or more solutions will be carried out in more detail, resulting in one or more preliminary designs. Such preliminary designs are to be compared on technological as well as economic merits.

The definite design will be made after a final choice has been made, in which other factors may also play a role. They can be written on the basis of the definite design requirements. This should usually be provided separately for the mechanical and civil engineering constructions. The mechanical installations may also be broken up in larger installations; e.g. different electrical system requirements or specific products such as mechanical trash-rack raking equipment, generator sets, etc. Manufacturers and contractors may be invited to tender depending on the requirements. The specifications for mechanical installations in particular need to be very clear in terms of specifications ,quality standards and performance.

It would, of course, be possible to request deals on the basis of the concept alone, without any requirements. The major drawback of this process, however, would be the receipt of deals that cannot be measured in terms of quality and performance. Therefore, the practice should be discouraged. After comparison of all tenders, contracts may be made among suppliers, contractors and the administrator, buyer or principal. Once agreement has been reached on the contract (and on the assumption that funding has been arranged), production of pumps will start and construction will begin on the site.

Sufficient supervision on the part of the principal will be required during manufacture and construction. Not only would this oversight deal with the actual building of the station and the installation of the equipment, it will also be important to test the performance of the mechanical equipment in the various factories. Detailed drawings of the equipment would also need to be reviewed on behalf of the principal, before beginning the actual production. Finally, when the station is operational tests must be carried out to check the pump output, etc.

Meanwhile the company must be formed for the station's potential operation and maintenance, often with the assistance of the pump supplier, and/or a consultant. The above technique can be streamlined for small pumping stations, and/or when a reliable supplier uses regular pumps and drives.

In this case, however, the contract would also have to be precise about quality requirements, expected results, etc. At all of these design phases, good communication between all the disciplines involved in the design process is expected. This will in practice be performed by, or on behalf of, the commissioning agency or the project principal. Coordination includes prompt preparation of plans, manufacture and shipping of equipment, station construction and on-site installation of equipment. To this respect, creation of a time schedule, to which all construction members (including the principal) would have to be committed, is part of the process of planning.

2.6 WATER TREATMENT

General guidelines: Water treatment requires transformation of raw water into potable quality through physical, mechanical, bacteriological and biological means. Design of water treatment plant requires the information given below:

- The water demand: The treatment works should be designed for the peak day demand
- The raw water quality: It is required to decide the type of treatment.

The following methods of water treatment are considered to be suitable for Tanzania.

- Screening or straining
- Plain sedimentation
- Chemical coagulation, flocculation and settling
- Filtration
- Disinfection
- Control of algae
- Taste and odour control
- Softening
- Removal of iron and manganese
- Defluoridation of water

Based on the life span of the different structures the following recommendations are made:

- Pumps, steel tanks, mechanical and electrical equipment, and internal piping must be planned to meet the potential 15-year demand.
- All other structures must be designed to meet future 20-year demand.

Gravity main

In design, service, and maintenance, a gravity main is, of course, the most preferred in terms of economy. The key one must also be of such a size that the total amount needed for the future expected peak day demand will flow through the pipe within 24 hours.

The main gravity will be on a steady decreasing gradient as far as possible, avoiding high points and low valleys. When the static pressure exceeds the allowable pressure of the pipe a break pressure tank with ball valve should be mounted.

Excessive high points should be avoided and the pipeline should not be laid higher than the hydraulic gradient (negative pressure) at all. Where it is inevitable, the pipe must be mounted freely on concrete supports and kept in place by metal ring braces, which are placed in the concrete support, for the pipeline to be laid above ground. Inside the ring bracket the pipe requires to have free movement.

Galvanized steel or ductile iron can be used in the pipes above ground The pillars are to be mounted at pipe length.

Concrete anchors in steep gradients and much smaller distances in steeper gradients should be built on all gradients at every 200 m and at far less distances. Anchors should also be given at all horizontal and vertical direction shifts at all equivalent tees and at all valves. The concrete should be shaped in such a way that the pipe curvature follows.

Suitable air valves should be fitted with stop valves at all high points, and one air valve at every 1000 m in flat areas. Arrangements should be placed at all low points of wash out. They are not to have the same diameter as the central one.

The non-return or reflux valves can be located 3 to 5 km away, facilitating maintenance and repair, and helping to reduce water hammer. The repair and replacement unions or flanged joints should be installed in all pipelines every 350 to 500 m for inspection purpose.

Pipe will be 0.6 m long by 0.75 m in well equipped trenches, for medium pipes up to 100 mm; and for larger pipes, 0.8 m long by 1.0 m medium. The pipe should be placed on a level bed cushion of sand or dirt, free of stones under the pipe. It should also be free of stones around the pipe.

Special case should be taken to where mains are crossing roads and tracks, where a minimum of 1m cover should be provided. It is an advantage when the pipes can be laid inside a steel, D1 or C1 pipe with a larger diameter. Those protections and covers should be extended 3 m beyond road width on both sides. The gravity mains are built in such a way that the pressure head available is only lost in overcoming the frictional resistance to water flow. Therefore, the speed to be produced is so maintained that it is, neither too low to require a large diameter pipe; nor too high to cause unnecessary pressure and head loss.

2.6.1 Water Tanks

In order to align the constant supply rate from the water source or treatment plant with the fluctuating water demand in the distribution area, water tanks are provided for. The storage capacity should be sufficiently high to satisfy the combined variations between water supply and demand, and in case of breakdown between source and tank. If there were no water reserves in the distribution areas, the supply source and the treatment plant would have to be able to meet the variations of the community's water demand served. Generally, it is not economical and often not even technically feasible. For that purpose, a water tank is provided. It also helps to maintain adequate pressure in the distribution. Storage tanks may be classified in the following way:

(i) Classification based on the position of the tank. Under this category tanks may be:
   (a) Surface storage tanks
   (b) Elevated storage tanks

(ii) Classification based upon the material of construction.
   (a) R.C.C. tanks
   (b) Masonry tanks
   (c) Concrete tanks
   (d) Steel tanks

(iii) Classification based on the shape of the tank.
   (a) Circular tanks
   (b) Rectangular tanks
   (c) Intze tanks

2.6.2 Calculating Capacity of the Reservoir or Tanks

The total capacity of the reservoir can be divided into three categories. The total of all the three categories determines the storage capacity of the reservoir.

- **Balancing reserve:** Water demand always varies from hour to hour, but treated water often comes at a constant rate from treatment plants. Balancing reserve is the quantity of water required to be retained in the distribution network to balance the variable demands. Mostly, it is measured using mass curve or hydrographs.
- **Break-down reserve:** This storage is estimated to be 1½ to 2 hours of average daily supply.
- **Fire reserve** = $(F \cdot C)\, T$ (2.44)

where,

F = Fire demand

C = Reserve for pumping capacity

T = Fire duration 10 hours and (2-5) hour

Fire reserve is calculated in Eq. 2.44.

2.7 DISTRIBUTION SYSTEM

Having treated the raw water in water treatment plants to the appropriate specifications, the last and final step of water supply schemes includes water delivery to consumers. The main objective of the distribution systems is to establish adequate water pressure at different points. It depends on the topography of the distribution area and its elevation regarding the location of the water treatment plants. The distribution system may be classified into three categories.

- Gravity system
- Pumping system without storage
- Dual system with storage

Water from storage tanks can be distributed to consumers by arranging pipelines in two ways which is explained later on.

2.8 DESIGN OF THE SEWERAGE SYSTEM

Energy concept

The energy principle of sewerage should be taken into consideration before starting its design. Our system is intended to move waste from any section of the city or sewerage area to a designated point where it will be handled or discharged into the surface water. Energy is required for transporting water from one point to another. This energy is available in the form of potential energy, so that the point of entry of water into the system is higher in the lever than the point of discharge. The lost potential energy is equal to the friction of the sewer wall, multiplied by the distance traveled by the sewage.

Knowing that the friction is proportional to the square of the velocity, we see that we need to be thrifty with energy in flat areas and plan our sewers to run at the minimum allowable level. However, if there is insufficient energy available, a supplementary supply may be given by pumping. If there is more than the minimum energy requirement we will allow a higher flow speed. If it happens that the full allowable flow velocity does not use all the available energy, the surplus must be dissipated by, for example, drop manholes. Based on these considerations, it can be concluded that utmost care needs to be taken not to lose energy unnecessarily, particularly, in flat areas. This means that friction and significant flow velocity variations in the sewer network need to be avoided, manholes and bends need to be simplified, no application of drop manholes, overflows need to be planned with minimal energy loss for DWF and transitions need to be properly engineered.

2.8.1 Layout

The first move will be to design the layout. One set protocol cannot be followed. In general, there should be an effort to follow the trend of natural drainage. Working from the overall plan down to the specifics is advisable. Next, mark on a map the boundary lines of the area to be sewered (catchment area), the watersheds and the major valleys. The sewer lines then need to be tentatively designed to match the most efficient flow pattern and the area's main sewer line needs to be decided. It should be borne in mind that the shortest route must always be chosen. Figure 2.37 gives some examples of flow patterns.

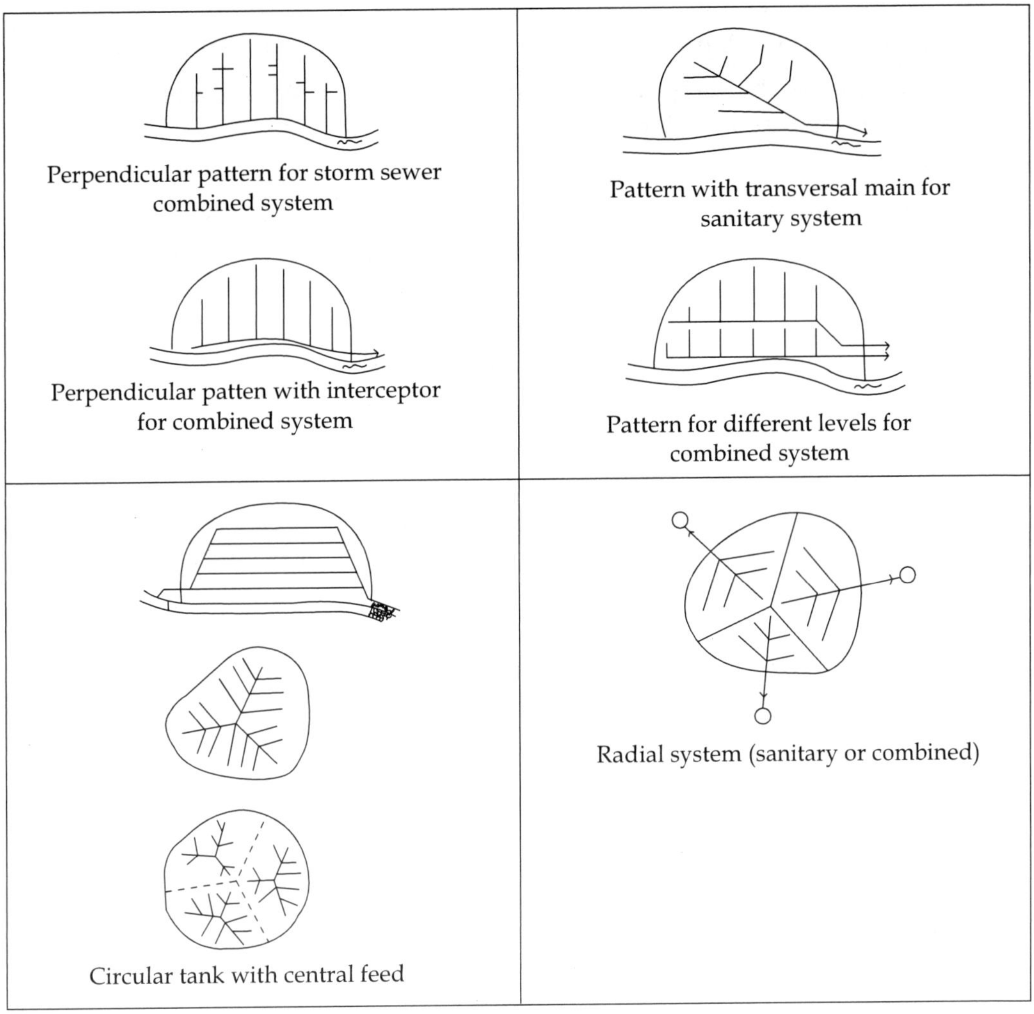

Fig. 2.37: Patterns of sewerage systems (Operation and Maintenance of Water and Sewerage Systems, 1999)

Some general rules to follow during the design of the layout are as follows:

- No sewer can pass underneath the building
- Prevent intersections with railroads, canals, rivers, etc.
- Manholes must be plotted at all junctions and slope or direction shifts. The overall distance to the manhole will be between 35 and 50 metres. A drain has to

be straight in between manholes. The sewer can be examined from these manholes to find obstructions, and a cleaning system can be passed through them with buckets and steel wire. In the case of curved sewers, this is not likely. Sometimes, these have to be used in winding and narrow streets.

- Wide open sewers can be curved and have manhole intervals of between 100 and 200 metres.
- The sewer locations depend on the sewer type and the width of the street.

Now, on the layout, the sites for the pumping stations, the storm water overflows and the treatment plants can be indicated.

Calculations

By providing this full model and knowing the population at the end of the design cycle, it is possible to measure the sewage output and storm water flow, and the whole network. This is usually achieved with the aid of a calculation list. The procedure is as follows.

- Mark the boundary lines of the community or of the main sewer catchment area in that area, including potential extension areas to be connected.
- Mark the internal border lines of the tributary areas for each sewer line computation portion. This includes the part tributary areas along the main sewer as well as the tributary areas connected to the main sewer through secondary sewers. These areas form the flow pattern for both household and industrial wastewater and storm water.
- Find area's surface by planimetry and calculate this value using the run-off coefficient of that area (average reduction factor at the end of the design period) to find the surface of the reduced tributary area.
- Compute the tributary population by multiplying the surface of the area with the population density.
- Evaluate the cumulative average and maximum overall dry weather flow (DWF) – domestic flow, industrial flow, including infiltration water.
- Estimate the time of concentration for the segment under consideration and measure the cumulative storm flow (RWF).
- Minimum sewer sizes: 0.15 m for connection to a house; 0.20 m. 0.30 m for storm sewers, for sanitary sewers.
- Measure the flow and combined flow data for real sewage flow and storm or waste flow (combined system). Check estimated flow time with actual flow time.
- Build and measure facilities such as overflows, pumping stations, storage basins, etc. In the case of sanitary sewer or storm drain, the same measurement list can be used by simply eliminating redundant columns.

2.8.2 Responsibility of Operation and Maintenance

Responsibility of operation and repair is typically split between the household and one or more entities. The most important point is that everyone involved should know his role and duties unambiguously as early as possible in the project. It is easier to spell out obligations in written papers, signed by all stakeholders, than to rely on the work spoken up.

Property owners: In normal circumstances, the householder is responsible for the operation and maintenance of all facilities inside the plot boundary and possibly up to the house sewer and collector sewer junction.

Institutions: Institutions are typically responsible for the entire network of societies. Where more than one organization is involved (for example, where one is responsible for collecting sewage and the other for handling sewage), the most important need is to clearly identify areas of responsibility.

When choosing which institutions to be involved, there is generally very little discretion so it is important to review them to ensure that they have adequate expertise, resources and funds to carry out their duties properly.

Institutional reinforcement can be required prior to or during the project. The organization itself should not carry out administrative responsibilities; they can be subcontracted to private companies. Such a strategy has the advantage of reducing organization size and holding maintenance costs down. This strategy has worked successfully on large schemes like Orangi in Pakistan where the demand for routine maintenance is sufficient to help a number of private firms that compete for business with each other. On schemes where the amount of work is minimal, it is unlikely that the privatization of maintenance will bring significant benefits.

There may be circumstances where institutions should be responsible for maintenance of components on private property. An obvious example would be the emptying of interceptor tanks.

Community groups: There are a number of examples of communities that are responsible for collector sewer maintenance, particularly where they run across household parcels. These duties may be on a household basis, where the owner of the plot is responsible for the sewer crossing length or adjacent to his / her plot. Alternatively, a group of plot owners can cooperate to maintain the collector sewer they use together. Joint responsibility seems to be more suitable for structures where the collector sewer runs under public land and individual responsibility functions best when the sewer crosses plots of users. Attempts to keep societies responsible for trunk sewers met with no success. Many societies also believe systemic responsibility for eliminating waste from large numbers of people will remain.

The community's stability has an impact on the degree of obligation that it can sustain. In a shifting population, or where the sewers are built before properties are occupied, it is more difficult to implement maintenance duty. For these regions, holding consumer responsibility within the plot boundary is more fitting.

The private sector: Private sector use for service and, in particular, maintenance is frequently ignored but has a range of benefits. Work-to-work competition between companies aims to slash unit costs. The push to make bigger money speeds up the work, getting jobs completed quicker. The implementing organization may be able to reduce over-inflated and under-used labour power resulting in wage and management savings. Contractors are typically less affected by political intervention and union demarcations, resulting in a stable and rapid problem-reaction.

The use of contractors, however, has its disadvantages, as the push for profit can lead to the use of untrained workers and short cutbacks, both of which can produce unsatisfactory standards of work.

Contractors can use their resources to distort the process of selection. Contributions to the re-election funds of politicians and government employees who are underpaid are not unknown.

Therefore, service and repair must be adequately handled using contractors. Minimum performance levels need to be set and tracked. Processes for the selection of contractors shall be open to public inspection and considered fair and appropriate.

Supervision, operation and maintenance: All operation and maintenance activities must be supervised to ensure that they are performed in a secure manner, at an approved standard, at a reasonable cost and within an appropriate period. Effective management is focused on understanding what research needs to be done, why it needs to be done, who is responsible for doing it and what observable expectations need to be met. That needs qualified supervisors. Supervisory responsibility usually lies with the implementing agency, which is also a government entity. Day operations are usually contracted to private companies.

Although this may be much cheaper than using in-house workers, it is crucial that the agency responsible maintains full control over management. This will set and track standards and accept responsibility for the implementation of new work practices where applicable.

Implemented sewerage management, neighbourhood committees frequently disband and leave the routine maintenance to individual residents. This could be sufficient for minor problems like local blockages but it is difficult to deal with major issues. Problems that affect a large number of properties allow residents to regroup, collect funds and carry out repairs.

It all takes time, and can generate major social tensions. Localized maintenance without overall oversight by residents may generate variable standards of workmanship. Residents with limited emergency funds should spend as little as possible on sewerage repair which can contribute to the network's long-term deterioration.

Overall, it is an asset to supervise workers with local information. It makes continuous interaction between the administrators, the network and their users. Any issues that arise can be solved immediately and without the need for extensive bureaucracy.

However, supervision of on-plot construction is one area where the use of local staff can be disadvantage. Personal knowledge of the families constructing the sewers, and community pressure, may cause the supervisor to accept a standard of construction lower than that, which would normally be acceptable.

Inadequate operation and maintenance is the most common reason for sewerage schemes to fail and persistent attention is needed to achieve long-term success. Whoever is responsible for carrying out the operation and maintenance of a scheme must be managed by an organization with the skills and legal duty to ensure that it is carried out effectively. However, on-plot building supervision is one area where local workers can be disadvantageous. Personal knowledge of the sewer building families and community pressure can result in the supervisor accepting a construction standard that is lower than that usually appropriate.

The most common cause for sewerage schemes to fail is insufficient operation and maintenance, and constant attention is required to achieve long term success. Whoever is responsible for the implementation and maintenance of a scheme needs to be handled by an agency with the expertise and legal responsibility to ensure that it is successfully carried out.

EXERCISE

1. Differentiate between wet and dry intake towers. Which would you suggest will be the best for use in supply from River Yamuna for a metropolitan city like Delhi?
2. Design a river intake with respect to (i) number and size of openings in the intake well (the diameter and height of the intake well, and (iii) the gravity pipe for raw water connecting the intake and the jack well. The required data is:

   RL of the river bed : 100 m
   RL of lowest water level : 102 m
   RL of high flood level : 120 m
   Population to be served : 50,000
   Average water demand : 200 Lpd

   Assume suitable data required. Draw sketches to show the designed details.
3. Draw a neat sketch showing the intake structure which you suggest for drawing water for a city water supply plant being direct from a river under sluices.
4. Write brief notes with neat sketches on the following:
   (a) Simple submerged intake
   (b) Dry intake tower and wet intake towers
   (c) Intake well for canal
   (d) Intake for sluice-ways of dams

5. Design a bell mouth canal intake for a city of 1,75,000 persons drawing water from a canal which runs only for 10 hours a day with a depth of 1.5 m. Also calculate the head loss in the intake conduit if the treatment works are ¼ km away? Draw the neat sketch of canal intake. Assume the velocity through the screens and bell mouth to be less than 16 cm/sec and 32 cm/sec respectively.
6. Explain the Darcy's law. What is its limitation? Discuss its validity.
7. Explain site selection criteria for artificial recharge. Elucidate how to monitor groundwater quality?
8. Explain in detail interference among wells. Explain induced recharge method with their flow pattern.
9. Explain the pumping test to estimate the safe yield from an open well.
10. A well penetrates fully 10 m thick water bearing stratum of medium and having coefficient of permeability of 0.05 m/sec the well radius is 10 cm and is to be worked under a drawdown of 5 m at the well face. Calculate the discharge from the well. What will be the percentage increase in the discharge if the radius of the well is doubled? Take $R = 300$ m in each case. Describe a case study of your choice for a groundwater contaminated site? Differentiate between direct runoff and base flow.
11. Draw a neat sketch showing the intake arrangements for drawing water from river to a water treatment plant located in a city situated on higher elevation than of river.
12. Describe the design criteria required to construct an intake tower to draw water from river to WTP.
13. Describe the design criteria required to construct an intake tower to draw water from a canal to WTP.

REFERENCES

1. Deep Wells. (n.d.). Retrieved from Groundwater Engineering: https://www.groundwatereng.com/dewatering-techniques/deep-wells
2. Engineers, M. S. (2020). Useful information on centrifugal pumps. Retrieved from https://www.michael-smith-engineers.co.uk/resources/useful-info/centrifugal-pumps
3. Source: Engineering Notes. "Water Engineering, Intake Types." Sengupta. (http://www.engineeringenotes.com/water-engineering-2/surface-water/intakes-types/types-of-intakes-water-collection-water-engineering/44456
4. (1999). Operation and Maintenance of Water and Sewerage Systems. Ministry of Water - Tanzania - Rwegarulila Water Resources Institute.
5. Reciprocating Pump – Components, Working and Uses. (n.d.). Retrieved from https://theconstructor.org/practical-guide/reciprocating-pump-components-working-uses/2914/

3
CHAPTER

Quality and Characteristics of Water

3.1 INTRODUCTION

Water is important for human life. Safe drinking water influences the quality of health and productivity. Unsafe water kills more people each year than war and all other forms of violence combined. Meanwhile, our drinkable water sources are finite and are less than 1 per cent of the earth's freshwater is actually accessible to us. Without action, the challenges will only increase by 2050, when global demand for fresh water is expected to be one-third greater than it is now.

Completely pure water is rarely present in nature. The originally pure rainwater often absorbs different gases, dust, and other impurities when flowing. This water contains salt, carbon, and inorganic impurities as it travels deeper on the ground. Thus, this water should be treated and filtered for the safety of public health, economy and security of various industrial processes before supplying to the public. It is most important for the water work engineer to thoroughly test, analyse and treat the sources collected from the raw water prior to its distribution. The water supplied to the public should be strictly according to the standards laid down from time to time.

3.2 OBJECTIVES OF WATER QUALITY MANAGEMENT

Water treatment and purification is mainly aimed at extracting water from the best available sources and subjecting it to processing that ensures water of good physical quality, free from bad taste or odour and containing nothing that may be harmful to health. Chemically pure water is the one that contains two hydrogen parts and one oxygen portion. Water drops only coming out of the clouds can be known as chemically pure water, but they absorb certain gases and other matter as the drops fall down. Yet having chemically pure water is not necessary sometimes, the presence of other minerals gives water a pleasant taste. Water sources can be collected directly from rivers, lakes or springs. The different waters thus obtained that vary greatly in purity and suitability for the purposes for which they are needed, depending to a large extent on the sources from which they are extracted. Underground water may contain much dissolved matter from the state it is passing through. Shallow wells are much more likely to pollute the bacteria than deep wells. Via the discharge of sewage aand industrial effluents, streams and rivers absorb impurities from the soil and are carried to wetlands, lakes and reservoirs.

3.3 WATER IMPURITIES

Impurities in water contain solids which are dissolved and suspended. Calcium bicarbonate is a salt which has solubility. Calcium bicarbonate solution is transparent, as calcium and bicarbonate are present as ions of atomic size that are not large enough to reflect light. Some soluble minerals give the solution a colour. Soluble iron salts create pale yellow or green solutions; some of the copper salts form blue solutions. These solutions are simple, though coloured. Suspended solids are substances not fully soluble in water and exist in the form of particles. Usually, these particles add visible turbidity to the water. In most surface waters dissolved and suspended solids exist. Seawater is high in soluble sodium chloride; suspended sand and silt make it slightly cloudy.

3.4 SURFACE WATER

The final course of rain or melting snow depends on the landscape it is flowing through. Very little water penetrates the ground in areas consisting of hard packed clay. The water creates "runoff" in those situations. The rivers drain into bays and estuaries, and the water eventually returns to the sea, completing one important step of the hydrological process. When water runs off along the shore, it agitates and suspends sand and soil particles, producing silt in the surface water. The streaming activity additionally erodes rocky surfaces, creating more sand. It is aerated as the surface water cascades over rocks. The combination of oxygen, inorganic nutrients leached from the earth, and sunlight supports a large variety of water lifeforms including algae, fungi, bacteria, small crustaceans, and fish.

Riverbeds are often lined with trees and drainage areas are forested which feed the rivers. Leaves and pine needles make up a major percentage of the water's sbiological material. This substance is a significant cause of fouling ion exchange resin used in water treatment until it dissolves in the water. Over time the physical and chemical properties of contact with surface water differ considerably. A sudden storm can cause the composition of a water source to change drastically over a short term. The surface water chemistry varies with the seasons over an prolonged period of time. Strong runoff occurs during times of heavy rainfall. Depending on the geochemistry and biology of the landscape this can have a beneficial or unfavourable effect on the characteristics of water. Over multi-year or multidecade periods of drought and rainfall, surface water chemistry varies too. Water supply for industrial use is seriously impacted by long periods of drought. Where rivers flow into the ocean there are additional problems with the incursion of salt water into the river during times of drought. When designing water treatment plants and systems, industrial users must take into account the quality of the surface water.

3.5 GROUNDWATER

Water falling on porous soils, such as sand or sandy loam, drains or percolates into the soil. The water meets a wide variety of mineral species arranged in complex layers, or strata, in such situations. The minerals can comprise granite, gneiss, basalt, and shale. In certain situations, under impermeable clay there might be a layer of very permeable sand. Water in the ground often follows a complex, three-dimensional path. Groundwater hydrology research involves monitoring the motions of water. Table 3.1 shows the comparison of surface water and groundwater characteristics.

Table 3.1: Comparison of surface water and groundwater characteristics

Characteristic	Surface water	Groundwater
Turbidity	High	Low
Dissolved minerals	Low-moderate	High
Biological content	High	Low
Temporal variability	Very high	Low

Unlike surface sources, groundwater is largely free from suspended pollutants, since it is filtered as it passes through the strata. Filtration also removes much of the biological contaminants. Some groundwaters with a high content of iron contain sulphate which reduces bacteria. In industrial water systems these are a cause of fouling and corrosion. Chemistry at groundwater appears to be very stable over time. A groundwater may contain an excessive amount of solid formation, but it can be treated effectively because of its relatively stable chemistry.

3.6 MINERAL REACTIONS

When groundwater meets minerals it dissolves them according to their characteristics of solubility. Chemical reactions occur in some cases which enhance the solubility of the minerals. A clear example of this is the interaction with calcareous groundwater. Surface percolating water contains greenhouse gases. One of such gases is carbon dioxide, which when dissolved in water, produces carbonic acid. A further source of carbon dioxide is the decomposition of organic matter under the soil. Limestone is a combination of calcium carbonate and magnesium. The basic mineral is only slightly soluble in neutral water. The slightly acidic groundwater reacts in a neutralization reaction with simple limestone that creates a salt and neutralizing gas. The salt formed by the reaction is a mixture of the bicarbonate calcium and magnesium. The two bicarbonates are relatively soluble. This reaction is the cause of the most common problems that industrial users face with deposition and corrosion.

If the groundwater is not treated until use in industrial cooling and boiler systems, the calcium and magnesium (hardness) type scale on heat transfer surfaces. In boiler feedwater applications bicarbonate thermal breakdown in the boiler results in high levels of carbon dioxide in condensate return systems. This can cause serious corrosion to the network. Structural, the porous calcareous is, that is, it includes tiny holes called "interstices" and channels A broad calcareous formation can retain massive quantities of groundwater in its structure. Limestone formations containing these large quantities of water are referred to as aquifers. If a well is drilled into a calcareous aquifer, it can continuously extract the water for decades. Unfortunately, because of the neutralization/ dissolution reactions mentioned above the water is very hard. For most uses, that requires extensive treatment.

3.7 CHARACTERISTICS OF WATER

For the purpose of classification, the impurities present in water may be divided into the following three categories. The water quality characteristics can be determined by assessing physical, chemical and microbiological or biological properties of water.

3.7.1 Physical Characteristics

The physical characteristics of the water are discussed below.

3.7.1.1 Turbidity

Turbidity is caused by the presence of colloidal and suspended matter in the water. The character and quantity of turbidity depend on the amount of soil through which the water has passed groundwater is less turbid than the surface water. Turbidity is a measure of water's resistance to light moving through it. Turbidity is expressed as either NTU (nephelometric turbidity units) or mg/L (parts per million) or mg/L. Turbidity is measured by

- Turbidity rod or tape
- Jackson's turbidimeter
- Bali's turbidimeter

The sample to be tested is poured into a test tube and put in the meter and turbidity units are read by a needle or digital monitor directly onto the scale. Drinking water does not have turbidity exceeding 10 NTU. This test is useful for determining the detention time in raw water settling and for the dosage of coagulants required to eliminate turbidity.

Turbidity rod or tape: Turbidity rod is used to calculate in-field water turbidity. It consists of a graduated aluminium tube, approximately 20.3 cm long, with a graduated non-stretchable tape, approximately 12.2 cm long, attached at the top end. A screw containing a platinum needle and a nickel ring is inserted at the lower end of the aluminum pin. The graduated tape, at its top end, has a mark indicating the eye location during the examination. To find the turbidity, the lower end of the rod is slowly submerged in water the turbidity of which is to be established. At the indicated spot, the eye is constantly held and the platinum needle is looked up. Owing to water turbidity the rod is pulled slowly in water until the platinum needle just disappears from the view. Reading the graduated tape near the water surface directly gives the rod turbidity at mg/L and only rough value of water turbidity is provided.

Jackson's turbidimeter: This is a laboratory tool used to measure turbidity up to 100 mg/L. It consists of a metal stand containing in it a metal container and a graduated glass tube. A regular candle is placed underneath the stand. The water sample is poured into the container, and the illustration of the typical candle flame is seen in the glass tube through the turbid water. The water level in the glass tube rises slowly until the picture of the flame ceases to be visible as shown in Fig. 3.1.

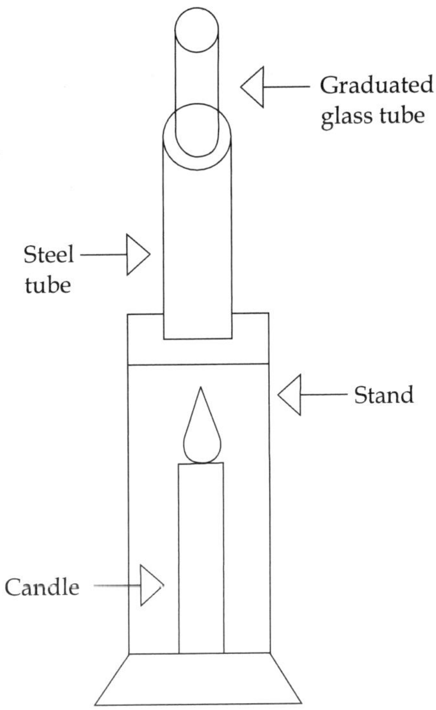

Fig. 3.1: Jackson's turbidimeter (Sengupta)

The water column height, measured in the graduated glass tube, provides measurement of the water's turbidity. The longer the 21.5 cm light path corresponds to 11 JTU while the 10.8 cm light path corresponds to 200 JTU where 1 JTU= 1mg/L.

Bayli's turbidimeter and nephlometer: This is a very reliable and is suitable with sample turbidity is less than 5 units. Both of the above experiments are based on colour matching techniques in which standard samples whose turbidity is known and test samples whose turbidity is to be determined are matched to light and colour intensity sources produced in both samples. It consists of a galvanized iron box in which two glass tubes are held at one end and a reflector 250 watt bulb is placed at the other end as shown in Fig. 3.2. One tube contains normal known turbidity solution, while the water sample is stored in the other tube. The tube is firmly placed in a frame at its bottom, with beveled holes. The tubes are surrounded by blue cobalt plates on all four sides, and a white opal glass plate at its centre. Because of blue cobalt plates, both tubes emit blue light, and a comparison is made. If the light varies, another tube that contains standard solution of specific turbidity is inserted in place of the first until the colour matches in both tubes. At this point the standard solution gives the turbidity of the given water sample. The turbidity is either expressed as being equal to mg/L or BTU.

In order to improve the efficacy of this test, intensity of current, tests are now being done using semi-conductor system called "photometer" instead of colour intensity. Photometers are mounted behind the samples in Bayli's turbidimeter and hence it only measures the intensity in the direction of incidence. It is implemented upon the concept of absorption. In Nephlometer, photometers are located at the right angle to measure samples as shown in Fig. 3.3.

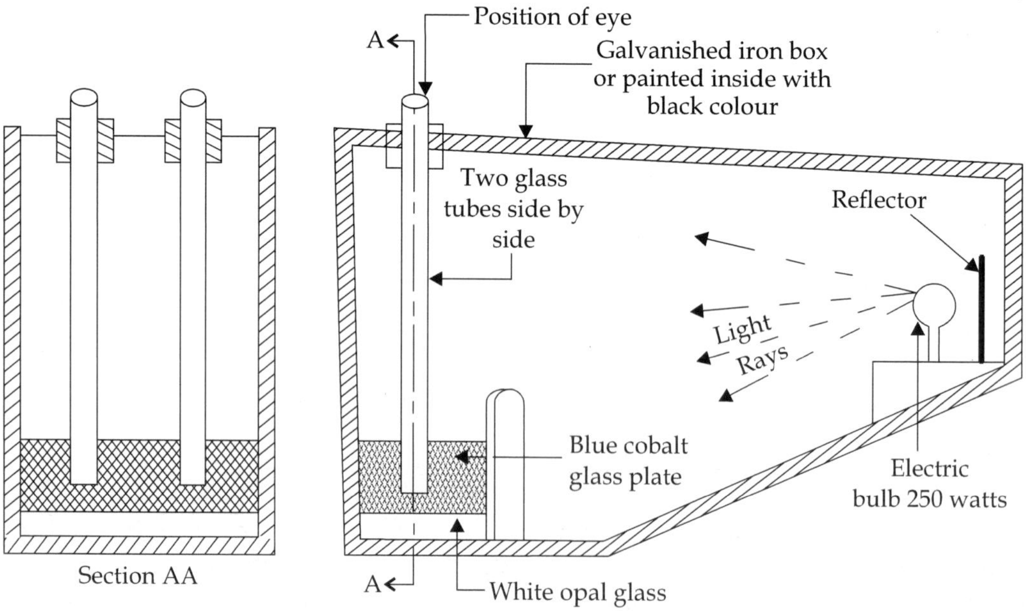

Fig. 3.2: Bayli's turbidimeter (Sengupta)

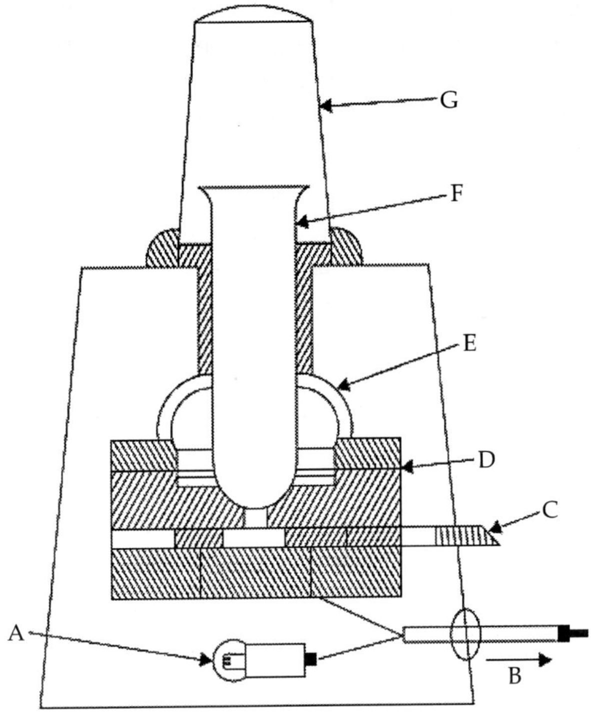

Fig. 3.3: Nephlometer

The different parts of a nephlometer can be described as under :

A = Light source

B = Sensitive micro-ammeter

C = Filter wheel with a series of colour filters

D = An annular photocell

E = A reflector to collect the scattered light

F = A test tube

G = A metal test tube cover to exclude extraneous light

The pressure at the right angles to the direction of incidence and is based on the principal scattering. Turbidity is represented as a normal unit obtained when 1 mg formazine dissolved in 1 L of pure water instead of silica and is termed as 1 NTU (nephlometeric turbidity unit) or 1 FTU (formazine turbidity unit). The Indian standard has specified the acceptable turbidity unit for drinking water to be 1 NTU and permissible limit to be 10 NTU.

3.7.1.2 Colour and Temperature

The presence of colour in water is generally due to the presence of colloidal organic matter, but often it is also due to mineral and dissolved organic impurities. The colour may be defined as one milligram of cobalt-platinum dissolved in 1 L of water and is said to be 1 colour unit. As per Indian standard the water to be used for domestic consumption should not contain colour value greater than 20 mg/L on a platinum cobalt scale. Although the presence of colour in water is not harmful but it is unacceptable.

Measurement of colour: Presence of colour in water can be determined by using colour matching technique. One mg of chloroplatinate ions when dissolved in 1 L of water will produce colour of 1 unit on platinum-cobalt scale. Colour intensity is expressed in terms of TCU (True colour unit, ppm, mg/L, unit) Colour can also be determined by using tintometer, a tool with Nessler's tube. Testing of colour should be done within 72 hours after collection of water samples. The Indian standard prescribes the acceptable colour unit of 5 TCU and permissible limit of 25 TCU for domestic water consumption.

Water temperature is measured by ordinary thermometers. The surface water temperature is usually taken or considered at ambient temperatures, while groundwater temperature can be more or less than ambient. For public water supply the acceptable temperature lies between 4.4°C and 10°C as per Indian context. The temperature above 35°C is not appropriate for public supplies, because such temperature makes water unpalatable.

3.7.1.3 Taste and Odour

The presence of taste and odour in water can be due to the presence of dead or live micro-organisms, dissolved gasses such as hydrogen sulphide, methane, carbon dioxide or oxygen mixed with organic matter, minerals such as sodium chloride, iron compounds and carbonates, and other substances such as sulphates. The taste and odour tests can be performed by the sense of smell and by tasting the water because they are present in such small amounts that chemical analysis makes it difficult to detect them. Water with bad odour is unacceptable and should not be distributed to the public. Odour intensities are calculated by using method of threshold odour number. This number is numerically equal to the amount of water needed with the number of dilutions being applied till the water becomes odourless for a given known amount of water sample.

Measurement of taste and odour: The taste measurement can be carried out using a tool osmoscope, and odour is measured by diluting the water sample to an extent that the taste is barely visible and its severity is expressed as standard units. TON (threshold odour number) reflects the dilution ratio that is barely noticeable in taste and odour as indicated in Eq. 3.1 & Eq. 3.2.

$$\text{Dilution ratio} = \frac{\text{Final volume}}{\text{Original volume}} \tag{3.1}$$

$$\text{Threshold odour number} = \frac{A + B}{A} \tag{3.2}$$

where,

A = Volume of odourous water (mL)

B = Volume of odour free water (mL) required to produce a 200 mL mix water sample.

The Indian standard prescribes the odour limit to be 1 TON and permissible limit to be 3 TON for domestic water to be used for drinking purpose and other usages.

3.7.2 Chemical Characteristics

In the chemical analysis of water, these tests are done that will reveal the sanitary quality of the water. Chemical tests involve the determination of:

- Total solids
- pH value
- Hardness of water
- Chloride content
- Metal and other chemical substances
- Nitrogen and its compounds
- Dissolved gases
- Alkalinity

3.7.2.1 pH Value of Water

The pH value denotes the concentration of hydrogen ions in the water and it is a measure of acidity or alkalinity in water.(Eq. 3.3)

$$pH = -\log_{10}[H^+] \text{ or } \frac{1}{\log_{10}[H^+]} \tag{3.3}$$

pH measurement is important because it provides a means to identify and compare other characteristics or behaviours such as corrosive activity or other interrelated factors that regulate biological function in a waterbody. The standard pH value as shown in Fig.3.4 varies from 0 to 14, depending on the quality of the dissolved salts and minerals present in the water. The pH value of normal water is considered to be 7. However, the pH value from 0–7 is considered to be acidic and from 7–14 is considered to be alkaline. The Indian standard recommends pH value of 6.5 to 8.5 for public water supplies.

The lower pH values can lead to tuberculation and corrosion in the distribution system or its pipelines. However, the high pH values greater than 8.5 can lead to incrustation, deposits of sediments and other bad depositions in the distribution systems. Water pH value is usually calculated by the use of pH papers or pH metre. Using pH meter, pH can be read directly on scale or by digital display. The determination of pH value is also helpful in water treatment management during softening and coagulation processes.

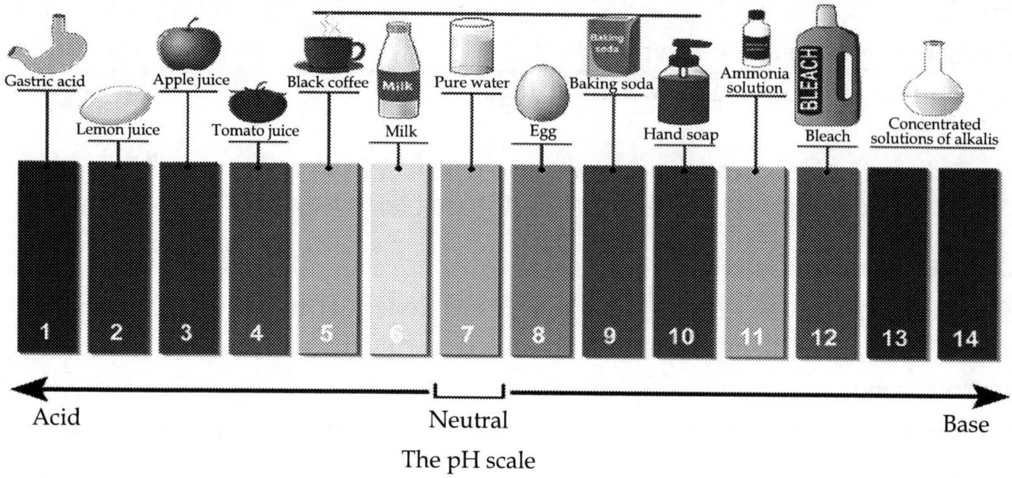

Fig. 3.4: pH scale representation and its values

The pH of a water can be measured by using two types of methods:

(i) Colorimetric method

(ii) Electrometric method

(i) Colorimetric method: This is perhaps the most unusual approach in which any indicator is applied to the water sample, and the colour of the known pH value solution. These basic colours can be issued by the manufacturers with each indicator in the form of coloured liquids in glass tubes, coloured glass disks or coloured maps. Colorimetric method is shown in Fig. 3.5.

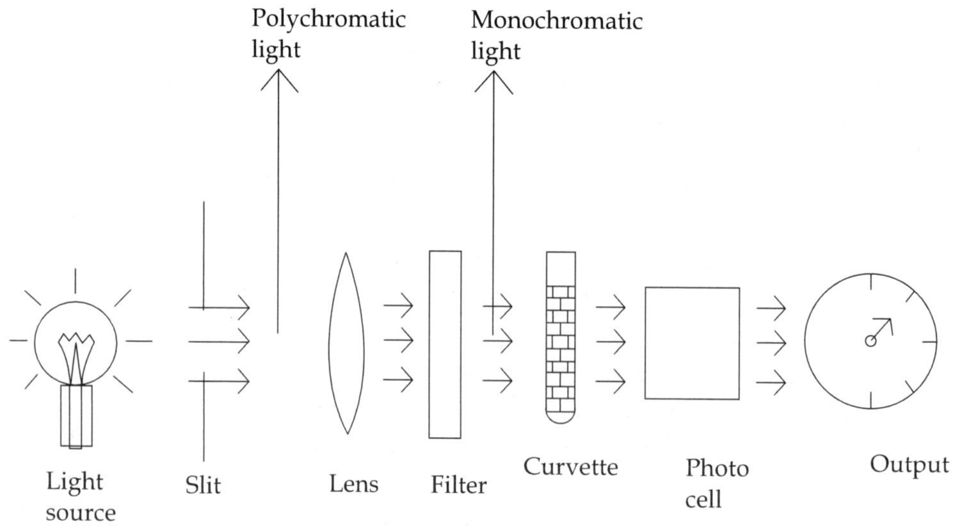

Fig. 3.5: Colorimetric method

The indicators for acidic range are bromophenol blue, methyl red, and chlorophenol red, thymol blue (acid range). The indicators for alkaline range are thymol blue (alkaline range), phenolphthalein red, and toyl-red.

(ii) **Electrometric method:** A pH-meter is used in this process. The water specimen is placed in a beaker and the instrument's two electrodes are dipped in water that is connected to a dry cell as shown in Fig. 3.6. The current which passes through the circuit indicates the pH value read directly on the dial. Commonly used electrodes consist of calomel, glass, or quinhydrone. The pH value is shown electronically on the dial in some of the latest pH metres, up to the second decimal position of the value.

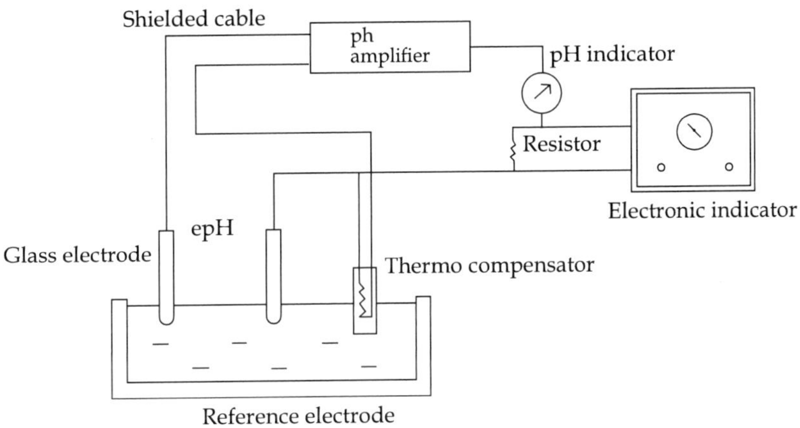

Fig. 3.6: Circuit for electronic pH measurement

3.7.2.2 *Total Solids and Suspended Solids*

Total solids include the solids in colloidal suspension and dissolved form as shown in Fig. 3.7. Solid materials in wastewater may consist of organic and/or inorganic materials and organisms. The solids must be significantly reduced by water treatment before public water supplies.

Total solids (*TS*): Total solids include both the suspended solids and the dissolved solids which are obtained by separating the solid and liquid phase by evaporation. Total solids (*TS*) in water is the amount of all solids, which are determined by drying a known volume of the sample in a pre-weighed crucible dish at 105°C. After cooling in a desiccator, the crucible dish is again weighed. *TS* is determined by using Eq. 3.4:

$$TS = \frac{(M_1 - M_2)}{V} \tag{3.4}$$

where,

M_1: Mass of crucible dish after drying at 105°C (mg)

M_2: Mass of initial crucible dish (mg)

V: Volume of sample (L)

Total suspended solid (TSS): Total suspended solids are the solids in a filter and is usually determined by filtration using glass fibre filters. In all analytical procedures for determination of suspended solids, weighed filters are used for sample filtration, the filters are dried at about 105°C after filtration, cooled in a desiccator to room temperature and the weight of the loaded filter is determined. TSS is determined by using the Eq. 3.5:

$$TSS = \frac{(M_4 - M_6)}{V} \tag{3.5}$$

where,

M_4 = Mass of filter after drying at 105°C (mg)

M_5 = Mass of initial filter (mg)

V = Volume of sample (L)

Total dissolved solids (TDS): However, the solids passing through the above mentioned filter paper consist of colloidal and dissolved solids. Therefore, the TDS can be determined by subtracting the amount of suspended solids from the amount of total solids in a given water sample.

Volatile suspended solids (VSS): Solids are volatilized and burnt off in a muffle furnace with the same TSS sample at 550 ± 50°C for 25–30 minutes. The amount of volatile solids represent an estimate of organic matter in the solids, while the amount of non volatile solids (fixed) represent the inorganic or mineral matter in a given water sample and VSS can be calculated by using Eq. 3.6:

$$\text{VSS} = \frac{(M_4 - M_6)}{V} \tag{3.6}$$

where,

M_4 = Mass of filter after drying at 105°C (mg)

M_6 = Mass of filter after ignition at 550°C (mg)

V = Volume of sample (L)

Similarly, the amount of volatile dissolved solids can be calculated by subtracting the amount of VSS obtained using the above method from the amount of total volatile solids in a given water sample. The total volatile solids in a water sample can be measured directly by burning the sample in a muffle furnace at 550 ± 50°C for 25–30 minutes as the difference in weight provides the amount of TVS in a water sample. The desirable limits of TDS as prescribed in Indian standard for drinking water is 500 mg/L with permissible limit to be 2000 mg/L. No pronounced effects on human beings due to high concentration of solids in water has been mentioned in the literature. However, the presence of suspended solids in water may accelerate the biological growth or the presence of such solids may indicate the presence of organic matter in the water.

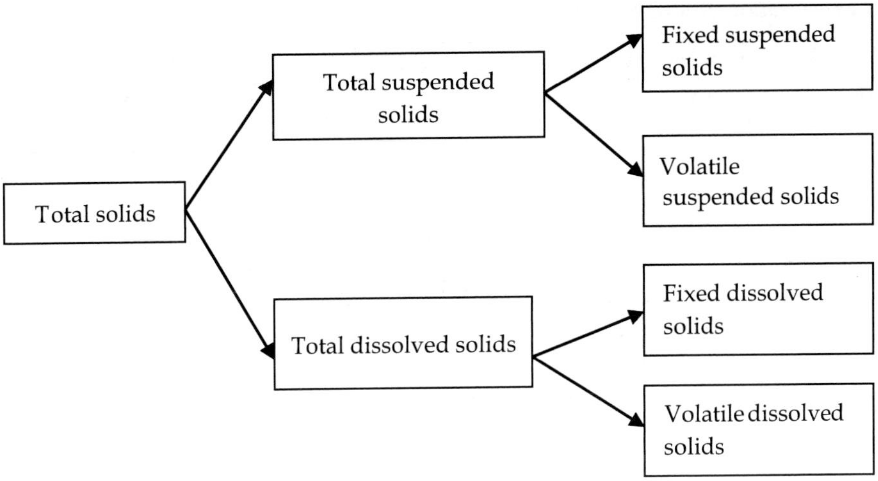

Fig. 3.7: Characterization of solids in water

Ionic balance: The TDS concentration in water is due to the presence of ions adsorbed on the surface of solids and cannot be removed easily by simple filtration or sedimentation process. The ions that are generally associated with TDS concentration in water are (i) major constituents (1–1000 mg/L) include sodium, calcium, magnesium, bicarbonate, sulphate and chlorides, and (ii) minor constituents (0.01-10 mg/L) include iron, strontium, potassium, carbonates, nitrate, fluoride, boron and silica. These are common ions in water and are measured independently. Thereafter, on equivalent basis all the determined ions are added thus representing the total concentration of TDS. In order to check the ionic analysis accuracy the sum of cations should be equal to anions on equivalent basis and therefore electrical neutrality must be maintained. Any inequality or error indicates that either the additional constituents might be present in water or any error has been occurred during analysis of ions.

Example 3.1: In water analysis major ions has been determined in the laboratory procedure and the results for the same is mentioned below as in Table 3.2.

Table 3.2: Concentration of ions(mg/L)

Ca^{2+} = 60 mg/L	HCO_3^- = 275 mg/L
Mg^{2+} = 20 mg/L	SO_4^{2-} = 75 mg/L
Na^+ = 100 mg/L	Cl^- = 100 mg/L

If 10% error is acceptable should the analysis be considered accurate or authentic?

Solution: The first step is to convert the concentration of all ions as cations and anions separately from mg/L to milli-equivalent per litre and sum all of them as shown in Table 3.3.

Table 3.3: Concentration of ions (mg/milli-equivalent)

Ion	Cations			Ion	Anions		
	mg/L	mg/ milli-equivalent	meq/L		mg/L	mg/ milli-equivalent	meq/L
Ca^{2+}	60	40/2 = 20	3.00	HCO_3^-	275	61/1 = 61	4.51
Mg^{2+}	20	24.3/2 = 12.15	1.64	SO_4^{2-}	75	96/2 = 48	1.56
Na^+	100	23/1 = 23	4.34	Cl^-	100	35.5/1 =35.5	2.81
Total cations			**8.98**	**Total anions**			**8.88**

The error (%) can be calculated as:

$$\text{Error (\%)} = \frac{8.98 - 8.88}{8.88} \times 100 = 1.12\% < 10\% \text{ (hence, analysis is O.K.)}$$

In order to arrange the cations and anions in the order for convenience these common ions can be displayed in Fig. 3.8 as:

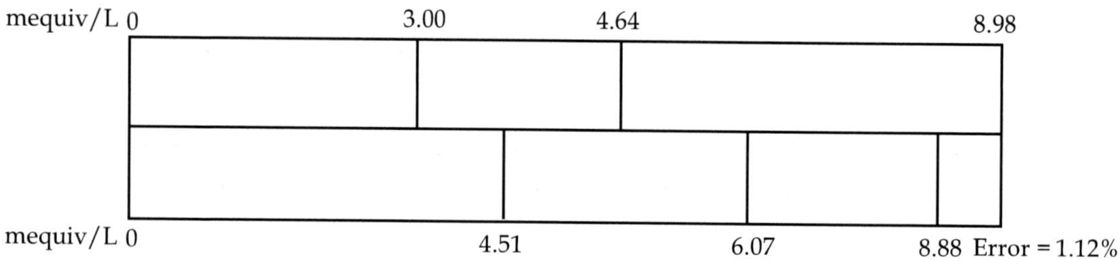

Fig. 3.8: Bar diagram showing the concentration of ions (mequiv/L)

3.7.2.3 Alkalinity

It is known as the concentration of ions in water capable of neutralizing the hydrogen ions. Alkalinity is therefore a measurement of the ability of water to neutralize acids. It is caused by the presence of bicarbonates of calcium and magnesium. It may also be caused by carbonates or hydroxides of sodium, potassium, calcium and magnesium. The alkalinity caused by bicarbonates is known as *bicarbonate alkalinity* and the one which is caused by carbonates is known as *carbonate alkalinity*. However, hydroxide alkalinity or caustic alkalinity is caused by hydroxides. Also all the compounds that cause alkalinity are not responsible for causing hardness in water. It is produced in water by anions caused by inorganic mineral salts, dissolved gasses and dissolved organic matter in water. However, the major cause of acidity in water is caused due to the presence of mineral acids, carbon dioxide, sulphates of iron and aluminium in water. The acceptable limits being prescribed by Indian standard for alkalinity in drinking water to be 200 mg/L while the permissible limit is 600 mg/L.

Constituents of alkalinity in natural water systems include CO_3^-, HCO_3^-, OH^-, H^+, $HSiO_3^-$, $H_2BO_3^-$, HPO_4^-, $H_2PO_4^-$, HS^- and NH_3. Most common constituents of alkalinity are CO_3^- and HCO_3^- with little amount of OH^- ions. The reaction equations (Eq. 3.7–Eq.

3.10) of emergence and dissolution of these minerals from atmosphere to the water system are:

$$CO_2 + H_2O \rightleftharpoons H_2CO_3 \text{ (dissolved } CO_2 \text{ and carbonic acid)} \quad (3.7)$$

$$H_2CO_3 \rightleftharpoons H^+ + H_2CO_3^- \text{ (bicarbonate)} \quad (3.8)$$

$$HCO_3^- \rightleftharpoons H^+ + CO_3^{2-} \text{ (carbonate)} \quad (3.9)$$

$$H_2O + CO_3^{2-} \rightleftharpoons HCO_3^- + OH^- \text{ (hydroxide)} \quad (3.10)$$

The reaction equations (Eq. 3.7–3.10) are weak reactions chemically. However, utilization of bicarbonate ion as a carbon source by algae can derive the reaction towards right side that ultimately results in excessive enrichment of OH^- ions in water. Therefore, the water with heavy algal growth is found to be highly alkaline with pH value ranging around 9 to 10. Therefore, the relative quantities of alkalinity species are pH dependent and can be shown through Fig. 3.9.

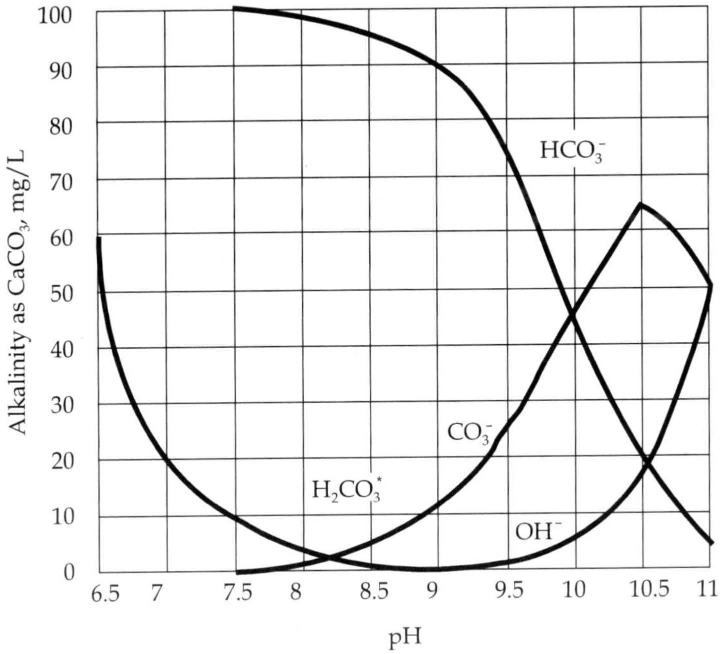

Fig. 3.9: Alkalinity and pH dependency at 25°C at total alkalinity concentration of 100 mg/L (source: Sawyer and McCarty)

The high concentration of alkalinity can impart bitter taste to water. It can also leads to foul pipes and incrustation of pipes in distribution system.

Measurement: Alkalinity is determined by titrating the water sample with an acid and computing the hydrogen equivalents of acid and is expressed as mg/L as $CaCO_3$. The titration can be carried out using 0.02 N H_2SO_4 indicating that 1 mL of acid will neutralize 1 mg of alkalinity as $CaCO_3$. Hydrogen ion reacts with alkalinity and can be expressed with the following reaction equations (Eq. 3.11 – Eq. 3.13) and relationship between molecular weight, atomic weight, moles and number of gm equivalent weight is shown in equations (3.14–3.17).

$$H^+ + OH^- \rightleftharpoons H_2O \tag{3.11}$$

$$CO_3^{2-} + H^+ \rightleftharpoons +HCO_3^- \tag{3.12}$$

$$H\,CO_3^- \rightleftharpoons H_2CO_3 \tag{3.13}$$

$$\text{Molecular weight} = \Sigma \text{ Atomic weight} \tag{3.14}$$

$$\text{No. of moles} = \frac{\text{Given weight}}{\text{Molecular weight}} \tag{3.15}$$

$$\text{Equivalent weight} = \frac{\text{Molecular weight}}{\text{Valancy}} \tag{3.16}$$

$$\text{No. of gm equivalent weight} = \frac{\text{Given weight}}{\text{Equivalent weight}} \tag{3.17}$$

During alkalinity analysis and titration the acid is added slowly to the water sample and pH is noted continuously. The inflection curve as shown through Fig. 3.10 occurs at around pH value of 8.3 and 4.5. This indicates that conversion of carbonate to bicarbonates occurs at pH 8.3. However, because bicarbonate is also an alkalinity species an equal amount of acid must be added to complete neutralization. Therefore, neutralization of carbonate is only half complete at pH 8.3 and all hydroxides and one half of carbonate is measured at pH 8.3 only. At pH 4.5 all the bicarbonate has been converted to carbonic acid including bicarbonate resulting from the reaction of acid and carbonate. Therefore, the amount of acid required to neutralize the alkalinity sample to pH 4.5 is equivalent to the total alkalinity of a given water sample. If T is the amount of acid required during titration to reach pH 8.3 and Q is the total quantity of acid required to end point value of 4.5, following is consideration in order to assess the different forms of alkalinity:

If $T = Q$, all alkalinity is OH^-

$T = Q/2$, all alkalinity is CO_3^{2-}

$T = 0$, (Initial pH <8.3), all alkalinity is HCO_3^-

$T < Q/2$, predominant species are CO_3^{2-} and HCO_3^-

$T > Q/2$, predominant species are OH^- and CO_3^{2-}

It is supposed that OH^- sample with pH less than 9 is not important. Therefore the CO_3^{2-} scan be measured by 2T and HCO_3^- can be measured by remaining amount of (Q–2T).

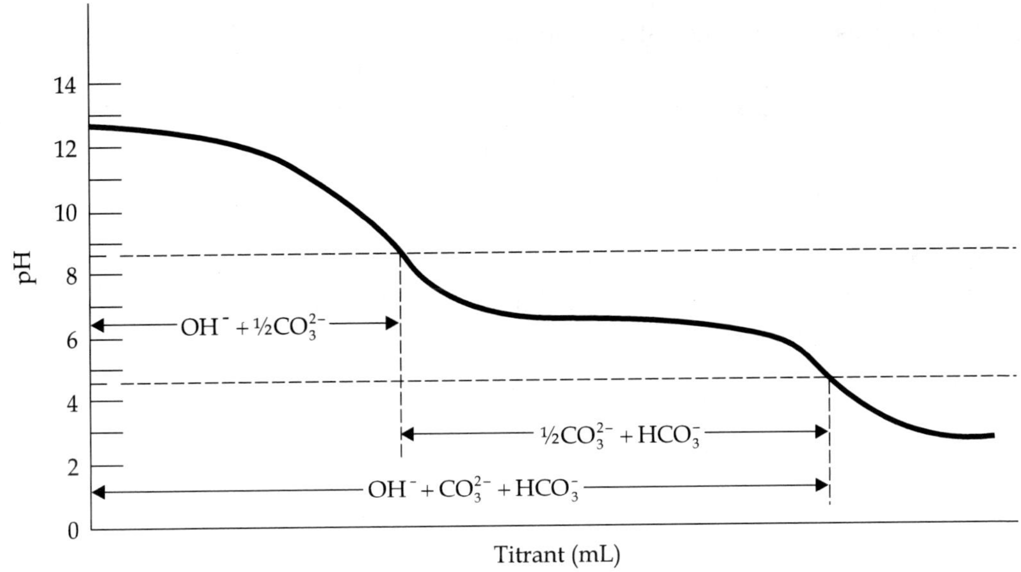

Fig 3.10: Alkalinity curve during titration process for determination of different alkalinity species in a water sample

Example 3.2: A 250 mL of water sample of water has initial pH value of 9.5 requires 25 mL of 0.02 NH_2SO_4 to titrate the sample with pH 4.5 as end point. Determine the alkalinity of water sample in mg/L of $CaCO_3$? Also calculate all the alkalinity species, if 10 mL acid has been used as equivalence point at pH 8.3?

Solution: As 1 mL of 0.02 NH_2SO_4 will neutralize 1 mg of alkalinity therefore, 25 mL of acid in a 250 mL sample will neutralize the alkalinity concentration of

$$= \frac{25 \text{ mL}}{250 \text{ mL}} \times \frac{1000 \text{ mL}}{1 \text{ L}} = 100 \text{ mg/L as CaCO}_3$$

At initial pH value of 9.5 the OH^- ion as pOH in the water sample is 4.5. The OH^- ionic concentration can be determined as:

$$[OH^-] = \frac{10^{-4.5} \text{mol OH}^-}{\text{L}} \times \frac{1 \text{ Equiv}}{\text{mol OH}^-} \times \frac{50{,}000 \text{ mg CaCO}_3}{1 \text{ Equiv}} = 1.6 \text{ mg/L as CaCO}_3$$

Therefore, 1.6 mL of acid would be required to measure OH^- in one L of sample. For a sample of 250 mL the volume of acid used can be calculated as:

$$= \frac{250 \times 1.6}{1000} = 0.4 \text{ mL}$$

Also when 0.4 mL acid measures the OH^- ion concentration in water sample, therefore 9.6 mL of acid has been used to measure half of the carbonate ions and another 9.6 mL of acid is required to measure the remaining one half of carbonate ions as CO_3^{2-}, Therefore, from a total of 25 mL of acid used 5.4 mL will be utilized to measure the HCO_3^- ions. The amount of each species is as under:

$$OH^- = 1.6\frac{mg}{L} \text{ as } CaCO_3 \text{ (calculated as above)}$$

$$CO_3^{2-} = \frac{19.2 \text{ mL}}{250 \text{ mL}} \times \frac{1000 \text{ mL}}{1 \text{ L}} = 77\frac{mg}{L} \text{ as } CaCO_3$$

$$HCO_3^- = \frac{5.4 \text{ mL}}{250 \text{ mL}} \times \frac{1000 \text{ mL}}{1 \text{ L}} = 21.6\frac{mg}{L} \text{ as } CaCO_3$$

Total alkalinity = 1.6 + 77 + 21.6 ≈ 100 mg/L as $CaCO_3$

Example 3.3: Water sample consists of 210 gm of CO_3^{2-}, 122 gm of HCO_3^- & 68 gm OH^-. Find the alkalinity of water sample and express it as $CaCO_3$ in mg/L.

Solution:

$$\text{gm equivalent of } CO_3^{2-} = \frac{210}{30} = 70 \text{ gm equivalent of } CaCO_3$$

$$\text{gm equivalent of } HCO_3^- = \frac{122}{61} = 2 \text{ gm equivalent of } CaCO_3$$

$$\text{gm equivalent of } OH^- = \frac{68}{17} = 4 \text{ gm equivalent of } CaCO_3$$

gm equivalent of $CaCO_3$ = 13

Therefore, total alkalinity of $CaCO_3 = 13 \times 50 \times 10^3 = 650 \times 10^3$ mg/L **Ans.**

3.7.2.4 Hardness of Water

As per WHO, the water hardness is the traditional measure of the capacity of water to react with soap, as hard water requiring more soap to produce lather. It is a property of water, which prevents lather formation in soap. It may also be defined as the amount of dissolved calcium and magnesium and, sometimes, other divalent and trivalent metallic elements in the water. Hard water is high in dissolved minerals, largely calcium and magnesium. The presence of these cations with a charge of +2, especially Ca^{2+} and Mg^{2+} do not pose any health threat, but they can engage in reactions that leave insoluble mineral deposits. These deposits can make hard water unsuitable for many uses, and so a variety of means have been developed to "soften" hard water in order to remove the calcium and magnesium ions. Dissolved calcium and magnesium salts are primarily responsible for scaling in pipes and water heaters and cause numerous problems in laundry and kitchen. The hardness may be classified of two types as:

Temporary hardness of water: It may be due to the presence of magnesium and calcium carbonates in water that makes it temporarily hard. The hardness in water can be removed by boiling the water. When the water is boiled the soluble salt of $Mg(HCO_3)_2$ present in it is converted to $Mg(OH)_2$ which is insoluble and the CO_2 escapes. The remaining salts gets precipitated and is removed by simple filtration.

Permanent hardness of water: When the soluble salts of magnesium and calcium are present in the form of chlorides, nitrates and sulphides in water, the hardness is termed permanent hardness and cannot be removed by boiling. Therefore, washing soda or salts of zeolite are added to the water. Insoluble carbonates are formed when washing soda reacts with the sulphide and chloride salts of magnesium and calcium and thus hard water is converted to soft water.

Impacts of hardness: Hard water results in less foam formation with soap while increasing its use while $MgSO_4$ hardness causes "Laxative effect" if its concentration is greater than 50 mg /L. To a large extent, individuals are protected from excess intake of calcium by a tightly regulated intestinal absorption mechanism through the action of 1, 25-dihydroxy-vitamin D, the hormonally active form of vitamin D. Although, calcium can interact with iron, zinc, magnesium, and phosphorus within the intestine, there by reducing the absorption of these minerals. On the other hand, the major cause of hypermagnesemia is renal insufficiency associated with a significantly decreased ability to excrete magnesium. Hard water can cause mineral buildup in plumbing, fixtures, and water heaters, and poor performance of soaps and detergents. The corrosion and incrustation in water pipes is also due to high concentration of hardness in water. Hardness of water sample is also expressed as $CaCO_3$ equivalent of multivalent cation present in water. For drinking water the quality the hardness concentration varies from 5 to 8 degrees in British unit as 1 British degree of hardness is equivalent to 14.25 mg/L as $CaCO_3$.

Measurement of hardness: Hardness is typically expressed as calcium and magnesium ions in water in terms of calcium carbonate and is expressed in mg/L or ppm as expressed in Eq. 3.18.Water hardness is determined by EDTA method of titration. There are three methods of determining total hardness of water.

$$\text{Total hardness as } CaCO_3 \left(\frac{mg}{L}\right) =$$

$$= \left(\frac{Ca^{2+}}{\text{Equivalent weight of } Ca^{2+}} \times \text{Equivalent weight of } CaCO_3\right) +$$

$$\left(Mg^{2+} \times \frac{\text{Equivalent weight of } CaCO_3}{\text{Equivalent weight of } Mg^{2+}}\right) \tag{3.18}$$

However, determining the calcium and magnesium ions concentration in water and considering the equivalent weight of calcium and magnesium, the hardness can be calculated as (Eq. 3.19):

$$\text{Total hardness (mg/L)} = \left(C_a^{2+}\left(\frac{mg}{L}\right) \times \left(\frac{50}{20}\right)\right) + \left(Mg^{2+}\left(\frac{mg}{L}\right) \times \left(\frac{50}{12}\right)\right) \tag{3.19}$$

Further the hardness can be expressed in terms of **carbonate and non carbonate hardness**. The **carbonate hardness** can be termed as the total hardness or alkalinity whichever is lesser. However, **non-carbonate hardness** may be defined as total hardness in excess of alkalinity of a sample. Also, if the alkalinity concentration of a water sample is greater than total hardness than there is no non-carbonate hardness present in the sample.

Carbonate hardness is sensitive to heat and precipitates at high temperature as CO_2 escapes out. (Eq. 3.20 & Eq. 3.21).

$$Ca(HCO_3)_2 \xrightarrow{\text{yields}} CaCO_3 + CO_2 + H_2O \tag{3.20}$$

$$Mg(HCO_3)_2 \xrightarrow{\text{yields}} Mg(OH)_2 + 2CO_2 \tag{3.21}$$

Example 3.4: Calculate the total hardness, carbonate and non carbonate hardness of water sample with the following data:

(i) Alkalinity = 240 mg/L as $CaCO_3$ (ii) Ca^{2+} = 155 mg/L (iii) Mg^{2+} = 90 mg/L

Solution: The total hardness can be calculated as:

$$\text{Total hardness (mg/L)} = \left(C_a^{2+}\left(\frac{mg}{L}\right)\times\left(\frac{50}{20}\right)\right)+\left(Mg^{2+}\left(\frac{mg}{L}\right)\times\left(\frac{50}{12}\right)\right)$$

$$\text{Total hardness (mg/L)} = \left(155\left(\frac{mg}{L}\right)\times\left(\frac{50}{20}\right)\right)+\left(90\left(\frac{mg}{L}\right)\times\left(\frac{50}{12}\right)\right)$$

$$= 388 + 375 = 763 \text{ mg/L}$$

Alkalinity is 240 mg/L and is less than total hardness value of 763 mg/L, therefore the carbonate hardness is found to be 240 mg/L only.

The non-carbonate hardness is calculated as = 763 – 240 = 523 mg/L as $CaCO_3$.**Ans.**

However, the total alkalinity is caused by the presence of CO_3^-, HCO_3^- and OH^- ions in water. A very low amount of alkalinity may be caused by H^+ ions of water.

Total alkalinity (mg/L as $CaCO_3$) = CO_3^- + HCO_3^- +OH^- – H^+

Example 3.5: Calculate the total alkalinity of water sample containing 100 mg/L of CO_3^- ions and 75 mg/L of HCO_3^- ions?

Solution: The total alkalinity can be calculated as:

$$CO_3^- \text{ as } CaCO_3 \text{ (mg/L)} = \left(100\left(\frac{mg}{L}\right)\times\left(\frac{50}{30}\right)\right) = 167 \text{ mg/L}$$

$$HCO_3^- \text{ as } CaCO_3 \text{ (mg/L)} = \left(75\left(\frac{mg}{L}\right)\times\left(\frac{50}{61}\right)\right) = 62 \text{ mg/L}$$

Therefore, total alkalinity is calculated as = 167 + 62 = 229 mg/L as $CaCO_3$

The alkalinity caused by the presence of OH^- and H^+ ions is ignored if the pH value of the water sample is not known to us. Le us consider the pH value of a water sample to be 9.5, therefore the presence of ions is:

H^+ ions = $10^{-9.5}$ moles/L and OH^- = $10^{-4.5}$ moles/L

H^+ ions = $10^{-9.5}$ moles/L = $10^{-9.5}$ (moles/L)(1 gm/mole)(10^3 mg/gm) = $10^{-6.5}$ (mg/L) × (50/1) mg/L as $CaCO_3$

$= 5 \times 10^{-6.5}$ mg/L as $CaCO_3$

OH^- ions $= 10^{-4.5}$ moles/L $= 10^{-4.5}$ (moles/L) (17 gm/mole)(10^3 mg/gm) = 0.53 mg/L
$= 0.53 \times (50/17)$ mg/L

$= 1.58$ mg/L as $CaCO_3$.

Therefore, from the above calculations it can be inferred that the alkalinity contribution of H^+ ions is negligible and OH^- ions is also very low as compared to the alkalinity value of 240 mg/L due to the presence of carbonate and bicarbonate ions. However, various methods of hardness determination are as under:

(a) **Clarke's method:** This approach is based on the assumption that the hardening substances react with insoluble compounds of soap before lather is formed. Hence, total hardness is measured by determining the standard soap solution needed to achieve a permanent lather with the known volume water sample with constant shaking as shown in Fig. 3.11. The approach has become obsolete.

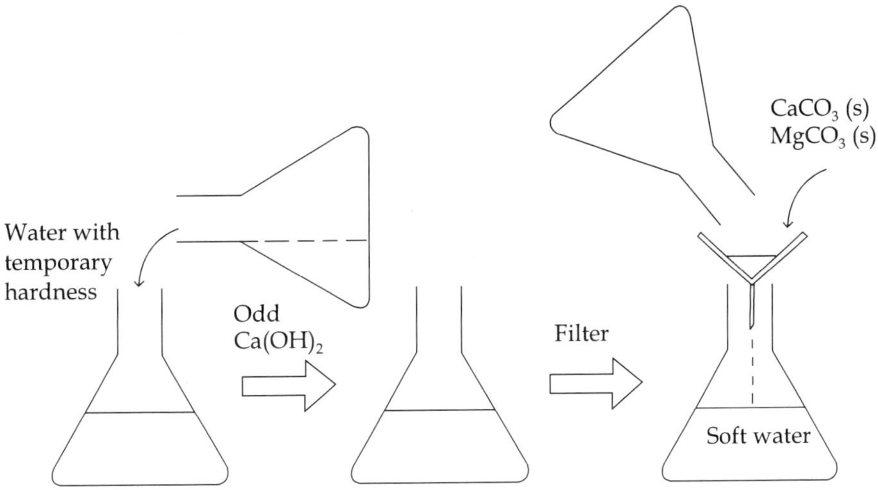

Fig. 3.11: Clarke's method

(b) **Hehner's method:** Temporary hardness is determined by titration using a regular sulphuric acid solution using a methyl orange indicator. Normal sodium carbonate solution is added to the water sample and evaporated to dryness to assess the consistency of the permanent hardness. The excess quantity of sodium carbonate required to turn the sulphates and chlorides into carbonates gives the durability.

(c) **Versenate method:** Titrating to di-ethylene di-amine tetra-acetic acid (EDTA) solution using Erichrome black- T as indicator at with pH 8.5 to 11 decides hardness in this step. Hardness is objectionable for drinking water because much soap is required for washing clothes, etc., (ii) scale is formed in boilers and hot water heating systems, and (iii) corrosion and incrustation of pipelines and plumbing fixtures is encouraged. The Indian standard has prescribed the acceptable limit of 200 mg/L and permissible limit of 600 mg/L for the water used for domestic purpose. However, the prescribed hardness limit for water to be supplied for public usage is 75-115 mg/L. The hardness index specified for water consumption I domestic purpose on the basis of hardness as $CaCO_3$ is shown in Table 3.4.

Table 3.4: Degree of hardness of water analysis sample

S. No.	Degree of hardness	Hardness as $CaCO_3$ (mg/L)
1.	Soft	0-55
2.	Slightly hard	56-110
3.	Moderately hard	101-200
4.	Very hard	201-500

Example 3.6: Water contains the following dissolved ion

$[Na^{2+}] = 56$ mg/L

$[Ca^{2+}] = 40$ mg/L

$[Mg^{2+}] = 30$ mg/L

$[Al^{3+}] = 3$ mg/L

$[Cl^{-}] = 165$ mg/L

$[HCO_3^{-}] = 165$ mg/L

Atomic weights Ca = 40, Mg = 24 , Al = 27, H = 1, C = 12, O = 16, Na = 23 & Cl = 35.5

Compute the total hardness and non-carbonate hardness of the sample as $CaCO_3$ in mg/L.

Solution: Total hardness $CaCO_3\left(\frac{mg}{L}\right) =$

$$\left(Ca^{2+} \times \frac{\text{Equivalent weight of } CaCO_3}{\text{Equivalent weight of } C^{2+}}\right) + \left(Mg^{2+} \times \frac{\text{Equivalent weight of } CaCO_3}{\text{Equivalent weight of } M^{2+}}\right) +$$

$$\left(Al^{3+} \times \frac{\text{Equivalent weight of } CaCO_3}{\text{Equivalent weight of } Al^{3+}}\right)$$

$$= \left(40 \times \frac{50}{20}\right) + \left(30 \times \frac{50}{12}\right) + \left(3 \times \frac{50}{9}\right) = 242 \text{ mg/L as } CaCO_3$$

Since hardness occurs due to presence of bivalent and trivalent cation salt.

Non-carbonate hardness = Total hardness – Carbonate hardness

$$\text{Carbonate hardness} = \left(HCO_3^{-} \times \frac{\text{Combining weight of } CaCO_3}{\text{Combining weight of } HCO_3^{-}}\right)$$

$$= 190 \times \frac{50}{61} = 165 \text{ mg/L as } CaCO_3$$

Hence, non-carbonate hardness = 242 – 156 = 86 mg/L

3.7.2.5 Chloride Content

Chloride in natural waters near the mines and in seawater is present in the form of dissolved sodium chloride and chlorides may also be present due to the combination of salt water and sewage in the water.s However excessive concentration of chloride is harmful and unsafe for human consumption and usage. The acceptable limits of chloride concentration in water as prescribed in Indian standard is 250 mg/L for drinking water. The chloride concentration is measured using titration process by argentometric method using potassium chromate as an indicator with normal silver nitrate solution.

3.7.2.6 Fluoride

Fluoride ions have dual importance in water sources. Short range of fluoride concentration is required in drinking water. As strong concentration of F- induces dental fluorosis (the teeth are disfigured) and at the same time, "dental caries" result in a concentration of less than 0.8 mg/L. Hence, maintaining the F- concentration in drinking water between 0.8 and 1.0 mg/L is important. Fluorides are added to drinking water if sufficient fluoride is not present in water. However, large concentration of fluoride is toxic to humans and other animals. Excessive fluoride concentration of greater than 2 mg/L causes discolouration of teeth called mottling of teeth. Excessive fluoride intake may also result in disease called fluorosis that causes bone problems and weakening of bones.

Among the many methods suggested for determining fluoride ion in water, the most effective and applicable to variety of samples are the colorimetric method (SPANDS) and the ion specific electrode method. Because all colorimetric methods are prone to errors due to the existence of interfering ions, the sample may need to be purified before making the fluoride estimation. When the fluoride electrode is dipped in the sample whose concentration is to be calculated, a potential is determined by the existence of fluoride ions by any modern pH meter having an extended millivolt-scale. Organic matter may intervene positively, but variably. The degree of interference depends on the type and concentration of the sampled organic matter. Clean all glasswares thoroughly and rinse to minimize the error that could arise from streaks or particles outside the curves, as well as from traces of surfactants or dichromate cleaning solution that may adhere to the internal glass surfaces

3.7.2.7 Nutrients

A number of minerals and constituents in water can affect the domestic and drinking water quality of water used for the human consumptions. Owing to the excessive or low concentration of fluoride outside a range of 0.8-1.5 mg/L in water can cause problems with cavities, mottling in teeth also can cause bone disease such as fluorosis. A laxative effect is induced in the human body due to the excessive presence of sulphate in the water. Nitrates and nitrites are considered together, as conversion occurs in the atmosphere from one type to another. The presence of nitrogen and phosphorus in water indicates the presence of organic matter or contamination in water. The health effects of nitrate in the body are usually a result of its ready conversion from nitrite.

Nitrogen content: The presence of nitrogen in the water indicates the presence of organic matters in the water. The nitrogen may be present in the water may be in one or more of the following forms.

(a) Free ammonia
(b) Albuminoidal nitrogen
(c) Nitrites
(d) Nitrates

(a) Free ammonia: The free ammonia indicates the initial stage of aerobic degradation of organic matter and indicates the recent pollution in a stream. Organic-nitrogen (organic-N) is found in proteins, amino acids, urea, living or dead organisms (algae and bacteria) and decaying plant material. The microbiological degradation of organic nitrogenous matter creates ammonia. Therefore, it occurs both in many groundwaters and in surface waters. Ammonia concentration0.26 or 1.8 mg/L (for long-term exposure) in water at high temperature and pH value is toxic and chronic to fish.

The proportions of the two sources of ammonia nitrogen in surface water depend on pH. In general, distilling ammonia from the sample and storing it in boric acid is preferred for accurate tests. Organic-N is usually determined from the laboratory method called total Kjeldahl nitrogen (TKN), which measures a combination of organic N and ammonical nitrogen. Since N can transform from one form to another, it is often considered in its totality as total nitrogen (TN). Ammonical nitrogen can be calculated using Nessler's reagent by titration or colorimetric methods as with alkaline Nessler's reagent, ammonia produces a yellow coloured compound. Addition of EDTA before addition of Nessler's reagent or adding Rochelle salt solution in the presence of alkaline Nessler's reagent prevents precipitation of residual Ca and Mg.

(b) Albuminoid nitrogen: Albuminoid nitrogen indicates the concentration of nitrogen before the start of decomposition process. The albuminoid nitrogen indicates amount of organic pollution in the water as it comes into existence from animal and vegetable matter and finally from the nitrogenous substances in solution in various stages of decomposition. The organic matter in itself is not much harmful to health, but is undesirable as it putrefies and gives water a disagreeable taste and odour. It enhances the microbial growth as it offers food for them. The organic matter from animal origin contains high amount of albuminoid nitrogen as compared to same amount of vegetable matter.

(c) **Nitrites:** Nitrite in water is caused either by oxidation of ammonium compounds, or by nitrate reduction. This is unstable as an intermediate step in the nitrogen cycle as well as of aerobic or anaerobic process. Nitrites in water are regarded as a special danger signal as it indicates the active putrefaction of nitrogenous organic matter is going on as the result of bacterial activity. The presence of nitrites indicates organic pollution and represents the transitional stage in the oxidation of organic matter between ammonia and nitrates as shown in Eq. 3.22. Nitrites are generally present in water in small amounts as they are very soon converted to stable nitrates.

The normal pure water should not contain any nitrites or traces. The absence of nitrites also does not mean that the water is necessarily safe, while their presence in any but the smallest measurable amounts shows pollution. A normal concentration of nitrite in natural water is around tenths of mg/L. However, the higher concentration can be found in human waste, sewage, or biologically treated effluents and in contaminated streams. Nitrite concentration of 0.005 mg/L NO_2-N is the limit for water supply systems. Very high levels of nitrite are typically related to water with unsatisfactory microbiological activity.

$$2Kr_2HgI_4 + NH_3 + 3KOH \xrightarrow{\text{yields}} (NH_2Hg_2IO) + 7KI + 2H_2O \tag{3.22}$$

As per APHA, 2005 nitrites can be determined by the methods (i) colorimeter or spectrophotometer that can be operated at wavelength of 543 nm, and (ii) Nesslerization method

(d) **Nitrates:** Nitrate is a nutrient ingested by plants and animals in surface water and is stored in cells as cell protein. Plant growth stimulation, particularly of algae, can cause objectionable eutrophication. Chemical fertilizers, decayed vegetables and animal matter, domestic effluents, land contamination of waste sludge, industrial pollution, leachates from refuse dumps, and surface washouts are major sources of nitrate. Nitrate is the most oxidised type of compounds of nitrogen typically found in natural waters. It is an end product of final conversion of ammonia to nitrate in aerobic oxidation process. The nitrate found in pure well water extracted from a vast catchment is primarily the product of biological activity in the soil's surface layers, intensified by agriculture and manure application. These sources can contaminate streams, rivers, lakes and groundwater.

The excessive nitrate concentration ultimately leads to "blue baby disease" particularly, in infants known as infant methemoglobinemia. The Indian standard has prescribed the permissible limit of nitrate concentration in drinking water to be 45 mg/L. Nitrate is calculated either by ammonia reduction or by matching phenol-disulphonic acid-produced colours. Determination of nitrate (NO_3^-) is difficult due to the fairly complex procedures needed, the high likelihood of presence of interfering constituents and the small concentration ranges of the different techniques. Interferences include dissolved organic matter, nitrite, hexavalent chromium, and surfactants.

3.7.2.8 Hydrogen Sulphide (H_2S)

The presence of offensive or "rotten egg" smell or odour and taste in water is an indicator of hydrogen sulphide gas in water. Hydrogen sulphide is a by-product in anaerobic decomposition of sulphate present in the organic matter. However, sulphates are end products of aerobic decomposition of sulphur present in the organic matter. Sulphur-reducing bacteria, which use sulphur as an energy source, are the primary producers of large quantities of hydrogen sulphide. These bacteria chemically change natural sulphates in water to hydrogen sulphide. It also occurs as natural mineral in some soil and rock formations that dissolves over time and is released into groundwater. Sulphur-reducing bacteria flourish in oxygen-deficient environments such as deep wells, plumbing systems, water softeners and water heaters. These bacteria also live in the hot water side of a water distribution system.

The magnesium corrosion control rod present in many hot water heaters can chemically reduced by naturally occurring sulphates to hydrogen sulphide. The effects of H_2S gas when present in water is its corrosiveness to metals such as iron, steel, copper and brass. It can tarnish silverware and discolour copper and brass utensils. It can also cause yellow or black stains on kitchen and bathroom fixtures. Also the presence of H_2S in water used for preparation of coffee, tea and other beverages not only discolours its appearance but it can also change the taste of foods cooked in such type of waters. In water treatment units the high concentrations of dissolved hydrogen sulphide also can foul the resin bed of an ion exchange water softener as it provides a convenient environment for these bacteria to grow. A "salt-loving" bacteria, that uses sulphates as an energy source, may produce a black slime inside water softeners. The concentration of H_2S in water as stipulated in Indian standard should not be more than 0.05 mg/L as the same can cause taste and odour problem.

Hydrogen sulphide is traditionally determined using an acid displacement procedure, followed further analysis using gas chromatography flame photometric detector. The procedure has been used for water, sewage, and effluents containing 0–2.0 m/L of sulphide. The methylene blue colorimetric method is another standard analytical procedure for hydrogen sulphide determination, at concentrations ranging between 0.1 and 20 mg/L.

3.7.2.9 Sodium

Owing to the high solubility of its salts and the abundance of mineral deposits the sodium ion is pervasive in the water. Wind-borne sea spray may make an significant contribution in the vicinity of coastal areas, either by dropping on to land surfaces where it flows to water supply or by rain washout. A further source of sodium in water is domestic, commercial, and industrial discharges. Sodium salts are usually not highly toxic because of the potency with which mature kidneys excrete sodium. In comparison to adults, the effects on children are different due to the immaturity of the infant kidneys. With drinking water a limit of 200 mg/L is required. The concentration of iron and manganese in water can be determined by using colorimetric method using UV-VIS spectrophotometer.

3.7.2.10 Sulphates

Sulphates are commonly distributed in nature, and excessive quantities of sulphates dissolved in drinking water contribute to hardness problems. Sulphate minerals can cause scale to build up in water pipes similar to other minerals and may be associated with a bitter taste in water that can have a laxative effect on humans and young livestock. Elevated sulphate levels in combination with chlorine bleach can make cleaning clothes difficult. Sulphur-oxidizing bacteria produce effects similar to those of iron bacteria. They convert sulphur into sulphate, producing a dark slime that can clog plumbing and/or stain clothing. Blackening of water or dark slime coating toilet tanks may indicate a sulphur-oxidizing bacteria problem. Sulphur-oxidizing bacteria are less common than sulphur-reducing bacteria. The well may have a biofouling problem and a corrosion related problem. For the most prevalent sulphate salts the taste threshold concentrations are from 200 to 500 mg/L for sodium sulphate, 250 to 900 mg/L for calcium sulphate and 400 to 600 mg/L for magnesium sulphate. However, the dissolved sulphate concentration range as stipulated in Indian standard for potable water is 200–400 mg/L.

3.7.2.11 Petroleum Hydrocarbons

Water-soluble elements of crude oils and processed products contain a number of toxic compounds. High molecular weight tars in the water are less harmful than compounds of medium-molecular weight such as petrol. Low molecular weight compounds are usually of little value because they are very volatile and evaporate rapidly. Human beings have a very low threshold of taste for petroleum hydrocarbons, whose taste is particularly repellent. All components of crude oil are degradable by bacteria but at varying levels, and petroleum hydrocarbons may also be metabolized by a variety of yeasts and fungi. The spillage of diesel at sea is even more harmful than crude oil (very tarry) or spillage of petrol. The tolerance for dissolved hydrocarbons in harbour basin water does not exceed 0.30 mg/L. When used for washing, polluted harbour water can impart an unpleasant taste to fish. Commercial fishermen may also risk spoiling a whole catch if their nets are filled with diesel or oil (Sources and Impacts 2020). As per Indian standard the presence of polynuclear aromatic hydrocarbons (as PAH), should not be greater than 0.0001mg/L in any case (APHA 2005)

3.7.2.12 Asbestos

Asbestos is a general term for the amphibole and serpentine mineral groups of fibrous silicate minerals. They are organic or inorganic compounds chosen on the basis of carcinogenicity, mutagenicity, teratogenicity or high acute toxicity known or suspected by them. Asbestos is generally renowned by six minerals aschrysotile, crocidolite, anthophyllite, tremolite, actinolite, and amosite. Asbestos is popular in household water sources. The use of asbestos cement (170 g of asbestos per kg with 80% chrysotile and 20% crocidolite) for pipes in distribution systems may add to the drinking water's asbestos content. It is estimated that the background rates range from less than 1 million to 10 million fibres per litre. The key issue of asbestos is its high carcinogenicity.

3.7.2.13 Dissolved Gases

Oxygen and carbon oxide are the gasses present mainly in natural water. The surface water contains significant quantities of dissolved oxygen as it is absorbed from the atmosphere. Algae and other minuscule water plant life also give the water oxygen. The presence of dissolved oxygen in the water makes it fresh and sparkle. But more oxygen induces material corrosion to the tubing. Water absorbs atmospheric carbon dioxide. If water overflows with calcium and magnesium salts, carbon dioxide reacts with the salts and turns them into bicarbonates, causing water hardness. It is easy to determine the presence of carbon dioxide by applying lime solution to water gives milky white colour.

3.7.2.14 Biodegradable Organics

Biodegradable organics are most commonly measured in terms of BOD (biological oxygen demand), consisting primarily of proteins, carbohydrates, and fats. BOD may be defined as the amount of oxygen required by the microorganisms to oxidize the biologically active organic matter. The higher the BOD in water the lesser will be dissolved oxygen in water to support the aquatic life thus indicating high strength of pollution in water. The urban sewage normally contains a BOD concentration of 200–500 mg/L. Harbour basin water may contain a BOD value between 50 and 150 mg /L.

Bio-chemical oxygen demand: If the water is polluted with sewage, the demand for oxygen in sewage exerted by microorganisms of organic matter is called biochemical oxygen demand of that water. The aerobic activity continues in sewage until the oxygen is present. As the oxygen exhausts the anaerobic activity starts with indication of foul smell. The biochemical oxygen demand (BOD) is a laboratory test that provides the oxygen demand for aerobic oxidation of decomposable organic matter and other inorganic materials in water and wastewater under controlled temperature and incubation time conditions. It is a direct measure of the amount of oxygen required by microorganisms to carry out the oxidation process. The test is applied to sources of fresh water (rivers, lakes), wastewater (domestic, industrial), contaminated water bodies, marine water (estuaries, coastal water) and also to determine contamination rates, water body assimilation ability and waste treatment plant efficiency in terms of organic matter removal.

The organic matter present in the wastewater as shown through Fig. 3.12 in terms of BOD can be categorized in two groups as (i) carbonaceous matter (ii) nitrogenous matter. The carbonaceous BOD is the amount of oxygen needed by the microorganisms in a sample to decompose biodegradable carbonaceous matter. This is the first stage of oxidation and is known as first stage BOD. In the second stage after carbonaceous matter exhausts, the autotrophic bacteria oxidize the nitrogenous matter present in the sample and is termed as nitrogenous BOD or the second stage BOD. Therefore, 5 days period of incubation at fixed temperature conditions is the period in which around 65 to 70% of the organic matter degrades.

However, it is assumed that overall 95% to 99% oxidation completes in around 20 days. Therefore, the sample results obtained in the laboratory are represented as BOD_5 or simply BOD. Another reason to consider 5 days as standard duration as thereafter the interference of nitrogenous bacteria starts the nitrification process that ends on sixth or seventh day. Sanitary engineers normally are interested in assessing the carbonaceous BOD, therefore, 5 days period of BOD as BOD_5 is chosen for study purpose. Kinetics of BOD is indicated in following Eq. 3.23–Eq. 3.26.

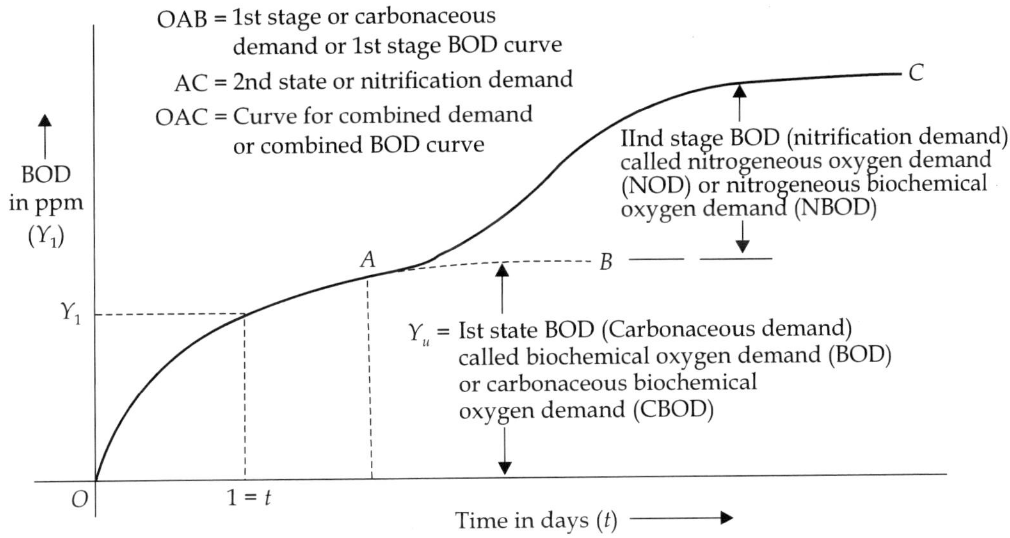

Fig. 3.12: The BOD curve indicating various demands during the BOD test with respect to time (Rahman 2012)

(a) Kinetics of BOD: The BOD at any time t can be calculated using the relationship of organic matter biodegradation as

$$\frac{dL_t}{dt} = -kL_t$$

where, L_t oxygen equivalent of carbonaceous oxidisable matter present in sewage after t days from start of oxidation (mg/L).

t = time in days

k = rate constant (per day)

$$\int_{L=L_0}^{L=L_t} \frac{dL_t}{L_t} = \int_{t=0}^{t=L_t} -k\,dt$$

$\log_e (L_t) - \log_e (L_0) = -kt$

$$\log_e \frac{L_t}{L_0} = -kt$$

or $\log_e \left(\frac{L_t}{L_0} \right) = -kt$

$$\boxed{L_t = L_0 \cdot e^{-k \cdot t}} \quad (3.23)$$

where, k is known as BOD rate constant or de-oxygenation rate constant (per day). In literature the $k_D = \frac{k}{2.3}$ has the same meaning. Now, the quantity of organic matter oxidized during t days = organic matter present at (t = 0) – organic matter left at (t = t)

$$y_t = L_0 - L_t$$
$$= L_0 (1 - e^{-k \cdot t})$$

Since y_t is the amount of organic matter oxidized during t days, hence it is also same as oxygen absorbed in t days.

$$\boxed{BOD_t = y_t = L_0 (1 - e^{-k \cdot t})} \quad (3.24)$$

However, BOD ultimate equivalent to amount of organic matter present in the wastewater can be calculated as:

$$\boxed{y_u = \frac{BOD_5}{(1 - e^{-k \cdot t})}} \quad (3.25)$$

As the fact that ultimate BOD (y_u) will occur at $t = \infty$

$y_\infty = L_0 (1 - e^{-k \cdot t}) = L_\infty$, independent of temperature and is affixed quantity

However, the BOD rate constant (k) is a temperature dependent parameter and can be computed by the following relation:

$$\boxed{k_{(T°)} = k_{(20°)} [1.047]^{T - 20°}} \quad (3.26)$$

Also, if $T > 20°$ then $C = k_{(T°)} > k_{(20°)}$ and when $T < 20°$ then $C = k_{(T°)} < k_{(20°)}$

The test is carried out in a 300 mL BOD bottle incubated at 20°C temperature conditions for 5 days. It is ensured that the light must be excluded from the incubator to prevent algal growth that can produce oxygen in the bottle and may deviate the results of the samples. The nitrification intervention may be removed by sample pretreatment by using inhibitors such as methylene blue. Since the test is primarily a bioassay method, standard conditions of temperature, nutrient supply, pH value of 6.5-7.5, sufficient microorganism population as seed and absence of microbial growth inhibiting substances need to be considered.

The low solubility of oxygen in water requires dilution of strong wastes to ensure demand does not increase the oxygen available. This dilution water is therefore prepared in the laboratory by aerating the tap water for 4-8 hours duration till it attains the saturated concentration of DO with addition of nutrients, pH buffer solution along with addition of little seed as microorganisms. The prepared nutrients solutions are added at the rate if 1 mL/L of dilution water prepared for the analysis. However, other reagents are added as per the requirements and recommendations in BOD determination method specified by APHA, 2005.

The test can also be carried out at different temperature conditions and correction of time and *k* values need to be considered to obtain the desired results. This indicates that the BOD test will provide the same result when the sample is incubated at 27°C for 3 days as compared to the test carried out at 20°C for 5 days incubation period. Therefore, incubation time in days and temperature and are important parameters for reporting the test results.

(b) Calculations: In calculations, the final BOD concentration in the laboratory can be determined by assessing the dissolved oxygen in a water sample as Eq. 3.27-Eq. 3.28:

$$BOD_5\left(\frac{mg}{L}\right) = \frac{\text{Volume of diluted sample}}{\text{Volume of undiluted sewage sample}} \times (DO_{initial} - DO_{final}) \tag{3.27}$$

or

$$BOD_5\left(\frac{mg}{L}\right) = \text{Dilution factor} \times (\text{DO consumed after 5 days}) \tag{3.28}$$

All required reagents used for DO estimation and BOD determination are:

- **Phosphate buffer:** Dissolve 8.5 g KH_2PO_4, 21.75 g K_2HPO_4, 33.5 g $Na_2HPO_4 \cdot 7H_2O$, and 1.7 g NH_4Cl; dilute to 1000 mL in distilled water. Any further modification the pH will be 7.2. When there is some indication of biological growth, discard reagent.
- **Magnesium sulphate:** Dissolve 22.5 g $MgSO_4 \cdot 7H_2O$ in approximately 700 mL distilled water and dilute to 1 litre.
- **Calcium chloride:** Dissolve 27.5 g $CaCl_2$ in approximately 7000 mL of distilled water and dilute to 1 litre.
- **Ferric chloride:** Dissolve 0.25 g $FeCl_3 \cdot 6H_2O$ in a distilled water of around 700 mL and dilute to Sodium sulphate 0.025 N solution: Dissolve 1.575 g Na_2SO_3 in distilled water and dilute to 1000 mL. There should be a routine solution to this.
- **Acid and alkali solutions 1N:** Prepare 1N H_2SO_4 and 1N NaOH or neutralize the caustic or acidic sample • Nitrification inhibitor: 2-chloro-6-(trochloromethyl) pyridine [Nitrification inhibitor 2570-24 (2.2% TCMP)].

Standard BOD reagent as glucose-glutamic acid solution: glucose grade dry reagent and glutamic acid at 103°C for 1h. With distilled water, dissolve 150 mg of glucose and 150 mg of glucose acid and dilute to 1000 mL. Prepare fresh for immediate use. The standard BOD value of the solution lies in the range of 200 ± 20 mg/L.

Chemical oxygen demand (COD): Chemical oxygen demand (COD) is another test method that calculates the amount of equivalent oxygen required to oxidize the organic matter present in water by means of chemical method. Therefore, COD may be defined as the oxygen required in oxidizing the amount of total organic matter present in the water. As a means of measuring organic strength for streams and contaminated water bodies, it is a significant fast measured parameter that can provide results in 2-4 hours duration as compared to BOD results that are obtained in 3-5 days. The test is useful in the study of wastewater treatment plants efficiency assessment and the monitoring of fairly contaminated water bodies. In addition, the test is fairly simple, accurate, and uninfluenced by interferences as in the BOD test.

The inherent weakness of the test is its inability to differentiate between the biologically oxidizable and biologically inert content and to find out the aerobic biological stabilization system rate constant. The test is carried out in the presence of $K_2Cr_2O_7$ as a strong oxidant along with H_2SO_4 as digestion reagent. The chemical silver sulphate ($AgSO_4$) is added to the acid reagent in order to eliminate the chloride interferences during the digestion and oxidation process. The sample after mixing of above chemicals is digested for 2-4 hours at 150°C in a digester. Thereafter, the sample is cooled to room temperature and the amount of oxygen required can be determined by titration method or by using colorimetric procedure. However, if the concentration of organic compounds present in water is known the theoretical oxygen demand of the sample can be calculated comfortably and accurately. But it is not always possible to determine all the organic compounds in wastewater easily. However, organic matter in terms of theoretical COD can be calculated by balancing the reaction equation in terms of theoretical oxygen required to oxidize the known organic compound with end products as CO_2 and H_2O and other species (if present) such as NO_3^- as shown in Eq. 3.29.

$$x.\text{Organic matter} + y.\,O_2 + z.\,NH_3 \xrightarrow{\text{yields}} a.CO_2 + b.H_2O + c.NO_3^- \quad (3.29)$$

In order to determine the COD of given water sample the test is carried out using $K_2Cr_2O_7$ as a strong oxidant solution that assesses the amount of molecular oxygen utilized in oxidizing the organic matter in the presence of strong acid. The open reflux and close reflux methods are generally used to calculate the COD of a water sample. However, the closed reflux method is preferred over the open reflux method as low quantity of oxidizing agents and acids are generally used and the test result takes less time as compared to open reflux method. Also the closed reflux colorimetric method is preferred as it utilizes lowest concentration of all chemicals and the results can be obtained in 2 hours. Oxidation of most organic compounds is up to 95-100 per cent of the theoretical value.

The organic matter is completely oxidized by potassium dichromate and silver sulphate as a catalyst in the presence of concentrated H_2SO_4 for CO_2 and H_2O growth.

In open reflux method the remaining or excess $K_2Cr_2O_7$ is titrated with ferrous ammonium sulphate [Fe $(NH_4)_2(SO_4)_2$] after the reaction. The dichromate consumed gives the oxygen (O_2) required for oxidation of the organic matter. The chemical reactions involved in the method are as under(Eq. 3.30-Eq. 3.32):

$$2K_2Cr_2O_7 + 8\,H_2SO_4 \longrightarrow 2\,K_2\,SO_4 + 2Cr_2(SO_4)_3 + 8\,H_2O + 3O \quad (3.30)$$

$$C_6H_{12}O_6 + 6O_2 \longrightarrow 6CO_2 + 6H_2O \quad (3.31)$$

$$Cr_2O_7^- + 6Fe^{2+} + 14H^+ \longrightarrow 6Fe^{3+} + 2Cr^3+ + 7H_2O \quad (3.32)$$

(a) Preparation of reagents and standards*

- Standard potassium dichromate solution, 0.25N (0.04167 M): Dissolve 12.259 g $K_2Cr_2O_7$ dried at 103°C for 24 h in distilled water and dilute to 1000 mL. Add about 120 mg sulphamic acid to take care of 6 mg/L NO_2 –N.
- Sulphuric acid reagent: Add 10 g of Ag_2SO_4 to 1000 mL concentrated H_2SO_4 and let it stand for one to two days for complete dissolution.
- Standard ferrous ammonium sulphate approx. 0.25N (0.25M): Dissolve 98 g $Fe(NH_4)_2(SO_4)_2{\cdot}6H_2O$ in about 400 mL distilled water. Add 20 mL concentrated H_2SO_4 and dilute to 1000 mL.
- Ferroin indicator: Dissolve 1.485 g 1, 10-phenanthroline monohydrate and 695 mg $FeSO_4{\cdot}7H_2O$ in distilled water and dilute to 100 mL.
- Mercuric sulphate: $HgSO_4$, crystals, analytical grade.
- Potassium hydrogen phthalate (KHP) Standard: Dissolve 425 mg lightly crushed dried potassium hydrogen phthalate ($HOOC{\cdot}C6H_4{\cdot}COOK$) in distilled water and dilute to 1000 mL. This solution has a theoretical COD of 500 µg O_2/mL. This solution is stable when refrigerated, up to 3 months in the absence of visible biological growth.

*preparation of $K_2Cr_2O_7$ and H_2SO_4 may be different with open and closed reflux methods as the requirement of oxidizing agent and acid will be different and therefore should be prepared accordingly.

(b) Calculations: However, practically using titration method the COD concentration is determined as in Eq. 3.33:

$$COD\left(\frac{mg}{L}\right) = \frac{(a-b)x\ Nx\ 8000}{\text{mL of sample}} \quad (3.33)$$

where, a = mL FAS used for blank

b = mL FAS used for sample

N = normality of FAS

8000 = mille equivalent weight of $O_2 \times 1000$

(c) ThOD (Theoretical Oxygen Demand): The theoretical COD concentration can also be assessed, if the exact formula and concentration of the organic matter present in the system is known. Therefore, the amount of oxygen required for its decomposition can be computed theoretically as in Eq. 3.34:

$$\underset{(180)}{C_6H_{12}O_6} + \underset{(192)}{6\,O_2} \longrightarrow 6\,CO_2 + 6\,H_2O \tag{3.34}$$

It infers that 180 g of $C_6H_{12}O_6$ requires 192 g of O_2 for its decomposition and therefore,

300 mg/L of $C_6H_{12}O_6$ requires $= \frac{192}{180} \times 300 = 320 \frac{mg}{L}$ as ThOd or COD

(d) BOD/COD ratio: BOD as ultimate BOD_u represents the amount of biodegradable organic matter present in the water whereas COD represents the total biodegradable and non-biodegradable organic matter present in water. Therefore, the difference of the two provides the amount of non biodegradable matter in the sample. The BOD_u/COD ratio indicates the biodegradability of any organic matter present in the water sample. For a excellent amount of biodegradable matter in a water sample the BOD_u/COD ratio should approach towards 1. However, in any case the value is always less than 1.0 or approaches towards 1.0. BOD can be equal to COD but cannot be greater than it in any case.

Hence, $\frac{BOD_u}{COD} < 1.0$

Also, if the ratio is found to be between 0.92 to 1.0, the water is said to be practically biodegradable. The ratio also helps in fixing the various dilutions for assessing the BOD of water sample in laboratory as the early results of COD in 2-4 hours and knowing the BOD/COD value of particular wastewater from available literature one can be able to assess the approximate BOD concentration and the dilutions may be fixed accordingly. When the BOD_5 is evaluated and BOD_5/COD ratio is referred to as BOD/COD, then the ratio may be taken as 0.63 to 0.68 in place of 0.93 to 1.0 as indicated in Eq. 3.35.This may be due to the fact that BOD_5 is around 68% of BOD_u and therefore,

$$\frac{BOD_5}{COD} = 0.63 - 0.68 \tag{3.35}$$

Example 3.7: A 2% solution of sewage sample is kept at incubation temperature of 20 degree Celsius. If the initial DO and final DO values after 5 days incubation period are 8.5 mg/L and 5.5 mg/L respectively, compute the BOD.

Solution: As we know that

$$BOD_5\left(\frac{mg}{L}\right) = \frac{\text{Volume of diluted sample}}{\text{Volume of undiluted sewage sample}} \times (DO_{\text{initial}} - DO_{\text{final}})$$

$$= \text{Dilution factor} \times \text{DO consumed}$$

where, volume of diluted sample = 100 mL

Volume of undiluted sample = 2% = 2 mL

Initial DO = 8.5 mg/L

DO after 5 days as final DO = 5.5 mg/L

Therefore, putting the values in the above equation, we get

$$=\frac{100}{2}\times(8.5-5.5)=150 \text{ mg/L}$$ **Ans.**

Example 3.8: If the of waste water sample is 75 mg/L and the reaction rate constant K (base e) is 0.345 per day. Find the amount of BOD remaining in the given sample after 10 days.

Solution: Given that: $k = 0.345/\text{day}$

and we know that $BOD_3 = y_3 = L_0(1 - e^{-k \cdot t})$

$$75 = L_0(1 - e^{-k \cdot t})$$

Therefore,

$$L_0 = 116.245 \text{ mg/L}$$

Now, remaining BOD after 10 days can be calculated as

$$L_t = L_0\, e^{-k \cdot t}$$

$$= 116.245 \times e^{-0.345x} = 3.69 \text{ mg/L}$$ **Ans.**

3.7.2.15 Metals and Other Chemical Substances

Water contains specific minerals or metal substances such as iron, manganese, copper, lead, barium, cadmium, selenium, fluoride, arsenic and others. Owing to the deposition of ferric hydroxide and manganese oxide, excess clothing can cause discolouration during washing and incrustation in water mains due to the presence of iron and manganese. Lead and beryllium are extremely toxic, and therefore the recommended concentration should be extremely with low. Arsenic and selenium are also highly toxic and it is recommended that the same should completely be eliminated from the system. Humans are affected by the presence of high quality copper in drinking water. Many waters make up essential constituents of trace amounts of several metals. Some of those metals are listed as contaminants as well. Owing to their toxicity discharged or released into the atmosphere, the presence of either of these metals in large concentrations can interfere with other beneficial water uses.

Arsenic (As): Arsenic is a notorious feature of toxicity. However, its toxicity depends on the compound's chemical (valence) and physical form, the path by which it reaches the body, the exposure dose and length and many other biological parameters. Arsenic is generally associated with lead solder, lead shot, battery grids, cable sheaths, and boiler piping as an alloying additive. Most arsenic now comes from paints or pharmaceuticals, and is commonly found in sewage treatment. Chronic symptoms can be the result of low intake accumulation of arsenic compounds in the body. Arsenic compounds also have been shown to have carcinogenic properties.

The maximum concentration of As_3^+ in water does not exceed 72 µg/L for protection of aquatic life. And the average should be no more than 140 µg/L. The maximum concentration of As recommended by FAO for water to be used for irrigation is 100 µg/L. The USEPA and Indian standard has prescribed minimum acceptable As concentration limit of 0.05 mg/L for water used for domestic purposes. The As concentration in water can be determined using spectrophotometric method by using atomic absorption spectrophotometer (AAS). It is recommended that an effort to determine the valence and chemical types of the element should be made when water is found to contain arsenic at levels of 0.05 mg/L. Arsenic content in seawater is around 0.002 mg/L. The primary concerns are carcinogenicity and mutagenicity.

Cadmium (Cd): Cadmium occurs in minerals of sulphide which also contain zinc, lead, or copper. The metal is used with various other metals in galvanizing, batteries, paint pigments and in alloys. Commonly, cadmium is correlated with zinc. In biological processes, cadmium does not show signs of being an important trace element; on the contrary, it is extremely toxic to the human body. Cadmium is highly toxic, and has been involved in food poisoning in some cases. Minute amounts of cadmium are thought to be responsible for adverse reactions in human kidney arteries. Cadmium also commonly induces cancers in laboratory animals, and is epidemiologically related to some human cancers.

A 200 µg/L concentration of cadmium is harmful to certain fish. Owing to industrial discharges or the deterioration of galvanized steel, cadmium can reach water. The permissible amount of cadmium recommended by the FAO for water to be used for irrigation is 10µg/L. The USEPA has prescribed the cadmium concentration limit to be 0.005 mg/L and Indian standard ahs prescribed the cadmium limit to be1 mg /L for drinking water. Cadmium and its components are blacklisted products which cannot be discharged or deposited into the atmosphere under international agreement.

Lead (Pb): Lead in a water supply can come from factory, smelter discharges and mine, or from plumbing and plumbing fixture dissolution. While it is not a nutritionally necessary item, it is important to control due to its toxicity to human health. Tap water which is inherently not corrosive or is not properly handled can contain lead resulting from an attack on lead service pipes, lead interior plumbing, brass fittings and fittings on solder pipe joints primarily from galena (PbS). This lead is produced in metallic form in batteries, munitions, soldering, welding, pigments, insecticides, and alloys.

For several years lead was also used as an anti-knock agent in the form of tetraethyl lead in gasoline. The administration of food and drugs controls the amount of lead in food and in house paints. The USEPA has recommended the lead concentration for drinking water level to be 15 µgL, under the lead-copper law. The permissible limit specified in Indian standard is 0.05 mg/L for drinking water (Sources and Impacts 2020).The lead in batteries, cable sheathing, sheets and pipes can be recovered and recycled but much of the lead used in compound form, such as paints and petrol additives, is lost to the atmosphere and ultimately ends up in the aquatic environment.

Chromium (Cr): Many rocks and soils have small quantities of chromium them. Chromium is a highly insoluble form in its naturally occurring state; however, most of the more common soluble forms found in soils are mostly the result of industrial pollution. Chromium's main applications include chromium alloys, chromium plating, oxidizing agents, corrosion inhibitors, decorative glass and ceramic pigments, as well as photography. Hexavalent (soluble) chromium as Cr(VI) compounds are carcinogenic, with a guideline value of 0.05 mg/L.

Mercury (Hg): Mercury is a comparatively rare element in the world, it is the product of natural geological activity and man-made pollution. Natural source mercury can enter the aquatic environment through weathering, dissolution and biological processes. While extremely useful to humans, mercury is also highly toxic to the human body in all forms, especially in the form of methyl mercury, because it cannot be excreted and thus acts as a cumulative poison. The potential for long-term human health risks from consumption of mercury contaminated fish has led many nations to lay down laws and guidelines for acceptable levels of mercury in the seafood. Nearly all levels of water above 1 ppb are due to industrial effluents linked to the manufacture of chlorine and caustic soda, pharmaceuticals, mirror coatings, mercury lamps and other fungicides. Methyl mercury is 50 times more toxic than inorganic mercury. The Indian standard has recommended the mercury concentration in drinking water should not be more than 0.001 mg/L as Hg.

Nickel (Ni): Nickel is all-rounder in the environment. Nickel is almost definitely important for animal nutrition and is also possibly necessary for humans. Nickel is a fairly non-toxic element; however, animal studies have shown that some nickel compounds are carcinogenic. The Indian standard prescribed the Ni concentration limit should not be more than 0.02 mg/L.

Tin (Sn): Tin and its compounds are biologically important and controversial chemicals. Not all chemical forms of tin are equally biologically active as is the case with other elements. Some organic tin compounds, also known as organotins, are toxic, as opposed to the low toxicity of inorganic tin (derived from consuming canned foods). The constituents of anti-fouling paints tributyltin and triphenyltin are extremely toxic and their existence in harbour waters is typically limited to 0.002 and 0.008 μg/L, respectively. Organotin anti-fouling paints are not allowed in many countries on vessels less than 25 metres long, and the beginning of a fishing season usually sees an increase in this compound in the water as freshly painted vessels are launched back into the water.

Barium (Ba): Barium is found in the earth's crust at a concentration of 0.50 g/kg and the most common source is the mineral barytes, barium sulphate. Traces of barium exist in most soils. The sources of barium include mineral deposits, disposal of drilling wastes, smelting of copper and motor vehicle parts manufacturing industries effluents. Barium is also used in various industrial processes, for example, in vacuum tubes, spark-plug alloys, Getter alloys, Fray's metal and in X-ray machines as a lubricant for anode rotors. The major health effects of barium results in difficulties in breathing, increased blood pressure, changes in heart rhythm, stomach irritation, brain swelling, muscle weakness, and damage to the liver, kidney, heart, and spleen. As per Indian standard the barium concentration in drinking water should not be more than 0.7 mg/L.

Beryllium (Be): Beryllium is usually present as part of mineral deposits of feldspar, and can occur in small localized deposits as the mineral beryl. Burning fossil fuels is the primary source of beryllium in the atmosphere, although pollution is usually mild. Beryllium can join the water system by weathering rocks in ground acquifers, causing atmospheric effects on rainwater collection systems and industrial and municipal discharges. Beryllium is used in alloys of metal and in some electrical devices. The Indian standard has no evidence of any limits for beryllium in drinking water. However, literature evidences shows it to be not more than 0.20 μg/L.

Selenium (Se): Another determining factor in their existence in drinking water is the chemical composition of selenium, and hence its solubility. Selenium is commonly used as a colorant for glass in the electronics, TV cameras, solar batteries, machine cores, grinders, xerographic plates and ceramics industries. It is also used as a trace factor for feeding stuffs. Several animal species, including the man, have recognized selenium as an essential nutrient. Dietary selenium levels of 5 mg/kg or more of food may cause chronic intoxication, and this value has been the dividing line between toxic and non-toxic feeds in seleniferous areas. The amounts of selenium in soil and vegetation vary within large limits as a result of geochemical variations. It is released from natural deposits and copper smelting process. The potential health effects of consumption of selenium with drinking water is hair and fingernail changes, damage to the peripheral nervous system and fatigue and irritability. There is a range of human intake of selenium that is compatible with health, and deficiencies or toxicity may occur outside of this range. As per Indian standard the selenium concentration of 0.01 mg/L is acceptable for potable waters.

Silver (Ag): Silver occurs naturally in the form of elements, often as separate ores. This also refers to the ores of gold, copper and zinc. Since some metals like lead and zinc are used in delivery systems and also since silver oxide is used to treat water sources in some countries, silver levels in drinking water may often be elevated. Silver is used in the production of silver nitrate, silver bromide and other photographic compounds, water distillation devices, mirrors, silver plating machines, special batteries, table cutlery, jewellery, dental medical and research devices including amalgams. The Indian standard prescribes the concentration of silver in drinking water should not be more than 0.1 mg/L.

Aluminium: Aluminium along with its compounds comprise about 8% of the earth's surface. As the occurrence of aluminium naturally in form of silicates, cryolite, and bauxite rock makes it important. Natural processes account for most of the redistribution of aluminium in the environment. Acidic precipitation mobilizes aluminium from natural sources, and direct anthropogenic release of aluminium compounds associated with industrial processes occur mainly to air. In pure water, aluminium has a minimum solubility in the pH range 5.5–6.0; concentrations of total dissolved aluminium increase at higher and lower pH values.

Use of aluminium salts as coagulants in water treatment may lead to increased concentrations of aluminium in finished water. Where residual concentrations are high, aluminium may be deposited in the distribution system. Disturbance of the deposits by change in flow rate may increase aluminium levels at the tap and lead to undesirable colour and turbidity (WHO, 1996). Concentrations of aluminium at which such problems may occur depend on a number of water quality parameters and operational factors at the water treatment plant. The Indian standard stipulates the permissible limit of aluminium in drinking water to be 0.2 mg/L.

Copper: Copper is used in manufacturing of electrical wiring, pipes, valves, fittings, coins, cooking utensils and building materials. It is present in munitions, alloys (brass, bronze) and coatings. Copper compounds are used as or in fungicides, algicides, insecticides and wood preservatives and in electroplating, azo dye manufacturing, engraving, lithography, petroleum refining and pyrotechnics. Copper compounds can be added to fertilizers and animal feeds as a nutrient to support plant and animal growth (Landner & Lindestrom, 1999; ATSDR, 2002). Copper's presence in water supply is not harmful to health but it can interfere with the intended domestic water uses. Copper improves aluminum and zinc fittings corrosion, fabric stains and plumbing fittings. Copper is used as a catalyst in alloys, in anti-fouling paints and as a preservative for wood. Urban wastewater contains large amounts of copper. The level of human taste for copper is 5.0-7.0 mg/L and the taste is repulsive. As per Indian standard the permissible limit of copper in drinking water should not be more than 1.5 mg/L.

Iron and manganese: Iron and manganese are the common metallic elements found in the earth's crust. The elements dissolve in water percolating through soil and rock containing iron and manganese and hold these elements in the form of solution. Iron and manganese are not much hazardous but their presence can cause problems in drinking water supply. In deep wells, with low oxygen content, the dissolved iron and manganese containing water is clear and colourless. Of the two elements iron is considered to be a major contaminant in drinking water supplies and manganese normally exists in iron-bearing water. Sometimes the iron pipes used in water supply and distribution may be a source of iron in water. Iron and manganese can stain laundry, plates, and fixtures.

The tap water containing iron and manganese when exposed to air changes from colourless to a coloured dissolved solid form of these elements due to oxidation. Iron that does not form particles large enough to settle out and that remains suspended (colloidal iron) leaves the water with a red tint. Manganese usually is dissolved in water, although some shallow wells contain colloidal manganese (black tint). These precipitates or sediments may be severe enough to plug water pipes. Iron generally causes reddish-brown staining of laundry, porcelain, dishes, utensils, glasswares and manganese causes a brownish-black stain. Such stains cannot be removed by soaps and detergents and therefore chlorine bleach and sodium carbonates are used to remove such stains. Iron and manganese accumulation creates an economic problem in water supply systems and also increases the energy costs from pumping water through constricted pipes or heating water with heating rods coated with iron or manganese mineral deposits.

The problem that exists is presence of iron or manganese bacteria in water. Ferrous salts are brittle under the pH conditions in drinking water sources and precipitate as insoluble ferric hydroxide, which settles out as rusty silt. Such water tastes unpalatable, allows "iron bacteria" to expand and the silt slowly decreases water flow in the piping. The Indian standard has recommended the permissible limit of iron in water used for drinking and other domestic usages is 0.3 mg/L. However, as per Indian standard the allowable range of manganese for domestic water supplies is 0.1–0.3 mg/L.

Zinc: Owing to the leaching action of zinc from galvanized pipes, brass and other zinc alloys, the concentration of zinc in the drinking water can be considerably higher than that in surface water. Zinc imparts an unpleasant astringent taste to water, especially in concentrations above 5 mg/L. The water can seem opalescent, and on boiling produce a greasy image. The zinc levels should be held far below that amount. The Indian standard ahs recommended the concentration range in drinking water to be 5-15 mg/L. The concentration of all the above mentioned heavy metals in water can be determined applying atomic emission techniques using equipment atomic absorption spectrophotometer (AAS).

Refractory organics: This category of contaminants is large and consists of chlorinated alkanes (carbon tetrachloride), chlorinated ethenes (polyvinyl chloride or PVC), polynuclear aromatic hydrocarbons (naphthalene, coal tar), pesticides, herbicides and fumigants (DDT, Endrin, Aldrin, Lindane, Methoxychloride, Toxaphene and Silvex), mono-Dichlorobenzenes (solvents), Benzenes (Benzene, Toluene), Phenols and Chlorophenes(Sources and Impacts 2020).

Chlorinated alkanes and chloroethanes: The chemical industry mainly uses the chlorinated alkanes as an intermediate in the manufacture of certain organochlorine compounds. Therefore, they are manufactured in large amounts and hence, many are contained in raw and finished drinking water. Carbon tetrachloride is a haloalkaneis found to be an occasional chlorine contaminant used in drinking water disinfection but is not created in drinking water as a consequence of the chlorination process. Carbon tetrachloride has been widely used as an aerosol propellant. The WHO has established the prescribed limits of the chemical to be used for domestic water supplies should not be greater than 0.003 mg/L.

Chloroethanes are another group of compounds commonly used as solvents, softeners, paint thinners, dry-cleaning fluids and intermediates in a number of industrial processes. They are also found in raw and processed drinking water, due to their large use. Owing to their high volatility, they are typically lost from surface water to the atmosphere, and thus generally occur at lower concentrations. Vinyl chloride is primarily used in the production of PVC resins which, in effect, make up the most widely used plastics worldwide. High concentrations of PVC were found in effluents discharged by the manufacturing plants of chemicals and latex and in drinking water as a result of substandard leaching (improper cured) PVC pipes used in water distribution systems. Cancer associated vinyl chloride is mutagenic in a variety of biological systems, including Salmonella and E. Coli. Coli. Many chlorinated ethenes include 1,1-dichloroethane (used in the packaging industry), trichloroethane (used as a dry-cleaning solvent) and tetrachloroethane (used in dry-cleaning and as a metal degreaser).

Polynuclear aromatic hydrocarbons (PAH): PAH is a wide collection of organic compounds from both natural and industrial sources which are present in the environment. PAH is rarely present in the atmosphere alone and several encounters with PAH mixtures can occur whereby the potency of known carcinogenic. In certain instances interaction with linings centred on coal-tar during distribution is known to lead to an increase in water concentration of PAH. Due to the close association of PAH with suspended solids, the application of the treatment to reach an appropriate turbidity level must ensure that minimum PAH levels are reached. Aromatic hydrocarbons that enter the harbour basin's aquatic environment from vessel discharges such as ballast water, bilge pumping, engine exhaust, coastal refinery effluents, crude oil power stations, terrestrial run-off (especially from urban storm water containing asphalt road particles) and leaching (creosoted components from jetties and wharves). The Indian standard has not provided any relaxation or limit of PAH in water used for domestic usage.

Pesticides: Pesticides and their derivatives are chlorinated hydrocarbons, persistent herbicides, plant insecticides, easily leachable pesticides from the plant and pesticides that are routinely applied to the water supply to control disease vectors, such as mosquito larvae (malaria and dengue fever). Only the chlorinated hydrocarbon insecticides occur frequently of these compounds and these are very persistent in the environment where they have become widespread. Typical pesticides include:

- DDT (dichloro diphenyl trichloroethane) is a persistent insecticide which is stable under most environmental conditions and resistant to complete breakdown of soil microorganisms by enzymes.
- Aldrin and dieldrin are the two similar pesticides that accumulate in the food chain and are very persistent. They are used around the roots of fruit trees for termite protection.

- Chlordane is a wide-spectrum insecticide often used for management of the termite and for homes and gardens.
- HCB or hexachlorobenzene, are commercially manufactured for their use as a fungicide.
- Heptachlor is another wide-spectrum insecticide used to control insects from agricultural soil. It is highly persistent.
- Lindane is a group of broad-spectrum insecticide called organochlorine insecticides. It is used in a wide range of applications including animal care, houses, water (for mosquitoes), plants, seeds and soil.
- Methoxychlor is an insecticide which is used to treat crops and livestock.

The Indian standard prescribes the limit of all such pesticides in drinking water should not be more than 0.001 mg/L.

Benzenes and mono-dichlorobenzenes: Benzene and toluene are mainly derived from petroleum, or as a by-product of gas processing. Both are commonly used as intermediates in the chemical industry, as well as for the manufacture of styrene, phenol, acetone, and cyclohexene (used in nylon processing).Significant amounts of toluene are used as fuel additives in the production of plastics, paints and detergents. The WHO has recommended the concentration of benzene in drinking water should not be more than 0.001 mg/L. Monochlorobenzene is commonly used as a solvent in manufacturing insecticides and phenols. Dichlorobenzenes are important intermediates for dyes, moth repellents, deodorants, dielectric fluids, fluids for heat transfer, and insecticides.

Phenols and chlorophenols: Chlorphenols are used as biocides, and are present in water as a result of phenol containing chlorinated water sources. Chlorophenols are well known for their low levels of taste and odour. As per Indian standard the concentration of phenol in drinking water should not be more than 0.002 mg/L. Chlorophenols main solution to reducing emissions is to avoid phenol (from petrochemical industries) and chlorinated phenolic pesticides (agriculture) from contaminating the water sources.

Trihalomethanes: Trihalomethanes as chloroform and bromoform exist primarily as reaction products of chemicals used in oxidative treatment reacting with naturally occurring materials found in water. Their formation is particularly linked to the use of chlorine to disinfect water sources. Nevertheless, it is important to note that chlorine is an efficient water disinfectant and that there are considerable risks of disease from microbiological pollutants resulting from inadequate disinfection.

3.7.2.16 Bacterial and Microscopical Characteristics of Water

Water-borne diseases: The most important aspect in terms of water quality is the presence of various biological microorganisms in water. World Health Organization has reported that 80 percent of communicable diseases spread by water. The microorganisms are known as pathogens as they posses very high capability of infecting or transmitting various diseases to humans. The microorganisms are not the native aquatic species and require an animal host for their growth and flourishing. Diseases such as cholera, gastroenteritis, typhoid, amoebia, diarrhoea, measles, hepatitis (jaundice), leptospirosis, dracontiasis are caused by water-borne bacteria and microorganisms or pathogens. The most common and widespread threat associated with drinking water is contamination, directly or indirectly, by sewage, other pollutants, or excrement from humans or animals. If such contamination is recent, and carriers of communicable enteric diseases are among the contributors, some of the living causative agents may be present. Drinking such polluted water or its use in the preparation of food will lead to more cases of infection. Microbiological consistency of natural and processed waters varies. Drinking water would preferably not contain any microorganisms considered to be pathogenic to humans.

(a) **Bacteria:** The word *bacteria* derives from its shape and characteristics as in Greek, it means "rod" or "staff". Bacteria (singular: bacterium) are microscopic, single-celled organisms classified as prokaryotes that can flourish in versatile environments as the organisms can be found in soil, water and inside the human gut. The internal structure of bacteria lacks a nucleus, and contains DNA that either floats freely in a twisted, thread-like mass called the nucleoid, or in separate, circular pieces called plasmids. Bacteria can be beneficial as well as detrimental to human health. Several bacteria of group *Streptococcus, Clostridium perfringens* (C. perfringens), *E. coli* and *S. aureus* can cause severe soft tissue infection called necrotizing fasciitis (sometimes called flesh-eating bacteria).

Drinking water faecal contamination may introduce a variety of intestinal pathogens – bacterial, viral, and parasitic – whose existence is linked to microbial diseases and carriers present in the community at that time. Bacterial intestinal infections are widely spread across the globe. Others reported to occur in polluted drinking water include strains of *Salmonella, Shigella, Escherichia coli, Vibrio cholerae, Yersinia enterocolitica* and *Campylobacter fetus*. These species can cause diseases that range in severity from mild to serious and often fatal dysentery, cholera, or typhoid gastroenteritis. Bacterial pathogens as modes of transmission include the ingestion of infected water and food. Table 3.3 shows the list of infectious agents potentially present in drinking water contaminated by sewage. Most of the bacteria present in water requires oxygen for their survival and based on oxygen requirement may be classified as (i) aerobic bacteria that require oxygen for their survival, (ii) anaerobic bacteria that flourish and can survive in absence of free oxygen however, such bacteria utilize bounded molecular oxygen in NO_3^-, SO_4^-, etc., and (iii) facultative bacteria are those that can survive and flourish both in absence and presence of oxygen.

Table 3.5: List of infectious agents potentially present in drinking water contaminated by sewage (Sources and Impacts 2020)

Microorganisms	Diseases	Remarks
Bacteria		
Escherichia coli (enteropathogenic)	Gastroenteritis	Diarrhoea
Legionella pneumophila	Legionellosis	Acute respiratory illness
Leptospira (150 spp.)	Leptospirosis	Jaundice, fever
Salmonella typhi	Typhoid fever	High fever, diarrhoea
Salmonella (~1700 spp.)	Salmonellosis	Food poisoning
Shigella (4 spp.)	Shigellosis	Bacillary dysentery
Vibrio cholerae	Cholera	Extremely heavy diarrhoea, dehydration
Yersinia enterocolitica	Yersinosis	Diarrhoea
Viruses		
Adenovirus (31 types)	Respiratory disease	
Enteroviruses (67 types, e.g., polio, echo, and Coxsackie viruses)	Gastroenteritis, heart anomalies, meningitis	
Hepatitis A	Infectious hepatitis	Jaundice, fever
Norwalk agent	Gastroenteritis	Vomiting
Reovirus	Gastroenteritis	
Rotavirus	Gastroenteritis	
Protozoa		
Balantidium coli	Balantidiasis	Diarrhoea, dysentery
Cryptosporidium	Cryptosporidiosis	Diarrhoea
Entamoebahistolytica	Amoebic dysentery	Prolonged diarrhoea with bleeding
Giardia lamblia	Giardiasis	Mild to severe diarrhoea, nausea
Helminthes		
Fasciola hepatica	Fasciolasis	Sheep liver fluke
Dracunculusmedinensis	Dracunculosis	Guinea worm
Ascarislumbricoides	Ascariasis	Roundworm
Enterobiusvericularis	Enterobiasis	Pinworm
Hymenolepis nana	Hymenolepiasis	Dwarf tapeworm
Taeniasaginata	Taeniasis	Beef tapeworm
Taeniasolium	Taeniasis	Pork tapeworm
Trichuris trichiura	Trichuriasis	Whipworm

Bacteriological testing is the most important method for the identification of recent and therefore potentially hazardous faecal contamination. It is helpful in hygienic evaluation of the water quality with a sensitivity and accuracy that is missing from regular chemical analysis. It is vital that water is routinely and thoroughly inspected, as contamination can be sporadic and may not be identified by a single sample test. For this reason, it is critical that drinking water is regularly tested through a simple test rather than occasionally through a more complicated test or series of tests. Therefore, bacterial analysis should be carried out on regular basis and ensured that water for domestic and drinking usage should be free from harmful bacteria. The involvement of iron bacteria contributes to tuberculation and pitting of vessels. Pitting is usually localized corrosion, and practically incrustation is tuberculation. The presence of sulphur bacteria in water affects the durability of concrete structures as the release of corrosive gasses being produced during the metabolism. These gases when react with concrete results in the formation of compounds whose volume is much more than that of the original constituents which leads to the creation of cracks on the surface of concrete.

(b) Viruses: A virus is a submicroscopic infectious agent that replicates only inside the living cells of an organism. It consists of only one piece of genetic material either RNA or DNA and a protein coat. It is not considered to be "living" because it requires a host cell to survive long-term for energy. Also it reproduces by "hijacking" a host cell, and using its ribosomes to make new viral proteins. A virus can infect all types of lifeforms, from animals and plants to micro-organisms, including bacteria and archaea. Viruses of particular concern in relation to water-borne infectious disease transmission are primarily those that replicate in the intestine and are excreted in large numbers in the faeces of infected individuals.

A concentration of as high as 108 viral units per gram of faeces has been reported. The enteric viruses have a remarkable ability to live in the aquatic environment, and can remain viable for days or months. Viruses primarily enter the water system by sewage discharges. With the currently available techniques, large variations in the number of sewage viruses are generally reported. Many of the approximately 100 recognized enteric viruses may be isolated from sewage on any given day, with the particular forms being those prevalent in the population at the time. When any waste comes in contact with drinking water, viruses are passed on and remain active depending on temperature and other variables, for varying periods of time.

The primary route of access to enteric viruses is commonly assumed to be through direct contact with infected individuals or through interaction with faecally contaminated items. However, the viruses have the ability to survive even at low infectious dosage and exposure. The infection may be high in ingestion of contaminated water. The infectious viral hepatitis and gastroenteritis disease occurs from such water source contamination of the sewage and is well documented in various epidemiologically catastrophic outbreaks. Water supplies can be highly contaminated in some developed countries, and water treatment systems can be less efficient and effective. Therefore, in such countries a large number of people are at high risk as drinking water is a medium for the environmental transmission of enteric viruses and therefore must be regarded as a very significant health hazard potential.

Enteric viruses can also be transmitted, as with other microbial infections, by infected food. Enteric viruses can cause a wide variety of syndromes, including rashes, fever, gastroenteritis, myocarditis, meningitis, respiratory disease, and hepatitis, among others. Nevertheless, two diseases of epidemic proportions – gastroenteritis and infectious hepatitis – can occur when drinking water is polluted with sewage. Many viruses as identified are small particles with a diameter of 270-350 microns in stools of diarrhoea infected individuals. Viral gastroenteritis, typically lasting for 24-72 hours with nausea, vomiting, and diarrhoea, occurs in individuals of all ages. It is most dangerous in very young or very old people, where dehydration and electrolyte imbalances can occur rapidly and life threatened if not corrected without delay. Hepatitis, if mild, may need only rest and restricted activity for one or two weeks, but when serious it can cause death from liver failure, or lead to chronic liver disease. However, the water contamination due to faulty and leaky joints in sewage and water supply systems has increased in recent years thus making the huge population vulnerable to such diseases (Sources and Impacts 2020).

(c) **Protozoa:** Protozoa are eucaryotic single-celled microorganisms without cell walls as many of them are aerobic. Protozoa are free living or parasitic as many of them feed on other small micro-organisms and bacteria. These are highly adaptable and widely distributed in natural waters, although some of the aquatic protozoa are pathogenic. The three world-wide distributed important pathogenic intestinal protozoa found in human beings are *Entamoeba histolytica*, *Giardia spp.*, and *Balantidium coli* and are generally transmitted by drinking water. Such bacteria, respectively, are the etiological agents of amöbian dysentery, giardiasis, and balantidiasis, and have all been associated with drinking water outbreak. As species, a large number of the intestinal pathogenic protozoa exists in the faece of man-infected individuals and also present in wide range of domestic and wild animals. *Coliform* species do not seem to represent a strong predictor of *Giardia* or *E. Histolytica* in processed water due to their improved resistance to disinfection by these protozoans. Under adverse environmental conditions aquatic protozoa forms cysts that are difficult to remove by disinfection process and therefore requires special treatment process such as microfiltration or membrane separation.

(d) **Helminths:** Helminths, worm-like parasites, are hermaphroditic and bisexual species. The classification is based on morphology of egg, larvae and adult stages. The adults are classified as flukes (trematodes), tapeworms (cestodes), roundworms (nematodes). In drinking water a great variety of helminth eggs and larvae exists and should be eliminated before water to be supplied for domestic usage. The two classes of helminths associated with contamination in water supply systems are copepod and cercariae. The first group of helminths includes dracunculus and spirometra that grow in aquatic copepods and are acquired by the humans with water containing the intermediate host crustacea. Tapeworms of the genus spirometra have also a stage in aquatic copepods and much rarer in humans. The second group of helminths (group of flukes) includes schistosoma, ancylostoma, necator with roundworms whose infectious larvae can penetrate the human skin and mucous membranes. The third helminths group includes man-infective resistant eggs or cysts as the exposure to them when swallowed with drinking water, makes the individual sick. However, *Ascaris lumbricoides* (roundworm) and *Trichuristrichuria* (whipworm) are the most commonly used intestinal helminths. The human tapeworms of the genera Hymenolepis and Echinococcus, have the potential risk in drinking water when man is infected by ingesting eggs usually acquired from dogs.

(e) **Malaria and dengue as disease:** Both malaria and dengue are not spread directly by drinking water, but by fish-breeding vectors or carriers, in this case, mosquitoes. In order to avoid the spread of these diseases, it is of utmost importance that drinking water reservoirs within the port area are sufficiently protected in all endemic areas to prevent mosquitoes from gaining access to the free water surface where they can breed unimpeded. These reservoirs comprise elevated water distribution tanks (access manholes), reinforced concrete reservoirs, header tanks of any size and water cisterns in all toilet types. Therefore, all types of manholes should be covered with purpose-made manhole covers and all vents should be equipped with filters or mosquito nets.

Examining water for the presence of bacteria is critical from a public health perspective for the water supply engineers. The bacteria may be harmless to humans, or harmful to humans. The former category is referred to as non-pathogenic bacteria and the latter category is referred to as pathogenic. A lot of the bacteria present in water come from rain, soil and plants. Some of these multiply and continue their life while the rest die out in due course of time. The selective medium which promotes the growth of particular bacteria and integrates the growth of other organisms is used in the laboratory to detect the presence of the bacteria needed, typically coliform bacteria. Subsequent experiments are carried out for bacteriological research.

Tests on pathogenic indicators in water: The analysis of assessing the pathogenic microorganisms in water is very time consuming and costly. However, the tests are required to be carried out in special circumstances of suspicion of any such pathogenic microorganisms in water. Therefore, an indicator can be used to identify the presence or absence of such pathogens in short time that can save both time and cost. The ideal pathogen indicator should be applicable to all types of water and always be present when pathogens are there. It should also be absent when pathogens are not in the sample and can be easily detected without interference of any other microorganisms or species in a sample. Most of the pathogens in water originates from faecal contamination in water. Therefore such microorganism residing in human tract or intestine can be a good indicator of confirming the pathogens in water.

The microorganisms meeting all the above criteria and requirements belong to fecal coliform group classified as total coliforms and faecal coliforms. The group comprising of various bacteria but *Escherichia coli* also known as *E. coli* are found exclusively in intestinal tracts of warm blooded animals and excreted in feces largely. The *E. coli* in itself is non-pathogenic but their presence in water indicates the recent fecal contamination in water. These microorganisms survive much longer time in outside atmosphere as compared to other pathogens. The *E. coli* can be further sub divided into biotype and serotype. The microorganisms are also found in water moving through its native soil and decaying vegetation. The members exclusively found in faeces comprise of 96% of all coliforms of human feces. The microorganism has been recently termed as thermotolerant coliforms by WHO.

Also the feces of humans and animals also contains other type of bacteria known as fecal streptococci (enterococci), clostridium, prefringens, bacteroides and lactobacilli. The coliform bacilli called as *B. coli* is known as lactose fermenters, non sore forming and Gram-negative microorganisms that can ferment lactose with gas formation in 48 hours at 35°C. The methods generally used in determination of pathogen indicators are plate count test, most probable number also known as MPN test and membrane filtration technique test. The above-mentioned various tests required to determine the presence of pathogenic microorganism in water are described here.

(a) **Plate count test:** In this technique the population of bacteria is allowed to flourish in a broth by plating a small and dilute portion of water sample and the number of bacterial colonies can be counted after a defined period of time. Complete number of bacteria present in a 1 mL of water are counted in this process as shown in Fig. 3.13. 1 mL of water sample is diluted into 99 mL of sterilized water and 1 mL of diluted water is combined with 10 mL of gelatin agar. This mixture is then held for 24 hours at 37°C in the incubator, or 48 hours at 20°C. Thereafter the sample is taken out of the incubator and bacteria colonies are counted using microscope. The WHO guidelines modified in 1993 states that no faecal coliform should be present in water entering the distribution system up to the end users.

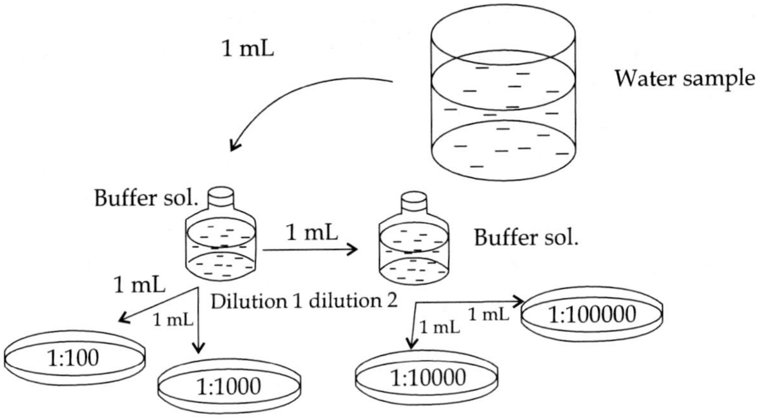

Fig. 3.13: Plate count test for coliform determination in water sample (Baldaniya 2016)

(b) Multiple tube fermentation technique for coliform bacteria (MPN test): In this test lactose containing broth (MacConkey broth) and other substances that inhibit the growth of non-coliform organisms in placed in groups in series with 5 test tubes in each group. All the 5 tubes of each group are inoculated with 10 mL, 1 mL and 0.1 mL of water sample as Group I, II and III, respectively. All the tubes are incubated at 35°C for 24 hours and formation of gas in the tubes is observed. The presence of gas formation in the tubes indicates the positive test and hence presence of coliform group is confirmed. However, if the gas formation does not take place, the tubes are allowed for further incubation of 24 hours as 48 hours is the maximum time for the release of gas by coliform microorganisms.

A more selective culture medium (brilliant green bile broth) with material taken from the positive tubes is inoculated for the confirmatory test. If the gas formation does not take place even after 48 hours the test is considered to be negative and absence of any coliform in a water sample. The former test as shown in Fig. 3.14 is called as presumptive test and the latter is known as confirmatory test. The most probable number (MPN) of present bacteria can then be determined from the number of inoculated tubes and the number of positive tubes collected in the confirmatory test by using a specially designed statistical tables. The table provides MPN index in terms number of microorganism per 100 mL of water sample when the testing is carried out using the above-mentioned 15 tubes of different dilutions.

For example, If a test is being carried out and result indicates that in 10 mL sample 3 tubes are positive with 1 mL sample 2 tubes are positive and 0.1 mL sample only 1 tubes is positive. The final combination in a test comes out to be 3-2-1. Using this combination of 3-2-1 the presence of number of coliform microorganism can be read out from Table 3.8 as 17per 100 mL. This indicates that 17 number of coliforms are present in 100 mL of water sample.

However, the presumptive MPN test can be further extended to confirmatory and faecal coliform MPN test. For performing the confirmatory test the growth from each of the positive tubes of the presumptive MPN test are transferred technically using a wire loop to brilliant green bile lactose broth and incubated for 48 hours at 35°C. The gas formation is an indication of positive test. For performing the faecal coliform presence test the growth from each of the positive tubes of presumptive MPN test is transferred aseptically using a wire loop to E.C. medium broth and incubated for 24 hours at 44.5°C. The gas formation indicates the presence of faecal coliform group.

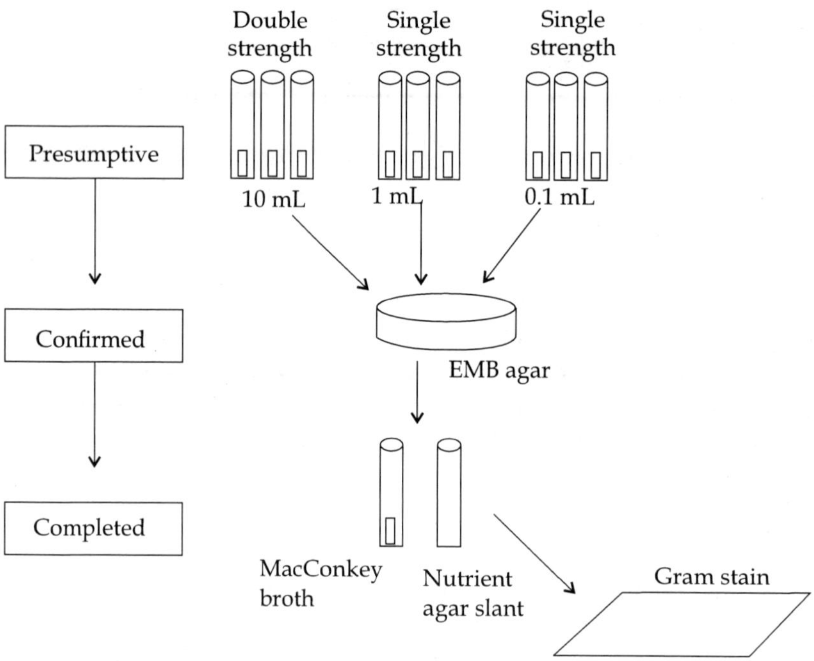

Fig. 3.14: MPN test used for *E. coli* determination in water (Akeju, Awojobi, and State 2015)

Coliform index is another term used to measure the coliform bacteria present in the water sample and is obsolete nowadays as more accurate and advanced methods such as membrane filtration technique is preferred.

(c) **Membrane filtration method for total coliform and thermotolerant (faecal) coliforms:** The simplest and most advanced technique mostly used in determination of pathogens or presence of coliform bacteria in water is membrane filtration technique. In this technique water sample is filtered through a specially designed sterile membrane with 80% porosity and pore size of 5-10 μm. The bacteria during filtration of water sample retains on the membrane. Thereafter the filter is rinsed with suitable sterile buffer solution which is placed on pad saturated with required nutrient medium and incubated at desired temperature. The bacteria flourishes on the nutrient medium can be visible in colonies retained on the membrane filter surface and the same is counted as each colony represents the bacterium in original sample. When the membrane containing the bacteria is incubated with a selective differential culture medium in a sterile container at a suitable temperature, characteristic colonies of coliforms/ thermotolerant coliforms grow, which can be counted as direct.

For determination of coliform group M-Endo broth or LES Endo agar is used and the sample is incubated at 35°C for 20-22 hours and total coliform is counted. Coliform bacteria produce characteristic colonies with pink to dark red colour with golden metallic sheen and mostly with greenish tint. Non-coliform bacteria lack the characteristic of sheen when used grown in this medium. However, for determination of faecal coliforms count M-FC broth is used as a medium with incubation at 44.5°C for 22 hours. The coliform colonies of this group appear in blue colour while other bacteria that grow on this medium are grey to cream in colour. Also for the determination of faecal streptococci count, the test is carried out using *M. enterococcus* agar or KF-Streptococcus and the sample is incubated at 35°C for 48 hours. The colonies of light pink to dark red with pink margins indicate the presence of faecal streptococci and are counted.

Because the filtration area is relatively small, it can accommodate the growth of only a limited number of colonies: the optimum number is between 20 and 80, with a maximum of 200. If this number is exceeded, very small atypical colonies or superimposed colonies may develop, or the overpopulation may inhibit development. The choice of sample volume to be filtered depends on the form of water. To ensure sufficient dispersal through the membrane filter surface, volumes less than 10 mL should be applied to the filtration apparatus after addition of at least 10 mL of sterile diluent.

Although the use of all of these media to detect presumptive coliform species is based on lactose fermentation, the characteristic reaction varies with each medium. The number of coliforms/ thermotolerant coliforms per 100 mL is then given by Eq. 3.36.

Coliform or thermotolerant coliforms per 100 mL

$$= \frac{\text{No. of coliforms or thermotolerant coliforms per 100 mL}}{\text{No. of mL of sample filtered}} \times 100 \qquad (3.36)$$

Drinking water standards: For any waterbody to function adequately in satisfying the desired use, it must have corresponding degree of purity. Drinking water should be of highest purity. As the magnitude of demand for water is fast approaching the available supply, the concept of management of the quality of water is becoming as important as its quantity. Table 3.6 shows the drinking water specifications (IS 10500:1991)

Table 3.6: Drinking water specifications (IS 10,500:1991)

S. No.	Characteristics	Desirable	Max. permissible limit
	Essential characteristics		
1.	Colour, Hazen Units, Max	5	25
2.	Odour	Unobjectionable	
3.	Taste	Agreeable	
4.	Turbidity, NTU, Max	5	10
5.	pH value	6.5 to 8.5	
6.	Total hardness (as CaCO3), mg/L, Max	300	600
7.	Iron (as Fe), mg/L, Max	0.3	1
8.	Chlorides (as Cl), mg/L, Max	250	1000
9.	Residual-free chlorine, mg/L, Max	0.2	
	Desirable characteristics		
10.	Dissolved solids, mg/L, Max	500	2000
11.	Calcium as (Ca), mg/L, Max	75	200
12.	Magnesium (as Mg), mg/L, Max	0.1	0.3
13.	Copper (as Cu), mg/L, Max	0.05	1.5
14.	Manganese (as Mn), mg/L, Max	0.1	0.3
15.	Sulphate (as SO_4), mg/L, Max	200	400
16.	Nitrate (as NO_3), mg/L, Max	45	100
17.	Flouride (as F), mg/L, Max	1	1.5
18.	Phenolic compounds (as C_6H_5OH), mg/L,	0.001	0.002
19.	Mercury (as Hg), mg/L, Max	0.001	
20.	Cadmium (as Cd), mg/L, Max	0.01	
21.	Selenium (as Se), mg/L, Max	0.01	
22.	Arsenic (as As), mg/L, Max	0.05	
23.	Cyanide (as CN), mg/L, Max	0.05	
24.	Lead (as Pb), mg/L, Max	0.05	
25.	Anionic detergents (as MBAS), mg/L, Max	0.02	1
26.	Chromium (as Cr_6^+), mg/L, Max	0.05	
27.	PAH, mg/L, Max		
28.	Mineral oil, mg/L, Max	0.01	0.03
29.	Pesticides, mg/L, MAX	Absent	0.001
30.	Alkalinity, mg/L, Max	200	600
31.	Aluminum (as Al), mg/L, Max	0.03	0.2
32.	Boron, mg/L, Max	1	5

The acceptable and permissible limits of water to be used for domestic and drinking purposes has been prescribed by various agencies. In India it is Bureau of Indian Standard (BIS), Central Pollution Control Board (CPCB) and Indian Medical Research Council (ICMR) while other international organizations include American Public Health Association (APHA), World Health Organization (WHO). The water usage limits for various parameters is practised by public health officials, physicians and researchers as shown in Table 3.7.

Table 3.7: Permissible limit of drinking water by different agencies (Communication 2012)

S. No.	Parameters	USEPA	WHO	ISI	ICMR	CPCB
1.	pH (mg/L)	6.5-8.5	6.5-8.5	6.5-8.5	6.5-9.2	6.5-8.5
2.	Turbidity NTU			10	25	10
3.	Conductivity (mg/L)					2000
4.	Alkalinity (mg/L)					600
5.	Total hardness (mg/L)		500	300	600	600
6.	Iron *mg/L)		0.1	0.3	1	1
7.	Chlorides (mg/L)	250	200	250	1000	1000
8.	Nitrate (mg/L)			45	100	100
9.	Sulphate (mg/L)			150	400	400
10.	Residual (mg/L) free chlorine			0.2		
11.	Calcium (mg/L)		75	75	200	200
12.	Magnesium (mg/L)		50	30		100
13.	Copper (mg/L)	1.3	1	0.05	1.5	1.5
14.	Fluoride (mg/L)	4	1.5	0.6-1.2	1.5	1.5
15.	Mercury (mg/L)	0.002	0.001	0.001	0.001	No relaxation
16.	Cadmium (mg/L)	0.005	0.005	0.01	0.01	No relaxation
17.	Selenium (mg/L)	0.05	0.01			No relaxation
18.	Arsenic (mg/L)	0.05	0.05	0.05	0.05	No relaxation
19.	Lead (mg/L)		0.05	0.1	0.05	No relaxation
20.	Zinc (mg/L)					15
21.	Chromium (mg/L)	0.1		0.05		No relaxation
22.	E. coli (MPN/100 mL)					No relaxation

Effluents water quality standards by Ministry of Environment is indicated in Table 3.8.

Table 3.8: Effluent water quality standards (Ministry of Environment)

S. No.	Parameter	Inland surface water	Public sewers	Land for irrigation	Marine/coastal areas
1	Colour and odour	All efforts should be made to remove colour and unpleasant odour as far as practicable		All efforts should be made to remove colour and unpleasant odour as far as practicable	All efforts should be made to remove colour and unpleasant odour as far as practicable.
2	Suspended solids mg/L, max.	100	600	200	(a) For process waste water (b) For cooling water effluent 10 per cent above the total suspended matter of influent.
3	Particle size of suspended solids	Shall pass 850 micron IS sieve	-	-	(a) Floatable solids, solids max. 3 mm (b) Settleable solids, max 856 microns
4	pH value	5.5 to 9.0	5.5 to 9.0	5.5 to 9.0	5.5 to 9.0
5	Temperature	Shall not exceed 5°C above the receiving water temperature			Shall not exceed 5°C above the receiving water temperature
6	Oil and grease, mg/L max,	10	20	10	20
7	Total residual chlorine, mg/L max	1	-	-	1
8	Ammonical nitrogen (as N), mg/L, max.	50	50	-	50
9	Total Kjeldahl Nitrogen (as N); mg/L, max. mg/L, max.	100	-	-	100
10	Free ammonia (as NH_3), mg/L,max.	5	-	-	5
11	Biochemical oxygen demand (3 days at 27°C), mg/L, max.	30	350	100	100

12	Chemical oxygen demand, mg/L, max.	250	-	-	250
13	Arsenic (as As).	0.2	0.2	0.2	0.2
14	Mercury (as Hg), mg/L, max.	0.01	0.01	-	0.01
15	Lead (as Pb) mg/L, max	0.1	1	-	2
16	Cadmium (as Cd) mg/L, max	2	1	-	2
17	Hexavalent Chromium (as Cr^{+6}), mg/L, max.	0.1	2	-	1
18	Total chromium (as Cr) mg/L, max.	2	2	-	2
19	Copper (as Cu) mg/L, max.	3	3	-	3
20	Zinc (as Zn) mg/L, max.	5	15	-	15
21	Selenium (as Se)	0.05	0.05	-	0.05
22	Nickel (as Ni) mg/L, max.	3	3	-	5
23	Cyanide (as CN) mg/L, max.	0.2	2	0.2	0.2
24	Fluoride (as F) mg/L, max.	2	15	-	15
25	Dissolved phosphates (as P), mg/L, max.	5	-	-	-
26	Sulphide (as S) mg/L, max.	2	-	-	5
27	Phenolic compounds (as C_6H_5OH) mg/L, max.	1	5	-	5
28	Radioactive materials:	10^{-7}	10^{-7}	10^{-8}	10^{-7}
	(a) Alpha emitters micro curie mg/L, max.				
	(b) Beta emittersmicro curie mg/L	10^{-7}	10^{-7}	10^{-6}	10^{-7}

29	Bio-assay test	90% suivival of fish after 96 hours in 100% effluent	90% suivival of fish after 96 hours in 100% effluent	90% suivival of fish after 96 hours in 100% effluent	90% suivival of fish after 96 hours in 100% effluent
30	Manganese	2 mg/L	2 mg/L	-	2 mg/L
31	Iron (as Fe)	3 mg/L	3 mg/L	-	3 mg/L
32	Vanadium (as V)	0.2 mg/L	0.2 mg/L	-	0.2 mg/L
33	Nitrate nitrogen	10 mg/L	-	-	20 mg/L
These standards shall be applicable for industries, operations or processes other than those industries, operations or process for which standards have been specified in Schedule of the Environment Protection Rules, 1989.					

Summary of common impurities found in fresh water and their treatment is tabulated in Table 3.9.

Table 3.9: Summary of common impurities found in fresh water and their treatment type

Constituent	Difficulties caused	Chemical formula	Means of treatment
Turbidity	Imparts unsightly appearance to water; deposits in water lines, process equipment, etc.; interferes with most process uses.	-	Coagulation, settling, and filtration.
Hardness	Chief source of scale in heat exchange equipment, boilers, pipe lines, etc.; forms curds with soap, interferes with dyeing, etc.	Salts of Ca & Mg, as $CaCO_3$	Softening, demineralization, internal boiler water treatment, surface active agents.
Alkalinity	Foam and carryover of solids with steam; embrittlement of boiler steel; bicarbonate and carbonate produce CO_2 in steam, a source of corrosion in condensate lines.	HCO_3^-, CO_3^{2-}, OH^-, as $CaCO_3$	Lime and lime-soda softening, acid treatment; hydrogen zeolite softening, demineralization dealkalization by anion exchange.
Free mineral acid	Corrosion	H_2SO_4, HCl as $CaCO_3$	Neutralization with alkalies
Carbon dioxide	Corrosion in water lines, particularly steam and condensate lines	CO_2	Aeration, deaeration, neutralization with alkalies
pH	pH varies according to acidic or alkaline solids in water; most natural waters have a pH of 6.0-8.0	Hydrogen ion concentration defined as: $\frac{1}{\log 10[H^+]}$	pH can be increased by alkalies and decreased by acids

Sulphate	Adds to solids content of water, but in itself is not usually significant, combines with calcium to form calcium sulphate scale	SO_4^{2-}	Demineralization, reverse osmosis, electrodialysis, evaporation
Chloride	Adds to solids content and increases corrosive character of water	Cl^-	Demineralization, reverse osmosis, electrodialysis, evaporation
Nitrate	Adds to solids content, but is not usually significant industrially: high concentrations cause methemoglobinemia in infants; useful for control of boiler metal embrittlement	NO_3^-	Demineralization, reverse osmosis, electrodialysis, evaporation
Fluoride	Cause of mottled enamel in teeth; also used for control of dental decay: not usually significant industrially	F^-	Adsorption with magnesium hydroxide, calcium phosphate, or bone black; alum coagulation
Sodium	Adds to solids content of water: when combined with OH^-, causes corrosion in boilers under certain conditions	Na^+	Demineralization, reverse osmosis, electrodialysis, evaporation
Silica	Scale in boilers and cooling water systems; insoluble turbine blade deposits due to silica vaporization	SiO_2	Hot and warm process removal by magnesium salts; adsorption by highly basic anion exchange resins, in conjunction with demineralization, reverse osmosis, evaporation
Iron	Discolours water on precipitation; source of deposits in water lines, boilers, etc.; Interferes with dyeing, tanning, papermaking, etc.	Fe^{2+} (ferrous) Fe^{3+} (ferric)	Aeration; coagulation and filtration; lime softening; cation exchange; contact filtration; surface active agents for iron retention
Manganese	Same as iron	Mn^{2+}	Same as iron
Aluminum	Usually present as a result of floc carryover from clarifier; can cause deposits in cooling systems and contribute to complex boiler scales	Al^{3+}	Improved clarifier and filter operation
Oxygen	Corrosion of water lines, heat exchange equipment, boilers, return lines, etc.	O_2	Deaeration; sodium sulphite; corrosion inhibitors
Hydrogen sulphide	Cause of "rotten egg" odour; corrosion	H_2S	Aeration; chlorination; highly basic anion exchange
Ammonia	Corrosion of copper and zinc alloys by formation of complex soluble ion	NH_3	Cation exchange with hydrogen zeolite; chlorination; deaeration

Dissolved solids	Refers to total amount of dissolved matter, determined by evaporation; high concentrations are objectionable because of process interference and as a cause of foaming in boilers	None	Lime softening and cation exchange by hydrogen zeolite; demineralization, reverse osmosis, electrodialysis, evaporation
Suspended solids	Refers to the measure of undissolved matter, determined gravimetrically; deposits in heat exchange equipment, boilers, water lines, etc.	None	Subsidence; filtration, usually preceded by coagulation and settling
Total solids	Refers to the sum of dissolved and suspended solids, determined gravimetrically	None	See "dissolved solids" and "suspended solids"

3.7.2.17 Water Purification Processes in Natural System

Introduction to self-purification of stream: If from the point of disposal of sewage in stream, the stream water is examined towards down streams, it is observed that the quality of stream water successively changes. Near the place of disposal the water is polluted and it becomes purified after some travel towards downstream side due to natural forces of purification. When sewage is discharged into streams, the water becomes polluted. The suspended solids are gradually deposited in the stream bed in layers, depending upon the hydrographic characteristics of the stream. The organic matter of these so formed layers starts decomposing and stabilizing. In due course, the decomposed materials are washed away with the current or are converted into simple constituents.

Various organisms present in the sewage start the decomposition of the volatile organic matter and stabilize it. Nitrogen, sulphur, carbon and other inorganic elements present in sewage move under their natural bio-geo chemical cycles. Algae and other microscopic organisms eat mineralized food and supply oxygen to the stream for maintaining aerobic condition. The bacteria are eaten away by the protozoa, which further become the food of fish and other aquatic life. Thus, the stream in course of time becomes free of bacteria and protozoa. The natural purification process continues and makes the stream water pure after some time. Therefore, in practice, it is observed that the stream water after some travel from the point of disposal becomes free of sewage pollution and comes in its natural purified state.

Factors affecting self-purification of stream

(a) **Dilution:** When the sewage is mixed up with large volume of water or it is much diluted, the sewage always remains in aerobic condition and anaerobic condition never comes because dissolved oxygen always remains in the water.

(b) **Current:** The self-purification of stream directly depends on the current. When there is no current the sewage matter deposits near the outfall causing formation of sludge bank and foul odours. In heavy current, the sewage is thoroughly mixed up with the stream water preventing all nuisances. In slow current, the sedimentation takes place causing growth of algae resulting in the production of oxygen. Therefore, slow current is better for the self-purification.

(c) **Sedimentation:** With slow current, the heavier solids settle in the stream bed and start anaerobic decomposition. The products of decomposition are again mixed with water by the current. If the dilution is sufficient, the anaerobic condition will not develop and the scouring tendency of streams during flood will wash the deposits.

(d) **Temperature:** As the activities of organisms depend on the temperature, the self-purification also depends on temperature. At low temperature the organisms activities are slow due to which the rate of decomposition is slow and in warm temperature this action is reverse. Therefore, in summer, the stream will self-purify is less than in winter. But the quantity of dissolved oxygen is more in cold water than in hot water.

(e) **Sunlight:** The pathogens are killed if they are exposed to sunlight, therefore, sunlight helps in self-purification. Algae also grow in sunlight causing production of oxygen. Therefore, indirectly sunlight helps in the oxidation of unstable matter.

(f) **Oxidation:** The organic matter, immediate after mixing with the stream or large quantity of water, starts getting oxidized due to the development of the oxidizing organisms present in the water. This process prevails till the complete oxidation of the organic matters. The oxygen demand is satisfied and the stream becomes purified. The streams get their oxygen demand by its aeration by wind or by the microscopic organisms.

(g) **Reduction:** The reduction occurs in the streams due to the hydrolysis of the organic matters biologically or chemically, Anaerobic organisms start the splitting complex organic matter present in the sewage. This action produces odours and gases and the stabilization starts.

Effect of oxygen demanding wastes on rivers: The amount of dissolved oxygen (DO) in water is the most commonly used indicator of a river's health. The solubility of oxygen depends on temperature, pressure, salinity and the dissolved oxygen concentration in a healthy stream and it ranges from 7–9 mg/L. As the DO value in stream drops below 4 or 5 mg/L the forms of life that can survive begin to be reduced. In an extreme case, when anaerobic conditions exist, most higher forms of life are killed.

(a) **Factors affecting amount of DO available in rivers**

- Oxygen demanding wastes affect available DO.
- Tributaries bring their own oxygen supply.

- Photosynthesis adds DO during the day but the same plants remove oxygen at night.
- Respiration of organisms living in water as well as in sediments remove oxygen.
- In the summer rising temperatures reduce solubility of oxygen.
- In the winter oxygen solubility increases, but ice may block access to new atmospheric oxygen.

(b) mass balance approach: The mass balance approach is applied to a river by considering it to be a plug-flow reactor. If volumetric flow rate and concentration of a given pollutant in stream and waste discharge is known the concentration after mixing can be calculated. It is assumed that the concentration of any contaminant or pollutant in sewage or in any industrial wastewater entering in river or any natural stream can be determined by considering the fact that resultant mixing may take pace and the contaminants' concentration after dilution in a stream can be determined as shown in Fig. 3.15. Considering that the initial conditions of a river stream flowing with flow rate vale of Q_R and pollutant concentration at that instant of time as C_R, the resultant concentration of any pollutant after mixing can be determined as in Eq. 3.37:

$$Q_m C_m = Q_R C_R + Q_w C_w \tag{3.37}$$

or

$$C_m = \frac{Q_R C_R + Q_W C_W}{Q_R + Q_W}$$

and

$$Q_m = Q_R + Q_w$$

where, C indicates the concentration of any pollutant in terms of mg/L or g/m^3 with flow rate value of Q in L/d or m^3/d. The subscript m, R and w designate the mixing condition, river or stream and waste, respectively.

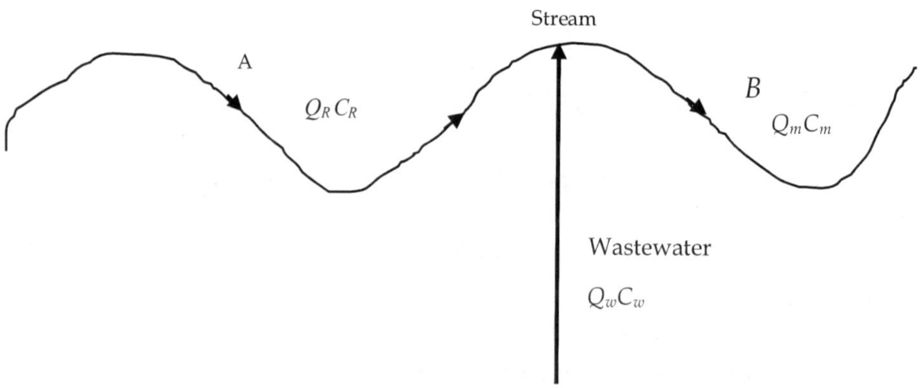

Fig. 3.15: Instantaneous mixing of pollutants in a stream and resultant concentration.

Process of self-purification of stream: The self-purification process of streams polluted by sewage can be grouped in the following four zones of divisions as shown in Fig. 3.16.

(a) Degradation zone: This zone is situated just near the outfall of sewage. The water is turbid with dark colour. Decomposition of solid matters takes place in this zone and the anaerobic decomposition prevail.

(b) Active decomposition zone: This zone is just after the degradation zone. The water is greyish and darker than the previous zone. The objectionable odours of hydrogen sulphide and other sulphur compounds prevail and scum may also be seen on the surface in this zone.

(c) Recovery zone: In this zone, the stabilization of organic matters takes place and BOD of water is reduced. The contents of dissolved oxygen start rising up above 40% of the saturation value. The bacterial load decreases as the food supply of bacteria diminishes.

(d) Clear water zone: In this zone, the stream attains normal conditions as were prevailing before sewage was discharged into it

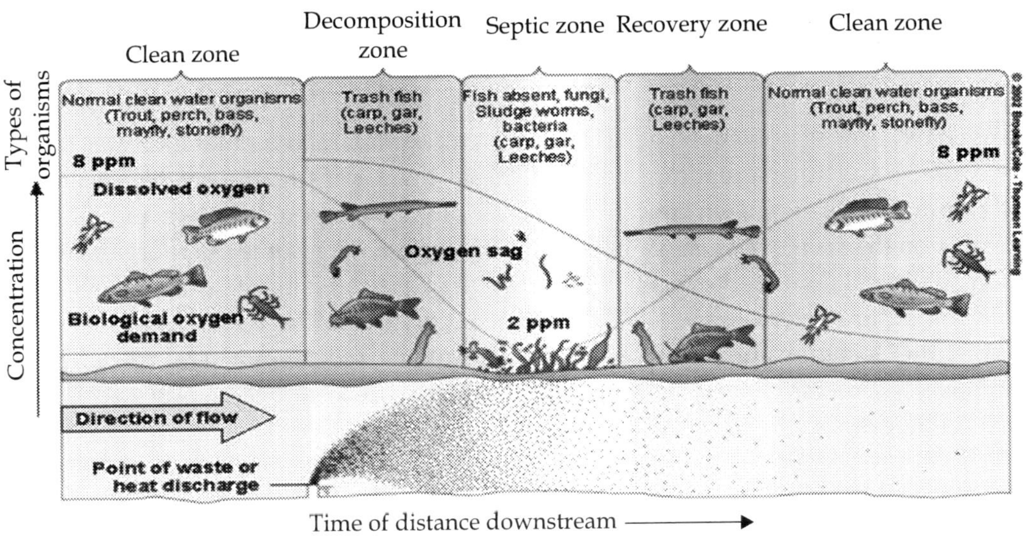

Fig. 3.16: Various zones in self-purification process of river stream

Example 3.9: Treated wastewater enters the stream as shown in the Fig. 3.17. The concentration of sodium at point *A* in stream is 20 mg/L with flow rate 10 m^3/s. The concentration of sodium in wastewater is 200 mg/L with flow rate 1.6 m^3/s. Determine the final concentration of sodium at point B assuming that complete mixing has occurred.

Solution:

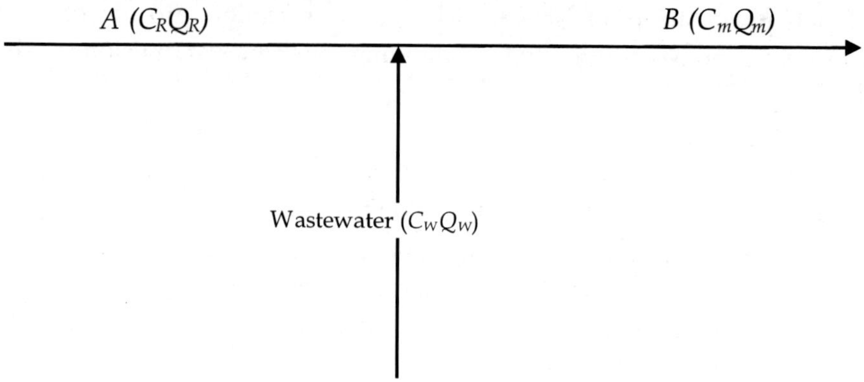

Fig. 3.17: Line diagram showing the entry of wastewater

We know that

$$C_m = \frac{Q_R C_R + Q_W C_W}{Q_R + Q_W} \text{ and } Q_m = Q_R + Q_w$$

where, $Q_R = 10\ \text{m}^3/\text{sec}$ and $C_R = 20\ \text{mg/L or g/m}^3$

$Q_w = 1.6\ \text{m}^3/\text{sec}$ and $C_w = 200\ \text{mg/L or g/m}^3$

Therefore,

$$C_m = \frac{10 \times 20 + 1.6 \times 200}{10 + 1.6} = \frac{200 + 320}{11.6} = \frac{520}{11.6} = 44.83 \frac{\text{g}}{\text{m}^3} = 44.83\ \text{mg/L}$$ **Ans.**

Dissolved oxygen: Dissolved oxygen (DO) is a significant parameter that indicates the survival of fish and other aquatic microorganisms in any water stream or water system. A minimum dissolved oxygen concentration of 4 mg/L is required for fish and other concentration to survive in such water streams. However, even some microorganisms can still survive at dissolved oxygen concentration of around 2 mg/L. Also oxygen is an important parameter on any system that generally helps the microorganisms to carry out the decomposition process and carry out the biochemical reactions. The decomposition process can get rid of any odour or any problem if such processes are carried out in presence of oxygen.

When a waste containing biodegradable organic matter is disposed of in a stream containing dissolved oxygen, the biodegradation process starts with utilization of dissolved oxygen present in the stream water by the microorganisms of the waste to convert the organic material into stable end products including new cells. The amount of oxygen required by the microorganisms to degrade the biologically active organic matter into end stable products or wastes is known as biochemical oxygen demand of water in terms of BOD. The rate at which the dissolved oxygen used or depletes in a stream generally depends upon the amount of organic matter, amount of microorganisms, and dilution capacity of stream including physical parameters such as flow rate, velocity, temperature, width and depth. The natural mechanisms involved to contribute dissolve oxygen to surface water includes (i) reaeration also known as dissolution of oxygen in stream from atmosphere, and (ii) oxygen production by algal photosynthesis.

From the principle of equilibrium, when the concentration of DO drops below the equilibrium value, the net transfer of oxygen molecules from atmosphere to water starts till it attains the equilibrium concentration, also known as saturation value. If the actual value of DO when determined on the field or in laboratory is observed to be less than the saturated value, the water is said to be undersaturated. However, when the DO concentration value is greater than the saturated value in a water stream the water is said to be supersaturated with dissolved oxygen. However, the concentration of dissolved oxygen in water is temperature dependent and may increase or decrease with respect to change in temperature. It has been observed that in a waterbody or a stream as the temperature of water increases the DO concentration in water decreases and vice versa.

The difference between the saturation concentration and actual concentration of dissolved oxygen in water is known as oxygen deficit and can be calculated by using Eq. 3.38.

$$D = C_s - C \tag{3.38}$$

where, D = Dissolved oxygen deficit (mg/L)

C_s = Saturation DO concentration of (mg/L)

C = Actual DO concentration (mg/L)

and the rate of change of DO deficit can be represented by Eq. 3.39.

$$\frac{dD}{dt} = -\frac{dC}{dt} \tag{3.39}$$

The negative sign indicating the decrease in DO in a stream increases the deficit indicating the deficit to be driving force in reaeration. The increase in rate of reaeration increases the dissolved oxygen concentration in water and ultimately decreases the deficit.

The presence of algae in water metabolizes the inorganic compounds using sunlight as an energy source. During the daytime the oxygen produced during the algal photosynthesis is available to replenish the dissolved oxygen in water. The availability of excessive nutrients and sunlight to algae thus imparts enough oxygen to water and therefore the water becomes supersaturated with $C > C_s$ and in this case the deficit value becomes negative. Algae utilize the waste products from bacterial decomposition and major algal activity usually occurs at the downstream side. It is an area of greatest bacterial activity where the oxygen is needed the most in a stream. However, during night the light is not available, therefore the algae obtain energy from endogenous catabolism with utilization of dissolved oxygen being imparted to the water during daytime and can be represented by Eq. 3.40.

$$CH_2O + O_2 \rightarrow CO_2 + H_2O \tag{3.40}$$

This reaction involves utilization of DO from water by algae thus demanding the oxygen rather than supplying it to the stream. This difference in algal catabolism during day and night thus results in diurnal variations in the dissolved oxygen in streams with heavy algal growth. The dissolved oxygen concentration often peaks around 2 to 4 P.M. during daytime with the lowest concentration levels occurring just before sunrise. In order to maintain the equilibrium concentration of DO in water the excessive oxygen produced during the daytime cannot be stored for its use during the night and is released to the atmosphere. Also due to high algal growth resulting in high endogenous catabolism may deplete the dissolved oxygen to a point of maximum extent where aquatic animals including fish die. Due to the inconsistency of oxygen produced photosynthetically, the reaeration is considered the most reliable source of dissolved oxygen. The oxygen produced photosynthetically may be considered for dissolved oxygen model in waters with high algal growth.

Oxygen sag curve model

The oxygen sag curve model, also known as dissolved oxygen model, was developed by Streeter and Phelps in 1925. The model can be used to predict the changes in the oxygen deficit as a function of oxygen depletion as BOD and reaeration in a stream.

Rate of oxygen removal

The depletion rate of dissolved oxygen from the stream can be assumed to be at the same rate of BOD exertion. Therefore;

$$\frac{dy}{dt} = -\frac{dC}{dt} \text{ and similarly } \frac{dy}{dt} = \frac{dD}{dt} \tag{3.41}$$

Equation 3.41 indicates the increase in oxygen deficit rate. There is an increase in BOD exertion rate which can be calculated as shown in Eq. 3.42;

$$y = L_0 - L \tag{3.42}$$

where, L_0 = Ultimate BOD (mg/L)

The rate of change in the dissolved oxygen deficit at time t with respect to the BOD as oxygen demand of a waste is proportional to the oxygen equivalent of the remaining organics. It can be represented by using Eq. 3.43.

$$r_D = K_1 L_1 \tag{3.43}$$

where, r_D may be defined as the rate of change in the oxygen deficit during oxygen consumption. The term k_1 is reaction rate constant of oxygen utilization and can be derived from laboratory tests on the wastewater. The rate constant k_1 should always be calculated for temperature changes and is not dependent on other parameters that affect dilution of stream water.

Rate of reaeration

The rate of reaeration with respect to reduction in oxygen deficit can be calculated by using Eq. 3.44.

$$r_k = -k_2 D \tag{3.44}$$

where, r_k is the rate of atmospheric oxygen addition to the waterbody indicating the term D as the oxygen deficit with k_2 as a reaction rate constant at which oxygen is added to the water system. The negative sign indicates the fact that an increase in oxygen supply due to reaeration thus reduces the oxygen deficit in the waterbody or stream. The various factors affecting k_2 value include stream turbulence, velocity, surface area, water depth, and temperature. The k_2 value is obtained at standard temperature conditions at 20°C while any change in temperature required correction in k_2 values with a value of 1.016.

The oxygen sag curve model

The oxygen deficit in a stream is a function of both oxygen utilization and reaeration. However from the above equations it is inferred that the two processes have opposite effects on the deficit and can be represented graphically through Fig. 3.18. The rate of change in the deficit can be calculated as (Eq. 3.45)

$$\frac{dD}{dt} = r_D + r_R = k_1 L_1 - k_2 D \tag{3.45}$$

The actual oxygen concentration can be calculated by (C_s – D) with the sag curve shown in Fig. 3.18 resulting in oxygen sag curve model most commonly used to describe the process. The oxygen of deficit equation can be further modified by integrating the above equations for available DO concentration in a stream with respect to BOD exertion and reaeration can be represented by Eq. 3.46.

$$D = \frac{k_1 L_0}{k_2 - k_1}\left(e^{k_1 t} - e^{-k_2 t}\right) + D_0 e^{-k_2 t} \tag{3.46}$$

where, the term t is the time in days required by the pollutant to travel in the stream from the point of discharge and is considered an independent variable in the above equation (Eq. 3.46). The time travelled by the pollutant from the point of discharge at any given downstream point can be calculated as:

$$t = \frac{x}{u} \tag{3.47}$$

where, x is the distance travelled by the pollutant along the stream in metres and u is the stream velocity in m/sec. The units of t must always be kept in days. Substituting values for t, or x/u, the value of D at that point of stream can be calculated. The point at which DO concentration is found to be lowest in a stream is the most important point of *oxygen sag curve* as this point represents the maximum impact of wastewater discharge on the dissolved oxygen present in a stream. The point of lowest DO concentration in a stream is called *critical deficit concentration* (D_c), and the point of time of travel at which this critical DO deficit occurs is termed the *critical time* (t_c). Considering that the rate of change of the deficit is zero at the maximum deficit condition, the term D_c can be further deduced to Eq. 3.48.

$$D_c = \frac{k_1}{k_2} L_0 e^{-k_1 t_c} \tag{3.48}$$

Similarly, the term t_c can be further deduced and represented by conventional Eq. 3.49 as:

$$t_c = \frac{1}{k_2 - k_1} \ln\left[\frac{k_2}{k_1}\left(1 - D_0 \frac{k_2 - k_1}{k_1 L_0}\right)\right] \tag{3.49}$$

Equation 3.48 and Eq. 3.49 can be used to determine the critical oxygen level in a stream and the position at which this maximum deficit occurs. Table 3.10 shows reaeration rate constants for different water types and Table 3.11 shows variation in DO concentration (mg/L) in water with variation in temperature.

Table 3.10: Reaeration rate constants for different waters

Waterbody	Range of k_2 at 20°, base *e*
Small ponds and backwaters	0.1-0.23
Sluggish streams and large lakes	0.23-0.35
Large streams of low velocity	0.35-0.46
Large streams of normal velocity	0.46-0.69
Swift streams	0.69-1.15
Rapids and waterfalls	More than 1.15

Source: Metcalf & Eddy, Inc.

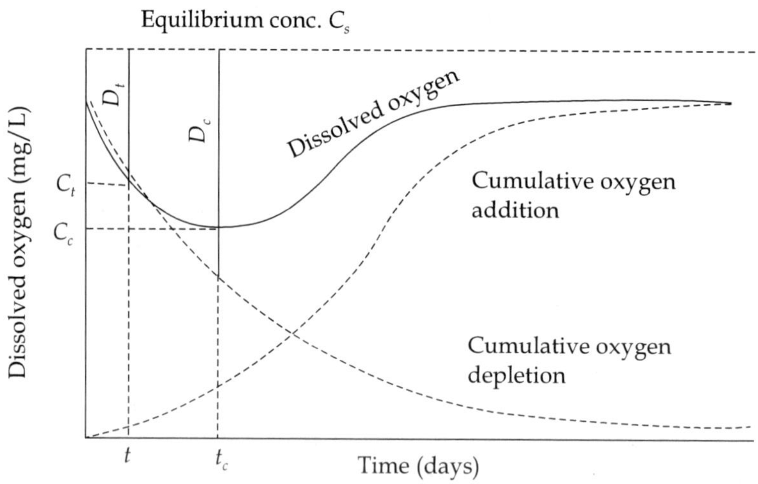

Fig. 3.18: Graphical representation of oxygen sag curve model

Table 3.11: Variation in DO concentration (mg/L) in water with variation in temperature

	Chloride concentration in water (mg/L)			
Temperature (°C)	0	5000	10,000	15,000
0	14.62	13.73	12.89	12.10
5	12.77	12.02	11.32	10.66

10	11.29	10.66	10.06	9.49
15	10.08	9.54	9.03	8.54
20	9.09	8.62	8.17	7.75
25	8.26	7.85	7.46	7.08
30	7.56	7.19	6.85	6.51

Source: Thomann and Muller (1987)

Example 3.10: Using the DO model equations: The treated sewage with maximum flow rate of 12,000 m^3/day with BOD_5 of 50 mg/L, dissolved oxygen concentration of 1.5 mg/L, and a temperature of 30°C has been discharged in a running stream. The stream has minimum flow rate of 0.65 m^3/s with BOD_5 of 4 mg/L and dissolved oxygen concentration of 7.5 mg/L with water temperature of 25°C. The complete mixing of sewage and stream is almost instantaneous, and the velocity of the mixture is 0.2 m/s. From the flow regime, the reaeration constant is estimated to be 0.4 $day^{-1}$for 20°C conditions. Sketch the dissolved oxygen profile of 100-km reach of the stream below the discharge. Assume BOD rate constant to be 0.23 d^{-1} at 20°C. Assume any other data suitably.

Solution:

1. Determine characteristics of wastewater-stream mixture.

(a) Resultant flow as Q_{mix}:

$$Q_w = \frac{12{,}000\ m^3}{d} \times \frac{1\ d}{24\ h} \times \frac{1\ h}{60\,min} \times \frac{1\,min}{60\ s} = 0.14\ m^3/s$$

$$Q_s = 0.65\ m^3/s$$

$$Q{\cdot}C = \frac{Q_w C_w + Q_s Q_s}{Q_{mix}}$$

where, $Q_{mix} = Q_s + Q_w$

$$Q_{mix} = 0.14 + 0.65 = 0.79\ \frac{m^3}{s}$$

(b) BOD mix:

$$y_{mix} = \frac{Q_s y_s + y_w Q_w}{Q_s + Q_w} = \frac{0.65 \times 4 + 0.14 \times 50}{0.79} = 9.6\ mg/L$$

Convert to ultimate BOD. (assume $k_1 = 0.23$ for mixture.)

$$y_u = L_0 = \frac{y}{1 - e^{-k.t}} = \frac{y}{1 - e^{(-0.23 \times 5)}} = 14\ mg/L \text{ as ultimate BOD}$$

(c) Dissolved oxygen:

$$DO_{mix} = \frac{0.65 \times 7.5 + 0.14 \times 1.5}{0.79} = 6.43\ mg/L$$

(d) Temperature:

$$T_{mix} = \frac{0.65 \times 25 + 0.14 \times 30}{0.79} = 25.88°C$$

2. Correct reaction constants for temperature

(a) BOD reaction rate correction (k_1):

$k_1 = k_{25.88} = k_{20}(1.047)^{25.88 - 20} = 0.23 \times 1.31 = 0.30 \text{ d}^{-1}$

(b) Stream reaeration rate correction (k_2)

$k_2 = k_{25.88} = k_{20}(1.016)^{25.88 - 20} = 0.4 \times 1.098 = 0.44 \text{ d}^{-1}$

3. Determine initial oxygen deficit, D_0.

(a) At T = 25.88, the equilibrium concentration of oxygen in fresh water is 8.3 mg/L; therefore

$D_0 = 8.3 - 6.43 = 1.87$ mg/L

4. Determine the critical deficit and its location.

(a) $$t_c = \frac{1}{k_2 - k_1} \ln\left[\frac{k_2}{k_1}\left(1 - D_0 \frac{k_2 - k_1}{k_1 L_0}\right)\right]$$

$$= \frac{1}{0.44 - 0.30} \ln\left[\frac{0.44}{0.30}\left(1 - 1.87 \frac{0.44 - 0.30}{0.30 \times 14}\right)\right] = 2.27 \text{ d}$$

(b) $$D_c = \frac{k_1}{k_2} L_0 e^{-k_1 t_c} = \frac{0.30}{0.44} 14 e^{-0.30 \times 2.27} = 4.83 \text{ mg/L}$$

(c) This condition will occur at a distance of

x = 0.2 m/s × 86,400 s/d × 2.27 d = 39.22 km downstream from point of discharge.

5. Determine the deficit at points 20, 75, and 100 km from the point of discharge.

(a) $$t = \frac{x \text{ km}}{u \text{ km/d}}$$

$$u = 0.2 \frac{\text{m}}{\text{s}} \times \frac{86{,}400 \text{ s}}{\text{d}} \times \frac{1 \text{ km}}{1000 \text{ m}} = 17.3 \text{ km/d}$$

$$t_{10} = \frac{10}{17.3} = 0.58 \text{ d}$$

$$t_{20} = \frac{20}{17.3} = 1.16 \text{ d}$$

$$t_{75} = \frac{75}{17.3} = 4.3 \text{ d}$$

$$t_{100} = \frac{100}{17.3} = 5.8 \text{ d}$$

$$t_{150} = \frac{150}{17.3} = 8.67 \text{ d}$$

(b) The deficits at these times are:

Using equation $D = \frac{k_1 L_0}{k_2 - k_1}(e^{-k_1 t} - e^{-k_2 t}) + D_0 e^{-k_2 t}$

Therefore,

$$D_{10} = \frac{0.30 \times 14}{0.44 - 0.30}(e^{-0.30 \times 0.58} - e^{-0.44 \times 0.58}) + 1.87 e^{-0.44 \times 0.58} = 3.56 \text{ mg/L}$$

$$D_{20} = \frac{0.30 \times 14}{0.44 - 0.30}(e^{-0.30 \times 1.16} - e^{-0.44 \times 1.16}) + 1.87 e^{-0.44 \times 1.16} = 4.12 \text{ mg/L}$$

$$D_{75} = \frac{0.30 \times 14}{0.44 - 0.30}(e^{-0.30 \times 4.3} - e^{-0.44 \times 4.3}) + 1.87 e^{-0.44 \times 4.3} = 4.04 \text{ mg/L}$$

$$D_{100} = \frac{0.30 \times 14}{0.44 - 0.30}(e^{-0.30 \times 5.8} - e^{-0.44 \times 5.8}) + 1.87 e^{-0.44 \times 5.8} = 3.02 \text{ mg/L}$$

$$D_{150} = \frac{0.30 \times 14}{0.44 - 0.30}(e^{-0.30 \times 8.67} - e^{-0.44 \times 8.67}) + 1.87 e^{-0.44 \times 8.67} = 1.67 \text{ mg/L}$$

6. The dissolved-oxygen concentrations at each point are found to be:
   C_{10} = 8.3 - 3.56 = 4.74 mg/L
   C_{20} = 8.3 - 4.12 = 4.18 mg/L
   $C_{39.22}$ = 8.3 - 4.83 = 3.47 mg/L (critical point)
   C_{75} = 8.3 - 4.04 = 4.26 mg/L
   C_{100} = 8.3 - 3.02 = 5.28 mg/L
   C_{150} = 8.3 - 1.67 = 6.63 mg/L

Considering the above points in view a dissolved oxygen sag curve profile with respect to distance in km and time in days can be sketched out as shown in Fig. 3.19 and the point of critical oxygen deficit and point of DO recovery can be determined.

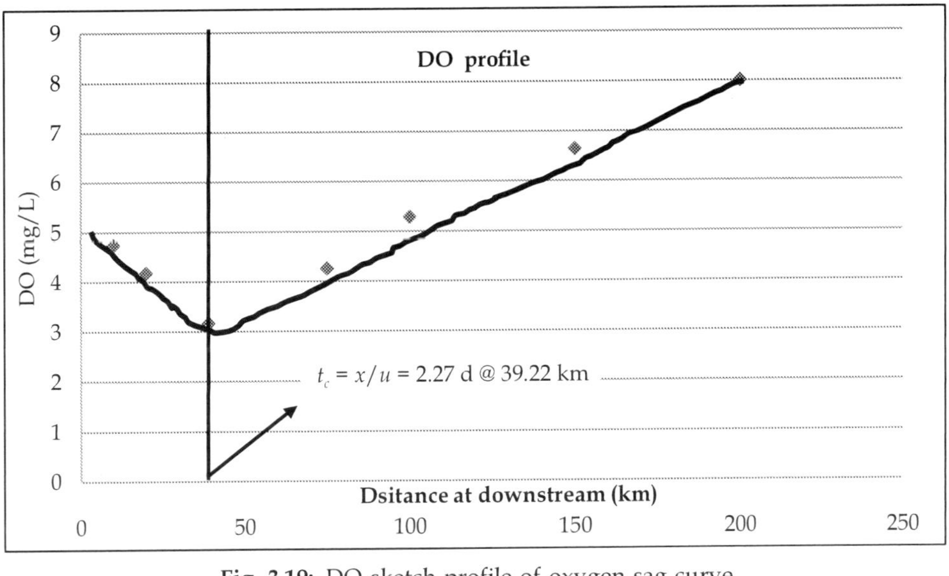

Fig. 3.19: DO sketch profile of oxygen sag curve

3.7.2.18 Water Quality Index for Drinking and Bathing

Water quality indices are tools to determine conditions of water quality. It is a well known method of expressing water quality that offers a stable and reproducible unit of measure which responds to changes in the principal characteristics of water. In this study Canadian Water Quality Index was used for evaluating the status of water quality. It consists of three measures of variance from selected water quality objectives: Scope (F_1), the number of variables not meeting water quality objectives; Frequency (F_2), the number of times these objectives are not met; and amplitude (F_3), the amount by which the objectives are not met. The index produces a number between zero (worst water quality) and 100 (best water quality).

Conceptual framework of CCME water quality index: The CCME WQI was originally developed as the Canadian Water Quality Index (CWQI). It is comprises 3 factors:

(a) **Scope:** It represents the percentage of variables that do not meet their objectives at least once during the time period under consideration ("failed variables"), relative to the total number of variables measured. It is represented by 'F_1'and can be calculated as (Eq. 3.50):

$$F_1 = \left(\frac{\text{Number of failed variables}}{\text{Total number of variables}}\right) \times 100 \tag{3.50}$$

(b) **Frequency:** It represents the percentage of individual tests that do not meet objectives ("failed tests"). It is represented by 'F_2'and can be calculated as (Eq. 3.51):

$$F_2 = \left(\frac{\text{Number of failed tests}}{\text{Total number of tests}}\right) \tag{3.51}$$

(c) **Amplitude:** It represents the amount by which failed test values do not meet their objectives. It is represented by 'F_3'and can be calculated in three steps as (Eq. 3.52 to 3.53):

(i) The number of times by which an individual concentration is greater than (or less than, when the objective is a minimum) the objective is termed an "excursion" and is expressed as follows. When the test value must not exceed the objective:

$$\text{Excursion} = \left(\frac{\text{Failed test value}}{\text{Guideline value}}\right) - 1 \tag{3.52}$$

For the cases in which the test value must not fall below the guideline:

$$\text{Excursion} = \left(\frac{\text{Guideline value}}{\text{Falied test value}}\right) - 1 \tag{3.53}$$

The collective amount by which individual tests are out of compliance is calculated by summing the excursions of individual tests from their objectives and dividing by the total number of tests (both those meeting objectives and those not meeting objectives). This variable, referred to as the normalized sum of excursions (NSE) is calculated as in Eq. 3.54:

$$\text{NSE} = \left(\frac{\sum_{i-1}^{n} \text{excursion}}{\text{No. of tests}} \right) \tag{3.54}$$

The term F_3 is then calculated by an asymptotic function that scales the normalized sum of the excursions from objectives (NSE) to yield a range between 0 and 100 as shown in Eq. 3.55.

$$F_3 = \left(\frac{\text{NSE}}{0.01 \text{ nse} + 0.01} \right) \tag{3.55}$$

Once the factors have been obtained, the index itself can be calculated by summing the three factors as if they were vectors. The sum of the squares of each factor is therefore equal to the square of the index. This approach treats the index as a three-dimensional space defined by each factor along one axis. With this model, the index changes in direct proportion to changes in all three factors. The CCME water quality index (CCME WQI) can be calculated as (Eq. 3.56).

$$\text{WQI} = 100 - \left(\frac{\sqrt{F_1^2 + F_2^2 + F_3^2}}{1.732} \right) \tag{3.56}$$

The divisor 1.732 normalizes the resultant values to a range between 0 and 100, where 0 represents the "worst" water quality and 100 represents the "best" water quality as shown in Table 3.12.

Table 3.12: Canadian Council of Ministers of the Environment Water Quality Index

WQI score	Rating of water quality
95-100	Excellent water quality
80-94	Good water quality
60-79	Fair water quality
45-59	Marginal water quality
0-44	Poor water quality

Corrosivity indices: The corrosivity index can be determined using water quality parameters and can be represented with Langelier index (LI), Ryznar index (RI), aggressive index (AI), Larson-Skold index (LS) and Puckorius index (PI).The Ryznar index (RI; Ryznar 1944)is a mechanism that is utilized to foresee the probability of calcium carbonate scale to form in a given specimen of water. The Langelier index (LI; Langelier 1946) is used by water operators in order to stabilize water for controlling internal corrosion as well as the scale deposition. The water supply operators can improve their water supply systems by identifying the leakage potentials using the Langelier index. Leakage is a common issue in water distribution systems in many parts of India due to the nature of natural water available for drinking and other domestic usages. Langelier index (LI) depends on dissolvability attributes of compounds including calcium carbonate. It serves as an indicator for corrosivity. The aggressive index is developed by a relation of pH, calcium and total alkalinity.

Count of RI utilizes LI as a segment in predicting the scaling or corrosion susceptibility of water. In Puckorius index (PI), the measured pH utilizes in LI and RI as a substitute by the balance of pH as pH_{eq}. Larson Skold file includes groupings of chlorides, sulphates, carbonates and bicarbonates in its numerical equation. The scientific conditions used to compute different records and elucidations dependent on list esteems are summarized in Table 3.13 (Yousefi et al., 2016).

Table 3.13: Mathematical equations and expressions used to calculate various indices in water

Index	Formula	Interpretation based on the index value
LSI	$LSI = pH - pH_s$	LI < 0: Water is not saturated and has a corroding tendency
	$pH_s = (9.3 + A + B) - (C + D)$	LI = 0: Water is saturated and has no scaling tendency
	$A = (\log_{10}(TDS) - 1)/10$ $B = -13.12 \times \log_{10}(T + 273) + 34.55$ $C = \log_{10}(Ca^{2+}) - 0.4$ $D = \log_{10}(Alk)$	LI > 0: Water is supersaturated and has scaling tendency
AI	$AI = pH + \log_{10}(Alk \times Ca^{2+})$	AI < 10: Water is severely corrosive (highly aggressive)
		10 < AI < 12: Water is moderately corrosive
		AI > 12: Water has scaling tendency and has a non-aggressive tendency
RI	R I= 2pHs - pH	RI < 5.5: Water has a rigorous scaling tendency
		5.5 < RI < 6.2: Water has scaling tendency
		6.2 < RI < 6.8: Water is balanced and has no scaling or corrosive tendencies
		6.8 < RI < 8.5: Water has corrosive tendency
		RIC 8.5: Water has a rigorous corrosive tendency
PI	PI = 2pHs - pHeq $pHeq = 1.465 \times \log_{10}(Alk) + 4.54$	PI < 6: Water has scaling tendency
		6 < PI < 7: Water has little scaling and corrosive tendencies
		PI > 7: Water has a significant corrosive tendency
LS	$LS = (Cl^- + SO_4^{2-})/(B_{alk} + C_{alk})$	LS < 0.8: Water has scaling tendency
		0.8 < LS < 1.2: Higher corrosion rates can be obtained
		LS > 1.2: High rates of localized corrosion can be expected

Terms involved: pH actual pH of water, pH_s pH at saturation state of $CaCO_3$ TDS total dissolved solids (mg/L), T temperature (°C), Ca^{2+} calcium hardness of $CaCO_3$ (mg/L), Alk alkalinity of $CaCO_3$ (mg/L), pH_{eq}pH at equilibrium, Cl^- chloride (mg/L), SO_4^{2-} sulphate (mg/L), B_{alk} bicarbonate alkalinity of $CaCO_3$ (mg/L), C_{alk} carbonate alkalinity of $CaCO_3$ (mg/L).

(a) **Langelier index:** The Langelier index indicates the saturation degree of calcium carbonate in water. It is determined using pH, temperature, alkalinity, calcium concentration and total dissolved solids of water sample collected from the tap. The negative value of Langelier index indicates water to be undersaturated with calcium carbonate which has tendency to corrosivity in the distribution system. However, the positive value of Langelier index indicates the water to be oversaturated with calcium carbonate and having a tendency to deposit calcium carbonate forming scales in the distribution system.

The Langelier index (LI) value closer to zero indicates the water has just begun to saturate with calcium carbonate and can neither be strongly corrosive nor scale forming. For LI value less than 10 the water sample becomes unsaturated and corrosive. It has also been reported that the Langelier index in the range of –1 to +1 is found to have a relatively less corrosion impact on the metallic parts of the distribution system. The values of Langelier index beyond this range may result in leaks or laundry stains.

(b) **Ryznar stability index:** The Ryznar index developed by John Ryznar is a tool used to predict the likelihood of scale formation of calcium carbonate as well as its saturation level in a given water sample. The Ryznar stability index (RSI) is a parameter that is used to link an empirical database of scale thickness detected in municipal water systems. RSI indicates the link between calcium carbonate saturation state and scale formation. At RSI << 6, the index decreases with the increase in scale tendency whereas, at RSI >> 7 the formation of calcium carbonate probably does not lead to a protective corrosion inhibitor film. However, at RSI>> 8 the mild steel corrosion becomes a major problem.

(c) **Aggressive index:** The aggressive index (AI) is another important tool for observing water in the asbestos pipe and due to simplicity and convenience is sometimes better substitute for the Langelier index as an indicator towards corrosivity of water. This aggressive index is determined using pH, total alkalinity and calcium hardness concentration in a given water sample. The aggressive behaviour of the water can be determined using the aggressive index values and also a common indicator of tendency for corrosion. The AI value of 12 or above represents non-aggressive (non-corrosive) water. The AI values below 10 indicate extremely aggressive (corrosive) conditions while the AI values ranging between 10–11.9 suggest the water to be moderately aggressive.

(d) **Puckorious scaling index (PSI):** Developed by Puckorious 1991 (PSI) is an important tool for quantifying the relationship between formation of scale and saturation state by incorporating an estimated value of buffering capacity of the water into the index. Since, other indices do not account for the critical parameters such as buffering capacity of the water and the maximum quantity of precipitate that can form in bringing water to equilibrium. Water containing higher concentration of calcium, but low in alkalinity and buffering capacity can have a higher calcite saturation level. The higher calcium concentration increases the ion activity product. The minuscule decrease in carbonate concentration in the water would radically decrease the ion activity due to the presence of small quantity prior to the initiation of precipitation. Such water might have high tendency towards scale formation due to the driving force, however, the scales formed might be in very lower amounts. The water possesses the driving force but it does not have the capacity and ability for maintaining pH with the formation of precipitates.

(e) **Larson skold index (LSI):** The Larson-Skold Index (LSI) is an index developed on the basis of in situ measurements of corrosion in steel lines conveying Great Lakes water and is generally used to explain the corrosivity of water concerning mild steel (Larson and Skold 1958). The LSI is the ratio of the concentration of chloride and sulphate ions to the concentration of bicarbonate and carbonate ions. The parameters – chloride, sulphate, carbonate alkalinity and bicarbonate alkalinity – are used to evaluate the Larson-Skold index in the present study. The LSI value of less than 0.8 indicates the sulphate and chlorides possibly will not interfere in the formation of natural film.

The LSI value in between 0.8-1.2 designates the chlorides and sulphates interference with formation of natural film. However, the LSI value higher than 1.2 indicates greater corrosion rates than desired and the tendency towards high corrosion rates should be expected with the increase in the index values.

(f) **Chloride sulphate mass ratio (CSMR):** CSMR, developed by Edward et al. in 2007, is a basic parameter that defines the ratio in water as a significant cause for leaching of lead in water. The larger values of CSMR indicate release of higher levels of lead through galvanic corrosion due to lead-tin solder in copper pipe joints. The agencies providing water must inadvertently alter the CSMR of water through coagulants addition. The ratio is usually calculated by determining chloride and sulphate concentration in a given water sample and is expressed in terms of mg/L.

(g) **Relative stability (S):** This term indicates ratio oxygen available in the effluent as DO to the total oxygen required to satisfy its first stage BOD demand as Eq. 3.57 and Eq. 3.58.

$$S = 100[1 - (0.794)^{t_{20}}] \tag{3.57}$$

$$S = 100[1 - (0.630)^{t_{37}}] \tag{3.58}$$

where, t_{20} and t_{37} are time in days for a sewage sample to decolourise a standard volume of methylene blue soultion, when incubated at 20°C and 37°C, respectively.

Methods for hydrogeochemical and water quality evaluation: Parameters such as $Ca^{2+}$and Mg^{2+} were considered to assess the water quality and geochemical process. The equation prescribed by Todd (1980); Ragunath (1987); and Hem (1991) has been used to determine the total hardness (TH) in ppm (Eq. 3.59)

Total hardness (TH) = 2.497 Ca^{2+} + 4.115 Mg^{2+} (3.59)

(a) **Irrigation suitability:** Using the assessed water quality parameters the following other parameters are determined in order to check the quality of the water used for irrigation:

(i) Soluble sodium percentage (SSP)

(ii) Sodium adsorption ratio (SAR)

(iii) Residual sodium carbonate (RSC)

(iv) Magnesium adsorption ratio (MAR)

(v) Kelley's ratio (KR)

(vi) Permeability index (PI)

Soluble sodium percentage (ssp): Sodium is normally expressed in terms of percent sodium or soluble sodium percentage (% Na). Irrigation water is classified based on the reaction of sodium with soil which reduces its permeability. For assessing the suitability of water for irrigation purposes percentage of Na^+ is widely used (Wilcox 1955).To evaluate the sodium hazard in soil, the term soluble sodium percentage (SSP) or Na % is used. Todd (1980) explained soluble sodium percentage (SSP) or Na % as (Eq. 3.60):

$$\text{SSP or Na\%} = \frac{(Na^+ + K^+) \times 100}{(Ca^{2+} + Mg^{2+} + Na^+ + K^+)} \quad (3.60)$$

Sodium adsorption ratio (SAR): SAR is a term used to express sodium or alkali hazard of the water quality for irrigation purpose (Bhuiyan et al. 2015; Islamet al. 2016a, b). Excess amount of Na+ and low value of Ca^{2+} destroys the soil structure (Todd 1980). The SAR value of irrigation water expresses the relative proportion of Na^+ to Ca^{2+} and Mg^{2+} (Alrajhi et al. 2015) and is calculated as (Eq. 3.61):

$$SAR = \frac{Na^+}{\frac{(Ca^{2+} + Mg^{2+})}{2}} \quad (3.61)$$

However, the concentrations of Na, Ca and Mg ions in water, are expressed as Na^+, Ca^{2+} and Mg^{2+}, respectively (Ayers and Westcot 1985).

Residual sodium carbonate (RSC): Considering the alkaline earth and weak acids, the residual sodium carbonate (RSC) is computed as per Ragunath (1987); Rao et al. (2012) is (Eq. 3.62):

$$RSC = (CO_3^{2-} - + HCO_3^-) - (Ca^{2+} + Mg^{2+}) \quad (3.62)$$

Magnesium adsorption ratio (MAR): MAR also recognized as magnesium hazard (MH) and is calculated as per method suggested by Ragunath 1987 (Eq. 3.63):

$$MAR = \frac{(Mg^{2+}) \times 100}{(Ca^{2+} + M^{2+})} \tag{3.63}$$

Kelley's ratio (KR): Excess amount of sodium over calcium and magnesium is determined by Kelley's ratio (KR). To find out the suitability of groundwater for irrigation, Kelley's ratio equation (Kelley 1963) can be used (Eq. 3.64):

$$KR = \frac{Na^{+}}{(Ca^{2+} + M^{2+})} \tag{3.64}$$

(b) **Permeability index (PI):** By the long-term use of irrigation water, the permeability of soil is influenced by sodium, calcium, magnesium and by carbonate contents in the soil. Permeability index (PI) of the groundwater samples are determined by using the formula given by Doneen 1964 (Eq. 3.65).

$$PI = \frac{Na^{+} + (HCO_3^{-})^{\frac{1}{2}}}{(Ca^{2+} + Mg^{2+} + Na^{+})} \times 100 \tag{3.65}$$

All ionic concentrations are represented in milli equivalent per litre (meq/L). Table 3.14 shows sample classification as per specified standards for different water quality parameters.

Table 3.14: Sample classification as per specified standards for different water quality parameters

Parameters	Range	Class
SSP	<200	Suitable
	>200	Unsuitable
SAR	<20	Excellent
	20-40	Good
	40-60	Permissible
	60-80	Doubtful
	>80	Unsafe
RSC	<1.25	Safe
	1.25-2.50	Marginally suitable
	>2.50	Unsuitable
MAR (Kacmaz and Nakoman 2010)	<50	Suitable
	>50	Unsuitable
KR	<1	Suitable
	>1	Unsuitable
PI	<80	Good
	80-100	Moderate
	100-120	Poor

EC	<250	Excellent
	250-750	Good
	750-2000	Permissible
	2000-3000	Doubtful
	>3000	Unsuitable
TH	<75	Soft
	75-150	Moderate
	150-300	Hard
	>300	Very hard
Chloride	<0.14	Extremely fresh
	0.14-0.84	Very fresh
	0.84-4.23	Fresh
	4.23-8.46	Fresh brackish
	8.46-28.21	Brackish
	28.21-282.1	Brackish salt
	282.1-564.3	Salt
	>564.3	Hyperthaline

EXERCISE

1. What are the common impurities found in natural sources of water and explain their effects on its quality along with desirable and maximum permissible limiting values for obtaining potable municipal supplies?
2. Tests for common ions are run on a sample of water and the results obtained are shown below. If 10 per cent error in balance is acceptable, should the analysis be considered complete? Constituents are: Ca_2^+ = 55 mg/L, Mg_2^+ = 18 mg/L, Na^+ = 98 mg/L, HCO_3^- = 250 mg/L, SO_4^{2-} = 60 mg/L, Cl^- = 89 mg/L
3. Explain in detail different sources responsible for groundwater pollution with causes.
4. A 200 mL sample of water has an initial pH of 10. 30 mL of 0.02 N H_2SO_4 is required to titrate the sample at pH 4.5. Determine the total alkalinity of water in mg/L as $CaCO_3$? Also determine the quantity of each species of alkalinity if the 8.3 equivalence point is reached at 11 mL of acid.
5. Determine the carbonate and non-carbonate hardness of a water sample having a total alkalinity of 200 mg/L as $CaCO_3$ and 120 mg/L of Ca^{++}, and 60 mg/L of Mg^{++} as ions. Also determine the H^+ and OH^- ions concentrations in terms of $CaCO_3$ if the pH value of the water sample is 10.

6. Enumerate the common tests that should be carried out in the examination of water of laboratories attached to water treatment plant? Explain the significance of each of them?
7. Define MPN and pH value of water.
8. Write short notes on any five of the following in terms of characterization, instruments or methods for measurement, desirable limits and acceptable limits:
   (a) pH value of water
   (b) Hardness and alkalinity
   (c) *E coli*
   (d) M.P.N.
   (e) Turbidity
   (f) Conductivity
   (g) Solids in water
   (h) Colour
   (i) BOD
   (j) COD
   (k) Alkalinity
   (l) Hardness
   (m) Acidity
9. In a water treatment plant the pH values of incoming and outgoing waters are 7.2 and 8.4, respectively. Assuming a linear variation of pH with time determine the average pH value of water.
10. Why are some diseases called water-borne? Which of them come under this classification and type of bacteria or virus associated with it (a) typhoid (b) malaria (c) jaundice (c) influenza, and (d) cholera?
11. Enumerate the various impurities of physical, chemical and microorganic and radiological origin in raw water supplies and state their desirable and maximum permissible limiting values for obtaining potable municipal supplies.
12. What do you understand by the term sampling? Describe briefly the method of collecting samples of water from different sources for its physical and chemical examination. What special precautions will be required to be observed if biological examination is also to be conducted?
13. Enumerate the common tests that should be carried out in the examination of water of laboratories attached to water treatment plant? Explain the significance of each of them.
14. Describe how the samples of water should be collected and sent to the laboratory for routine testing indicating necessary precautions to be taken.

REFERENCES

1. Akeju, Tolutope O, Kehinde O Awojobi, and Osun State. 2015. "Enumeration of coliform bacteria and characterization of *Escherichia coli* isolated from Staff Club swimming pool in Ile-Ife, Nigeria" 6. https://doi.org/10.4081/mr.2015.5972.
2. Patel, Amit & Trambadiya, Dipen & Baldaniya, Lalji. (2013). Water quality for pharmaceutical use: A review. *Pharma Science Monitor - An International Journal of Pharmaceutical Sciences*. 4: 29-63.
3. Communication, Brief. 2012. "A Review of Permissible Limits of Drinking Water" 16 (1): 1–6. https://doi.org/10.4103/0019-5278.99696.
4. Rahman, Mirza & Akter, Karima & Rahman, Syed. (2012). Evaluation of 5 days BOD measurement of surface water in Jahangirnagar University Campus, Bangladesh. *Bangladesh Journal of Environmental Research*. 10: 51-57.
5. Potential Pollutants, Their Sources and Their Impacts. (2020). https://www.fao.org/3/x5624e/x5624e04.htm.
6. Alrajhi A., Beecham S., Bolan N. S., Hassanli A. (2015), Evaluation of soil chemical properties irrigated with recycled wastewater under partial root-zone drying irrigation for sustainable tomato production, *Agricultural Water Management*, 161: 127-135.
7. Bhuiyan MAH, Ganyaglo S., Suzuki S. (2015) Reconnaissance on the suitability of the available water resources for irrigation in Thakurgaon District of northwestern Bangladesh. *Appl Water Sci* 5(3): 229–239.
8. Doneen, L.D. (1964). Notes on water quality in agriculture. Davis, CA: Department of Water, Science, and Engineering. University of California.
9. Islam M.A., Zahid A., Rahman M.M., Rahman M.S., Islam M.J., Akter Y., Shammi M., Bodrud-Doza M., Roy B. (2016a) Investigation of groundwater quality and its suitability for drinking and agricultural use in the south central part of the coastal region in Bangladesh. Expo Health.
10. Islam S.M.D., Majumder R.K., Uddin M.J., Khalil M.I., Alam M.F. (2016b) Hydrochemical characteristics and quality assessment of groundwater in Patuakhali district, southern coastal region of Bangladesh. Expo Health.
11. Kelly W.P. (1963). Use of saline irrigation water. *Soil Sci*. 95(4): 355-39.
12. Langelier W.F. (1946) Chemical equilibria in water treatment. *Journal of American Water Works Association* 38: 169–178.
13. Larson, T.E., & Skold, R.V. (1958). Laboratory studies relating mineral quality of water to corrosion of steel and cast iron. *Corrosion*, 8: 215-218.
14. Puckorius, R.; Brooke, J.M. (1991) A new practical index for calcium carbonate scale prediction in cooling systems. *Corrosion*, 47: 280–284.
15. Raghunath, I.I.M. (1987). *Groundwater*. Second edition; Wiley Eastern Ltd., New Delhi, India, 344- 369.

16. Rao N.S., Subrahmanyam A., Kumar S.R., Srinivasulu N., Rao G.B., Rao P.S., Reddy G.V. (2012) Geochemistry and quality of groundwater of Gummanampadu sub-basin, Guntur District, Andhra Pradesh, India. *Environ Earth Sci* 67(5): 1451–1471.

17. Ryznar J.W. (1944) A new index for determining the amount of calcium carbonate scale formed by water. *Journal of American Water Works Association* 36: 472–486.

18. Todd, D.K. (1980). *Groundwater Hydrology* , Second Edition, Wiley, New York.

19. Wilcox L., (1955) Classification and use of irrigation waters. US Department of Agriculture.

20. Yousefi Z., Kazemi F., Mohammadpour R.A. (2016) Assessment of scale formation and corrosion of drinking water supplies in Ilam city (Iran). *Environ Health Eng Manag* J 3: 75–80.

4

CHAPTER

Water Treatment

With rapid industrialization and urbanization, it is projected that water requirements for energy and industrial usage will rise to around 18% of the total requirements in 2025 .Weak environmental management systems have led to the discharge of highly toxic and organic wastewater into the ecosystem by industries such as thermal power plants, chemicals, metals and minerals, leather manufacturing, and sugar mills. This caused contamination of the sources of surface and groundwater that is mostly used for irrigation and domestic purposes. Management of water quality in India is carried out under the rule of Water (Prevention and Pollution Control) Act, 1974, which was amended in 1988. This Act's fundamental purpose is to protect and restore the wholesomeness of national aquatic resources through pollution prevention and control. In 1977 the Water (Pollution Prevention and Control) Cess Act was enacted to provide for the levy and collection of a cessation on water consumed by individuals operating and carrying on other types of industrial activities.

The daily requirement of water for general public in various areas such as domestic, commercial and industrial usages makes it essential to be free from different types of impurities. In order to make it potable the water should be free from bad taste, odours and should have sparkling appearance. However in order to supply the safe drinking water to a community it is ensured that the water must be free from any germs or pathogenic microorganisms so that transmission of any water-borne disease may not occur. Therefore, the water available in any form for public supplies should be sufficiently treated before distributing to the consumers in terms of quantity and quality of required standards. The surface sources of water generally contain large amount of impurities which requires to be removed through the engineering systems of purification such as sedimentation, filtration and chlorination. The aeration process is also applied to the water that contains algae or other microorganisms, and can also be used for the removal of taste and odour causing dissolved gasses such as CO_2 and H_2S. The surface water turbidity caused by colloidal impurities can be high enough during the rainy or high flood season and therefore, can be removed by using sedimentation aided with coagulation.

The advancement in water treatment systems has eradicated many water-borne diseases from the community that have caused epidemics in the past. However, in recent years chemical contamination in water is a major concern. The chief source of these chemical pollutants are the rapid growth in industries in the last few decades and the effluents being discharged directly either to the surface waterbodies or into the ground or on land. The elevated use of chemicals in fertilizers, pesticides and insecticides and their usage in agricultural area has deteoriated the groundwater quality in recent years. The presence of nitrates and phosphates in fertilizers and presence of these insecticides and pesticides and their compounds in water causes detrimental health effects as they are toxicants, mutagents or carcinogenic. Numerous studies have shown that these chemicals are persistent and enter the food chain thereby accumulate in human body by the process of bioaugmentation and biomagnification. The process of removing such toxic and hazardous pollutants from water is therefore a tedious job and requires special treatment processes and systems.

These contaminants reaches the groundwater or surface water and exist in various non-identifiable sources in the ecosystem through point and non-point sources of water pollution. The examples of such water pollutant sources are landfill run-off and municipal waste and many a times may be in the form of organic and inorganic compounds dissolved in water. Sometimes the pollutants in the form of insecticides and herbicides, organohalides from agricultural waste, sewage and livestock produced pathogenic bacteria, and other biological volatile compounds may be present in fresh water.

The available water must be safe from any disease causing bacteria or microorganisms, should be pleasant in taste and attractive in appearance. Therefore, in order to provide such a potable water to the public for domestic and other usages, it should be thoroughly treated before being supplied to the community. The various methods used for water treatment generally depend upon the quality of raw water and the degree of its treatment required for the water supply to the consumers. Therefore, the nature and extent of water treatment, in fact, depends on the nature and extent of contaminants in it. As most of the groundwater is free from pathogens and does not contain significant amount of organic material in it. Therefore, such type of water can be directly used for public water supplies after a lower dose of chlorination. But sometimes the groundwater may contain high concentration of dissolved solids and gases in it thus imparting taste and odour problem in it. This may result in excessive enrichment of iron and manganese or high degree of hardness in water and requires physical treatment processes and systems.

Surface water on the other hand contains a large variety of contaminants and organic matter as compared to groundwater and therefore requires complex treatment methods. Most of the surface water possesses turbidity value greater than the concentration limits prescribed by Indian standard for drinking water. However, the fast moving streams may contain suspended solids or colloidal solids that can be removed by sedimentation or sedimentation aided with coagulation process. However, the surface water may also contain some hardness thus depending upon the geological conditions of the watershed through which it is flowing. The process of adsorption is carried out if the water contains colour, organic matter or heavy metals in it.The pathogenic microorganisms and viruses can be destroyed by using the disinfection process.

Water treatment methods: The different water treatment methods can be combined to make it a single system and is termed as water treatment plant comprising of different units of treatment systems. The various units of a typical treatment process are described below as:

- Screening
- Plain sedimentation
- Sedimentation aided with coagulation (SAC)
- Filtration
- Disinfection
- Aeration and desorption
- Water softening includes lime and soda process, ion exchange, membrane processes, desalination, etc.
- Miscellaneous treatment methods including fluoridation and defluoridation, recarbonation, etc.

The coarse material, floating debris, vegetation branches, sticks, aquatic life, fish, animals present in surface water source can be restricted or traced by screening. Thereafter, it can be removed and disposed of easily. Further, the suspended impurities present in water can be removed by the action of gravity through simple process of sedimentation. However, the water may sometimes contain the colloidal impurities or solids which cannot be removed by plain sedimentation process. Therefore in order to remove the suspended colloidal impurities in water the sedimentation process is aided with coagulation by adding certain chemicals that enhance the sedimentation process and is known as *chemical coagulation*.

However, after the sedimentation and coagulation, the water still contains some very fine suspended impurities which do not settle easily by gravitational settling and are therefore removed by filtering the water or passing it through a bed of filter media. The fine impurities present in water are caught in the voids present in the filter bed material thereby, the passing water gets devoid of such impurities and the process is known as *filtration*. The filtered water may still contain some organic impurities and pathogenic bacteria and therefore needs to be removed before supplying water to the public usage. The process of killing pathogenic microorganism and trace organic matter removal is known as *disinfection*.

The final treated water may still contain disagreeable taste and some odour problem. The water may be odorous due to the presence of dissolved gases or other impurities in it. The process of removing such gases from water by exposing it to the atmosphere is known as desorption, while addition of oxygen or air into water to remove gases by oxidation process is known as aeration. The odour and taste can also be removed by passing water through activated carbon media and is known as adsorption process. The hardness of treated water increases during such treatment processes and therefore can be removed by softening processes such as lime and soda process, ion exchange process, demineralization process and desalination process.

Fluoride concentration in drinking water lies in a very narrow range of 1-1.5 mg/L. The treated water may contain low and high concentration of fluoride and therefore needs to be controlled before being supplied to the public. The process of fluoride removal from water is known as *defluoridation* and the process of addition of fluoride in water is known as *fluoridation.* However, sometimes the water may be acidic in nature and can cause corrosiveness in the metallic pipes and systems. The addition of lime is carried out to remove acidity from water and the process is known as *liming.* The water may be alkaline in nature and can cause deposition of calcium carbonate scales, therefore the addition of carbon dioxide is done to neutralize it to normal pH and the process is known as *carbonation.*

The details of above-mentioned processes and systems can be used as a single system or in a combination of two or more than two units depending upon the degree of treatment required for the raw water. The same has been explained with suitable processes along with principles of each and with suitable examples. The water treatment schemes pertaining to groundwater source and surface water source is depicted through Fig. 4.1 and Fig. 4.2, respectively.

Layout of water treatment plant (WTP)

- The WTP will be as close to the city as possible in order to prevent pollution.
- All plant units should be placed in order of sequence and flow from one unit to another by gravity.
- All units are designed in such a way as to allow a minimum area to reduce construction costs.
- For potential expansion 5 appropriate areas should be reserved. Staff quarters and office should be provided in the vicinity of treatment plants so that operators can observe the plant seasonally.
- The treatment plant site should be very clean and should give a very good appearance of aesthetics.

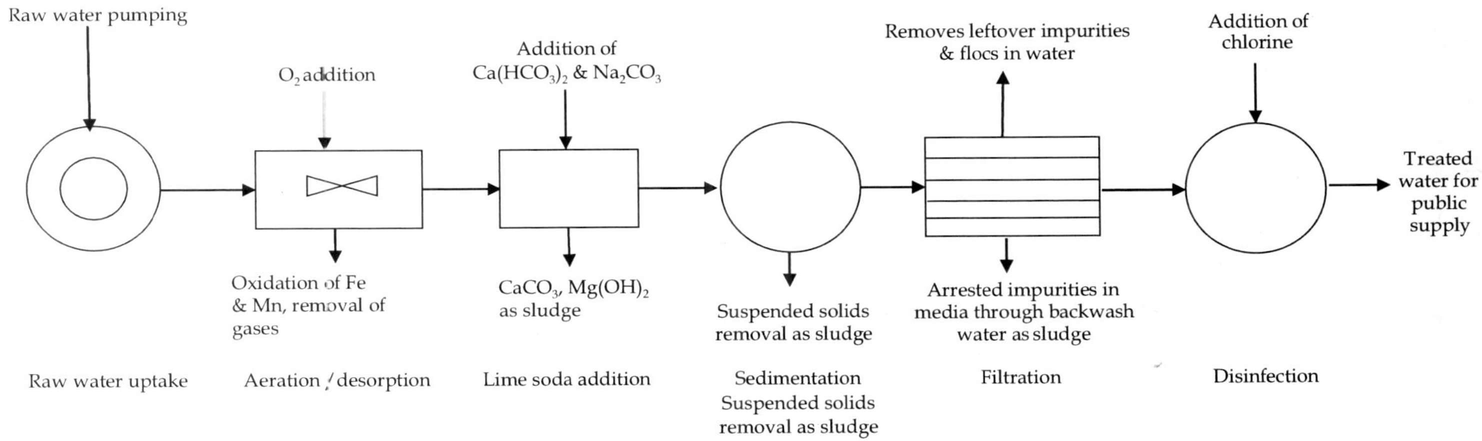

Fig. 4.1: Typical layout for groundwater treatment scheme for public water supplies

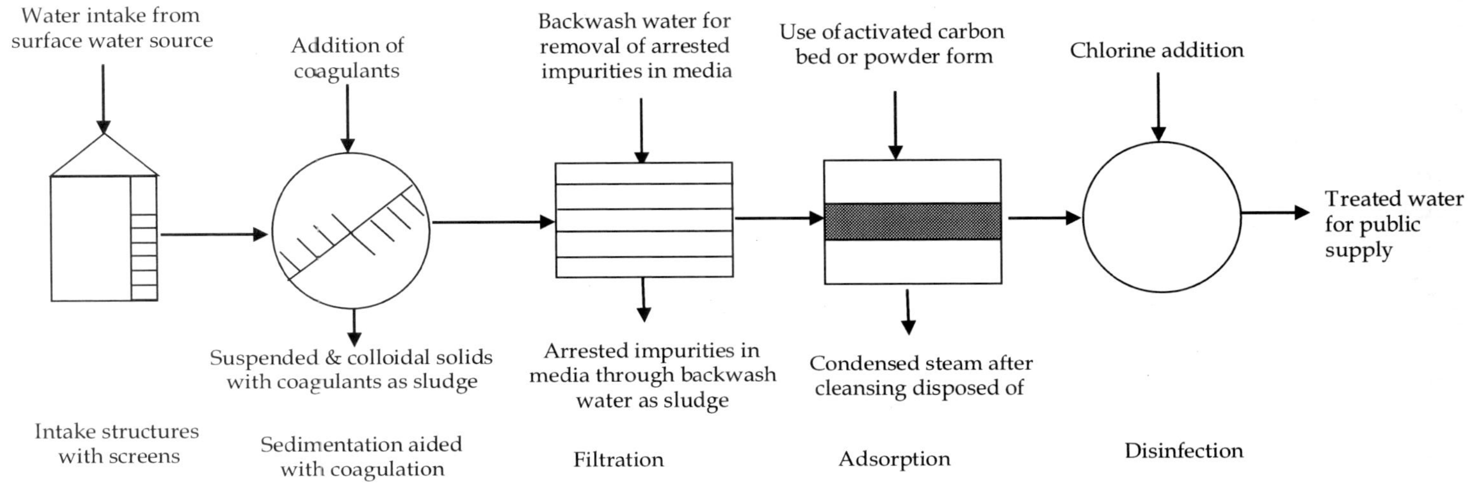

Fig. 4.2: Typical layout for surface water treatment scheme

The complete water treatment plant requires the following processes starting from the source of water up to the distribution zone in order of sequence as shown in Table 4.1.

Table 4.1: Water treatment processes and function of individual units

S. No.	Name of the unit	Purpose
1.	Intake work including pumping plant	Raw water from the source for treatment
2.	Plain sedimentation	To remove suspended impurities such as silt, clay and sand
3.	Sedimentation with coagulation	To remove the suspended matter
4.	Filtration	To remove microorganisms and colloidal matter
5.	Water softening plant	To remove hardness of water
6.	Miscellaneous treatment plants	To remove dissolved gases, tastes and odours
7.	Disinfection	To remove pathogenic bacteria
8.	Clear water reservoir	To store the treated water
9.	Pumps for pumping water in service reservoirs	If town or city is situated at higher elevation then pumping is required.
10.	Elevated or underground service reservoir	For distribution of treated water

The various techniques used for the water treatment and as an individual units of a water treatment plant are explained below.

4.1 SCREENING

Screening is carried out for removing heavy suspended impurities like plants, trees, animals, etc., that remains floating in water. Screening devices removes materials which would damage equipment or interfere with a process or piece of equipment. These bulky materials should therefore be removed before the pumps in order to avoid any clogging in the pumps and otherwise can damage them, if not removed. These types of screens are very common when surface water is used as a source of supply. Based on the opening size and flow through velocity the screens can be classified as (i) coarse screens and (ii) fine screens.

Coarse screens, also called racks, are typically bar screens, made up of vertical or inclined bars spaced at equal intervals around a channel into which water flows. Bar screens with fairly wide 10 mm to 100 mm openings are provided in front of pumps, whereas those ahead of sedimentation tanks have smaller 50 mm openings. In order to make the screen more effective and to reduce the head loss frequently the screens are placed about 45° to 60° angle with the horizontal. This angular position of screens increases the surface area against flow thus making it more effective. The bars used for the screens are typically around 10 × 40 mm rectangular in cross section, positioned parallel to the flow with a larger dimension. The clear openings should have ample total area so that flow through velocity is not greater than 0.8 to 1 m/sat average flow conditions.

However, the approach flow velocity should not be less than 0.3 m/s so as to allow all the impurities to remain in suspension. Bar screens are typically hand washed, and mechanical tools are often provided. These cleaning devices are rakes which sweep the entire screen periodically, removing the solids for further processing or disposal.

Fine screens are mechanically cleaning devices that use perforated plates; woven wire cloth or very closely spaced bars with 1.5 to 3 mm clear openings. These are generally placed after the coarse screens. These screens are generally used in water treatment units or are mainly used for the treatment of industrial wastewaters or for integrated systems. The velocity of flow ahead of and through the screen varies and affects its operation. The lower the velocity through the screen, the greater is the amount of screenings that would be removed from water. The lower the velocity, however, the greater the volume of solids accumulated into the channel upstream. The design speed should therefore be such as to allow 100 per cent removal of material of a certain size without unnecessary depositions. The velocity in the range 0.6 to 1.2 m/s can be used effectively under peak flow conditions. Additionally, the approach channel velocity at low flows should not be less than 0.3 m/s to prevent strong deposition.

Head loss: Head loss in the flow varies with the quantity and the nature of the screenings which is allowed to accumulate between screenings. The head loss from the screens can be calculated by using the flow values and the effective area of the screen openings. The sum of all the openings along the vertical projections placed in the channel is used as an effective area. The head loss is determined from the Eq. 4.1 from clean flat-bar screens:

$$h_L = \frac{1}{C}\left(\frac{V_s^2 - V_a^2}{2.g}\right) \tag{4.1}$$

or,

$$h_L = 0.0729(V_s^2 - V_a^2) \text{ (for new screens)}$$

where, h_L = Head loss (m)

V_s = Velocity through the screen (m/s)

V_a = Velocity before the screen (m/s)

C = An empirical discharge coefficient to account for turbulence and eddy losses, typically 0.7 for a clean screen and 0.6 for a clogged screen

g = Acceleration due to gravity, 9.81 (m/s^2)

Another formula often used to determine the head loss through a bar rack is Kirschmer's equation (Eq. 4.2):

$$h_L = \beta\left(\frac{W}{b}\right)^{\frac{4}{3}} h_v \sin\theta \tag{4.2}$$

where, h_L = Head loss (m)

β = Bar shape factor (2.42 for sharp edge rectangular bar, 1.83 for rectangular bar with semicircle upstream, 1.79 for circular bar and 1.67 for rectangular bar with both u/s and d/s face as semicircular).

W = Maximum width of bar u/s of flow (m)

b = Minimum clear spacing between bars (m)

h_v = Velocity head of flow approaching rack (m) = $hv = \frac{v^2}{2.g}$

θ = Angle of inclination of rack with horizontal

The head loss through fine screen is given by Eq. 4.3.

$$h_L = \left(\frac{1}{2.g.C_d}\right)\left(\frac{Q}{A}\right)^2 \tag{4.3}$$

where, h_L = Head loss (m)

Q = Discharge (m^3/s)

C = Coefficient of discharge (typical value 0.6)

A = Effective submerged open area (m^2)

The quantity of screenings depends on the nature of the water and the screen openings. Figure 4.3 shows the fixed bar type coarse screens.

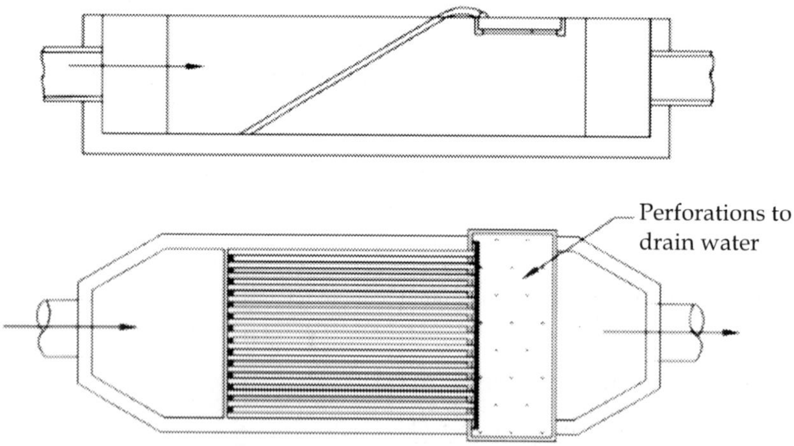

Fig. 4.3: Fixed bar type coarse or medium screens

Example 4.1: Design an inlet chamber and screens for a 50 million litres of daily water supply requirement for a city.

Design of Inlet chamber:

Average flow = 50×10^6 L/day = 50,000 m^3/d = 0.58 m^3/s

Peak flow = 50,000 × 1.8 m^3/day = 90,000 m^3/d = 1.04 m^3/sec

The approach velocity as calculated in the screen section may be taken as = 0.46 m/s

Therefore, the area of the chamber at peak flow can be calculated as $A = \frac{Q}{V} = \frac{1.04}{0.46} =$ 2.26 m^2

Now assume the depth of the chamber to be 1 m

Therefore, the width of chamber $= \frac{2.26}{1} = 2.26 \text{ m} \cong 2.3 \text{ m}$

The gross depth may be taken as = 1 m + 0.3 m (free board at the bottom) + 0.25 m(top free board) = 1.55 m

The dimensions of inlet chamber = 2.3 m × 1.55 m

The outflow from inlet chamber shall be taken to the screen chamber.

Design of screens

Basic assumptions:

Shape of bar = M.S. circular

Size = 10 mm facing flow

Clear spacing between the bars = 40 mm

Inclinations of bars with horizontal = 45° (cleaning manually)

Assuming velocity through velocity normal to screen = 0.8 m/sec

The area of screens, $A = \frac{Q}{V} = \frac{0.58}{0.8} = 0.725 \text{ m}^2$

Now assume the height of screen = 1 m

Therefore, the width of screen $= \frac{0.725}{1} = 0.725 \text{ m} \cong 0.73 \text{ m}$

Now assuming the screen bar of 10 mm diameter size with opening through screen is assumed to be 40 mm

Therefore, number of openings $= \frac{0.73}{0.04} = 18.25 \text{ m} \cong 18 = \text{n}$

As we know that with n number of opening the number of bars = $n - 1 = 18 - 1 = 17$

∴ Number of openings = 17

The gross width of the screen can be calculated as $= n \times 0.04 + (n - 1)0.01$

$= 18 \times 0.04 + (18 - 1)0.11 = 0.89 \text{ m}$

∴ The gross area now can be calculated as $(A') = 0.89 \times 1 = 0.89 \text{ m}^2$

However, in order to provide an effective surface area the screens are required to be placed at inclination against the flow, let us assume the inclination of 45°.

Therefore, effective surface area can be calculated as (Fig. 4.4):

$$(A_{\text{eff}}) = \frac{A'}{\sin 45°} = \frac{0.89}{0.71} = 1.25 \text{ m}^2$$

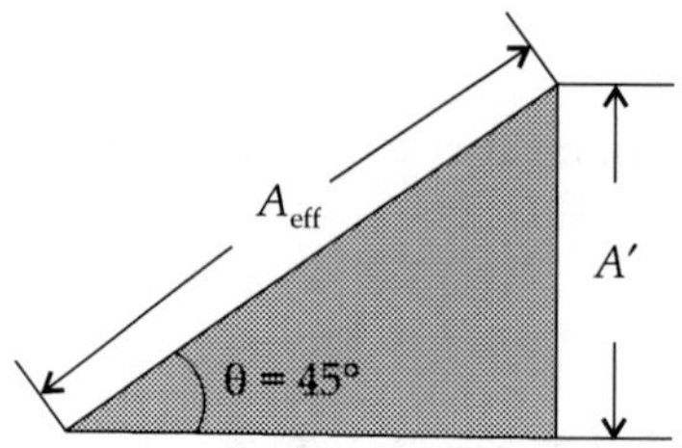

Fig. 4.4: Calculation of effective surface area (m^2)

The effective surface area of screens = 1.25 m^2

Based on the effective surface area the approach velocity V_a can be calculated as

$\therefore\ V_a = \frac{0.58}{1.25} = 0.46 \frac{\text{m}}{\text{s}}$ (O.K. as it is greater than minimum velocity of 0.3 m/s)

Provide 17 bars of 10 mm diameter @ 40 mm clear spacing

The width of the screen chamber shall be 89 cm

U/s of screens, 2 no's C.I. penstock gates shall be provided (one for each channel)

The depth of the screen shall be = 1 m + 0.3 m (free board at the bottom) + 0.25 m (top free board) = 1.55 m

Head loss calculation: The head loss through the cleaned screen and half cleaned screen can be computed as follows:

Velocity through screen: 0.8 m/s

Velocity at u/s of screen or approach velocity = 0.46 m/s

Head loss through the screen $= 0.0729(V_s^2 - V_a^2) = 0.0729\ (0.8^2 - 0.46^2)$

$= 0.0729\ (0.64 - 0.211) = 0.0312 \text{ m} = 3.12 \text{ cm}$

When the screen openings get half clogged, then

The area of the screens gets halved and the velocity gets doubled and C value to be taken as 0.6

Therefore, velocity through the screen = $V_s = 0.8 \times 2 = 1.6$ m/sec

Therefore, head loss $= 0.085\left(V_s^2 - V_a^2\right) = 0.085(1.6^2 - 0.46^2) = 0.085\ (2.56 - 0.211)$

$= 0.199 \text{ m} \approx 0.20 \text{ m} = 20 \text{ cm}$

This indicates that during initial flow conditions the head loss is minimal however with respect to time as the screens get clogged the head loss is found to be around 7 times higher than under normal conditions. Therefore, the screens are needed to be cleaned at frequent intervals in order to avoid the head loss and other operational and maintenance problems.

4.2 AERATION

Owing to the presence of toxic gases such as hydrogen sulphide and algae and associated species, in water imparts taste and odour problem and therefore aeration is required for their eliminations. The process may be used to remove the undesirable gases by the process of desorption and sometimes oxygen is added to water to oxidize the undesirable substances thereby managing the quality of water. Aeration also oxidizes iron and manganese, raises water content of dissolved oxygen, reduces CO_2 and decreases corrosion, and eliminates methane and other flammable gasses. In surface water it is a natural phenomenon as the water is exposed to atmosphere and the concentration of oxygen increases in it through the process of gas transfer continuously. However, the groundwater may contain gases such as hydrogen sulphide, ammonia and carbon dioxide. The gases such as carbon dioxide and hydrogen sulphide are corrosive in nature. These gases also generate taste and odor problem in water. These gases under atmospheric conditions possess very low solubility in water but at higher depths and under high pressure these gases get dissolved in water easily. Therefore, deep groundwater may contain high concentration of such gases.

The removal of phenols and humic acids from water can be carried out by aeration process but at a very slow rate therefore it proves to be ineffective. Iron and manganese are exits in surface water in hypolimnion lakes where anaerobic condition prevails. However, high concentration of iron and manganese is found in groundwater. But iron and manganese can be easily removed from water by aeration process. This is because upon contact with oxygen the iron and manganese oxidized to higher valances and forms ionic complexes in water in the form of precipitates that can be removed easily. However, in order to accelerate the oxidizing process sometimes potassium permanganate can also be added in water during aeration.

Principles of mass transfer: Mass transfer is transport of components under a chemical potential gradient. The component moves to the direction of reducing concentration gradient. The transport occurs from a region of higher concentration to lower concentration. Equilibrium is reached when the gradient is zero. The transport or migration of one constituent from a region of higher concentration to that of a lower concentration is known as mass transfer. Mass transfer operations depend on molecules diffusing from one distinct phase to another and are based upon differences in the physico-chemical properties of the molecules, such as vapour pressure or solubility. For inter phase mass transfer, there is a concentration gradient between bulk and interface, however under steady state, interface equilibrium is assumed. Mass transfer operation plays an important role in many industrial processes. A group of operations for separating the components of mixtures is based on the transfer of material from one homogeneous phase to another. These methods are covered by the term mass transfer operations which include techniques like gas absorption and stripping, liquid-liquid extraction, leaching, distillation, humidification, drying, crystallization and number of other separation techniques.

(i) **Absorption and stripping:** Absorption refers to an operation in which the transfer of material is from a gas phase to a liquid phase. A gas is absorbed by means of liquid in which the solute gas is more or less soluble from its mixture with an inert gas as well as more or less insoluble gas.

The liquid is essentially immiscible in the gas phase. The stripping is the separation of gas solute from the liquid phase. The separation of ammonia from an air-ammonia mixture by means of water is a typical example of absorption. The solute is then recovered from the solution by distillation. Depending upon situation, both the fluids (absorbent and absorbing fluid) are reprocessed and/ or reused for the operation.

(ii) **Adsorption and desorption:** It exploits the ability of transfer of mass from either a gas or a liquid to the surface of a solid. The adsorption is not a true inter-phase mass transfer operation because the fluid adheres to the solid surface instead of dissolving in the solid. Desorption involves the transfer of mass from the solid surface (adsorbents) to the gas or liquid medium (adsorbates). A few operations of practical applications are (a) elimination of toxic gases and deodorization of air, (b) recovery of solvents, (c) removal of ions from solution, as in demineralization of water, and (d) fractionation by selective adsorption of gases, vapours from gases, vapours from vapours and liquids from liquids and many other applications.

However, the aeration and desorption process are governed by the principle of mass transfer. In a liquid-gas transfer system under atmospheric pressure there is continuous transfer of molecules from the gaseous phase into the liquid phase till the condition of equilibrium is reached. Under such an equilibrium condition the number of molecules leaving the liquid is equal to number of molecules entering the liquid. This phase of dynamic equilibrium exists in surface water bodies that are fully saturated with oxygen and generally depends upon the exposed area of that water surface body. The process depends upon two characteristics (i) solubility, and (ii) transfer rate. This process continues until equilibrium is reached depending on the part pressure in the atmosphere of each specific gas.

Solubility: In 1803 William Henry formulated one of the gas law stating that: "At a constant temperature, the amount of a given gas that dissolves in a given type and volume of liquid is directly proportional to the partial pressure of that gas in equilibrium with that liquid." An equivalent way of stating this Henry's law is that the solubility of a gas in a liquid is directly proportional to the partial pressure of the gas above the liquid and is represented through Eq. 4.4.

$$C_s \propto p$$

or,

$$C_s = K_H \cdot p \tag{4.4}$$

where,

C_s = Saturation concentration of any gas (mL/L)

P = Partial pressure of any gas in contact with liquid or mL of gas dissolved in 1 L of liquid

K_H = Henry's constant (mL/L.atm) and at STP with temperature T = 0°C or 273 °K and 1 atmospheric pressure or 760 mmHg 1 mole of gas occupies 22.4 L of volume (Table 4.2) and Fig. 4.5 show the variation of Henry's law constant under different temperature conditions.

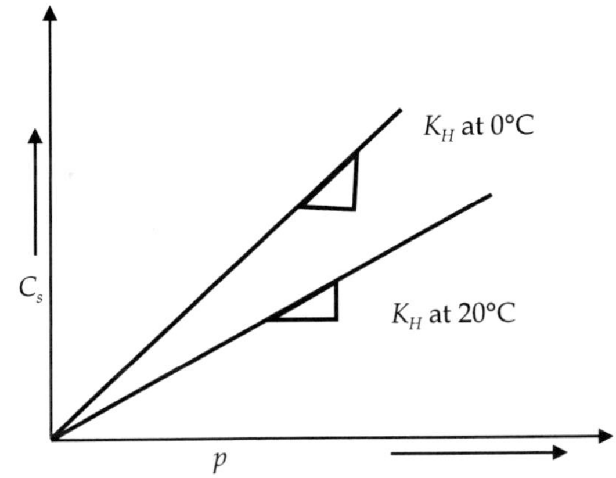

Fig. 4.5: Variation of Henry's law constant under different temperature conditions

Considering the line equation as

$$Y = mx + C$$

and comparing with Henry's equation, we get

$C_s = K_H \cdot p + 0$

Such that intercept = 0 and slope value = K_H

It states that as the temperature increases K_H value decreases

Table 4.2: Henry's constant values of gases under different temperature conditions

Temperature gas type	K_H value (mL/L) at 1 atmosphere & at STP			
	0°C	10°C	20°C	30°C
O_2	49.3	38.4	31.4	26.7
N_2	23	-	15.5	-
NH_3	1300	910	711	-
CO_2	1710	-	878	-
O_3	641	-	368	-
H_2S	4690	-	2670	-
CH_4	55.6	41.8	33.1	-

Example 4.2: Determine the solubility of O_2 in water at 0°C at 760 mm and report your answer in mg/L. Also calculate the solubility of O_2 at 20°C.

Solution: Composition of O_2 in atmosphere = 21%

$$C_s = K_H \cdot p$$

$$C_s = \frac{49.3\,\text{mL}}{l \times 760 \text{ mm Hg}} \times \frac{21}{100} \times 760 = 10.35 \frac{\text{mL}}{\text{L}}$$

We know that at STP (0° C, 760 mm Hg or 1 atmosphere) 1 mole of gas occupies 22.4 L volume

Also the molecular weight of O_2 = 32 g

Therefore, the O_2 concentration (mg/L)

$$= 10.35 \frac{\text{mL}}{\text{L}} \times \frac{32\,\text{g}}{22.4\text{ L}} \times \frac{1000\text{ L}}{100\text{ mL}} \times \frac{1000\text{ mg}}{\text{g}}$$

$$= 14.8 \text{ mg/L}$$

Now we know that under ideal gas law conditions:

$$\frac{P_1 \cdot V_1}{T_1} = \frac{P_2 \cdot V_2}{T_2}$$

Therefore,

$$\frac{22.4 \times P_1}{273} = \frac{P_2 \times V_2}{273 + 20}$$

And therefore V_2 = 24 L

Now the O_2 concentration at 20°C can be calculated as:

$$C_s = K_H \cdot P = 31.4 \times \frac{21}{100} \times \frac{32}{24} = 8.8 \frac{\text{mg}}{\text{L}}$$ **Ans.**

Mechanism of gas transfer: The rate of gas transfer is an important parameter in aeration. The resistance to the movement of molecules between two phases can be explained by using a two-film theory given by Lewis & Whiteman (1924). It states that the interface serves as barrier in between the two composed distinct films of which one is liquid and another is gas. The system can be shown through Fig. 4.6 and Fig. 4.7. Adsorption and desorption of gases is indicated in Fig. 4.8.

A molecule of a gas in bulk phase is to be transferred from the interior of gas bulk phase to the interior of bulk liquid phase. Therefore, it requires to cross the interface that exists between the gas-liquid phase. It requires to move from its own bulk gas

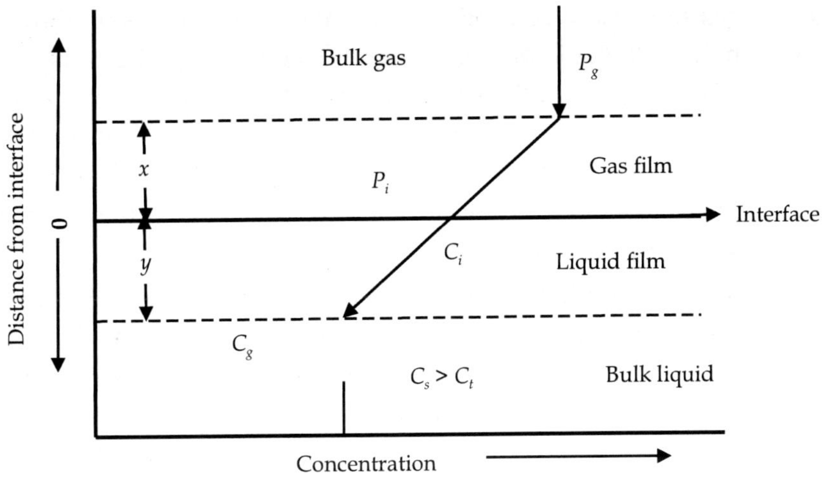

Fig. 4.6: Gas Transfer mechanism in gas-liquid interface (absorption: $C_s > C_t$)

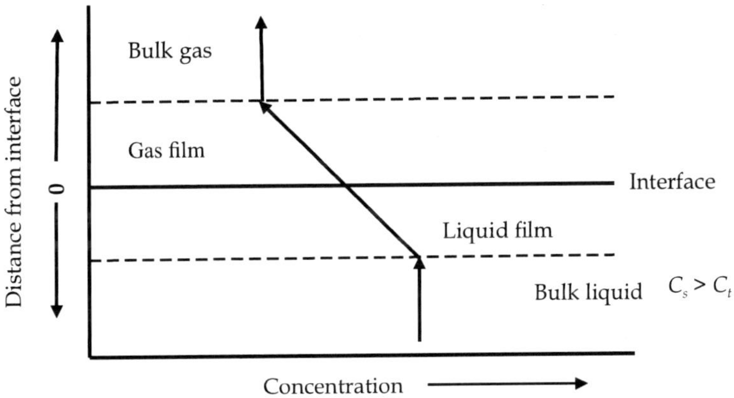

Fig. 4.7: Gas transfer mechanism in gas-liquid interface (desorption: $C_s < C_t$)

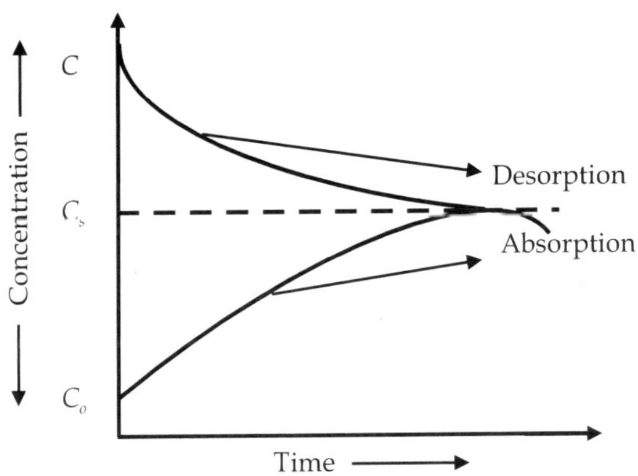

Fig. 4.8: Absorption and desorption process with time

where, C= Concentration at any time t and C_s = saturation concentration achieved during gas transfer

phase across the gas film and reach the interface and thereafter travel through the liquid film to reach the bulk liquid phase. However, when the liquid is fully saturated with the gas therefore under equilibrium conditions the gaseous molecules in the liquid starts escaping out from the liquid. The mass transfer is caused by the concentration gradient as $C_s - C$ which acts as the driving force for it. The resistance or acceleration to the movement of the molecules may be caused by certain factors such as diffusion, dispersion and turbulence in the liquid.

In most natural waters the sufficient agitation makes the existence of the bulk phase and these films are the limiting factors. Also when $P_i << P_g$, the gas film will offer more resistance than the liquid film and therefore the gas film is the controlling factor of the gas to be mixed. For an example in case of ammonia that offers more resistance when passing through a gas film and the process is known as gas-film controlled. Similarly, the oxygen and nitrogen offer more resistance to liquid film and therefore the process is called liquid film controlled. The gases of intermediate solubility such as hydrogen sulphide encounter approximately equal resistance through the two films and the process is said to be mixed-film controlled process.

The gas transfer is an important parameter can be described by the following expressions mathematically as in a two-film theory discussed above and as shown in Fig. 4.7, the following parameters can be explained as:

P_g = Partial pressure of gas in the main body of gas (bulk)

P_i = Partial pressure of gas at interface

C_i = Concentration of gas at interface

C_1 = Concentration of gas in the liquid phase (bulk)

The gas transfer mechanism can be explained by using the Fick's law of diffusion. It states that the rate of mass transfer of the gas $\left(\frac{dm}{dt}\right)$ is directly proportional to diffusion constant (D_L), area of the film (A) & concentration gradient.

or

$$\frac{dm}{dt} \propto -D_L\, A\frac{dc}{dx}$$

The negative (-) sign indicates that mass transfer takes place in decreasing concentration.

where,

$D_L = \text{Diffusion rate}\left(\frac{\text{m}^2}{\text{sec}}\right)$

$dc = \text{Concentration}\left(\frac{\text{mg}}{\text{L}}\right)$

A = Area (m^2)

y = Liquid film thickness (m)

Further the change in concentration of gas in gaseous phase with respect to liquid phase can be calculated as:

$$\frac{dc}{dx} = \frac{C_L - C_i}{y} = \frac{-(C_i - C_l)}{y} = \frac{-(C_s - C)}{y}$$

Concentration of gas in liquid phase (C_l & C are same)

$$\therefore \quad \frac{dm}{dt} = \frac{D_L\ A}{y}(C_s - C)$$

Divide the entire equation by volume, V

$$\therefore \quad \frac{d\left(\frac{m}{v}\right)}{dt} = \frac{D_L\ A}{V.y}(C_s - C)$$

$$\therefore \quad \frac{dc}{dt} = K_{LA}(C_s - C)$$

where, $K_{LA} = \frac{D_{L \cdot A}}{V \cdot y}$ overall mass transfer coefficient (1/time)

or the equation can be further modified by integrating both sides as

$$\int_0^c \frac{dc}{(C_s - c)} = \int_0^t K_{LA} \cdot dt$$

$$-\ln(C_s - C) = KL_A \cdot t + A$$

However, substituting the value of $t = 0$ and $C = C_0$ in the above equation, we get

$$A = -\ln(C_s - C_o)$$

Therefore, putting the value of A in the above equation, we get

$$-\ln(C_s - C) = K_{LA} \cdot t - \ln(C_s - C_o)$$

$$\ln\left(\frac{C_s - C}{C_s - C_o}\right) = -K_{LA} \cdot t$$

Therefore,

$$\boxed{\frac{C_s - C}{C_s - C_o} = e^{-kLA \cdot t}} \quad (4.5)$$

is the required gas transfer equation that denotes the absorption of gases as shown in Eq. 4.5.

Further, as per Eq. 4.6;

$$\frac{C_s - C}{C_s - C_o} = 1 - k$$

Therefore, $k = \frac{C - C_o}{C_s - C_o}$ is known as the coefficient of efficiency of a particular gas transfer system. (4.6)

Determination of mass transfer coefficient, K_{LA} by using graphical method as shown in Fig. 4.9.

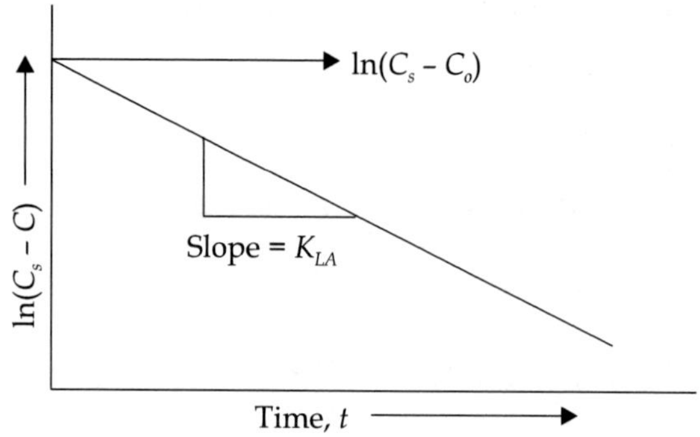

Fig. 4.9: Graphical representation of determining the K_{LA} value

The concentration of gas is measured with respect to time and the graph is plotted in between time and concentration (Eq. 4.7). The graphical equation can be represented as a straight line equation and can be compared with mass transfer equation.

We know that

$$\ln\left(\frac{C_s - C}{C_s - C_o}\right) = -K_{LA} \cdot t$$

Therefore,

$$\ln(C_s - C) = -K_{LA} \cdot t + \ln(C_s - C_o) \tag{4.7}$$

Therefore, can be compared with straight line equation as

$$y = mx + C$$

where,

$Y = \ln(C_s - C)$ with $X = t$

While slope = $-K_{LA}$ and Intercept = $\ln(C_s - C_o)$

Example 4.3: If H_2S is high in water and sprayer is used to decrease the concentration of H_2S. Suppose concentration decreases from 12 to 3 mg/L. Find the value of K_{LA}. Also initial concentration is 6.0 mg/L. Find the concentration in effluent at t = 3 sec.

Solution: Given that, C = 3 mg/L and C_o = 12 mg/L

Also, t = 3 sec

We know that $\frac{C_s - C}{C_s - C} = e^{-kLA \cdot t}$

Also, $C_s = K_H \cdot P$;

But partial pressure of H_2S in air = 0

Therefore, C_s value for H_2S = 0

Putting the C_s value in above equation, we get

$$\frac{0-3}{0-12} = e^{-kLA \cdot 3}$$

and $$0.25 = e^{-kLA \cdot 3}$$

or, $$\ln(0.25) = -K_{LA} \cdot 3$$

and $$-1.3862 = -K_{LA} \cdot 3$$

therefore, $$K_{LA} = 0.462\frac{1}{s}$$

Now using the K_{LA} value of 0.462 /s, the effluent concentration of gas in water can be calculated as:

$$\frac{0 - C}{0 - 6} = e^{-kLA \cdot 3}$$

and $$\frac{C}{6} = e^{-0.462 \times 3} \text{ and } C = 6 \times 0.250$$

Therefore, $$C = 1.5 \text{ mg/L}$$ **Ans.**

Example 4.4: Calculate the KLA value of a system with following data as shown in Table 4.3 with Cs value of 11.4 mg/L:

Table 4.3: DO Concentration (mg/L)

t (sec)	0	200	440	700	950
D.O. (mg/L)	3.5	6.0	8.2	9.5	10.2

Solution: Using the above given values time and D.O. concentration, convert the provided D.O. values in terms of Ln (C_s - C) and plot a graph between time and Ln (C_s - C) as below in Table 4.4;

Table 4.4: Conversion of DO in terms of Ln (C_s - C) (mg/L)

t (sec)	0	200	440	700	950
Ln (C_s - C) (mg/L)	2.07	1.69	1.16	0.64	0.18

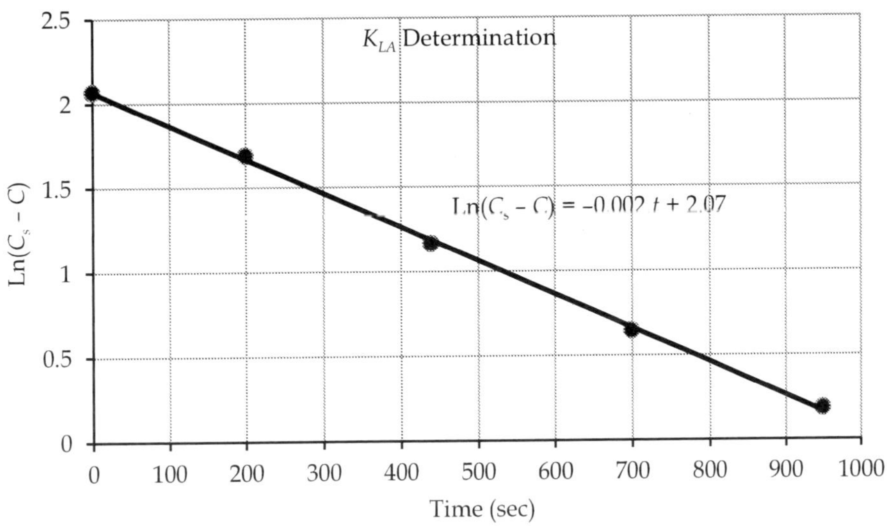

Fig. 4.10: Graph between time and Ln (C_s - C)

From the above plotted graph (Fig. 4.10), the slope of the graph is the K_{LA} value, therefore,

The computed value of $K_{LA} = \frac{0.002}{\text{sec}} = \frac{7.2}{\text{h}}$ Ans.

Example 4.5: Amount of sodium sulphite ($Na_2S_2O_3$) required to deoxygenate 5 m^3 volume of water having D.O. content is 9 mg/L.

Solution: The amount of oxygen required to dissociate $Na_2S_2O_3$ can be calculated by equation:

$$Na_2S_2O_3 + \frac{1}{2}O_2 \xrightarrow{\text{yields}} Na_2S_2O_4 \text{ (dissolved form)}$$

The molecular weight of $Na_2S_2O_3 = 2 \times 23 + 32 \times 2 + 16 \times 3 = 46 + 64 + 48 = 158$ g

Therefore, 158 g of $Na_2S_2O_3$ will consume = 16 g oxygen

But D.O. present in water = 9 mg/L = 9 g/m^3

D.O. in 5 m^3 of water = 9 g/m^3 × 5 m^3 = 45 g

But 16 g O_2 is required to dissociate = 158 g of $Na_2S_2O_3$

1 g O_2 is required to dissociate $= \frac{158}{16}$ g of $Na_2S_2O_3$

Therefore, 45 g of O_2 will dissociate $= \frac{158}{16} \times 45 = 444$ g of $Na_2S_2O_3$ ***Ans.***

Oxygen transfer in water: Oxygen in dissolved form is an important constituent of natural water systems as its presence provides life to the aquatic microorganisms. It is also important because in anaerobic biodegradation systems the end products of chemical and biochemical reactions often produce aesthetically foul odour, taste, and colour in water. The odour and colour can be easily removed by the process of adding of oxygen to the system. The rate at which dissolved oxygen is used depends on the quantity of organic constituents in the system. The rate of oxygen requirement and its transfer can be calculated by using the mathematical expressions as:

We know that the change in oxygen concentration in water with time can be calculated through Eq. 4.8.

$$\frac{dc}{dt} = K_{LA} \cdot (C_s - C) \tag{4.8}$$

$$\frac{dcV}{dt} = K_{LA} \cdot (C_s - C)V$$

where,

V = Volume of the tank (m^3)

The test is required to be carried out at 20°C. The systems are manufactured on the basis of tap water and therefore the D.O. of tap water is assumed to be 0 mg/L.

The rate of oxygen transfer can be calculated as

$$N_o = (K_{LA})_{20}(C_s)$$

where,

C_s = Saturation concentration of O_2 at 20°C = 9.1 mg/L

N_o = Standard oxygenation capacity

But the actual oxygen transfer under field condition depends on the temperature and therefore, the effect of temperature needs to be assessed and incorporated in Eq. 4.9,

$$K_{LA(T)} = K_{LA(20°C)(\theta)^{T-20}} \tag{4.9}$$

where,

θ = 1.015 to 1.04 (Typical value of θ = 1.024)

The mixing effect is also considered due to the fact that the oxygen transfer ratio of tap water to wastewater as the systems are manufactured on tap water basis, therefore the mixing effect is denoted by α and can be calculated as per Eq. 4.10;

$$\alpha = \frac{K_{LA}\ (\text{waste water})}{K_{LA}\ \ (\text{tap water})} \tag{4.10}$$

where, the value of α ranges between 0.3 to 1.2 with typical value of 0.75. However for diffused aeration system the value may range 0.4 to 0.8 and for mechanical aerators the value can be taken as 0.6 to 1.2.

The wastewater characteristic effect denoted by β is also required to be incorporated and can be calculated as in Eq. 4.11;

$$\beta = \frac{C_s\ (\text{waste water})}{C_s\ \ (\text{tap water})} \tag{4.11}$$

where, the wastewater characteristic effect value β may range in between 0.7 to 0.98 with typical value of 0.95.

Therefore, the oxygen transfer under field conditions can be calculated as shown in Eq. 4.12;

$$N_{\text{field}} = \alpha\, K_{LA\,20}\,(1.024)^{T-20}\,(\beta \cdot C_s - C)V\ \text{kg/h}$$

$$N_o = K_{LA(20)\cdot}\, C_{s20} \cdot V$$

$$K_{LA20} = \frac{\text{No}}{C_{s20} \times v}$$

$$N_{\text{field}} = \frac{\alpha\ \text{No}}{C_{s2} \cdot V} \times (1.024)^{T-20}(\beta \cdot C_s \cdot V - CV) \tag{4.12}$$

$$N_{\text{field}} = \frac{\alpha \cdot N_o}{9.1}(1.024)^{T-20}(\beta C_s - C)$$

where,

N_{field} = Oxygen transfer under field condition (kgO_2/kWH)

N_o = Oxygen transfer under standard conditions (kgO_2/kWh)

C_s = Saturation concentration of oxygen in water at operating temperature (mg/L)

C = D.O. level required in wastewater for biodegradation (1-2 mg/L)

T = Temperature of water (°C)

α = Correction factor for oxygen transfer to wastewater (0.80-0.85)

β = Correction factor for wastewater characteristics effect (0.70-0.98)

9.1 = SATURATION concentration of dissolve oxygen in water at 20°C

Example 4.6: Calculate the power requirement for aeration tank to remove 100 mg/L of chemical oxygen demand (COD) with wastewater flow rate of 1500 m^3/day under temperature condition of 10°C with following data:

Effluent D.O. = 1.5 mg/L, C_s = 11.3 mg/L, α = 0.85, β = 0.95 and N_o = 1.26 kg/kWh

Solution: Given that

Effluent D.O. = 1.5 mg/L

C_s = 11.3 mg/L

α = 0.85

β = 0.95

N_o = 1.26 kg/kWh

First calculate the oxygen requirement to oxidise 100 mg/L of COD in 1500 m^3/d of water as:

COD reduction required = 100 mg/L = 100 g/m^3

Therefore, O_2 consumption required = 100 g/m^3 = 0.1 kg/m^3

Given flow rate = 1500 m^3/d

Therefore, the O_2 transfer required = 1500 × 0.1 = 150 kg/d = 6.25 kg/h

Now we know that

$$N_{\text{field}} = \frac{\alpha \cdot N_o}{9.1}(1.024)^{T-20}(\beta C_s - C)$$

Therefore, putting the value in the above equation, we get

$$N_{\text{field}} = \frac{0.85 \times 1.26}{9.1}(1.024)^{10\quad 2}\ (0.95\times 11.3\quad 1.5)$$

$$= 0.118 \times 0.79 \times 9.235 = 0.86 \text{ kg/kWh}$$

Now O_2 requirement = 6.25 kg/h

Therefore, total power consumption $= \frac{6.25}{0.86} = 7.27$ kW

As we know that 1 H.P. = 746 watts, therefore the total power requirements in terms of HP can be calculated as

Total power required $= \frac{7.25 \times 1000}{746} = 9.75$ H.P. **Ans.**

Applications

- Addition of O_2 to oxidize Fe & Mn ions and for aerobic biological process such as aeration.
- Addition of Cl_2 for disinfection and removal taste and odour causing compounds such as O_3.
- Removal of H_2S and NH_3 for smell and odour problems.
- The dissolved CO_2 is added in water so as to decrease the pH value thus controlling the scaling problems in water systems.
- A high concentration of gases such as CH_4 and CO_2 are present in deep well water because of high pressure and low temperature, such gases therefore can be removed by desorption.
- Removal of volatile organics such as (THM).
- Removal of O_2 for boiler fed water.
- Removal of CH_4, H_2S (pH = 5) H_2S is highly soluble under alkaline condition
- Removal of NH_3 (pH = 10.5)
- Taste and odour compounds (chemical oxidation supply of Cl_2, O_3 and H_2O_2
- Carbon adsorption

The various applications of gas transfer systems as mentioned above are generally used and these equipment are designed for degasification purpose and aeration for oxidation process with an objective to drive the water-gas mix towards the equilibrium as fast as possible. The principle can be applied by either dispersing the water into atmosphere or by addition of oxygen in water. During the process when water is dispersed into the air thus maximizing the interfacial area per volume of water and ultimately minimizing the drop size. When water is supersaturated with any gas the rate of desorption will be fast and when water is undersaturated with gas the absorption rate will be high in the process. However, the principle can be best utilized for absorbing the gases in water as compared to oxygen while for undersaturated water the aeration proves to be a better technique.

4.2.1 Types of Aerators

Gravity aerators (Cascades)

Water is permitted to fall by gravity in gravity aerators, so that a wide area of water is exposed to atmosphere, often assisted by turbulence as shown in Fig. 4.11.

Cascade aerators

Cascade towers are another treatment systems consisting of a series of waterfalls that drop into small pools. This method is generally applied by exposing the water into the atmosphere not through the dispersed droplets but is exposed in thin sheets such that at every step it cascades down. The height of each step in such a cascade tower is

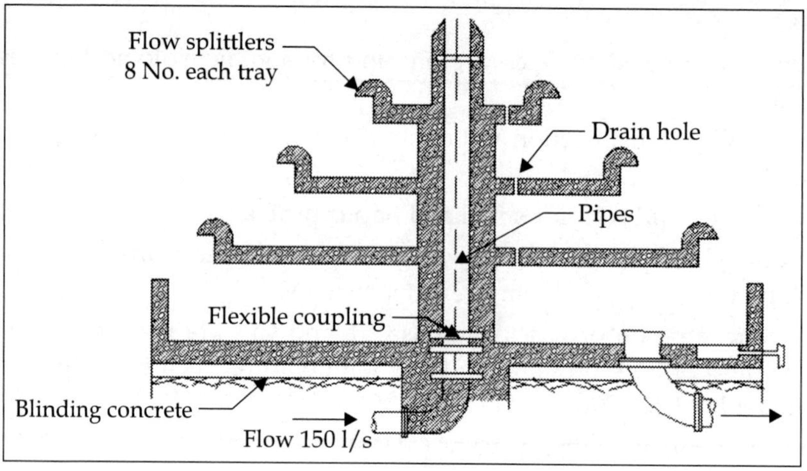

Fig. 4.11: Gravity aerators (cascades)

usually about 0.3 m and in total generally 10 steps are used. The contact time can be computed by using the number of steps between the water and the air. The height of the topmost step gives the head loss value through the system. The arrangements of cascade system is such that it may either be arranged longitudinally like the steps provided in staircase or may be arranged in a circle, with the every consecutive steps extending concentrically outward from top to bottom. The area requirement for cascade aerators depends upon the number of steps being provided and thus ranges from 4 to 9 m^2 per 50 L/s of water treated.

An example for computing the contact time can be explained below as in Eq. 4.13;

Splashing of water and most of the water will be in contact with air.

According to Newton's second law

$$t = \frac{1}{2} g \cdot t^2 \text{ or}$$

$$t = \sqrt{\frac{2h}{g}} \tag{4.13}$$

where,

n = The number of steps

t = Total time of exposure of water in air

h = Total height available for water head

h_{max} = 3–10 ft for 3 m single step/descending

$$t = \sqrt{\frac{2 \times 3}{9.8}} = 0.782 \text{ sec}$$

Fountain aerators

In water treatment systems for removal of gases from water by using water-in-air systems many types of equipment such as fountains, cascade towers or tray towers are generally used. Fountain cascade system consists of a grid of pipes suspended over a catch basin. The nozzles are fixed at the intersection of these pipes in such a position that flow of water is directly thrown upward in the atmosphere and return back to the catch basin when the kinetic energy of water is dissipated. The water from catch basin is recovered and a portion of this water is being recycled. For the design and other estimations the height of the spray along with water-air contact time, is being determined by using the pressure in the pipes. However, the dispersion pattern of the water in air depends on the nozzle characteristics and is calculated accordingly. The nozzle of size 20 to 40 mm diameter is generally used. The nozzles with low diameter with fine spraying provides higher surface-to-volume ratio but are not economical because of frequent clogging results in high operation and maintenance cost.

The parameters required while designing a spray aerator include pressure in the system, spacing between the nozzles and required flow rate per nozzle. At a flow rate of 5-10 L/s per nozzle the pressure of around 70 k-Pa is required and is very common. In order to prevent the excessive overlap of discharge from nozzles the grid spacing may vary from 0.6 to 3.5 m as the necessary distance. However, in a typical design at operating pressure of 70 k-Pa with 2.54 cm nozzle placed at centre-to-centre distance of 1.25 m comprising an area requirement of approximately 10 m^2 can treat the water at the rate of 50 L/s .

Spray towers

These spray aerators with special nozzles are often known to create a fine spray. Every nozzle has a diameter of between 2.5 and 4 cm and discharges between 18 to 36 L/h. Nozzle spacing will be such that, for one hour, every m^3 of water has an aerator area of 0.03 to0.09 m^2 as shown in Fig. 4.12.

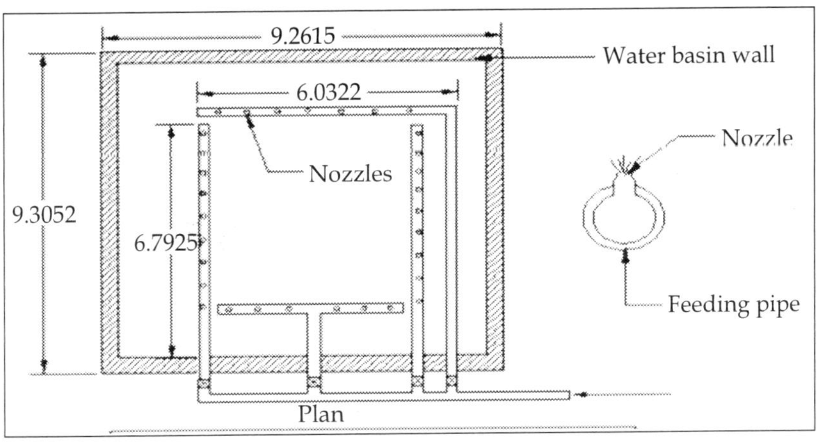

Fig. 4.12: Fountain aerators

Tray towers

Another equipment that works on the similar principle as cascade tower is tray towers in which the water is lifted and allowed to fall to a lower elevation. These tray towers intercept the flow with solid surface rather than intercepting in pools and over which the water must pass through its downward journey. The flow of water is broken down by solid surfaces—redwood slat trays placed in series or a series of porous-bottom trays containing stones, ceramic spheres, or other porous packing material. The tray works on the principle such that the porous material possesses large surface areas and helps the flow to spread through it in thin films. Therefore, the porosity of the system should be sufficient enough so as to ensure good circulation of air around the surfaces.

The application of tray towers is employed for removing iron and manganese from water through oxidation process. The tray towers are packed with large amount of coke pre-coated with potassium permanganate as a strong oxidant that initiates and accelerates the oxidation process. The deposited films of iron and manganese solids on the surface of the medium therefore serve as a catalyst for the precipitation reaction. In order to speed up the reaction and to precipitate the manganese ions slowly the pH value has to be raised around 9. Based on the above working principle many such equipment are available in the market under various proprietary names.

Injection or diffused aerators

Aeration is another most widely used method for dispersing and dissolving the air or oxygen into the water. Again, both absorption and desorption are enhanced by maximizing the interfacial area and thus minimizing the size of the air bubbles. When water is supersaturated with the gas the removal of such gas in known as *desorption* and when water is undersaturated with a gas the ingestion or dissolving the gas in water is known as *absorption*. However, the aeration process works better for absorption of oxygen in water and therefore generally used for the same. Such aeration systems consists of 2.5 to 5.0 m deep tank through which water flows and air is injected from the downside of the tank to upward using diffusers or tubes or by placing porous bottom or through spargers near the bottom and air is injected through them from outside source.

This air can be provided by using less complicated equipment such as blowers of respective capacity is sufficient to aerate the system. This capacity of the blower is good enough to deliver the required air volume at the pressure being determined by head loss through the distributing mechanism and the required depth of the water. The aeration devices are mostly used in wastewater treatment as compared to potable water treatment. These aeration systems can be employed with several variations. When the process is carried out in an enclosed tank with a positive pressure in the atmosphere above the liquid in the tank absorption rate will be increased while, the desorption rate may be decreased. Such an example is mechanical aerator consisting of an impeller when moves throws the water into atmosphere and thus breaks the air flow into smaller bubbles and enhances mixing patterns. Various equipment of such a type are available in market under patent name such as Simplex aerator and Dorocco aerator that make use of water-in-air and possess unique application of the basic principles discussed here.

A typical example of diffused aeration system as shown in Fig. 4.9 consists of a tank with perforated pipes, tubes or diffuser plates that are fixed at the bottom to release fine air bubbles from the compressor unit. The depth of the tank is maintained as 3 to 4 m, and the width of the tank is within 1.5 times its depth. If depth is more, the diffusers must be positioned under the water surface at a depth of 3 to 4 m. Aeration time is 10 to 30 min and 1 L of water requires 0.2 to 0.4 L of air as shown in Fig. 4.13. Number of bubbles is calculated as per Eq. 4.14.

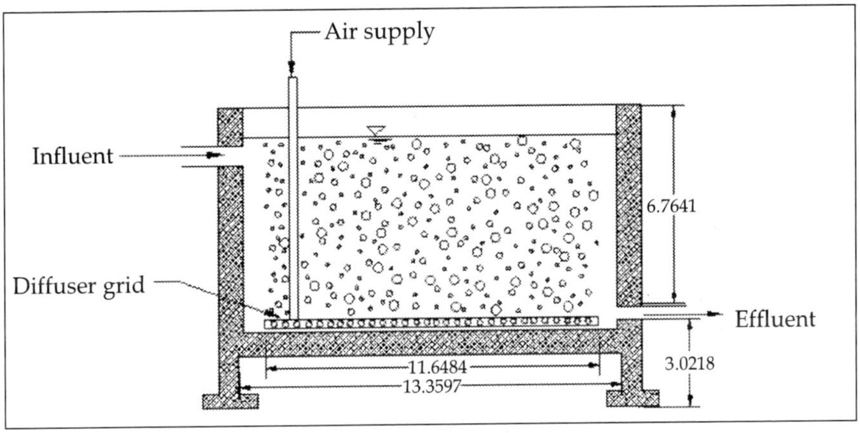

Diffused air system

Fig. 4.13: Injection or diffused aerators

Assume the bubbles are not breaking

Q_a = Air flow rate at contact time t (m^3/s)

Q = Water or wastewater flow rate (m^3/s)

h = Depth of the tank (m)

v_b = Rise velocity of the bubbles

V_b = Volume of the bubble (assume to be spherical) = $V_b = \frac{4}{3}\pi r^3$

r = Radius of bubble (m)

$$\text{Number of bubbles, } n = \frac{\text{Total air supplied (in form of bubble)}}{\text{Volume of a single bubble}}$$

$$\text{And } n = \frac{Q_a \times t}{V_b} = \frac{Q_a \times \frac{h}{v_b}}{V_b}$$

Therefore, $n = \dfrac{Q_a \times h}{V_b \cdot v_b}$ (4.14)

Example 4.7: Calculate the number of bubbles in an aeration tank with 30,000 m^3/d of water flow rate and depth of 2.8 m with detention time of 15 minutes. The bubble rising velocity is assumed to be 18 cm/s with compressed air flow rate of 0.6 m^3/m^3 of water/d. The radius of bubble may ne taken as 0.25 cm.

Solution: We know that

Q_a = Air flow rate = 30,000 × 0.6 $m^3/d \times m^3/m^3$ of water = 18,000 m^3/d

Q = Wastewater flow = 30,000 m^3/d

t = Detention time = 15 min = 15/24 × 60 = 0.01 d

v_b = Rise velocity of the bubbles = 18 cm/s = 0.18 m/s = 15,552 m/d

h = Depth of the tank = 2.8 m

r = Radius of bubble= 0.25 cm = 0.0025 m

Assume bubble to be spherical

Volume of the bubble, $V_b = \dfrac{4}{3}\pi r^3 = \dfrac{4}{3}\pi(0.0025)^3 = 6.5 \times 10^{-8}\ m^3$

We know that number of bubbles as $n = \dfrac{Q_a \times h}{V_b \cdot v_b}$

Therefore, putting the value in above equation, we get

$$n = \frac{18000 \times 2.8}{6.5 \times 10^{-8} \times 15{,}552} = 5 \times 10^7 \text{ bubbles}$$ **Ans.**

Mechanical aerators

The mechanical devices provided with paddles and aerator is fixed at the surface and splashes the liquid in the atmosphere all over the tank. This splashing liquid in the atmosphere comes in intimate contact of air and returns on the surface of tank. This makes the tank water continuously to remain in circulation from top to bottom. The aeration time depends upon the quality of water to be treated and process adapted for agitation. Various types of mechanical aerators under different patent names such Haworth system, Simplex aerators, Hartley system and Sheffield system are available in the market. The units consists of a driving unit motor that rotates the vanes or paddles of the aerator. The speed of the motor is fixed as per the amount of aeration required. The aerator will move around the surface range of 6-8 m diametrically. Mixing paddles are used here which may be either at the surface or submerged as shown in Fig. 4.9. However, the aerators of all such types are available in market with rating value ranging between 1.2–2.4 kg O_2/kWh.

4.3 SEDIMENTATION

The water when moving may contain suspended impurities having specific gravity higher than water. By the mere action of gravity these suspended impurities present in water may settle down if the turbulence in water is retarded. Therefore, the process of retarding turbulence by offering storage to water and settling down the impurities by action of gravity is known as sedimentation or plain sedimentation. The water is taken to a storage tank and detention time is provided where the physical separation of suspended solids from water takes place by settling down and clear water is passed away. The sedimentation with gravity as the driving force is the most common separation technique used in the water treatment for public supplies. However, when water is denser than solids the separation is carried out by the flotation process. Also, depending upon the characteristics and concentration of suspended solids the sedimentation process can be classified in different types.

The particles whose size, shape, and specific gravity do not change with time are known as discrete particles. However, the particles when comes in contact with other particles change their surface properties such as size, shape, and perhaps specific gravity with each contact, are called flocculating particles. Suspensions in which the concentration of particles is not sufficient to cause significant displacement of water as they settle or in which particles will not be close enough for velocity field interference to occur are called dilute suspensions. Suspensions in which the concentration of particles is too great to meet these conditions are called concentrated suspensions. These differences result in significantly different settling patterns and require separate analysis. The theoretical average time for which the water is detained in the tank is termed detention time. The tank used for the sedimentation process for offering storage to water is called sedimentation tank or clarifier or sedimentation basin.

Theory of sedimentation

The settlement of the particle in water is opposed by the various factors such as :

- The horizontal flow velocity that governs surface area of the tank at constant discharge value as less the velocity higher will be the surface area and more easily the particles will settle down.
- The viscosity is an important factor as the warm water is less viscous and helps in settlement of particles in the tank. The temperature cannot be controlled on such a large scale and its effect is not significant, therefore it is ignored.
- The shape, size and specific gravity of the particle are the important parameters that are required to be considered. This is because the fact that higher the specific gravity more easily the particles will settle down in the tank.
- The weight and volume of spherical particle is taken as $\frac{\pi d^3}{6}$ with surface area to be $\frac{\pi d^2}{4}$, therefore the small size particle will settle down slowly. Therefore, the shape and size of particle effects the velocities of the particles.

However, considering the above-mentioned factors the settling velocity of a spherical particle can be determined by using the Stokes' law and can be expressed mathematically in Eq. 4.15;

$$V_s = \frac{gp(G-1)d^2}{18\mu} \text{ for particle dia, } d < 0.1 \text{ mm and Re} < 0.5 \quad (4.15)$$

where,

V_s = Settling velocity of the particle (m/sec)

G = Specific gravity of the particle $= \frac{\rho_p}{\rho_w} = \frac{\text{Density of the particle}}{\text{Density of water}}$

μ = Dynamic viscosity of the fluid (N.s/m^2)

d = Diameter of the particle (m)

ρ = Density of water (kg/m^3)

g = Acceleration due to gravity (m/s^2)

Derivation of Stokes' Law

Consider the settling particle in fluid as shown in Fig. 4.14. When a particle moves in water the three forces acting on it are the gravitational force, F_g, which is equal to the mass of the particle multiplied by the gravitational constant, $m_p g$ and tries to settle it down. An upward buoyancy force (F_b) is equal to the mass of the displaced fluid times and is acting in opposite upward direction of the particle against settlement. During the sedimentation process when a solid particle settles down the downward settlement of the particle is opposed by the drag force of water. Therefore the remaining effective weight of the particle tries to accelerate it towards the downside of the tank till the drag force becomes equal to the effective weight of the particle. But by that time the particle attains a sufficient downward velocity by which it falls down at the bottom of the tank. This downward velocity is known as *settling velocity* of the particle, represented as V_s. The drag force by the fluid can be calculated by using Newton's law as per Eq. 4.16; gravitational force (Eq. 4.17) and buoyant force (Eq. 4.18);

$$\text{Drag force, } F_D = \frac{C_D \cdot \rho \cdot A \cdot V_s^2}{2} \quad (4.16)$$

where,

V_s = Settling velocity of the particle (m/s)

C_D = Coefficient of the drag depends on Reynolds number

A = Surface area of the particle (m^2)

ρ = Density of the water (kg/m^3)

Let us assume that the particle is a sphere of diameter d_p and density ρ_p (g/cm^3), and that the fluid density is ρ_w. Then the volume of the particle is $\pi d_p^3/6$ and we have

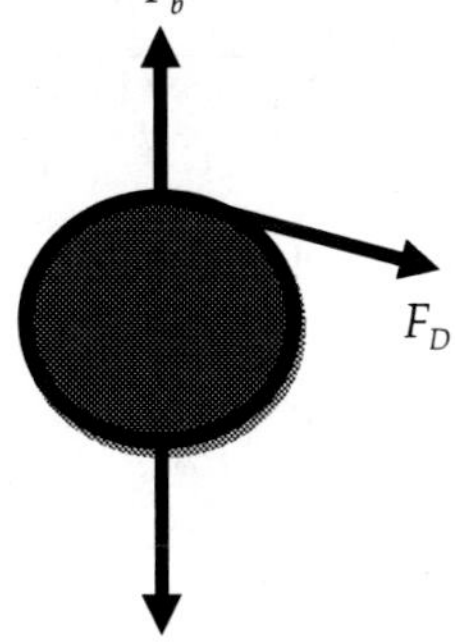

Fig. 4.14: Spherical particle in a tank

Therefore, gravitational force: $F_g = m_p \cdot g = \rho_p V_p \cdot g \; \frac{\rho_p \pi \cdot d^3}{6} \cdot g$ (4.17)

And buoyant force: $F_B = \rho_w V_p \cdot g = \frac{\rho_w \cdot \pi \cdot d^3}{6} \cdot g$ (4.18)

Also, the drag force is the result of frictional resistance to the flow of fluid past the surface of the particle. This resistance depends on the speed at which the particle is falling through the fluid, the size of the particle, and the viscosity, or resistance to shear of the fluid. Over a wide range of conditions, the friction force can be correlated with the Reynolds number. However, for most situations we are interested in, we can use Stokes' law, which states for a particle to settle down the drag force should be equal to the effective weight of the particle and therefore,

$$F_D = F_g - F_b$$

$$\frac{C_D \rho_w A V_s^2}{2} = \frac{\rho_p \pi d^3}{6} g - \frac{\rho_w \pi d^3}{6} g$$

Now putting the value of surface area of the particle $A = \frac{\pi d^2}{4}$, we get

$$\frac{C_D \rho_w \frac{\pi d^2}{4} V_s^2}{2} = \frac{\rho_p \pi d^3}{6} g - \frac{\rho_w \pi d^3}{6} g$$

$$\frac{\pi \cdot C_D \cdot \rho_w \cdot d^2 V_s^2}{8} = \frac{\pi d^3 g}{6} (\rho_p - \rho_w)$$

Now dividing the whole equation by ρ_w and πd^2, we get

$$\frac{C_D \cdot V_s^2}{8} = \frac{dg}{6}\left(\frac{\rho_p}{\rho_w} - 1\right)$$

Therefore, we get

$$V_s = \sqrt{\frac{4\left(\frac{\rho_p}{\rho_w} - 1\right) gd}{3 \cdot C_D}} = \sqrt{\frac{4(G-1)gd}{3 \cdot C_D}} \tag{4.19}$$

where G = Specific gravity of the particle $= \frac{\rho_p}{\rho_w}$ and,

C_D depends on flow regime, which in turn, depends on Reynolds number value R_e,

where $R_e = \frac{\rho \cdot V_s \cdot d}{\mu}$

where μ = Dynamic viscosity of the fluid ($gcm^{-1}s^{-1}$) and V_s is the velocity of the particle relative to the fluid as shown in Eq. 4.19. However, C_D depends on the flow conditions as under:

(i) When $R_e < 1$ (laminar flow condition) (Eq. 4.20 - Eq. 4.21);

The value of $C_D = \frac{24}{R_e} = \frac{24\mu}{\rho_w \cdot V_s \cdot d}$ (4.20)

Therefore, putting the value of C_D under laminar flow condition in the above equation, we get

$$V_s = \sqrt{\frac{4(G-1)\rho \cdot g \cdot V_s \cdot d^2}{3 \times 24 \times \mu}}$$

Therefore, we get

$$V_s = \frac{(G-1)gd^2}{18} \text{ for } d < 0.1 \text{ mm} \tag{4.21}$$

is the required equation and is known as *Stokes' law* and the settling velocity is often called the Stokes' velocity. A fundamental result of this equation is that the settling velocity increases as the square of the particle diameter, so that larger particles settle much faster than smaller particles.

(ii) When R_e: 1 to 1000 (transition flow condition)

The value of $C_D = \frac{24}{R_e} + \frac{0.3}{\sqrt{R_e}} + 0.34 = \frac{18}{R_e^{0.6}}$ (4.22)

The value in Eq. 4.22 can be substituted, therefore

$$V_s = \frac{0.153 g^{0.71} d^{1.14} (\rho_p - \rho_w)^{0.71}}{\rho_w^{0.29} \mu^{0.43}} \text{ for } d \text{ in between } 0.1 - 1.0 \text{ mm} \tag{4.23}$$

The above equation (Eq. 4.23) is known as **Allen's equation**

(iii) When $R_e > 1000$ (Turbulent flow condition) (Eq. 4.24-Eq. 4.25)

The value of $C_D = 0.4$ (4.24)

Therefore,

$$V_s = \sqrt{\frac{4(G-1)gd}{3 \times 0.4}} = \sqrt{3.33(G-1)gd} \text{ for } d > 1 \text{ mm} \tag{4.25}$$

The above equation is known as **Newton's equation**. Relation between drag force and Reynolds number is shown in Fig. 4.15.

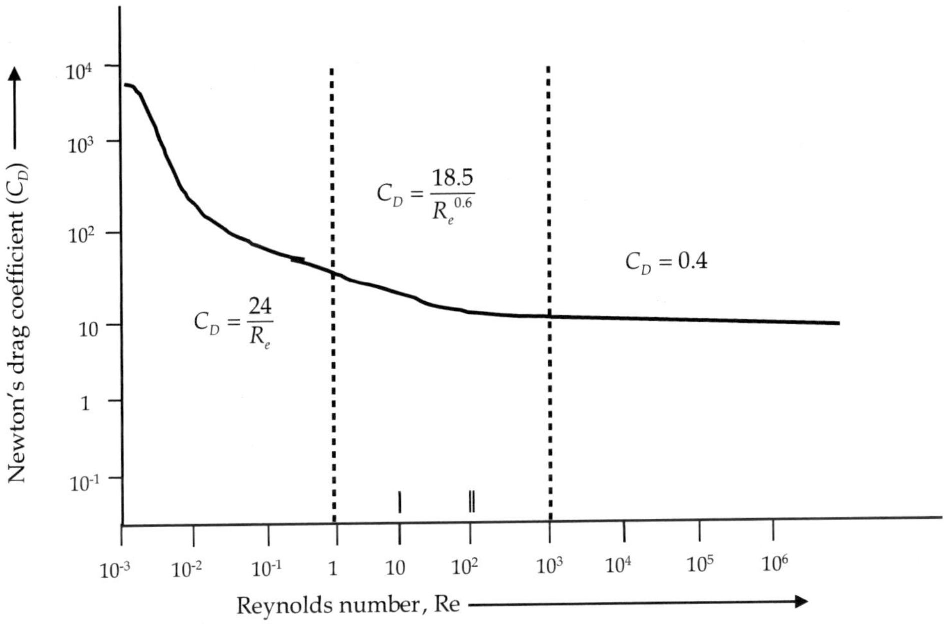

Fig. 4.15: Relationship between drag force (C_D) and Reynolds number (R_e)

The above-mentioned equations denote the settling velocities of discrete particles. The actual settling velocity is observed to be much lesser in sedimentation tank due to non-sphericity of the particles. The settling velocities of different sized particles can be further represented by modified Hazen's equation for transition zone as shown in Eq. 4.26;

$$V_s = 60.6\,d(G-1)\frac{(3T+70)}{100} \text{ for particles between 0.1 and 1 mm} \tag{4.26}$$

The above equation yields the following:

For inorganic solids: The value of G = 2.65

Therefore,

Settling velocity for inorganic solids (Eq. 4.27);

Therefore,

$$V_{s(in)} = 60.6\,d(1.65)\frac{(3T+70)}{100} = d(3T+70) \tag{4.27}$$

Similarly, settling velocity for organic matter: the value $G = 1.2$ (Eq. 4.28);

$$V_{s(o)} = 60.6\, d(0.2)\frac{(3T+70)}{100} = d(3T + 70) = 0.12\, d(3T + 70) \quad (4.28)$$

where,

T = Temperature of water (°C)

d = Diameter of the particle (m)

Example 4.8: Determine the settling velocity of the sand particles in a settling tank with size of diameter 100 μm and density 2.65 g/cm^3? (Note that the viscosity of water is 0.11g/cm/sec and the density of water is 1 g/cm^3.)

Solution: We will use our steady-state solution based on Stokes' law,

$$V_s = \frac{(2.65\,\text{g cm}^{-3} - 1.0\,\text{g cm}^{-3})(980\ \text{cms}^{-2})(100\times 10^{-4}\ \text{cm}^2)}{18\times 0.1185\ \text{g cm}^{-1}\text{s}^{-1}}$$

$$= 0.76\ \text{cm/sec or } 27\ \text{m/hr}$$

The settling velocity is relatively slow. For very small particles, the drag force is significant in comparison to the gravitational force acting on their small mass.

Sedimentation

The sedimentation process has many applications in water treatment and removal of suspended solids. However, the removal of suspended solids in normal tanks may be affected because of many factors such as density currents, eddies and turbulences caused by winds. In order to remove the concentration of suspended solids with high efficiency, therefore special types of tanks are constructed that prove to be quite effective in water treatment. However, the various factors that affect the settling properties of particles in the tank is the horizontal flow velocity in the tank that generally governs the surface area of the tank and ultimately the settling of the particles. The shape and size of the particles along with specific gravity are other important factors that govern the settling velocity of the particle. However, by decreasing the horizontal flow velocity and increasing the surface area of the tank the particles can be detained for a longer period thus increasing the removal efficiency of the tank.

The removal of discrete particles in such a tank is known as plain sedimentation process. The plain sedimentation process is a simple sedimentation on subsequent phase lightens the load and a possibility to monitor the activity of subsequent purification process in better way. Also due to no use of any chemicals in the process and no wastage of chemicals with produced sludge it is an economical process as compared to SAC. These sedimentation basins are circular or rectangular tanks of concrete. Such tanks are capable of removing 65-70% of suspended solids from water under normal conditions. In order to remove maximum percentage of particles from water such tanks are to be technically designed with suitable dimensions. The tanks are generally constructed based on flow and velocity conditions as horizontal flow type rectangular tanks and horizontal radial or spiral flow type circular tanks.

Types of sedimentation tanks

Depending upon various factors sedimentation tanks are classified as follows.

Based on methods of operation:

(a) Fill and draw type tanks

(b) Continuous flow type tanks

Based on shape

(a) Circular tanks

(b) Rectangular tanks

(i) **Fill and draw type sedimentation tank:** Such type of tanks are also known as intermittent sedimentation tanks where the raw water is stored for a certain period of time such as 12-24 hours. During this period of storage the suspended particles heavier than water settle down at the bottom of the tank and are deposited as sludge. The clear water is allowed to be decanted or taken out from the tank as shown in Fig. 4.16. The settled impurities in the form of sludge are cleaned and the tank is made ready for next filling. This cleaning operation takes around 6-12 hours. Therefore, it takes around 30-36 hours for the tank to make it ready for the next operation of filling raw water. Therefore, two tanks should be constructed in parallel so that if one is under cleansing the other could be operated. Due to more time and high labour such intermittent tanks are obsolete and not preferred nowadays.

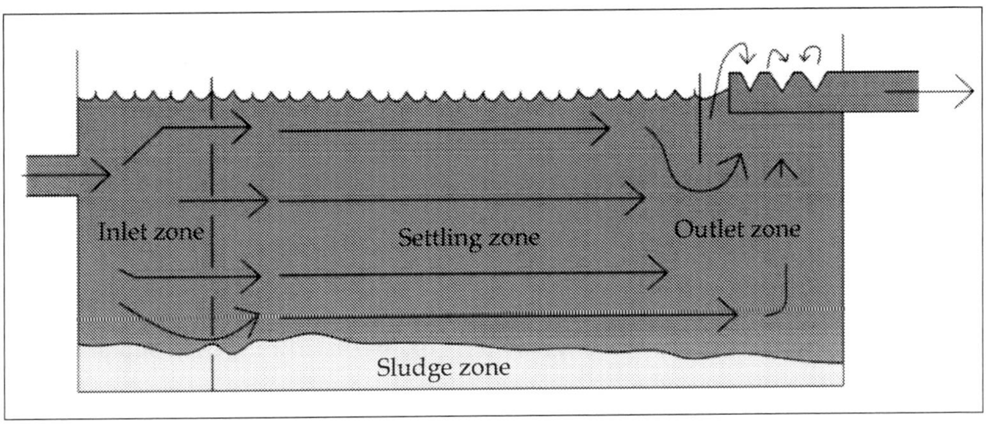

Fig. 4.16: Fill and draw type sedimentation tank

(ii) **Continuous flow type sedimentation tank:** The continuous type of sedimentation tanks are very common nowadays as the flow in the tank is in continuity and not stored as in case of intermittent tanks. The horizontal flow velocity of water in the tank is reduced to such an extent that allows the flow to travel up to sufficient length while entering the tank from one end and leaving the flow at other end. The tank is designed on such a flow velocity that the time taken by the particle to travel from inflow to outflow is slightly more than the time required for its settlement. The settled suspended particles at the bottom of the tank during the sedimentation process are removed continuously by using sludge withdrawal pipes from sludge hopper as shown in Fig. 4.17.

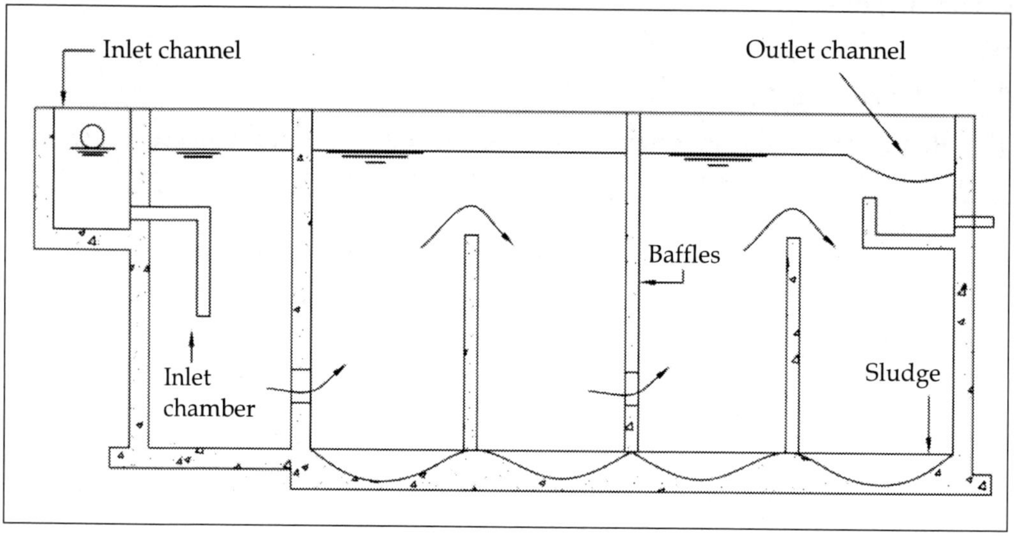

Fig. 4.17: Continuous flow type tank

Horizontal flow settling tanks: Based on the ideal flow conditions the tanks should have equal velocity at all points lying on each vertical line of the settling zone of the tank. In order to accomplish the above objective, in the horizontal flow settling tank the direction of flow of water is substantially horizontal. The horizontal flow settling tanks may be further classified into the following two types:

(a) Rectangular tanks with longitudinal flow

(b) Circular tanks with radial flow

Rectangular tanks with longitudinal flow: These are long rectangular tanks in which water is allowed to enter the tank through an inlet on one side of the tank and goes out with outlet on the opposite end of the tank after gradually flowing in horizontal direction as shown in Fig. 4.18. Inlet and outlet arrangements at the top edge of the tank are provided so that in the inlet proper instantaneous mixing of particles may take place and at the outlet the particles may not escape without detention. These inlet and outlet arrangements are provided with baffles, so that the flowing water can disperse uniformly and thereby avoid direct currents.

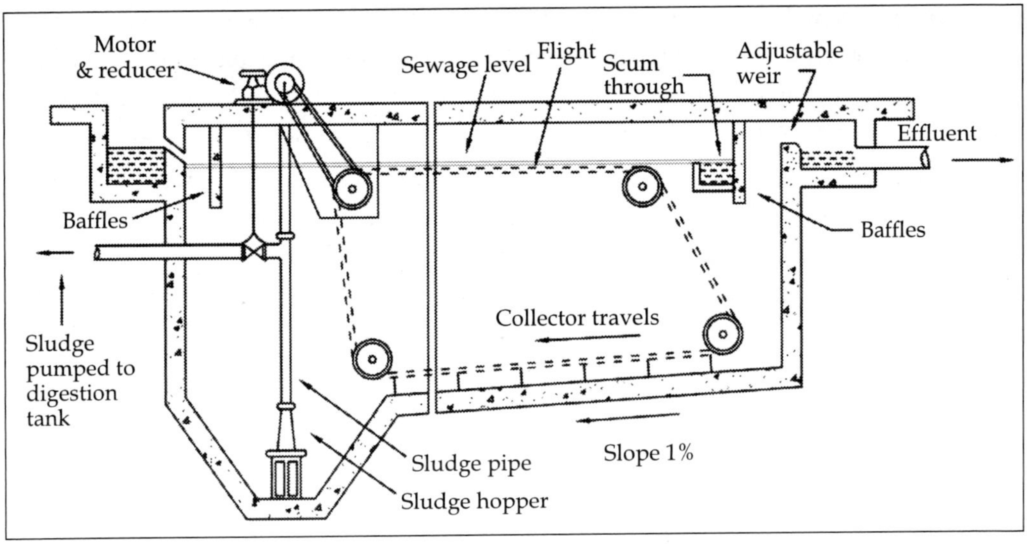

Fig. 4.18: Rectangular tank with longitudinal flow

Rectangular tanks have overall lengths and widths of 90 m and 30 m respectively with ratios of length to width of 1.5:1 to 7.5:1 and ratios of length to depth of 5:1 to 25:1, respectively. A minimum depth of 2.5 m is given for primary settling tanks, and 3.5 m for secondary settling tanks. The tank floor is provided with a 1 per cent slope from the end of the outlet towards the end of the inlet where a sludge hopper with a sludge removal pipe is installed. The sludge hopper's side slopes range from 1.2:1 to 2:1 (right to horizontal). In such tanks the average sewage flow rate is 1.5 mph. The accumulated sludge (or settled material) at the bottom of the tank is mechanically removed by a device controlled with electricity. It consists of two endless chains moving on sprocket wheels and supporting wooden cross-bars or flights which periodically sweep the sludge towards the hopper given near the inlet end, from which it is removed under hydrostatic head or pumped into separate sludge digestion tanks. Through means of a second blade, called skimmer, the scum produced at the surface is extracted by the mechanical scraper, via a scum trough.

Circular tanks with radial flow

The circular tanks with radial flow are of the following two types:

(a) Circular tank with central feed

(b) Circular tank with peripheral or rim feed

In circular tanks the water is allowed to enter from the centre of the tank and is allowed to flow radially towards the periphery of the tank equally in all directions. This decreases the horizontal velocity of the water continuously as it travels from the centre towards the periphery of the tank. This decrease in horizontal velocity of the particle changes its settling velocity continuously. During this duration of travelling the particle settles down at the bottom of the tank easily. The water thereafter flows over the periphery with equal discharge at constant rate through weirs. In such type of circular tank with raw water fed from centre (Fig. 4.19), a pipe suspended from a bridge or sealed in concrete under the tank floor is carried to the middle of the tank.

The raw water that flows gradually in the radial direction reaches the tank's edge, where the clarified water passes through a regulating weir and flows through an effluent channel (also called effluent laundering) and finally into the effluent pipe. A suspended circular baffle, a short distance from the tank wall, forms an annular space in which sewage is discharged in a tangential direction, in a circular tank with peripheral or rim feed. The water flows spirally across the tank and under the baffle, and the clarified water is skimmed over weirs on both sides of a weir trough that is centrally located. The sludge settled at the bottom of the tank is scrapped to the sludge sump mechanically and is continuously withdrawn through sludge pipes. The circular tanks are not economical as compared to rectangular tanks but have a high efficiency of removing suspended solids from water.

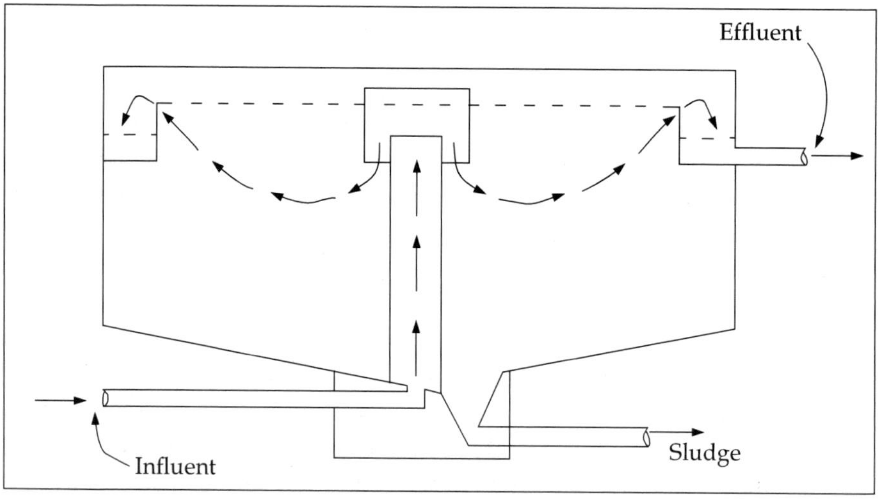

Fig. 4.19: Circular tank with central feed

Circular tank diameters vary widely from 3 to 60 m but the most common range is from 12 to 30 m. The water depth in the tanks ranges between 2 m for primary settling tanks and 3.5 m for secondary settling tanks. The floors of these tanks are equipped with a slope from periphery to the centre of 7.5 to 10 per cent to form an inverted cone at the bottom of the tank. A sludge removal system installed in the tank continuously extracts the sludge (or settled material) that is collected at the bottom of the tank. The tanks measuring 3 to 9 m in diameter are operated by the sludge removal system on beams spanning the tank. Tanks of 10.5 diameter and more have a central pier that serves the sludge removal system and is accessed via a passageway or bridge as shown in Fig. 4.20.

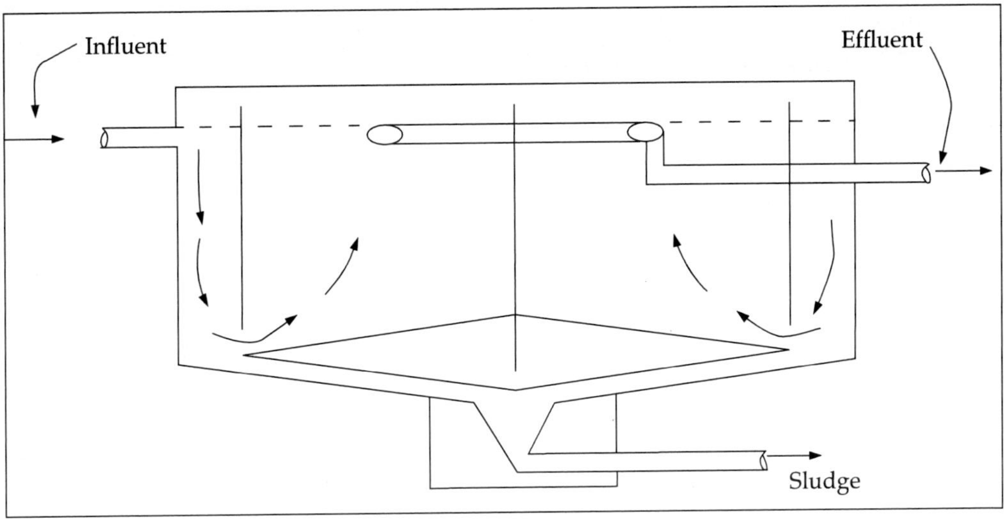

Fig. 4.20: Circular tank with peripheral feed

The sludge removal system consists of scraper blades that are placed on two or four collector arms or raking arms that also carry scum removal blades. The collector arms are rotated at a slow speed by a driving mechanism, so that in around 30 to 40 minutes one revolution is complete. The slow motion of the arms prevents the formation of unnecessary currents and also requires little operating power. As the collector's arms rotate, the sludge is scraped into a small hopper near the middle of the tank from which it is drained via a pipe.

Vertical flow settling tanks

Vertical flow settling tanks are square or circular in shape and may contain hopper bottoms. The sides of these tanks are vertical in the upper portion and inclined to form hopper bottom at 45° or more in the lower portion. The water enters from the bottom of the tank while the upflow velocity decreases as water moves on the cross-sectional surface of the tank. The tanks may be plain sedimentation tanks. Mostly the sedimentation is aided with coagulation. The flocculated masses quickly fall to the bottom of the tank creating a layer of sludge. The water entering the tank comes up through the layer of the sludge, and is clarified. The clarified water passes into a circumferential or central weir at the top of the tank and discharges into a channel draw off. The tank is built in such a way that the upward speed of water flow is less than the settling speed of the smallest particles to be extracted. Thus, waste can be moved at relatively higher velocity in such tanks. Shorter detention times are also expected for these tanks resulting in fewer capacities. The sludge is drained under hydrostatic pressure or by pumping from certain tanks.

Design criteria for plain sedimentation tanks

The design aspects of sedimentary tanks are as follows:

- Detention time of the tank
- Capacity of tank and Overflow rate
- Inlet and outlet arrangements
- Shapes of tanks
- Other miscellaneous considerations

Detention period

Detention time, also known as detention period, or hydraulic detention time of a sedimentation tank may be defined as the theoretical time taken by a particle of water while entering the tank and leaving the tank or the average time taken by water to flow through the tank. In case of intermittent tank it may be defined as the time required to fill the tank completely. Therefore, hydraulically it can be calculated as the volume of water in the tank per unit of discharge entering the tank.

Therefore, detention time or hydraulic detention time (HRT) of the tank can be calculated as in Eq. 4.29 for rectangular tank and Eq. 4.30 for circular tank.

$$\text{Detention time for rectangular tank, } t = \frac{\text{Volume of the tank}}{\text{Flow rate}} = \frac{L \times B \times H}{Q} \tag{4.29}$$

where, L = Length of the tank (m)

B = Width of the tank (m)

H = Height of the tank or water depth in the tank (m)

Q = Incoming flow rate (m^3/d or m^3/h)

$$\text{Detention time for circular tank, } t = \frac{\text{Volume of the tank}}{\text{Flow rate}}$$

$$= \frac{d^2(0.011\,d + 0.785\,H)}{Q} \tag{4.30}$$

where, d = Diameter of the tank (m)

H = Height of the tank or water depth in the tank (m)

Q = Incoming flow rate (m^3/d or m^3/h)

The volume of the circular tank is calculated based on the design aspects as the tank is constructed with a cone shaped bottom provided with 1:1 slope and therefore the volume of the tank is computed as $d^2(0.011\text{ d} + 0.785\text{ H})$. However, the unit of detention time or HRT is expressed in days, hours or minutes. The plain sedimentation tanks are generally designed on the basis of detention time of 4-8 hours, whereas in case of sedimentation aided with coagulation the detention time of 2-4 hours is generally used. In case of rectangular sedimentation tanks the width is kept in the range in between 10-12 m. The length of the tank should not be more than 4 times of the width normally but many times it may range in between 1 to 6 times of width. The horizontal velocity entering the tank lies in the range of 0.15 to 0.9 m/min, but the typical value of 0.3 m/min is very common. The design of such a tank is based on the maximum daily demand of water which is the total amount of inflow in the tank in 24 hours or in a day.

Short circuiting in sedimentation tank

For efficient removal of suspended material from a tank the material should be evenly distributed throughout the surface area cross section of the tank. However, in the actual detention time taken by the particle to travel from one end to other should be equal to the theoretical detention time of the tank. But due to some problems such as eddies, gusty winds and turbulence in the tank, moreover the formation of dead ends in the tank allows the escape of particles in the outlet of the tank without being detained or settle up. The process of escaping of particles from the tank without being detained is termed *short circuiting*. Therefore, under such practical conditions the actual time taken by the water to pass through the tank, also known as flow through period, is always less than the theoretical hydraulic detention time or simply the detention time. Another term known as *displacement efficiency* can also be calculated by using the ratio of flow through period and theoretical detention time or detention period as shown in Eq. 4.31.

$$\text{Displacement efficiency} = \frac{\text{Flowing through period}}{\text{Theoritcal detention time}} \tag{4.31}$$

In a plain sedimentation tank the value of displacement efficiency varies from 0.25 to 0.5 generally.

Particle scouring check in sedimentation tank

In a sedimentation tank the horizontal flow velocity should be maintained such that it should not be too slow so as to require high detention time and high area requirement for the tank. However it is also ensured that it should not be so high so as to cause the scouring or uplift of the deposited particles or materials that are settled in the bottom of the tank. Therefore the horizontal velocity in a sedimentation tank should always be less than displacement velocity or minimum scour velocity. This scour velocity can be computed by using the Eq. 4.32 given by Camp, 1946 as:

$$\text{Scour velocity } (v_d) = \sqrt{\frac{8\beta}{f'} \cdot g \cdot (G-1)d} \tag{4.32}$$

where, β − 0.04 for unigranular sand

= 0.06 or more for non-uniform sticky material

f' = Darcy-Weisbach friction factor = 0.025-0.03 for sedimentation tank

g = Acceleration due to gravity

G = Specific gravity of the particles

Tank capacity and overflow rate

Let the water entering the tank with containing uniform velocity V and Q is the discharge entering the basin, then capacity of the tank is determined by using the detention time and is also based on the overflow rate of the tank. The amount of water entering a tank at a given detention time is known as the capacity of the tank and can be computed by using Eq. 4.33 and surface area of tank is calculated by using Eq. 4.35.

$$\text{Volume } (V) = Q \times t \tag{4.33}$$

or

$$L \times B \times H = Q \times t$$

Because the term $\dfrac{H}{t} = v$

Therefore, $\text{Discharge } (Q) = A \times v \qquad (4.34)$

$\text{Surface area of tank } (A) = \dfrac{Q}{v} \qquad (4.35)$

where,

V = Volume or capacity of the tank (m^3 or L)
L = Length of the tank (m)
B = Width of the tank (m)
H = Height of the tank or water depth in the tank (m)
Q = Discharge or flow rate in the tank (m^3/d or L/d orm^3/h or so)
T = Detention period or detention time or HRT (d or h)
v = Horizontal inflow velocity of the tank (m/h or m/d or so)
A = Cross-sectional surface area of the tank (m^2)

The detention period depends on the quality of suspended impurities present in water. For plain sedimentation tanks, the detention period is found to vary from 4 to 8 hours.

However, the horizontal sedimentation tanks are designed based on the assumptions that:

- Particles and their velocity vectors are evenly distributed along the cross-sectional area of the tank; however it is a function of inlet zone.
- The liquid moves as an ideal slug down the length and along the depth of the tank.
- The particle hitting the bottom surface of tank is removed.

We know that in a horizontal tank of continuous type containing water with uniformly distributed solids is entering the tank with uniform horizontal velocity v and at a given discharge the velocity is computed through Eq. 4.36.

$$v = \frac{Q}{B \cdot H} \tag{4.36}$$

where, Q = Flow rate or discharge (m^3/d)
H = Depth of tank (m)

Let us consider in a rectangular tank with depth (H) and a discrete particle enters the tank with horizontal velocity (v) from inlet point A on the top. The particle will have settling velocity (v_s) which tends to make the particle settle down. The vector of horizontal velocity is governed by the length L of the tank and settling velocity is a vector function of depth (H) of the tank. The resultant path along the line AC is the vector sum of v_s and v in the tank and the particles following the path will settle down at the bottom of the tank before reaching the outlet zone of the tank. Therefore, the particle of this settling velocity will be removed 100% at whatever their height may be in the tank. This may be due to the fact that the trajectory AC will bring them down to settle at the bottom of tank. Every discrete particle is moving with horizontal velocity (v) and a downward velocity, v_s. The resultant path is given by the vector sum of its flow velocity (v) and its settling velocity (v_s) as shown in Fig. 4.21 and Eq. 4.37.

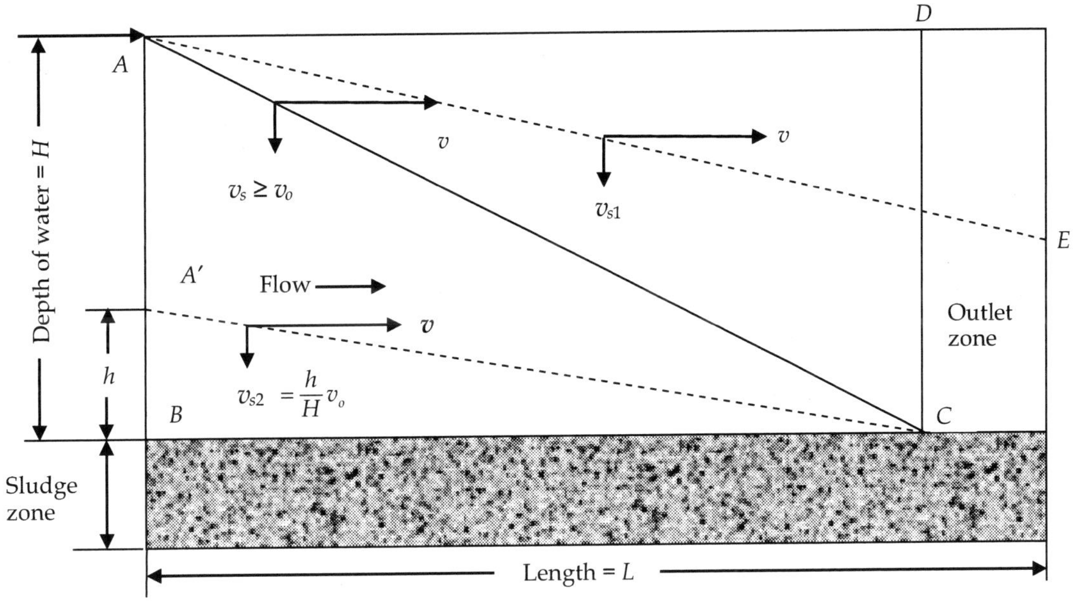

Fig. 4.21: Particle trajectory with respect to flow velocity and settling velocity in horizontal tank

Therefore,

$$\frac{v}{v_s} = \frac{L}{H}$$

$$v_s = \frac{v \cdot H}{L}$$

But
$$v = \frac{Q}{B.H}$$

Therefore,
$$v_s = \frac{Q}{B.H} \cdot \frac{H}{L} = \frac{Q}{BL} \qquad (4.37)$$

It shows that particles with settling velocity equal to or greater than Q/BL will settle down and are removed. If the smaller particles entering the other height h, if they have settling velocities lower than Q/BL will also settle down. All those particles having their settling velocities $\geq \frac{h}{H}\frac{Q}{BL}$ will settle down. If 70% of particles of a given size are proposed to be removed in a settling tank then settling velocity of that sized particles must be kept $\geq \frac{700}{100}\frac{Q}{BL}$. This approach assumes that the settlement of a particle at the bottom of the tank does not depend on the tank's depth but depends on the tank surface area.

In sedimentation tank the particles settle with a settling velocity (v_s) and the water in the tank rises upwards with a upward flow velocity denoted as v_o. For efficient and effective removal of all the particles of specified size and shape so as not to escape with clarified water the settling velocity of these particles should always be greater than rising velocity of the water. This indicates that the particles of specified size and diameter will be removed 100% when $v_s > v_o$ and will be 0% when $v_s < v_o$. This upflow velocity also known as *overflow velocity, overflow rate, surface overflow rate* (SOR) or *surface loading rate* represented as v_o is the basis of tank design and is a governing factor in overall removal of particles in a sedimentation tank. However, when the liquid is rising from the bottom to the top it is called *overflow rate* and when water is flowing over the top on the properly designed weirs of the tank it is called surface loading rate.

The overflow rate of the tank can be calculated by dividing the flow rate in the tank with the cross-sectional surface area of the tank. This may also be understood as the amount of water entering the tank per unit of surface area of it. Therefore, the overflow rate (v_o) value is kept low while designing a sedimentation tank and therefore should be kept about 80% of the settling velocity (v_s) of the particle. Therefore, the surface flow rate can be calculated using Eq. 4.38-Eq. 4.42;

$$\text{Over flow rate } (v_o) = \frac{Q}{A_s} \tag{4.38}$$

$$\text{However, for a rectangular sedimentation tank, } v_o = \frac{Q}{B \times L} \tag{4.39}$$

$$\text{Also, for a circular sedimentation tank, } v_o = \frac{Q}{\frac{\pi}{4}D^2} = \frac{4Q}{\pi D^2} \tag{4.40}$$

v_o = Surface overflow rate or overflow velocity ($m^3/d/m^2$ plan area or m/d)

$$\text{It can also be further modified as } v_o = \frac{\frac{V}{t}}{A_s} = \frac{\frac{A_s \times H}{t}}{A_s} \tag{4.41}$$

$$\text{Or simply the term } v_o = \frac{H}{t} = \text{velocity}\left(\frac{\text{m}}{\text{s}}\right) \tag{4.42}$$

Therefore, the surface over flow rate of a sedimentation tank should be less than the settling velocity of the particles in water and is independent of the depth of the tank. However, the surface area can be increased to make the sedimentation process more efficient and effective. Under normal conditions the plain sedimentation tanks in type-I settling are designed with overflow rate or SOR values in the range of 12-18 $m^3/d/m^2$ of plan area. However, the sedimentation aided with coagulation under type-II settling tanks are designed on the basis of overflow rate or SOR value of 20-30 $m^3/d/m^2$ of plan area. The weir loading rate of 300 m^3/day/m of weir length is quite satisfactory for controlling the overflow rate of the tank.

The accumulated sludge (or settled material) at the bottom of the tank is normally mechanically removed by a device controlled with electricity. It consists of two endless chains moving on sprocket wheels and supporting wooden cross-bars or flights which periodically sweep the sludge towards the hopper given near the inlet end, from which it is removed under hydrostatic head or pumped into separate sludge digestion tanks. Through means of a second blade called skimmer, the scum produced at the surface is extracted by the mechanical scraper, via a scum trough. For circular tanks the diameter of the tank should not be greater than 60 m.

While decreasing the overflow rate values even the particles with low settling velocities can also settle down easily. Therefore the depth of the sedimentation tank is not varied much and remains almost similar with respect to flow conditions. The depth has no significant effect on the efficiency of removal of particles in a tank. The depth of the sedimentation tank excluding the depth of sludge and silt deposition is generally kept as 3.5-4.5 m is considered to be fairly good enough while the depth value should not be less than 1.8 m and should not be greater than 6m in any case. However for providing the mechanical sludge removal mechanism a minimum depth of 0.8 to 1.2 m is generally considered additionally. Under normal practical consideration the additional sludge depth is considered to be 0.30 m and a free board of 0.30 m is also provided at the top of the tank.

Example 4.9: Determine the surface area of a settling tank for 0.5m^3/sec using design overflow rate as 32.5 m^3/m^2/day. Find the depth of the tank, if detention time is 95 min. Assume L/B ratio of 3:1 and length of tank should not exceed 100 m.

Solution: $Q = 0.5\text{m}^3/\text{sec}$

$\text{SOR} = v_o = 32.5\ \text{m}^3/\text{m}^2/\text{day}$

Detention time, $t = 95$ min

$$\text{Plan area} = A_s = \frac{Q}{SOR} = \frac{0.5 \times 86{,}400}{32.5} = 1329\ \text{m}^2$$

We know that horizontal velocity

$V = Q \times t = 0.5 \times 95 \times 60 = 2850\ \text{m}^3$

$$\text{Depth } (H) = \frac{V}{SOR} = \frac{2850}{1329} = 2.14 \text{ m}$$

Now assume width of tank $B = x$ m and therefore length $L = 3x$

The area of the tank = $A = x.3x = 1329$ m^2

Therefore, $3x^2 = 1329$ m^2

$$x = \sqrt{\frac{1329}{3}} = 21 \text{ m} = B$$

Therefore, $L = 3x = 3 \times 21 = 63$ m

Gross depth = 2.14 m + 0.5 m (sludge accumulation) + 0.3 m (free board top) = 2.94 m = 3.0 m

The tank may be designed with tank dimensions of 63 m × 21 m × 3 m.

$$\text{Weir loading rate} = \frac{Q}{2(L+B)} = \frac{10{,}000}{2(63+21)} = \frac{43{,}200}{168} = 257 \text{ m}^3/\text{d/m length of weir}$$

Example 4.10: Design a circular sedimentation tank to treat 6.5 million litres of raw water per day. The detention period may be assumed to be 4 hours (assume any other data suitably).

Solution: Given that

$$Q = 6.5 \times 10^6 \frac{L}{d} = 6500 \text{ m}^3/\text{d}$$

Detention time $t = 4$ hours

Assume depth of the tank = 3.0 m

We know that

Volume of the circular tank $V = d^2\,(0.011\,d + 0.785\,H)$

But $V = Q \times t$

Therefore, $V = 6500 \times \dfrac{4}{24} = 1084$ m^3

Now putting V value of 1084 m^3 and depth value as H to be 3.0 in above equation, we get

$1084 = d^2(0.011\,d + 0.785 \times 3.0) = 0.011\,d^3 + 2.35\,d^2$

$0.011\,d^3 + 2.35\,d^2 = 1084$

Solving by hit and trial method, we get

The diameter of the tank can be adapted, $d = 20.5$ m

Now the SOR of the tank can be checked by using equation as:

$$\text{Surface overflow rate (SOR)} = \frac{Q}{A_s} = \frac{Q}{\frac{\pi}{4}d^2} = \frac{6500}{\frac{\pi}{4}(20.5)^2}$$

$$= \frac{6500}{330} = 19.7\frac{m}{d} \text{ ((should be 12-18 m}^3\text{/d/m}^2\text{)}$$

Therefore increase the diameter, $d = 21$

$$\text{Now, SOR} = \frac{Q}{A_s} = \frac{Q}{\frac{\pi}{4}d^2} = \frac{6500}{\frac{\pi}{4}(22)^2}$$

$$= \frac{6500}{380} = 17.1\frac{m}{d} \text{ (between 12-18 m}^3\text{/d/m}^2\text{, hence o.k.)}$$

Now weir loading can be checked by using the equation as:

Weir loading rate $= \frac{Q}{\pi.22} = \frac{6500}{\pi \times 22} = \frac{6500}{69}$ = 94 m^3/d/m length of weir (< 300 hence, o.k.)

Therefore the tank should be designed for diameter of 22 m and height of 3.0 m.

Example 4.11: Design a rectangular sedimentation tank to treat 10 million litres of raw water per day. The detention period may be assumed to be 2 hours.

Solution: Raw water flow rate, $Q = 10 \times 10^6$ per day = 10,000 m^3/d

Detention time, t = 2 hours

Volume of the tank, $V = Q \times t = \frac{10{,}000 \times 2}{24} = 834 \text{ m}^3$

Assume depth of tank, H = 2.5 m

Therefore, surface area of the tank, $A = \frac{V}{H} = \frac{834}{2.5} = 334 \text{ m}^2$

Assume L:B ratio of 2:1

Now assume $B = x$ and therefore $L = 2x$

The area of the tank $= A = x.2x = 334 \text{ m}^2$

Therefore, $2x^2 = 334 \text{ m}^2$

$$x = \sqrt{\frac{334}{2}} = 13 \text{ m} = B$$

Therefore, $L = 2x = 2 \times 13 = 26$ m

Gross depth = 2.5 m + 0.5 m (sludge accumulation) + 0.3 m (free board top) = 3.30 m

The tank may be designed with tank dimensions of 26 m × 13 m × 3.3 m.

$$\text{Surface over flow rate (SOR)} = \frac{10{,}000}{26 \times 13} = 30 \text{ m}^3\text{/m}^2\text{/d} \quad < 40 \text{ m}^3\text{/m}^2\text{/d}$$

$$\text{Weir loading rate} = \frac{Q}{2(L+B)} = \frac{10{,}000}{2(26+13)} = \frac{10{,}000}{78} = 128 \text{ m}^3\text{/d/m length of weir}$$

Example 4.12: Calculate the size of a rectangular tank to treat 2.0 million litres of water per day. The detention period may be assumed as 3 hours and overflow rate less than 40,000 L. per m^2 of the surface area per day. (ESE Mains, Civil Engg. 2012)

Solution: Supply of water to be treated per day = 2×10^6 litres

Supply of water to be treated during the detention period of 3 hours i.e., the capacity of the tank;

$$= \frac{2 \times 10^6 \times 3}{24} = 0.25 \times 10^6 \text{ litres} = 250 \text{ m}^3$$

Let us assume the velocity of flow to be maintained through the tank = 20 cm/minute = 0.2 m/minute

The length of tank required = Velocity of flow × detention period = $0.2 \times 3 \times 60 = 36$ m

Cross-sectional area of the tank required

$$= \frac{\text{Capacity of tank}}{\text{Length of tank}} = \frac{250}{36} = 6.94 \text{ m}^3$$

Assuming an overflow rate of 36000 L/d/m^2

$$\therefore \frac{Q}{BL} = 36{,}000 \text{ L/d/m}^2$$

$$\frac{Q}{BL} = 1500 \text{ L/hr/m}^2$$

$$\text{But, } Q = \frac{2 \times 10^6}{24} \text{ L/hr} = 0.0833 \times 10^6 \text{ L/hr}$$

$$\therefore BL = \frac{Q}{1500} = \frac{0.833 \times 10^6}{1500} = 55.56 \text{ m}^2$$

$$\text{or, } B = \frac{55.56}{36} = 1.543 \text{ m}$$

$$\text{The depth} = \frac{250}{36 \times 1.543} = 4.5 \text{ m}$$

Using a free board of 0.5 m, the overall depth = 0.5 + 4.5 = 5 m

Hence , a rectangular sedimentation tank with an overall size of 36 m × 1.543 × 5 m can be used.

Type-I settling

This type of settling in which the particles during settling do not change their shape and size is known as type-I settling or discrete settling. The Stokes' law overall is applicable in water treatment for removal of definite size of particles and for proper sphericity of such particles a correction factor has to be accounted and needs to be determined. No such direct method is therefore available for determining the characteristics of settling particles in water. However, an indirect method also known as settling column analysis method can be used for measuring settling velocities of discrete particles in dilute suspensions. It also determines the settling characteristics of a suspension to be analysed and used. The prepared suspension to be tested should be completely mixed type in order to ensure the uniform distribution of the particles when placed in the column. The suspension to be tested is then allowed to settle without any disturbance.

Suppose that at time equal zero a particle just arrives at the surface and its settling velocity is such that it arrives at the lowest sampling port at a later time, say $t = t_o$ with its average settling velocity as shown in Eq. 4.43;

$$v_o = \frac{\text{Distance travelled}}{\text{Time taken by particle}} \tag{4.43}$$

where, H_o = distance travelled by the particle (m)

t_o = time travelled by the particle to reach the bottom of tank

Let us assume another particle initially suspended at a distance H_p above the sampling port with its terminal velocity lesser than v_o value but reaches at the port at same time as previous particle do.

The settling velocity of this particle therefore is calculated as per Eq. 4.44.

$$v_p = \frac{\text{Distance travelled}}{\text{Time taken by particle}} = \frac{H_p}{t_o} \tag{4.44}$$

Therefore,

$$t_o = \frac{H_o}{v_o} = \frac{H_p}{t_o}$$

or

$$\frac{v_p}{v_o} = \frac{H_p}{H_o}$$

However based on the above equations it can be understood that:

- It is presumed that all particles having settling velocities equal or exceed v_o with diameters equal to or greater than d_o, will arrive at or pass the sampling port in time t_o.
- However, particles having terminal settling velocity $v_p < v_o$ with diameter $d_p < d_o$, will pass the sampling port in time, t_o with the presumption that its original position was at, or below, a point H_p.
- It is further assumed that all the particles of all sizes are randomly distributed from top to bottom of the column and the suspension is homogeneous, then the fraction of these particles of size, d_p, with settling velocity (v_p) will arrive at the port at time (t_o) equal to $H_p/H_o = v_p/v_o$.
- Therefore, the removal efficiency of particle of any size in such a suspension is the ratio of the settling velocity of that particle to the settling velocity (v_o) represented by H_o/t_o.

The principle can be used to determine the settleability of particles by using a column of about 2 m filled with the suspension to be tested. The suspension type should be a completely mixed homogenously in order to ensure an initially uniform distribution of particles. It has already been discussed that the depth of the water column is not a factor in the analysis therefore under practical considerations a depth of about 2 m is suitable. Let us consider a completely mixed suspension containing initial suspended solids concentration as C_0 being determined before the start up of test. After time t_1 a sample drawn from the port and suspended solids concentration is determined as C_1. All such particles comprising of C_1 will have settling velocities less than H_o/t_1. Therefore, the mass fraction comprising of particles with $v < H_o/t_1$ can be calculated as shown in Eq. 4.45;

$$x = \frac{C_1}{C_o} \tag{4.45}$$

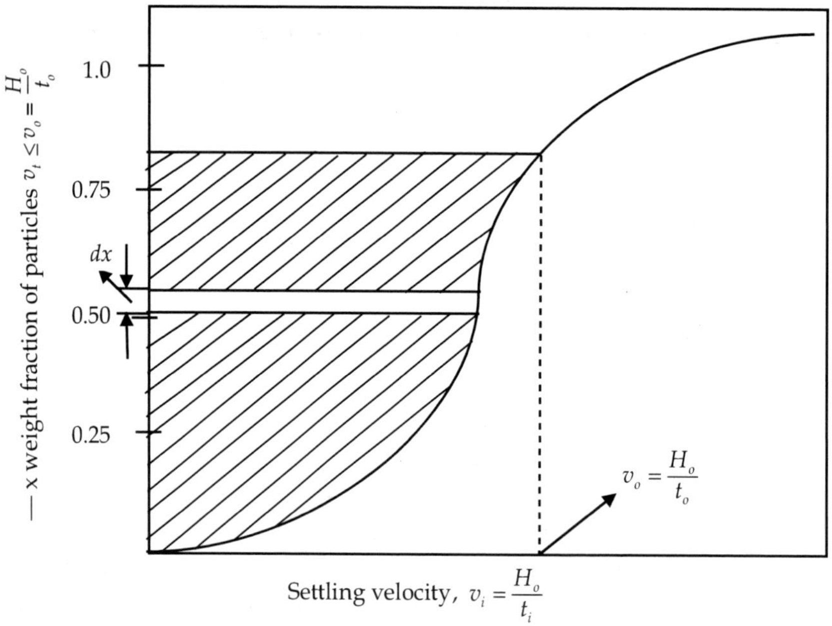

Fig. 4.22: Mass fraction of particles removed as function of settling velocity

During the test period a number of samples being drawn off with respect to time and x_i being a mass fraction of particles with settling velocities $v_i < H_o/t_i$. These vi and xi values are plotted on graph as shown in Fig. 4.22 and the fraction of particles corresponding to surface over flow rate or to any settling velocity can be obtained. Therefore, at any detention time t_0 the overall particle removal efficiency can be obtained. The particles having settling velocities greater than $v_o < H_o/t_o$ can be removed 100% effectively. Therefore, the mass fraction of particles as $1-x_0$ will be removed completely by time t_o. However, the remaining fraction of particle will be removed by a ratio v_i/v_o as shown through shaded area in Fig. 4.22. Thus knowing the x and v values the area can be found by integration by using Eq. 4.46.

$$X = 1 - x_0 + \int_0^{x_0} \frac{v_i}{v_o} \qquad (4.46)$$

where, X is the total mass fraction removed by sedimentation in the column.

The overflow rate or overflow velocity (v_o) is the controlling factor for removal of different types of particles in the settling tank. A general overflow rate value of 12-18 $m^3/m^2.d$ is generally adapted in type-I settling for design of sedimentation tank. However in case of type-II settling when sedimentation is aided with coagulation process the overflow rate value of 24-30 $m^3/m^2/d$ is generally adapted for the design of sedimentation tank. The lower overflow rate values ensures the settling of particles having low flow velocities. However, at a given Q value the plan area of the tank ($L \times W$) can be increased that will ultimately reduce the overflow rate of the tank thus enhancing the efficient removal of even particles having low settling velocities at a given detention time. Therefore, the depth has no effect on settling properties of particles during sedimentation process. Therefore the depth of sedimentation tank may be taken in the range of 1.8 m as minimum and 6 m as maximum with usual values ranging from 3 to 4.5 m.

Example 4.13: During settling column analysis test a sample was being run for type-I suspension with a 2 m column height and analysis data is shown as below in Table 4.5:

Table 4.5: Concentration of sample (mg/L)

Time (min)	0	30	60	90	120	150	200	250	320	400
Concentration (mg/L)	350	210	190	175	167	120	85	52	38	25

Calculate the theoretical removal efficiency of the settling tank with a over flow rate of 18 $m^3/m^2.d$ or 18 m/d?

Solution: Calculate the mass fraction remaining with respect to suspended solids initial concentration of 350 mg/L and also determine the settling velocity with respect to each mass fraction obtained. (Table 4.6). The time t and depth of column is used to determine the settling velocity of the particles for every sample taken.

Table 4.6: Calculation of mass fraction remaining

Time (min)	30	60	90	120	150	200	250	320	400
Mass fraction remaining	0.6	0.54	0.50	0.47	0.34	0.24	0.15	0.11	0.07
$v_i \times 10^{-2}$ (m/min)	6.6	3.3	2.2	1.6	1.3	1.0	0.8	0.6	0.5

$$\text{Overflow rate or overflow velocity, } v_o = \frac{18}{24 \times 60} = 1.25 \times 10^{-2} \text{ m/min}$$

Now determine the x_o and $\Delta x.\ v_i$ values from graph as shown in Table 4.7.

Table 4.7: Calculation of x_o and Δx

S. No. Column (1)	Difference in mass fraction strips (Δx) Column (2)	Settling velocity at intercept drawn from midpoint of Δx values as v_i (m/s) Column (3)	$\Delta x \cdot v_i$ (column 2 × column 3)
1	0.025	0.55×10^{-2}	0.000138
2	0.05	0.75×10^{-2}	0.000375
3	0.05	0.80×10^{-2}	0.0004
4	0.05	0.90×10^{-2}	0.00045
5	0.05	1.10×10^{-2}	0.00055
6	0.025	1.125×10^{-2}	0.000313
		$\Sigma\ \Delta x\ .v_i$	0.002225

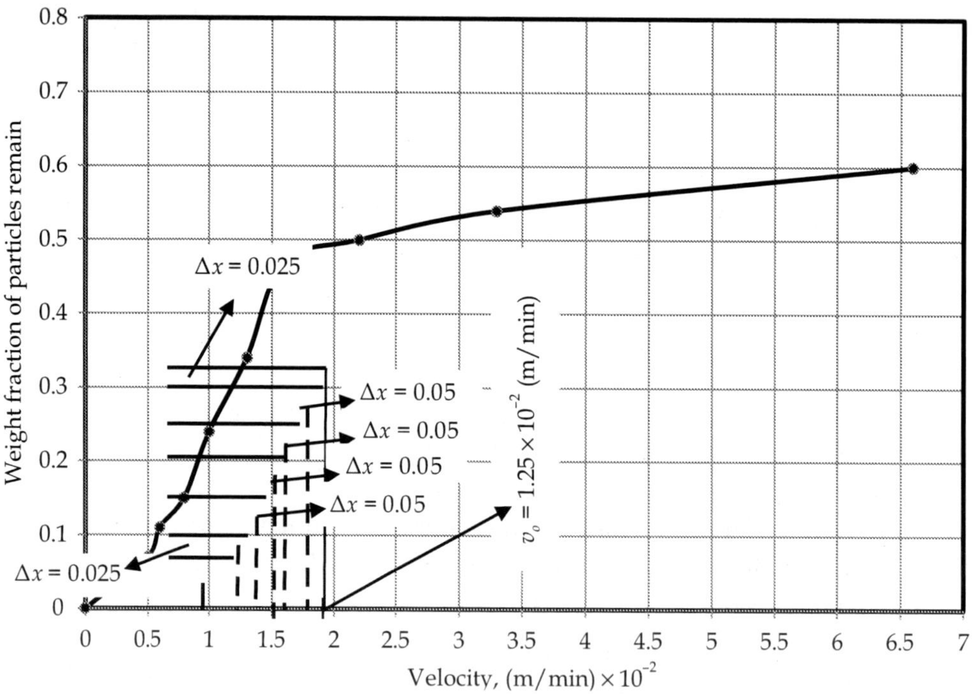

Fig. 4.23: Settling velocity curve for discrete particles in a settling column analysis

- The value as calculated above is $v_0 = \dfrac{18}{24 \times 6} = 1.25 \times 10^{-2}$ m/min (Fig. 4.23)
- The x_o value at above calculated v_o value is = 32.5%
- The summation of $\Delta x\ .v_i$ is computed to be = 0.002225

Therefore, the overall removal efficiency can be calculated as :

$$X = 1 - x_0 + \int_0^{x_0} \frac{v_i}{v_o}\, dx$$

or $$X = 1 - x_0 + \sum \frac{\Delta x \cdot v_i}{v_o} = 1 - 0.325 + \frac{0.002225}{1.25 \times 10^{-2}} = 0.675 + 0.178$$

$$= 0.853 = 85.3\%$$ **Ans.**

Inlet and outlet arrangements

The inlet is a system for spreading the water within the tank, and the outlet is a system for collecting outgoing water. Such systems should be carefully built and positioned in such a way that they do not block the flowing water or cause any damage to it. A sedimentation tank is a device that fills and retains water for some time to remove the suspended particles present in the water. These particles can settle down at the bottom and are removed using scrapers. If the particles suspended have a higher specific gravity than water, they settle at the top of the tank

Inlet and outlet arrangement

Inlet devices: Inlets are designed to disperse water at uniform velocities and equally. A baffle should be installed across the basin near the inlet and project several feet below the surface of the water to dissipate the inlet velocities and provide uniform flow as shown in Fig. 4.24.

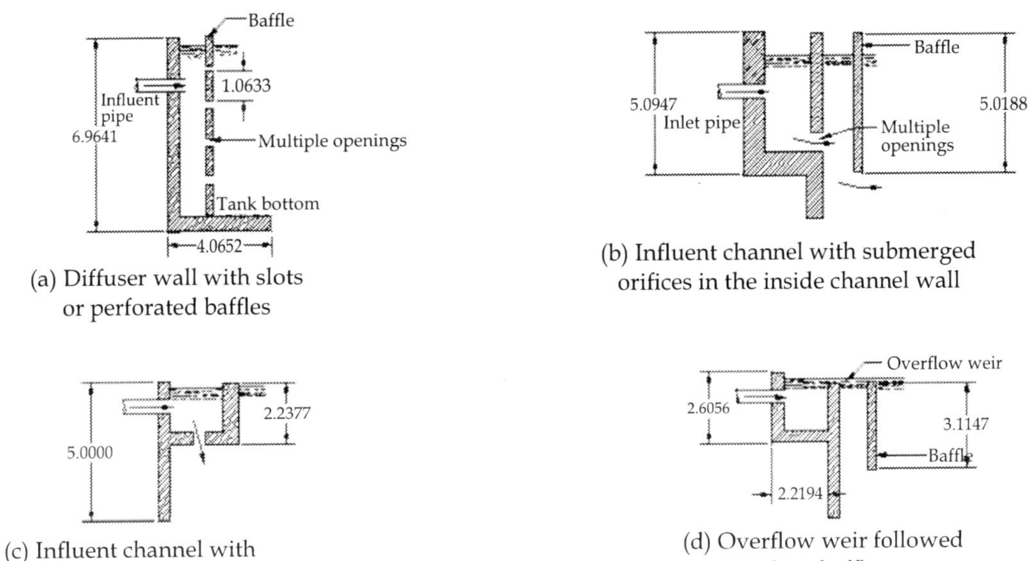

Fig. 4.24: Inlets for settling basins

Outlet devices: Outlet weirs or submerged orifices are designed to maintain speeds suitable for basin settlement and to minimize short circuiting. Weirs shall be adjustable in length, and at least equivalent to the tank perimeter. Peripheral weirs are, however, not appropriate because they appear to cause excessive short circuiting as shown in Fig. 4.25.

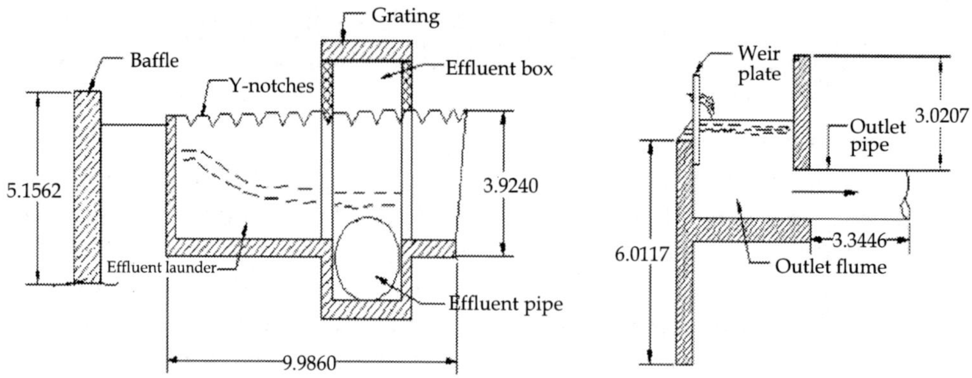

(a) Outlet consisting of V-notches, effluent box, effluent launder

(b) Outlet with rectangular weir

Fig. 4.25: Outlets for settling basins

Weir overflow rates

Big weir overflow levels lead to excessive outlet velocities. Such velocities reach backwards into the settling environment, contributing to the drainage of particles and flocks into the outlet. Usually, up to 300 m^3/d /m of weir loadings are used. Special inboard weir designs can need to be provided, as shown to minimize weir overflow levels.

Design details

- The detention period of 3 to 4 h for plain sedimentation tank and 2 to 2.5 h for when sedimentation aided with coagulation.
- The horizontal flow velocity should not be more than 30 cm/min (horizontal flow).
- The rectangular tank of size: L: B = 3-5:1 with maximum length of the tank should not exceed 100 m and generally the length of 30 m is quite common.
- However in case of circular tank the diameter should not be greater than 60 m and the circular tank with diameter range in between 20 and 40 m is quite common with depth of the tank in the range of 2.5 to 5.0 m.
- The plane sedimentation tank should be designed for a surface overflow rate of 12 to 18 m^3/d/ $m^2$of plan area and 24 to 30 m^3/d/$m^2$for sedimentation aided with flocculation.
- The slope of 1 per cent is generally kept towards the bottom of rectangular tanks and 8% towards the bottom of circular tank.

Sedimentation aided with coagulation

The various conventional water treatment processes as discussed earlier can treat the water up to certain extent but still the water contains some impurities and minerals that may harm the people during general water supplies. The shape and size of flocculent particles can be changed by adding suitable chemicals known as coagulants. These particles when come in close proximity of each other forms a heavier floc that settles easily under the action of gravity and is quite effective in settlement of fine colloidal particles. The process is known as sedimentation added with coagulation.

Particles of colloidal size possess certain properties which prevents them from agglomeration. This may be due to the large surface-to-volume ratio of the particles and because of the accumulation of electrical charges on the particle surface. However, flocculation is a slow mixing technique which promotes agglomeration of stabilized particles and coagulation is a chemical technique which is directed towards destabilization of charged colloidal particles. Therefore, the entire process of addition of chemicals (coagulants) and mixing (flocculation) is usually referred to as coagulation. It can be efficiently used for the water having turbidity more than 30-50 mg/L and is universally adopted at all the water treatment plants.

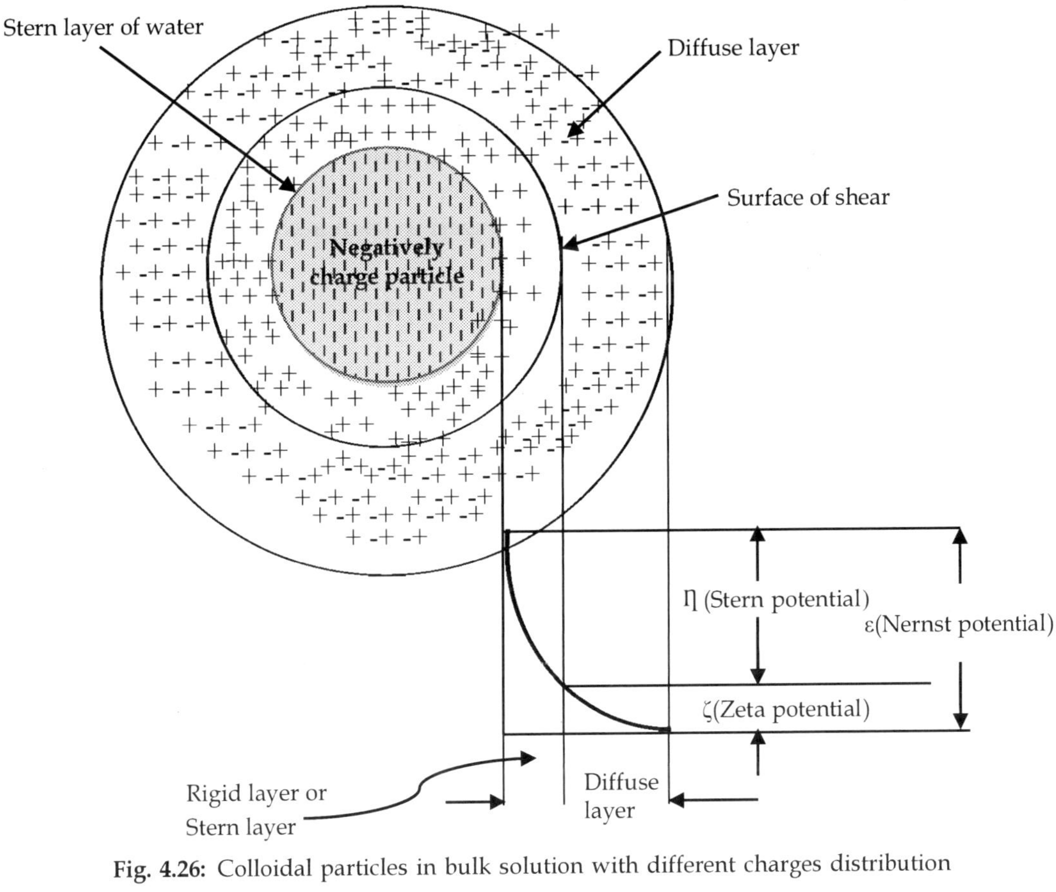

Fig. 4.26: Colloidal particles in bulk solution with different charges distribution

The ions of opposite charge being drawn from bulk solution is present in the bound layer also known as stern layer. It produces a rapid drop in potential known as stern potential denoted by η while further drop in stern potential is assessed and is known as zeta potential denoted by ξ. This zeta potential occurs between the shear surface of bound water layer and the point of electro neutrality in the solution as shown in Fig 4.26.

General properties of colloids

The particles that do not agglomerate naturally are called stable particles. This stability is due to the excessive large surface-to-volume ratio of the particles thus comprising very small size. Colloidal particles are charged electrically and so small that their surface area is very large in relation to the mass. However, most of the colloidal surfaces of the particles are negatively charged. They possess tyndall effect as these particles have dimensions and are heat-reversible such as organics in water. These colloids have a high surface area and therefore have plenty of active surfaces to adsorb. The colloid stability is due mainly to the preferential ion adsorption.

Electrical double layer

Although individual hydrophobic colloids have an electric charge, however there is no net electrical charge in a colloidal dispersion. While sufficient amount of counter ions are required to balance the electric charge in an atom which is produced by this diffuse layer in a colloidal dispersion. The charge distribution of a negatively charged colloid in the diffuse layer can be represented in Fig. 4.27 using the curve ABCD. The ions involved in this electro-neutrality are organized in such a way that they form a double layer electrical field.

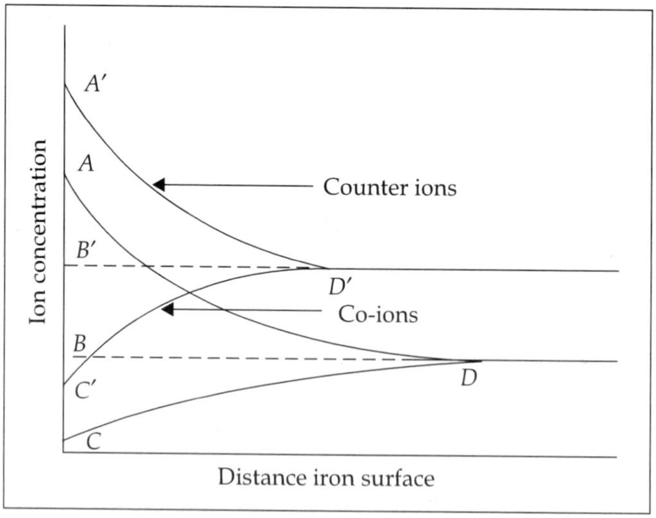

Fig. 4.27: Electric double layer

When two particles come in contact with each other the two main forces acting on them are electrostatic potential force also known as halo where the counter ions are surrounding the colloids and therefore unable to contact with each other. Another important force is van der Waals forces that is generally a positive force and favours the contact of particles. It is very strong at close distance of two particles and becomes weak rapidly if the distance between particles decreases. This is because this force is inversely proportional to the sixth power of distance between two particles. The sum of the resultant of these two forces can be shown through Fig. 4.26. Therefore, at greater distances it is weak however becomes attractive after passing the energy barrier at some distance between the colloids. As the force becomes attractive the agglomeration of particles takes place easily. This net repulsion force which acts as a barrier of energy must be resolved before aggregation of particles takes place. This magnitude of the energy barrier generally depends on (i) the concentration of particles in colloidal system, and (ii) the ionic water composition as shown in Fig. 4.26. Therefore, this energy barrier may be eliminated by applying a molecular bombardment of ions that may produce enough momentum for the particles to collide and contact. However, mechanical agitation of water also produces enough momentum in large size particles to cross the energy barrier and to agglomerate with each other. In water treatment systems by a slow mixing process such a process can be accomplished and thus proves to be efficient. In water purification processes this can be achieved by adding the chemical coagulants possessing high positive charge ions that agglomerates the particles resulting in floc formation that are heavier than water and therefore settle by action of gravity.

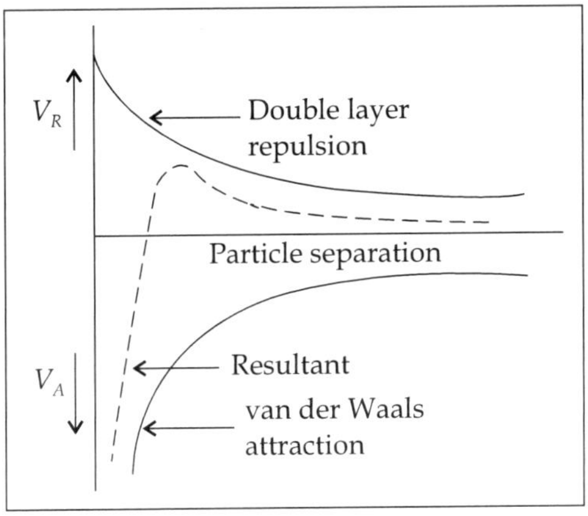

Fig. 4.28: Double layer repulsion

This double layer electrical theory can be well explained by using electric potential and charge on the particle or colloid with respect to Nerst and Zeta potential as shown in Eq. 4.47;

$$Z_p = \frac{4\pi\delta q}{D} \tag{4.47}$$

where,

Z_p = Zeta potential (mV)

q = Charge on particle (coulomb)

δ = Thickness of layer around the particle (zone of influence)

D = Dielectric constant of the media

For more zeta potential value we have to add more positive ions and therefore lesser chances of settling of colloids. However, with respect to stability factors it is stated that due to electrostatic force of repulsion (E) is directly proportional to the square of zeta potential and the distance between the particles as shown in Eq. 4.48;

Therefore,

$$E \propto Z_p^2, e^{-K \cdot H} \tag{4.48}$$

where,

E = Energy or dielectric force of repulsion

K = A constant

H = Distance of the particle

Z_p = Zeta potential (mV)

Coagulation process

The coagulation process can be carried out by using trivalent metallic salts such as alum or aluminium sulphate as $Al_2(SO_4)_3.18\ H_2O$ or ferric chloride as $FeCl_3$. These chemicals possess high positive charge ions and the coagulation is supposed to be carried out through four the mechanisms as (i) ionic layer compression, (ii) adsorption and charge neutralizations, (iii) sweep coagulation, and (iv) inter-particle bridging.

(i) Ionic layer compression: As discussed earlier the presence of high ionic concentration in a solution compresses the layers that are composed of predominantly of counter ions on the surface of colloid. In this case no energy barrier remains as the net force becomes attractive due to the predominance of van der Waals forces across the entire area because of the sufficiently compressed layer. The mechanism can be well understood by a natural example, when a turbid stream meets the ocean and delta formation takes place due to subsequent settlement of the silt particles. The coagulating power can be well explained by Hardy Schulz rule which states that the coagulation power of trivalent ions is more than 30-50 times than the divalent ions and is around 1500-2500 times greater than mono-valent ions.

(ii) Adsorption and charge neutralization: Some chemical species are able to adsorb colloidal particles on the surface. If the adsorbed species bear a charge compared to that of the colloids, such adsorption causes the colloidal particle to decrease its surface potential and results in destabilization. Reduction of surface load by adsorption is a very different process than double layer compression. The ionization of aluminium sulphateas $Al_2(SO_4)_3.18\ H_2O$ results in Al^{3+} cations and SO_4^- anions. The produced Al^{3+}cations immediately reacts with water forming an aquametallic ions in the form of $Al(OH)_3$ thus leaving H^+ ions.

These aquametallic ions have greater affinity for surfaces and becomes a part of ionic cloud surrounding the colloid. These ions when adsorbed on the surface of colloid thus result in surface charge neutralization. The electrostatic potential of the colloids disappears due to this charge neutralization, the ionic cloud disperses away and particles are free to contact each other. However, if the concentration of coagulants is increased in the system it will again result in restabilization, therefore an optimum dose of coagulants needs to be assessed before addition of such chemicals or coagulants in water. Coagulation can occur at a lower chemical dose at high concentrations of colloids as the colloids act as nuclei to promote precipitation formation.

Sweep coagulation

This mechanism is an integral part of the coagulation process because when aluminium sulphate reacts with water the aquametallic ion produced in the form of $Al(OH)_3$ is aluminium hydroxide. This $Al(OH)_3$ is a amorphous and gelatinous precipitate that attracts the particles and thus forms a floc. This floc formed is heavier than water and therefore settles down by the action of gravity. During the settling of this floc the other colloids may also get entrapped in it. Therefore the process of eliminating the colloids in such a manner is known as sweep coagulation.

Interparticle bridging

In carrying out the coagulation process in place of using chemicals and other salts as coagulants, the use of synthetic polymers is popular nowadays. These polymers possess the highly reactive surface and are available as linear or in branched structures. During the adsorption process a number of colloids are attached to a single polymers and likewise several such polymer groups come in proximity with each other thus agglomerates form a settleable mass. The charges on these polymers also supports and enhances the coagulation process. However, the chemicals used as coagulants contain positive charges but these polymers may have positive or negative charge on them or may be neutral. Therefore, such polymers may be used prudently for carrying out the coagulation process effectively. Active coagulating agents are considered to be several different natural compounds such as starch, cellulose, polysaccharide gums and protein materials as well as a broad range of synthetic polymeric compounds. Research has shown that both positive and negative polymers can destabilize negatively charged colloidal particles. However, optimum conditions of coagulation do not equate to a minimum zeta potential but depends on pH for that coagulant, depending on the solubility-pH relationship.

Flocculation

Flocculation is mechanical stimulation to agglomerate destabilized particles into small, fast-settling (or flocs) particles. Flocculation or gentle agitation results from differences in the velocity or gradients of the coagulated water which causes the fine moving, destabilized particles to come into contact and become large, readily settleable flocs.

Providing an initial quick (or) flash mix for the dispersal of coagulant or other chemicals into the water is a common practice. Then, slow mixing occurs, during which the floc develops. Rapid or flash mixing is the mechanism by which a coagulant is spread rapidly and evenly through the waterbody. This cycle typically takes place in a small basin immediately preceding the coagulation basin or at the head of it. The detention time of flash mixing is usually 30 to 60 seconds, and the head loss is 20 to 60 cm of water. Here colloids are destabilized, forming the nucleus for the floc. However, slow mixing provides a better contact between the particles that are finely split destabilization during rapid mixing.

Coagulation aided sedimentation

In order to carry out the coagulation process effectively and efficiently certain chemicals such as salts of aluminium sulphate, ferrous sulphate, copperas and sodium aluminate are added to the raw water. The effectiveness of these chemicals is generally enhanced if the water is slightly alkaline. Therefore, for efficient coagulation the water should be slightly alkaline or if it is not the alkalinity in the form of lime and soda ash should be added to the raw water.

This may due to the fact that the coagulation of metallic salts releases hydrogen ions along with coagulant species. These hydrogen ions neutralize the alkalinity such that the addition of 1 mg/L of alum can neutralize 0.5 mg/L of alkalinity. The low alkalinity in raw water reduces its buffering capacity during coagulation process and therefore pH drops sharply. Therefore, the optimum pH value must be maintained for best coagulation and some alkalinity must be present for the floc formation. However with respect to alkalinity in raw water and the coagulation process being carried out the water can be categorized in four different types of groups such as:

Group 1

This group comprises of high turbidity-low alkalinity in raw water and by using relatively small dose of coagulants the water of this type can be easily coagulated by the mechanism of adsorption and charge neutralization. The lowering in pH value makes the method more effective because the effectiveness of aquometallic ions increases at lower pH values. But the pH should not be lowered to excessive values so as to maintain the effectiveness of the system.

Group 2

This group comprising of high turbidity and high alkalinity in raw water. In such type of waters the pH will be relatively unaffected by coagulant addition as the mechanism of adsorption and charge neutralization will be less effective due to presence of high alkalinity in water. Such waters should be supplied with high coagulant dose and solids are removed by the mechanism of sweep coagulation.

Group 3

The water in this group consists of low turbidity with high alkalinity. In such type of water even if the particle charge has been neutralized by adding of coagulants, it becomes very difficult to remove small size colloids and therefore makes the coagulation process difficult. Therefore with moderate coagulant dose the colloids are removed through the mechanism of sweep coagulation. However, an addition of some turbidity may decrease the amount of coagulant dosage required.

Group 4

The water comprising of low turbidity along with low alkalinity belongs to this group. In this type of water the coagulation process becomes difficult due to low concentration of colloids and also low alkalinity concentration prevents the formation of $Al(OH)_3$. Therefore further addition of turbidity can convert this water into group 1 or addition of alkalinity in water can convert it to a group 3. Also addition of both turbidity and alkalinity in such water proves to be quite advantageous.

Coagulants for water treatment

(i) **Aluminium sulphate or alum:** Aluminium sulphate or alum with chemical formula $Al_2(SO_4)_3.18H_2O$ also known as filter alum is quite cheap and most commonly used universally accepted coagulant. It forms an excellent stable floc and quite clear treated water is achieved. It also helps in reducing taste and colour of raw water and addition of it does not require any skilled supervision for handling. However, the sludge produced using alum as a coagulant was troublesome and causing the main problem. This was because it becomes difficult to dewater the sludge and regenerate the alum thus ultimately causing the unsuitability for its disposal at low lying lands. But the recent study has been carried out on regenerating the alum from this sludge for its use in coagulation with a recovery cost of one-fourth of the cost of the recovered alum.

When alum is added to raw water it reacts with the bicarbonate alkalinity either present in the water or added to it thus resulting the formation of gelatinous precipitate or floc of aluminium hydroxide. The colloids and other fine particles attract this floc and therefore grows in size and become heavier than water resulting in settling down at the bottom of the tank.

The overall chemical reaction equation can be represented as Eq. 4.49;

$$\underset{\text{(Al hydroxide ppt.)}}{Al_2(SO_4)_3} \cdot 18H_2O + 3Ca(HCO_3)_2 \xrightarrow{\text{yields}} 3CaSO_4 + 2Al(OH)_3 \downarrow + 6CO_2 \uparrow + 18H_2O \tag{4.49}$$

The above equation indicates that addition of alum to raw water causes permanent hardness in it from the formation of calcium sulphate. Also, the formation of CO_2 gas can cause corrosiveness. It can also be seen from the equation that the alum reacts with bicarbonate alkalinity and this bicarbonate alkalinity exists in water at pH below 8.3 and above pH 4.5 or so. Therefore, the alum can work well in a very narrow pH range of 5 to 7.5 that shows good floc formations and sweep coagulation occurs efficiently.

The amount of alum required for coagulation generally depends on the turbidity and colour of the raw water. The proper optimum dose of coagulant can be calculated by using jar test procedure in the laboratory and the obtained optimum dose is implemented on real scale treatment plants. The alum dose lies in the range of 5 mg/L for relatively clear waters to 85 mg/L for highly turbid waters. However, the optimum average normal dose of alum to be added in raw water is about 17 mg/L. From the above equation the stoichiometry shows that for each mole of alum 3 moles of bicarbonate alkalinity are required to carry out the reaction. The molecular weight of alum is 600, and that of calcium bicarbonate is 300 and therefore it is obvious that for every 1 mg of alum 0.5 mg of alkalinity is required per litre of water to be treated.

However, at an optimum pH value of 5-7.5 around 5-10 mg/L of alkalinity must remain as residual after the reaction occurs. Therefore the raw water to be treated should contain at least 5-10 mg/L of alkalinity in it before adding the alum. When the raw water does not possess the sufficient alkalinity in the form of lime as $Ca(OH)_2$ or soda ash $Na_2CO_3$are generally supplemented and the reaction equation is as (Eq. 4.50-Eq. 4.51);

$$\underset{\text{(Al hydroxide ppt.)}}{Al_2(SO_4)_3} \cdot 18H_2O + 3Ca(OH)_2 \xrightarrow{\text{yields}} 3CaSO_4 + 2Al(OH)_3 \downarrow + 18H_2O \quad (4.50)$$

and

$$\underset{\text{(Al hydroxide ppt.)}}{Al_2(SO_4)_3} \cdot 18H_2O + 3Na_2CO_3 \xrightarrow{\text{yields}} 3Na_2SO_4 + 2Al(OH)_3 \downarrow + 3CO_2 \uparrow 15H_2O \quad (4.51)$$

However, addition of external alkaline salts such as lime and soda ash makes the process a little bit costlier and another drawback is that the effective pH range of 5 to 7.5 used in this process is very narrow.

Copperas as coagulant

Copperas is an iron salt with its chemical name is ferrous sulphate having chemical formula $FeSO_4.7H_2O$. The copperas salt is generally added to raw water in combination with lime. The lime may be added either to copperas or copperas may be added to lime. The stoichiometric of the reaction equation is shown in Eq. 4.52;

$$\underset{\text{copperas}}{FeSO_4 \cdot 7\,H_2O} + \underset{\text{lime}}{Ca(OH)_2} \longrightarrow \underset{\text{ferrous hydroxide}}{Fe(OH)_2} + CaSO_4 + 7\,H_2O \quad (4.52)$$

However, when copperas is added before the addition of lime the reaction equation is shown in Eq. 4.53 and Eq. 4.54:

$$FeSO_4 \cdot 7\,H_2O + \underset{\text{alkali}}{Ca(HCO_3)_2} \longrightarrow Fe(HCO_3)_2 + CaCO_4 + 7\,H_2O \quad (4.53)$$

and

$$\underset{\text{Lime ferrous hydroxide}}{Fe(HCO_3)_2 + 2\,Ca(OH)_2} \longrightarrow Fe(OH)_2 + 2CaCO_3 + 2H_2O \quad (4.54)$$

The produced ferrous hydroxide in any of the above processes further gets oxidized by air and water to form ferric hydroxide, the reaction equation is shown in Eq. 4.55 below.

$$4Fe(OH)_2 + O_2 + 2H_2O \longrightarrow \underset{\text{ferric hydroxide ppt.}}{4Fe(OH)_3\,(Floc)\downarrow} \quad (4.55)$$

This produced ferric hydroxide forms the flocs and enhances the process of sedimentation in the system. The copperas is a widely used coagulant for the treatment of water without colour and is economical as it is cheaper than alum. The coagulant works effectively in a limiting pH range of 8.5 and above. It does not provide satisfactory results when used for coloured waters. From the above stoichiometry it is clear that the quantity of copperas required is almost the same as that of alum.

Chlorinated copperas as coagulant

The addition of chlorine to a solution of copperas or ferrous sulphate, ends up in formation of ferric sulphate and ferric chloride as the final product of the reaction with water, the reaction equation can be represented as Eq. 4.56;

$$6(FeSO_4 \cdot 7\,H_2O) + 3Cl_2 \longrightarrow \underset{\text{Ferric chloride}}{2Fe\,Cl_3} + \underset{\text{Ferric sulphate}}{2Fe_2\,(SO_4)_3} + 42\,H_2O \quad (4.56)$$

This final end product obtained in combination of ferric chloride and ferrous sulphate is known as chlorinated copperas. It is quite effective in removal of colours from water having low pH values. However, both the end products of chlorinated copperas when used along with lime prove to be very effective in carrying out the coagulation process. The reaction equation with lime for both the end products is specified below in Eq. 4.57-Eq. 4.58;

$$\underset{\text{Ferric chloride lime}}{2FeCl_3 \quad + \quad 3Ca(OH)_2} \longrightarrow \underset{\text{ferric hydroxide ppt}}{2Fe(OH)_3\downarrow} + 3\,CaCl_2 \quad (4.57)$$

$$\underset{\text{Ferric sulphate lime}}{Fe_2\,(SO_4)_3 + 3Ca(OH)_2} \longrightarrow \underset{\text{ferric hydroxide ppt}}{2\,Fe(OH)_3\downarrow} + 3\,CaSO_4 \quad (4.58)$$

In the above reaction equation, the precipitates of ferric hydroxide produced thus helps in forming the flocs and therefore augment the process of sedimentation. The combined effect of these two end products is quite effective and works in wide range of pH variations. This may be apparent from the fact that ferric sulphate is quite effective in the pH range of 4 to 7 and above 9, whereas ferric chloride works well in the pH range of 3.5 to 6.5 and above 8.5.

Sodium aluminate as a coagulant

When water does not possess desired alkalinity the use of sodium aluminate salt in the form of $Na_2Al_2O_4$as a coagulant is quite effective. The chemical when dissolves in water it ultimately reacts with the salts of calcium and magnesium present in raw water. This results in the formation of calcium or magnesium aluminate as precipitates. The overall chemical reaction equation can be written as (Eq. 4.59-Eq. 4.61);

$$Na_2Al_2O_4 + Ca(HCO_3)_2 \longrightarrow \underset{\text{Calcium aluminate}}{CaAl_2O_4\downarrow} + Na_2CO_3 + CO_2\uparrow + H_2O \quad (4.59)$$

$$Na_2Al_2O_4 + CaCl_2 \longrightarrow CaAl_2O_4\downarrow + 2NaCl \quad (4.60)$$

$$Na_2Al_2O_4 + CaSO_4 \longrightarrow CaAl_2\,O_4\downarrow + Na_2SO_4 \quad (4.61)$$

This coagulant is not very common in treating the water used for general public supplies as it is 3/2 time costlier than alum. However, the coagulant is quite effective in treatment of water that does not hold sufficient alkalinity and cannot be treated by alum. Other important characteristics of using this salt, and is also evident from the equation that it reduces the temporary as well as the permanent hardness present in raw water. However, in case of alum the hardness increases due to formation of sulphates of calcium and sodium. Therefore, the coagulant is extensively used for boiler feed water where the hardness of water should be minimum.

However, sometimes synthetic polymers are also used as a coagulant in addition to alum. These polymers bind small flocs together to produce large flocs or mass that helps the settling process at a faster rate and thus makes the sedimentation process quite efficient in terms of colloid removal.

Merits and demerits of using alum and iron salts as coagulants

The use of alum and iron salts as coagulants for water treatment in general public supplies has its advantages and disadvantages are discussed below:

(i) The use of iron salts ends up with heavy floc formation and therefore high removal of colloids from the water can take place as compared to alum.

(ii) The use of iron salts not only removes suspended material but is effective in removing taste and odours from water due to its better oxidizing properties and thus removal of hydrogen sulphide from water.

(iii) The iron salts can work in wide pH range values as compared to alum.

(iv) The use of iron salts can cause staining in the system and also encourages the growth of iron bacteria in the water distribution system.

(v) The iron salts are more prone to enhance corrosiveness in water as compared to alum.

(vi) The iron salts due to corrosive nature therefore requires proper handling and more skilled supervision as compared to alum.

(vii) The iron salts are most commonly used in the treatment of sewage while alum is mostly sued for the treatment of raw water.

(viii) The following are the mostly used coagulants with normal dose and pH values required for best floc formation as shown in Table 4.8.

Table 4.8: Coagulant and dosage

S. No.	Coagulant	pH Range	Dosage (mg/L)
1.	Aluminium sulphate $Al_2(SO_4)_3 \cdot 18H_2O$	5.5 - 8.0	5 - 85
2.	Sodium aluminate, $Na_2Al_2O_4$	5.5 - 8.0	3.4 - 34
3.	Ferric chloride ($FeCl_3$)	5.5 - 11.0	8.5 - 51
4.	Ferric sulphate $Fe_2(SO_4)_3$	5.5 - 11.0	8.5 - 51
5.	Ferric sulphate $FeSO_4 7H_2O$	5.5 - 11.0	8.5 - 51

Example 4.14: Determine the quantity of alum required to treat 10 million litres of water per day in a treatment plant. From the jar test an optimum alum dose of 12 mg/L is required. Also assess the amount of carbon dioxide gas released per litre of water being treated.

Solution: Given that:

$Q = 10 \text{ MLD} = 10 \times 10^6 \text{ L/d}$

Alum dose required = 15 mg/L

Therefore, the total amount of alum required per day $= 10 \times 10^6 \times 15$ mg

$= 15 \times 10^7 \text{ mg} = 150 \text{ kg}$

From the chemical reaction equation it can be calculated as:

$$Al_2(SO_4)_3 \cdot 18H_2O + 3Ca(HCO_3)_2 \xrightarrow{\text{yields}} 3CaSO_4 + 2Al(OH)_3 \downarrow + 6CO_2 \uparrow 18H_2O$$

The molecular weight of alum = 2 (26.97) + 3 (32 + 64) + 18(18) = 666

Molecular weight of CO_2 = 12 + 32 = 44

Therefore the amount of CO_2 produced can be calculated on the basis as:

666 mg of alum will release $= 44 \times 6 = 264$ mg of CO_2

and 1 mg of alum will release $= \dfrac{264}{666}$ mg of CO_2

therefore, 15 mg of alum will release $= \dfrac{264}{666} \times 15 = 5.95$ mg of CO_2

Therefore, 5.95 mg of CO_2 will be released per litre of water treated with an alum dose of 15 mg/L of water.

Example 4.15: In a water treatment plant determine the quantity of copperas required to treat 15 million litres of water per day when copperas dose of 10 mg/L is required with lime.

Solution: Given that:

Amount of water treated $= 15 \times 10^6$ L/d

The quantity of copperas required = 10 mg/L

The total amount of copperas required to treat the 15 million liters of water per day $= 10 \times 15 \times 10^6 = 15 \times 10^7 \text{ mg} = 150 \text{ kg}$

The reaction equation when lime is added to water

$$\underset{\text{Quick lime}}{CaO} + H_2O \longrightarrow \underset{\text{Hydrated lime}}{Ca(OH)_2}$$

and,

$$\underset{\text{copperas}}{FeSO_4 \cdot 7H_2O} + \underset{\text{lime}}{Ca(OH)_2} \longrightarrow \underset{\text{ferrous hydroxide}}{Fe(OH)_2} + CaSO_4 + 7H_2O$$

The molecular weight of copperas = [56 + 32 + 4 × 16 + 7(18)] = 88 + 64 + 126 = 278

and molecular weight of quick lime (CaO) = 40 + 16 = 56

From the above stoichiometry it can be inferred that one mole of copperas requires one mole of lime,

278 mg of copperas requires 56 mg of quick lime

Therefore, 278 mg copperas requires = 56 mg of lime

or 1 mg of copperas requires $= \frac{56}{278}$ mg of quick lime

and therefore 150 kg copperas requires $= \frac{56}{278} \times 150 = 30$ kg of quick lime **Ans.**

Example 4.16: Coagulation sedimentation plant clarifies 40 MLD of water. The quantity of filter alum is required at plant is 80 mg/L. The raw water is having an alkalinity equivalent to 5 mg/L of $CaCO_3$. Determine the quantity of filter alum and quick lime having 85% of purity required per year in the plant.

Solution: Quantity of alum required $= 18 \times 10^{-6} \times 10^{-3} \times 40 \times 10^{6} \times 365$

$= 262.8$ tonnes/year

1 gm of alum reacts with 0.45 gm of $CaCO_3$ alkalinity.

18 gm of alum reacts with $0.45 \times 18 = 8.1$ mg/L

Additional $CaCO_3$ alkalinity to be provided in water $= 8.1 - 5.0 = 3.1$ mg/L

1 mole of $CaCO_3$ alkalinity is induced by 1 mole of 100% pure CaO

100 gm $CaCO_3$ alkalinity is induced by 56 gm of 100% pure CaO

3.1 mg/L $CaCO_3$ alkalinity is induced by $= \frac{56 \times 3.1}{100} = 1.736$ mg/L

3.1 mg/L $CaCO_3$ alkalinity is induced by $= \frac{1.736}{0.85} = 2.04$ mg/L of 85% pure CaO

Quantity of 85% pure CaO required $= 2.04 \times 10^{-6} \times 10^{-3} \times 40 \times 10^{6} \times 365$

$= 29.8$ tonnes/year

Example 4.17: If the alum dose required to coagulate is 10 ppm, find out the amount of alum (in quintals) needs to treat 10 MLD of water. (ESE Mains , Civil Engg. 2010)

Solution: Quantity of water to be treated $= 10 \times 10^{-6}$ l/day

Alum dose required = 10 ppm i.e. 10 mg/L

Amount of alum required per day $= 10 \times 10 \times 10^{-6}$ mg

$= 10^{8} \times 10^{-6}$ kg $= 10^{2} \times 10^{-2}$ quintal

= 1 quintal.

The sludge quantity of around 131 m^3 per day is too bulky and it is difficult to handle on daily basis. However, the removal of water content from such sludge can reduce its bulkiness to some extent. This dewatering of sludge can be carried out by various methods such as using clarifier thickeners, centrifugal thickeners or by filter press.

Also the treated effluent from the sedimentation tank still contains 12 mg/L of suspended solids, therefore it needs further treatment. In order to further reduce the suspended solids concentration the effluent can be taken to filtration unit. The filtration process can remove the suspended solids to a maximum extent depending upon the rate of filtration. It can reduce up to 99% of solids from water. This filtered water can be further taken for disinfection process where the bacteria or microorganisms along with remaining amount of organic matter can be eliminated completely, thus make it safe for drinking and other uses.

Rapid mixing or flash mixing

The optimum dose of coagulant can be determined by carrying a study using jar test apparatus. The water is violently agitated after addition of coagulants in order to ensure proper and uniform mixing. The process of mixing of coagulant salts with such a high stirring process is known as flash mixing. However, in order to keep the particles in suspension and to promote collisions between destabilized particles and in between these particles and flocs the water is stirred gently thereafter. This slow process of mixing of particles and flocs is known as *flocculation*. Thereafter the water is passed in sedimentation tank where the particles along with flocs are allowed to settle down and therefore removed from the water . This entire process of floc formation and solids removal from water is known as type-II settling.

Therefore, the rapid mix unit should be designed properly in order to promote thorough mixing and uniform coagulation to take place. Consequently, careful attention must be paid to the design of rapid-mix units. The time required for flash mixing (t) and velocity gradient (G) are the two important parameters required for the design of rapid-mix units. The velocity gradient is a measure of the relative velocity of two particles of fluid and the distance between them. The concept of velocity gradients can be explained in terms of power dissipation per unit volume of the tank.

Velocity gradient and power requirement in a flocculator

The slow mixing devices or flash mixing devices are used for coagulation and flocculation respectively. The velocity gradient (G) is a term that provides the intensity of mixing and is a function expressed as per unit of time can be used to assess the power dissipated (P) by a flocculator or mixer in flash mixing process and flocculation process. The relationship between velocity gradient and power dissipated by a flocculator can be calculated as shown in Fig. 4.29.

Consider a particle in water at width Δx, height Δy and to travel the length of Δz along a paddle flocculator. The drag force required by a paddle depends on the stress applied to the liquid with respect to the velocity of the particle and the distance travelled by it at height Δy. The stress can be defined in Eq. 4.62;

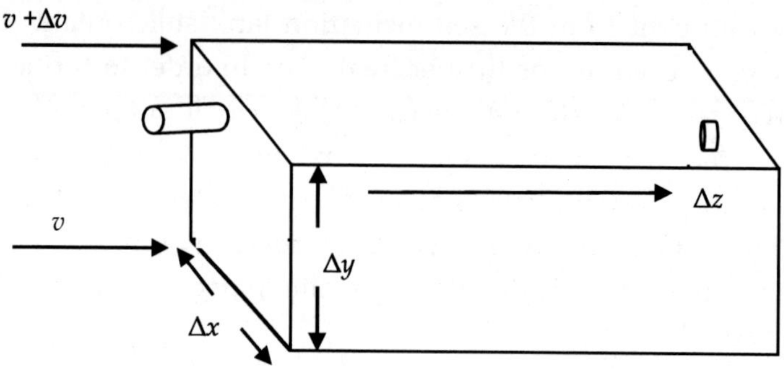

Fig. 4.29: Flocculator for mixing of coagulants

$$\tau = \mu \frac{dv}{dy} \tag{4.62}$$

where, μ = dynamic viscocity of water $\frac{N \sec}{m^2}$ or $\frac{kg}{m.\ s}$

τ = Stress produced by a particle in the liquid

F = Stress × Area = $\tau \cdot \Delta x \cdot \Delta z$

Torque = $F \times d = \tau \cdot \Delta x \cdot \Delta z \cdot \Delta y$

$$\text{Power input } (P) = \frac{\text{Work done}}{\text{Time taken}} = (\tau \cdot \Delta x \cdot \Delta z \cdot \Delta y)\frac{dv}{dy}$$

The power required per unit of tank volume or volume of water can be calculated by using Eq. 4.63;

$$\frac{\text{Power dissipated}}{\text{Volume}} = \frac{P}{V} = \left(\frac{\tau \cdot \Delta x \cdot \Delta z \cdot \Delta y}{\Delta x \cdot \Delta z \cdot \Delta y}\right) \cdot \frac{dv}{dy} = \tau \frac{dv}{dy} \tag{4.63}$$

$\frac{P}{V} = \mu \left(\frac{dv}{dy}\right)^2$ putting the value of velocity gradient $\left(\frac{1}{\sec}\right) = \frac{dv}{dy}$

We get

$$\frac{P}{V} = \mu G^2$$

or

$$\boxed{G = \sqrt{\frac{P}{\mu V}}} \tag{4.64}$$

where,

V = Volume of the tank (m^3)

P = Power dissipated (W)

μ = Dynamic viscocity of water $\frac{N\,sec}{m^2}$ or $\frac{kg}{m.\,s}$

G = Velocity gradient $\left(\frac{1}{sec}\right)$

The velocity gradient (G) is a parameter that provides the intensity of mixing of the coagulants, higher G values indicate higher mixing, vice versa and can be calculated using Eq. 4.64. In a flash mixing or rapid mixing the lesser G values are recommended. However, the G values in the range of 20-70 $sec^{-1}$and t values of 20-40 minutes are recommended by AWWA for slow mixing units.

Rapid mixing can be accomplished by using either mixing tanks with back-mix impellers or in-line flash mixers. The in-line flash mixers consist of mechanically driven impellers. The best G values for rapid mixing tanks operation varies from 700 to 1000, with detention time of approximately 2 min proves to be quite effective. The tanks and impellers of various configurations are generally used for flash mixing with the most popular units being square tanks consisting of back-mix impellers and others with baffles and flat bade impellers which are more efficient.

Sedimentation aided with coagulation tank

The sedimentation aided with coagulation tanks works on the principle of type-II settling suspension process, also known as clariflocculator or coagulation tanks, consists of different subunits as (i) feeding device for feeding of coagulants, (ii) flash or rapid mixing devices in a mixing basin (iii) flocculation basin equipped with mechanical or gravity flocculators, and (iv) settling tank or sedimentation basin. Such types of clariflocculators equipped with all the above subunits are manufactured by the Door Co. and other manufacturers under various patent names with configuration as shown in Fig. 4.16.

The chemical coagulants are added to the raw water either in dry form or in solution through the feeding device in the mixing basin. The added chemicals are then mixed homogeneously and agitated by rapid or flash mixers in the mixing basin. Thereafter, the water is agitated slowly by using mechanical flocculators or gravity floccculators. During this slow mixing process, also known as flocculation process, the floc formation takes place due to the inter-particle bridging in the tank by the chemical reactions taking place in the mixing basin and is then allowed to coagulate in the flocculation tank. After a certain period of given detention time this flocculated water is taken to the sedimentation tank. The flocculated particles as heavy flocs settle down in the sedimentation tank under the action of gravity. The effluent from sedimentation tank is therefore free from colloidal impurities or suspended solids to a certain extent and this treated water is taken to the filtration process for further removal of remaining impurities left over in the sedimentation process.

The sedimentation aided with coagulation process is quite effective in removal of suspended solids and colloidal matter as the effluent concentration with turbidity values as low as 10-20 mg/L can be achieved. It also helps in reducing the bacteria from the water such as 70% of the B-Coli index can be reduced efficiently. The design of sedimentation aided with coagulation tank comprises different subunits and the details of each subunit are described below:

Feeding devices for coagulants

The chemical coagulants used in the process can either be fed into raw water in a powdered form in a solution form. When such chemicals are added in dry form it is known as *dry feeding* however, when used in solution form is termed *wet feeding*. The wet feeding devices though costly but are more convenient in terms of operations, maintenance and synchronization with the dosages. The choice however depends on various factors such as the characteristics of the coagulants and the ease with which the dose can be added. The alum and activated carbon are such coagulants that are available in powdered form and in uniform size therefore can be added with dry feeding procedure. However, the copperas cannot be applied by dry procedure because it readily reacts with atmospheric air and thus with respect to temperature and moisture it turns into solid or sticky mass. Similarly, lime cannot be added with dry feeding method due to absorbing the atmospheric moisture and change in its characteristics of forming sticky mass in the orifices.

However, when the dose of coagulant is very small for proper control and precision wet feeding method is preferable. Another important factor that matters is the cost of the coagulants and the cost of feeding. This can be explained by the fact that when the water is to be treated on large scale the low cost coagulants should be used along with feeding devices or machines. This is because on such a large scale water treatment, the cost of feeding equipment is less important as compared to the cost of coagulants. However, on small scale feeding equipment plays an important role as compared to cost of coagulants, therefore dry feeding equipment and dry feed method are economical. Therefore, the wet feeding and dry feeding have their own advantages and therefore should be used accordingly, so that they can be utilized effectively, efficiently and should be economical too.

The two different methods used for feeding of coagulants are described below:

Dry feeding

Dry feeding is operationally easy and requires comparatively less space. The feeding machines are cheaper too. Dose management is difficult though. In this situation, the hopper-bottomed coagulant is contained in the tank in the form of powder. Agitating plates are mounted within the tank to avoid the arching of chemicals as shown in Fig. 4.30. The feeding is controlled by the speed of the toothed wheel connected to the venture system in the raw water stream. Dry feeding in powder form will directly feed the chemicals into water. Being relatively fine and uniform in size, alum can be easily fed by dry feeding. Copperas and lime can't be fed by dry feeding.

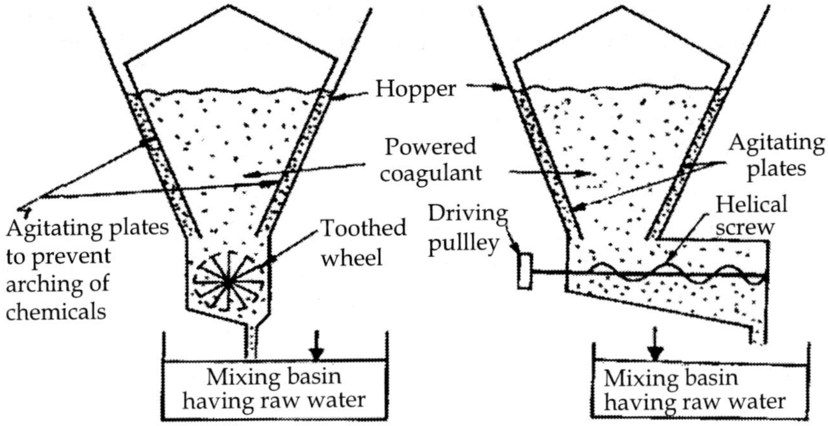

Fig. 4.30: Dry feeding device

Dry feeding systems are usually in the form of a hopper-bottomed tank. Within the tank, agitating plaques are mounted to prevent the coagulant from arching. The coagulant, filled in the tank in the powder form is allowed to fall in the mixing basin. The dose is regulated by the speed of a dented wheel or helical screw. In addition, the speed of the toothed wheel or helical screw is controlled by connecting it to a venture device mounted in the raw water pipes that feed water to the mixing basin. The amount of coagulant released is, therefore, regulated in proportion to the amount of raw water entering the mixing basin. In smaller plants, dry feeding systems are used.

Wet feeding

The dosage can be changed more readily in the wet feeding equipment and can be easily managed with automatic tools. However, in wet feeding, the chemicals that are corrosive in nature cause problems. In the case of wet feeding, the coagulant solution, the necessary strength is prepared and stored in the tank from which it can fall into the mixing pipe. The volume of the mixture falling in the raw water channel is balanced in proportion to the volume of flowing.

In wet feeding, the solution of the coagulant's necessary strength is prepared and stored in a tank, from where it can be trickled down through an outlet into the mixing tank. A float controlled valve keeps the level of coagulant solution in the feed tank constant to ensure a constant discharge rate for a constant feed rate of raw water flow into the mixing basin. When the inflow of raw water changes, a conical type arrangement adjusts the rate of coagulant feed. The working of an arrangement of a conical plug type is very simple. The mixing tank is combined with the float container, so that the water level stays the same in both. As the rate inflow of raw water raises the water depth in the mixing basin as well as the float chamber decreases, then the float is raised. The pinion and pulley rotate in the same direction as the float rises as this raises the conical plug and allows further quantity of coagulant solution to release into the mixing basin. As the flow rate decreases, the conical plug descends and lowers the coagulant feeding rate to a lower rate. Wet feeding systems are typically used by larger water treatment facilities, because they are more costly than dry feeding systems.

Rapid mixing

Rapid mixing or flash mixing is an important design aspect of coagulant dosing. After the addition of coagulants in water they should be mixed properly and distributed uniformly in the entire basin. This is because if the mixing is not done properly the same results in the unevenness distribution of these coagulants in water to be treated leading to poor process efficiency. After addition of the coagulant chemicals the water is violently agitated by using the flash mixing devices where these chemicals combine vigorously and disperse uniformly around the surface. The flash mixing is carried out for 30 to 60 seconds. If the water is mixed for less than 30 seconds, the chemicals will not be mixed correctly in the water. When the water is mixed for more than 60 seconds, however, then the blades of the mixer will shear the newly formed flocs back into tiny particles. The method of mixing chemical dispersal is known as the flash mixing or rapid mixing. The coagulation and flocculation processes take place after the flash mixing. The flash mixer impeller speed lies in the range of 100-120 rpm and the power requirement lies in the range of 2-5 kW/m^3/min. The G.t value in flash mixing processes range from 30,000 to 60,000 with the above-mentioned t values.

Flash mixer

In this system the solution of coagulants is thoroughly mixed in the water by means of a fan driven by the necessary drive of the electric motor. The water enters through the inlet, the deflecting wall deflects the water towards fan blades where chemicals can pass through chemical tubing. The spinning fan combines coagulants with water that ends up flowing out of the outlet as shown in Fig. 4.31.

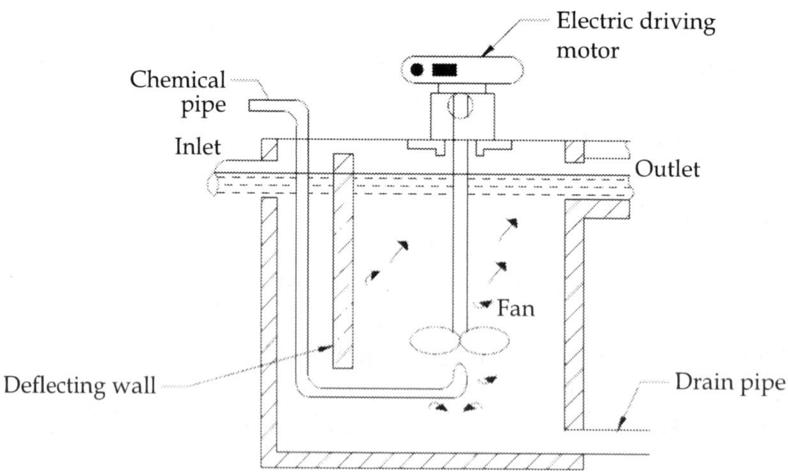

Fig. 4.31: Flash mixer

Mixing devices

As discussed earlier after adding coagulants to the raw water, the mixing of chemicals is carried out by rapid mixing devices where the water is violently agitated for 30-60 sec and thus allowing the uniform distribution of these chemicals in the entire water body. This mixing process is achieved by various mixing equipment such as pumps with blades, compressed air units, mixing units with baffle walls under gravitational flow. After flash mixing the water mixed with coagulants is then agitated gently for about 30 minutes, so that coagulants can react and attract each other resulting in the formation of flocs. The water flow velocity in the mixing basins is usually kept in the range of 15-30 cm/sec. The velocity should not be less than 10 cm/sec, in any case and should not be greater than 75 cm/sec. This is because low flow velocities may lead to the settlement of the coagulants and chemicals or other and high velocity can lead to the disintegration of the formed flocs. The two types of most commonly used mixing devices are can be discussed and shown in Fig. 4.32 and Fig. 4.33.

(i) **Mixing basin with baffle walls:** The mixing units are rectangular tanks provided with baffles at equal distances along the length of the tank. The water is allowed to move horizontally around the end of each baffle and the mixing is known as around the end baffles type. However, when these baffles are constructed vertically along the length of the tank and water is allowed to enter in vertical up and down movement under each baffle and the mixing is known as over and under baffle type. The obstructions caused by these baffle walls creates enough turbulence and agitation in the water that creates adequate mixing and enhances the formation of flocs. This 180° turn in direction of flow of water under gravity at each baffle provides enough momentum in water to mix up freely and evenly. The design basis in both the conditions remains the same however around the end baffle type of mixing is very common due to ease of cleanliness and operation.

Design criteria

For effective flocculation to take place the horizontal flow velocity in each channel of the tank in between the baffles lies in the range of 0.15 -0.45 m/sec. The high velocity values are generally adapted for water having high turbidity values and vice versa. The detention time of flocculation is generally taken in between 20-50 minutes. The length of the tank can be determined by using the velocity and detention time values. From the given flow rate value and detention time value the volume of the tank can be determined. Using the volume and length values the cross-sectional area can be determined by dividing the volume of the tank by the length of the tank. The depth of the basin can be fixed by dividing the cross-sectional area of the tank with the centre-to-centre distance between the baffles. The depth of the tank should not be less than 1 m in any case so as the process may not be affected under flow variations.

The number of channels can be determined by dividing the total length of the tank by flow length through each channel. The minimum distance between the two baffles should not be less than 45 cm. The clear opening between the end of each baffle and the wall of the basin should be kept around 1.5 times the distance between the baffles or should not be less than 60 cm. The tapered flocculation is sometimes carried out in order to reduce head loss in the tank such that the baffles spacing is increased gradually along the length of the tank. The flow is through gravity and therefore the flow of channels should slope towards the outlet of the tank. A head loss around the baffle in the channel lies in the range of 0.5 to 1.0 m.

However, in case of over and under baffles the width can be determined by dividing the cross-sectional area of each channel by the spacing between the baffles. The mixing devices contain no moving parts or equipment therefore due to the simplicity it is easy in operation and maintenance is also quite easy. The only disadvantage in the process is that due to the 180° bends of the flow the head loss is high at the bends and the flocculation process is affected at the straight way of the channel. It is estimated that the head loss at each 180°bends is almost equal to 3.2 times of velocity head. This head loss value is quite high as compared to the mechanical mixers. These baffle type mixing devices work on a very narrow range of velocity and flow values thus make the usage to be very rigid. These devices are less efficient in mixing the coagulants and therefore high amount of coagulant dosage is required as compared to mechanical flocculators. These baffled mixing devices are generally used for small plants where less skilled supervision is required and are not used for large plants.

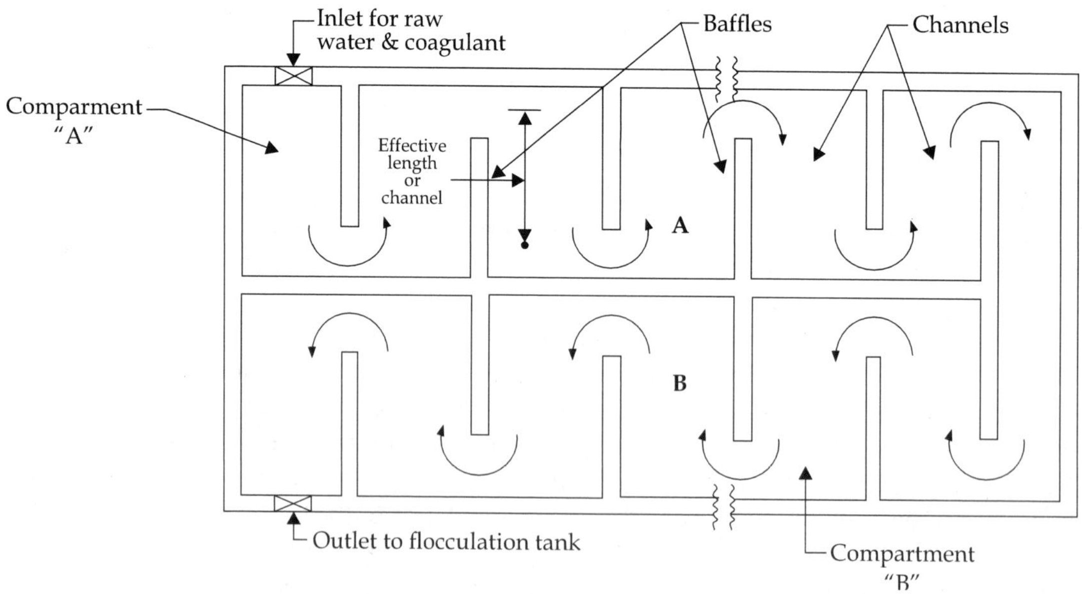

Fig. 4.32: End baffles type mixing basins

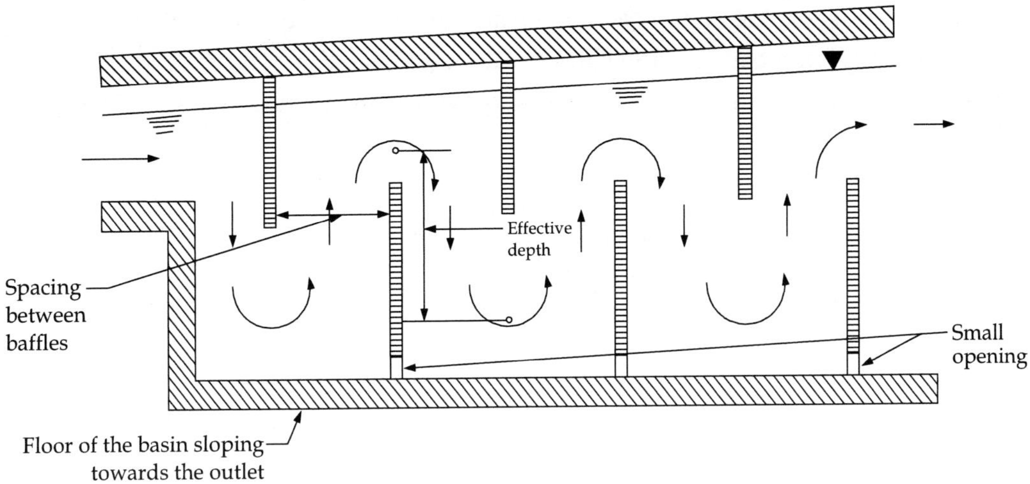

Fig. 4.33: Over and under baffles type mixing basin

Example 4.18: Design a around the end baffled flocculator to treat 30 million litres of water per day. The tank may be divided in two similar compartments with partition wall along the length of the tank and each half may have 6 m width. The clear distance between the baffle walls may be kept equal to minimum permissible. Assume detention time and flow velocity suitably. Detail out all the dimensions and baffles spacing in the overall length of each channel.

Solution: Given that:

Flow rate = Amount of water treated per day = 30 MLD = 20×10^6 L/d = 30,000 m^3/d

Number of compartments = 2

Width of each compartment = 6 m

Assume detention time, t = 30 minutes (20-50 minutes)

Therefore volume of the water to be treated per day =

$$V = Q \times t = \frac{30,000 \times 30}{24 \times 60} = 625 \text{ m}^3$$

Assume the horizontal velocity = 0.3 m/sec

The length of flow, $L = v \times t = 0.3 \times 30 \times 60 = 542$ m

The required cross-sectional area of each channel between the baffles

$$= \frac{\text{Volume of channel}}{\text{Length of flow}} = \frac{625}{540} = 1.16 \text{ m}^2$$

The spacing between the baffles may be assumed = 45 cm = 0.45 m

$$\text{Therefore, the depth of water in the tank} = \frac{\text{Cross-sectional area}}{\text{Spacing between the baffles}}$$

$$= \frac{1.16}{0.45} = 2.58 \text{ m}$$

The clear openings at the end of each baffle and the wall of the basin

$$= 1.5 \times \text{spacing between the baffles} = 1.5 \times 0.45 = 0.68 \text{ m} \approx 0.7 \text{ m}$$

The effective length of the each channel may be assumed to be average distance travelled by the water in channel.

Therefore, the effective length of flow in each channel

$$= \text{Clear width of either half of the tank} - 2 \times \frac{1}{2} \text{ of the opening}$$

$$= 6 \text{ m} - 2 \times \frac{1}{2} \times 0.7 = 6 - 0.7 = 5.3 \text{ m}$$

$$\text{Therefore, No. of channels required} = \frac{\text{Total flow length}}{\text{Effective flow length each channel}}$$

$$= \frac{540}{5.3} = 102$$

$$\text{Therefore, there will be} = \frac{102}{2} = 51 \text{ channels in each half of the tank}$$

The clear length of tank excluding the baffle walls and side walls = number of channels $\times$ 0.45 = 51 $\times$ 0.45 = 22.95 m

Assume thickness of each baffle wall = 7.5 cm

We know that number of baffles = No. of channels – 1 = n – 1 = 51 – 1 = 50

$$\text{The overall length of the tank} = 22.95 + \frac{7.5}{100} \times \text{No. of baffles} = 22.95 \times \frac{7.5}{100} \times (50)$$

$$= 22.95 + 3.75 = 22.7 \text{ m}$$

Ans.

Flocculation

After the addition of chemicals to water it is violently agitated by using flash mixing devices for uniform dispersion of chemicals in water rapidly. Thereafter, the water is transferred to the tank with slow mixing devices also known as flocculation tank and this slow mixing process is known as flocculation. The flocculation process relies on the mechanism of turbulence in order to promote collisions and can be measured by assessing the velocity gradients. Rectangular tanks are generally constructed and electric motors equipped with paddles are generally provided for mixing and stirring. During slow mixing process the paddles in flocculation tank rotate at the speed of 2-3 rpm. The clear distance between the paddles and the wall or the floor of the tank is taken as 15-30 cm. Time is another important factor required for design of flocculation unit by using a parameter known as G.t factor a dimensionless number. The most common G.t values in designing a flocculation tank is 10^4 to 10^5 with t values ranging in between 20 to 60 min with normal value of 30 minutes is considered to be effective. (Table 4.9). However, using high G values with short times tend to produce small, dense flocs, while low G values and long times produce larger, lighter flocs. This may be due to the fact that large, dense flocs are more easily removed in the settling basin.

Therefore varying the G values over the length of the flocculation basin proves to be quite advantageous.

However, at lower G values produced larger flocs combines with the small, dense flocs produced at high G values. Therefore, the G values are reduced by a factor of 2 from the influent end to the effluent end of the flocculator proves to be quite effective. Traditional flocculator design can be shown through Fig. 4.21. The unit comprising of long-rectangular basins equipped with mechanically operated paddles to provide power input. The paddles of these flocculators are made of either redwood or aluminum slats and can be operated either transverse or parallel to the longitudinal axis of the basin. Nowadays the process consists of rapid mixing, flocculation tank along with settling tank becomes one complete unit. Such a unit is shown in Fig. 4.22. The principles of mixing and flocculation in this unit are the same as those for the long-rectangular tanks as indicated in Table 4.10.

Table 4.9: Recommended G values in rapid mixing (as per AWWA)

Parameters	Recommended values			
Detention time (s)	20	30	40	>40
Velocity gradient G (sec^{-1})	1000	900	790	700

Table 4.10: General design criteria for mixing and flocculation

Parameters	Flash mixing	Flocculation (slow mixing)
Flocculator speed (rpm)	100–120	2–3
Mixing time (min)	1.5–02	20–60
Velocity gradient G(sec^{-1})	700–1000	20–80
Power required ($kW/m^3/min$)	2–5	-
G.t factor	$(3\text{-}6) \times 10^4$	10^4–10^5

Power requirement in mechanical flocculator

The G values and power requirement by mechanical driven flocculator can be calculated as follows:

We know that the power imparted by the paddle is directly proportional to the drag force exerted by the fluid.

Power imparted is directly proportional to drag force exerted by the fluid.

The power input of the flocculator can be determined by using Eq. 4.65 and Eq. 4.68.

$$P = F_D \cdot v_p \tag{4.65}$$

where,

P = Power input, W (N.m/s)

F_D = Drag force on paddles (N)

v_p = Velocity of paddles (m/s)

Drag force on paddle can be calculated by using Eq. 4.66;

$$F_D = \frac{1}{2} C_d \cdot \rho \cdot A_p \cdot v^2 \tag{4.66}$$

where,

C_d = Drag coefficient = 1.8 (for flat blades)

ρ = Density of fluid= 1000 kg/m^3 (for water)

A_p = Area of the paddle (m^2)

v = Relative velocity of the fluid with respect to the paddle (m/s)

Also, v_p= Peripheral velocity or velocity of paddle is shown in Eq. 4.67;

$$v_p = \frac{2\pi r N}{60} \tag{4.67}$$

where,

N = Number of revolutions

r = Radial distance of the outermost paddle from the centre of the shaft or lever arm

Also,

$v = v_p - v_{\text{fluid}} = v_p - kv_p = (1 - k)v_p$ with k value assumed to be 0.20 to 0.25
and therefore $v = (0.75 - 0.80)v_p$

But,

$$\boxed{\text{Power } (P) = F_D \cdot v = \left(\frac{1}{2} C_d \cdot A_p \cdot v^2\right) . v = \frac{1}{2} C_d . A_p . v^3} \tag{4.68}$$

where,

v_p = 0.09 to 0.9 m/s

A_p = total area of the paddles may be taken as 15-20% of the cross sectional area of the tank. This area of paddle (A_p) is the combined area of slats perpendicular to the cylinder of rotation. However, this area of paddles should not be more than 40% of total area encompassed by the paddles. The paddle velocity (v_p) relative to water and is about 75% of actual paddle speed. The velocity of the paddle should be kept less than 1 m/s. The distance between the paddle end and the walls of the tank from all dimensions should be kept 0.3 m in order to prevent any effect during high velocity gradients. The water should then be transferred from flocculation tank to the settling tank carefully so that no turbulence may take place to prevent the breaking of flocs. The design of mechanical flocculator can be well explained with the following steps.

Design steps for a mechanical flocculator:

1. At a given Q value assume the G values from 20 to 70 sec^{-1} and detention time t accordingly. Divide the tank in 3 chambers with decreasing G values so that area of one chamber = 1/3 of total tank area.
2. Designate the number of paddles per flocculator required.
3. Calculate the volume of flocculation tank using equation $V = Q \times t$

4. Assume the depth of the tank in the range of 2-3 m and calculate the surface area of the tank by dividing the area into three parts. The tank can be divided into three chambers with decreasing G values and set the dimensions of the tank accordingly. The length can be assumed three time the width of the tank ($L = 3W$)
5. After designating the dimensions in terms of width and length the chamber.
6. Set the minimum clearance of 40 cm between the outer paddle of flocculator and the floor of the tank and water surface.
7. Distance of the paddle end from the wall of the tank = 40 cm
8. Relative velocity of the paddle transferred to water (v) is taken as 75% the paddle velocity (v_p).
9. The coefficient of drag value for flat blade paddle (C_d) may be taken as 1.8.
10. Minimum spacing between the two paddles is taken as 50 cm.
11. Length of the paddle can be calculated based on the width of the chamber after deducting the minimum spacings from both the sides of the chamber wall using equation below as : $L = \dfrac{\text{Lenght or width of the tank} \quad 0.5 - 0.4 \times 2}{2}$
12. Also assume the area of paddle = 20% of total tank area, therefore the width of the paddle can be calculated as: width of the paddle = area of the paddle divided by length of the paddle.
13. Calculate the power requirement for individual chambers using the equation given as: $P_h = G_2 \times \mu \times v_h$
14. The power delivered by the paddle can be calculated using the Eq. 4.69:

$$P_p = \frac{1}{2} C_d \, \rho \, b \left(\frac{2\pi}{60}\right)^3 (1-k)^3 (r_0^4 - r_0^4) \tag{4.69}$$

where, P_p = Power requirement for the paddle (W)

r_o = Distance from the shaft to the outer end of the paddle (m)

r_i = Distance from the shaft to the inner edge of the paddle (m)

b = Width of the paddle (m)

N = Number of paddle rotations (RPM)

k = Velocity transfer factor with a value of 0.20-0.25

C_d = Coefficient of drag (value for flat blade = 1.8)

ρ = Density of water (kg/m^3)

V_p = Velocity of the paddle, $v_p = \dfrac{2\pi r N}{60}$ m/sec

15. However, the total power requirement for the chamber can be calculated by multiplying with the total number of paddles in a chamber as: $P = P_p \times$ Number of paddles in a chamber.

16. The total power requirement of the tank can be summarized in a tabular form as shown in Table 4.11;

Table 4.11: Power requirement of the tank

Chamber no.	G values (1/sec)	Velocity (m/sec)	N (rpm)	Power requirement P (W)

Power requirement in gravity flocculator

Works under the principle of gravity and categorized as

(a) Around the end baffles (horizontal flocculator)

(b) Over and under baffles (vertical flocculator)

Power dissipation or power loss or power requirement depends on the density of the liquid and can be represented as in Eq. 4.70;

$$P = \gamma \cdot Q \cdot h = \rho \cdot g \cdot h \cdot Q \qquad (4.70)$$

where,

γ = Specific weight of liquid (N/m^3)

ρ = Density of water (kg/m^3)

h = Head loss in flocculator (because of difference in velocities)

g = Acceleration due to gravity (m/s^2)

If there are n number of compartments

Therefore, number of baffles may be taken as = n – 1

Overall head loss in the chamber can be calculated by using Eq. 4.71;

$$h_L = \frac{nv_1^2}{2\,g} + (n-1)\frac{v_2^2}{2\,g} \qquad (4.71)$$

where, v_1 = Velocity between the baffles or initial velocity of the fluid

v_2 = Velocity at the end of baffles or baffle slot or shoot up velocity

Design steps for a gravity flocculator

1. At a given Q value assume the values from 20 to 70 sec^{-1} and detention time t accordingly. Divide the tank in 3 chambers with decreasing G values so that area of one chamber = 1/3 of total tank area. Calculate the volume of flocculation tank using equation $V = Q \times t$
2. The area of the tank can be calculated using the above volume calculated from step 1 and assuming the depth of the tank to be 2-4 m.
3. Assume length to width ratio of 3:1 and calculate the dimensions of the tank as length, width and depth.
4. The number of baffles to be provided per chamber can be calculated using Eq. 4.72:

$$n = \left[\frac{2\mu t}{\rho(1.44+f)}\left[\frac{HLG}{Q}\right]^2\right]^{\frac{1}{3}} \qquad (4.72)$$

where, μ = Dynamic viscosity of water (Ns/m^2)

ρ = Density of water (kg/m^3)

t = Detention time per compartment (sec) = total detention time/no. of compartments

f = Coefficient of friction = 0.3

Q = Water flow rate (m^3/sec)

L = Length of the channel

G = Velocity gradient (1/sec)

n= Number of baffles per compartment

5. The spacing between the baffles can be calculated by dividing the length of the chamber with number of baffles (L/n). The distance between the consecutive baffles ≥ 45 cm.
6. Clear distance between the end of each baffle and the wall is $1\frac{1}{2}$ times the distance between the baffles and it should be greater than 60 cm.
7. The water in the flocculator should not be less than 1m and must not be greater than 3 m.
8. In order to ensure better flocculation and to increase the floc size, the flocculator should be divided into three compartments with decreasing G values so as to such as $G_1 > G_2 > G_3$ such as G_1 = 50/sec, G_2 = 40/sec and G_3 = 30/sec. Such type of flocculator is known as tapered flocculator. The floc size of 1-2 mm is considered to be good enough for settling.
9. The head loss per chamber can be calculated by using the Eq. 4.73.

$$h_L = \frac{\mu\, t}{\rho\, g} G^2 \tag{4.73}$$

h_L = Head loss per chamber (m)

μ = Dynamic viscosity of water (Ns/m^2)

ρ = Density of water (kg/m^3)

t = Detention time per compartment (sec) = total detention time/no. of compartments

G = Velocity gradient (1/sec)

g = Acceleration due to gravity (m/sec^2)

Head loss per baffle can be calculated by using Eq. 4.74;

$$\text{Head loss per baffle} = \frac{h_L}{n} \tag{4.74}$$

where n is the number of baffles per compartment

10. Calculate the velocity through the chamber using the head loss equation which can be equated as shown in Eq. 4.75.

$$\frac{h_L}{n} = \frac{k}{2g} V^2 \tag{4.75}$$

where, K = A constant with value of 1.5

g = Acceleration due to gravity (m/sec^2)

V = Velocity through the baffles (m/sec)

11. The slot width can be calculated by calculating the area of the slot using velocity and discharge through the baffle slot. The area of the slot can be calculated using Eq. 4.76:

$$\text{Slot area} = \frac{Q}{V} \tag{4.76}$$

where, Q = Discharge through the slot (m^3/sec)

V = Velocity through the slot (m/sec)

Slot width (W) can be calculated using Eq. 4.77;

$$W \times H = \frac{Q}{V}$$

or

$$W = \frac{Q}{V \times H} \tag{4.77}$$

where, W = Slot width (m)

H = Height of the chamber

Q = Discharge through the slot (m^3/sec)

V = Velocity through the slot (m/sec)

The head loss and slot width in different sections can be calculated as shown in Table 4.12.

Table 4.12: Calculation of head loss and slot width

Chamber no.	G values (1/sec)	Number of baffles (n)	Spacing between the baffles (m)	Head loss (m)	Velocity (m/sec)	Slot width (m)

4.4 COAGULATION

The energy needed to neutralize the charge over suspended impurities is given by inducing rapid mixing in the coagulation cycle using either of the following methods:

(i) Mixing basin: Mixing is done by creating turbulence in water caused by the sudden change in the direction of water flow by providing the tank with baffle walls.

(ii) In round and end type mixing basin water flow happens in horizontal plane and in vertical plane in over and under type mixing basin water flow takes place as shown in Fig. 4.26 and Fig. 4.27.

Mechanical mixer

Mixing is caused by the vigorous agitation of coagulants in water with the aid of mechanical aerators and the mixing strength depends on the parameter called *temporal mean velocity gradient* which represents the relative velocity between the two particles at a specific distance in the tank in the coagulation tank as shown in Fig. 4.34 and Eq. 4.78-Eq. 4.79.

$$G = \frac{(v_1 - v_2)}{x} = \sqrt{\frac{P}{\mu v}} \tag{4.78}$$

P = Power supply to shaft (in watts)

μ = Dynamic viscosity of water $\left(\frac{\text{N} - \text{se}}{\text{m}^2}\right)$

v = Volume of the coagulation tank.

G is kept to be 300/sec in coagulation tank.

Detention time is 30-60 sec for the tank.

Deep circular and square tanks are adopted in this case , having $\left(\frac{H}{D \text{ or } B}\right) = 1{:}1\text{ -}3{:}1$

Power supply to the shaft is 1-3 watts/discharge in m^3/hr.

Speed of the shaft should be such that velocity generated at the tip of the paddle is more than 3 m/s.

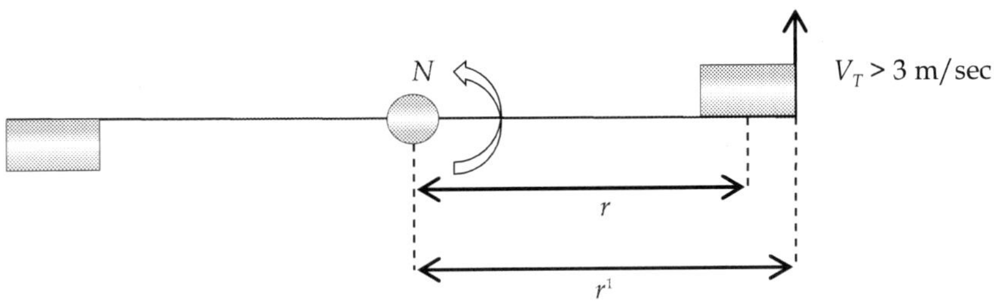

Fig. 4.34: Diagram showing the temporal mean velocity

$$V_T = W r^1$$
$$r^1 = r + b/2$$
$$V_T = \frac{2\pi N(r + b/2)}{60} r^1 \tag{4.79}$$

4.4.1 Sedimentation Aided with Coagulation

When water contains fine clay, and electrically charged colloidal impurities are continuously in motion and never settle due to gravitational force. Some chemicals are added to the water so that impurities which are not removed by plain sedimentation are removed. During its development and descent through water, the chemicals form insoluble, gelatinous, flocculent precipitate absorbs and entagles very finely suspended matter and colloidal impurities. Furthermore, these coagulants also eliminate colour, odour and taste from water. Turbidity of water is reduced upto 5-10 ppm and bacteria removed up to 65%. Coagulants are selected based on the water's pH. Owing to the low cost and ease of storage as solid crystals over long periods, aluminum or aluminium sulphate is commonly used in all treatment plants. The dosage of coagulants to be applied to the water depends on form of coagulant, water turbidity, water colour, water pH, water temperature, and time for mixing and flocculation. A jar test as shown in Fig. 4.40 specifies the optimum dose of coagulant needed for a water treatment plant.

First, all the water sample is taken in each jar to start the experiment and the coagulant is added in varying quantities in a container. It states the amount of coagulant added to each glass. Thereafter, all the paddles are rotated for around 10 minutes at a speed of 30-40 rpm with the help of an electric motor. The speed is decreased after this, and paddles are rotated for around 20-30 minutes. The paddle rotation is halted, and the flock established in each jar is noted and allowed to settle. The optimum dose of coagulants is the coagulant dose which gives the best floc.

Mechanical flocculation: It consists of rotating paddles with horizontal or vertical shafts or paddles suspended from horizontal oscillating plates, going up and down as shown in Fig. 4.35.

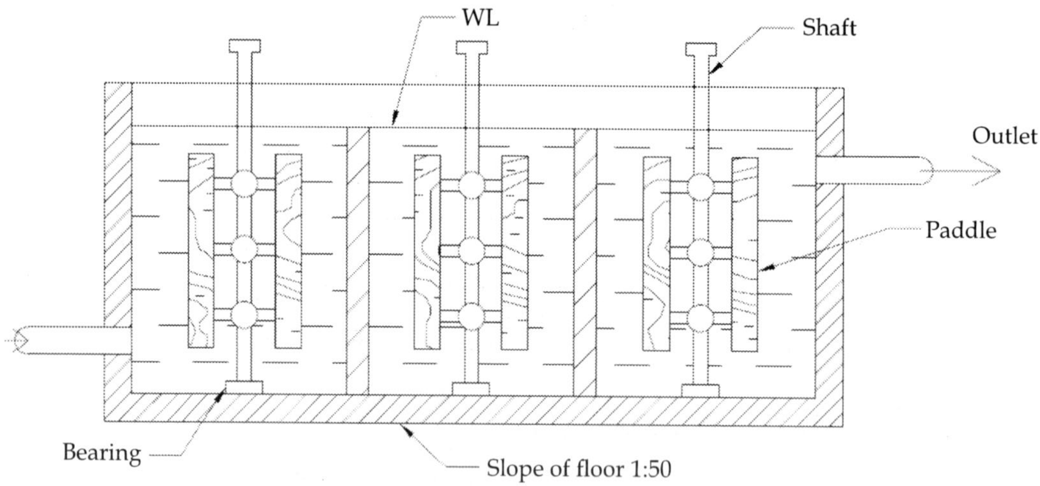

Fig. 4.35: Mechanical flocculator

Example 4.19: The flash mixing needs to be carried out in a water treatment plant with water at a flow rate of 50,000 m^3/d at 20°C. Determine the power requirement at velocity gradient of 900 sec^{-1} and t value of 30 sec?

Solution: Given that

$Q = 50000\ m^3/d$

$G = 900\ sec^{-1}$

t = 30 sec and

$$\mu = 10^{-3}\,\frac{\text{N sec}}{\text{m}^2}\ \text{at}\,20°\text{C}$$

Here volume of the water or tank volume, $V = \dfrac{50{,}000\ \text{m}^3}{24 \times 3600\,\text{sec}} \times 30\,\text{sec} = 17.36\ \text{m}^3$

$$G = \sqrt{\frac{P}{\mu V}}$$

$$900 = \sqrt{\frac{P}{10^{-3} \cdot 17.36}}$$

$$81000 = \frac{P}{10^{-3} \times 17.36}$$

Therefore, P = 14061.6 W = 14.061 W = 18.85 H.P.

Considering pump efficiency to be 70%, therefore, $P = \dfrac{18.85}{0.70} = 26.93$ H.P. **Ans.**

Example 4.20: Calculate the size of the basin and head loss in a gravitational type of flocculator with 16 baffles with a flow rate of 0.3 m^3/sec. The detention time may be assumed to be 25 minutes. The initial velocity and shoot up velocity may be taken as 0.2 m/s and 0.5 m/s, respectively. Also calculate the number of collisions in terms of G.t factor at the above given conditions?

Solution: $Q = 0.3\ \text{m}^3/\text{sec}$

Initial velocity through baffles = V_1 = 0.2 m/s

Final velocity at the end of baffles or shoot up velocity, V_2 = 0.5 m/s

Number of baffles = 16

Number of compartments, $n - 1 = 16 - 1 = 15$

Detention time, t = 25 minutes = 25 × 60 = 1500 sec

We know that the total head loss in a basin containing 16 baffles can be calculated as

$$h_L = \frac{nv_1^2}{2\,g} + (n-1)\frac{v_2^2}{2g} = \frac{1}{2 \times 9.81}(16 \times 0.2^2 + 15 \times 0.5^2) = 0.223\ \text{m}$$

Also the power requirement can be calculated as :

$P = \rho \cdot g \cdot Q \cdot h = 1000 \times 9.81 \times 0.3 \times 0.223 = 67.12 \times 9.81 = 658.44$ watts

Also t = 1500 sec

$$V = Q \times t = 0.3 \times 1500 = 452\ \text{m}^3$$

And we know that

$$G = \sqrt{\frac{P}{\mu V}} = \sqrt{\frac{658.44}{10^{-3} \times 450}} = 38.25/\text{sec}$$

Therefore, number of collisions, G.t = 38.75 × 1500 = 5.74 × 10^4 **Ans.**

Example 4.21: Calculate the power requirement for a mechanical flocculator with flow rate of 10,000 m^3/d containing 4 shafts with each shaft having 4 paddles and a detention time of 30 minutes rotating at 2 rpm. The radial distance of the outermost paddle may be taken as 1.7 m with C_d value of 1.8.

Solution: Given that

$$Q = 10{,}000 \frac{\text{m}^3}{\text{d}} \times \frac{1}{24 \times 60} = 6.94 \text{ m}^3/\text{min}$$

$N = 2$ rpm

$t = 30$ min

$r = 1.7$ m

$C_d = 1.8$

We know that

$$v_p = \frac{2\pi r N}{60} = \frac{2\pi \times 1.7 \times 2}{60} = 0.356 \text{ m/s}$$

$$v = 0.75\, v_p = 0.75 \times 0.356 = 0.267 \text{ m/s}$$

Assume length of the paddle, $L = 4.5$ m = maximum depth of the tank

Assume width of the paddle = 20 cm (10 – 25 cm)

Length of the paddle depends on the depth of the tank (Fig. 4.36);

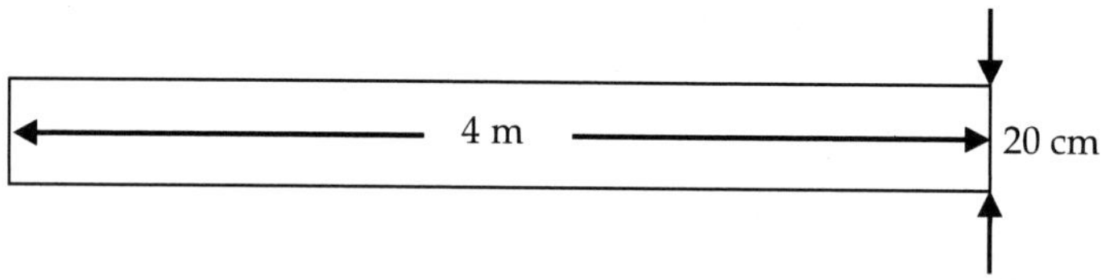

Fig. 4.36: Length of the paddle

The area of paddles in 4 shafts with each shaft containing 4 paddles as

$A = L \times W = [(4.5 - 0.15 - 0.15) \times 0.2] \times 4 \times 4 = 0.84 \times 16 = 13.44 \text{ m}^2$

Therefore, power requirement can be calculated as

$$P_p = \frac{1}{2} C_d\, \rho\, b\, w \left(\frac{2\pi N}{60}\right)^3 = \frac{1}{2} C_d\, \rho\, A (v_p)^3 = \frac{1}{2} \times 1.8 \times 1000 \times 13.44 \times (0.267)^3$$

= 230 watts

Volume of the basin can be calculated as

$V = Q \times t = 6.94 \times 30 = 208.3 \text{ m}^3$

$$G = \sqrt{\frac{P}{\mu . V}} = \sqrt{\frac{230}{10^{-3} \cdot 208.3}} = 33.23 \frac{1}{s}$$

Therefore, G.t. = $33.23 \times 30 \times 60 = 5.98 \times 10^4$ **Ans.**

Example 4.22: Calculate the flocculation basin area and paddle area having flow rate of 5000 m^3/d with paddle velocity of 0.6 m/s and detention time of 20 minutes?

Solution: Given that

t = 20 minutes

Velocity $v_p = 0.6$ m/s

Flow rate, $Q = \frac{5000}{24 \times 60} = 3.48 \text{ m}^3/\text{min}$

Assume water depth = 4 m (range of depth in the tank = 3-5 m)

Also, $v = 0.75\, v_p = 0.75 \times 0.6 = 0.45$ m/s

$$V = Q \times t = 3.48 \times 20 = 70 \text{ m}^3$$

Assume the length to width ratio of the tank to be: $L:W = 3:1$

Therefore area of the flocculator, $\frac{\text{Volume}}{\text{Depth}} = \frac{70}{4} 17.5 \text{ m}^2$

$$\frac{L}{W} = \frac{3}{1} \text{ or } L = 3\text{ W}$$

$$\text{Area} = L \times W = 3\text{W} \times \text{W} = 3\text{W}^2$$

$$3W^2 = 17.5 \text{ or } W = 2.41 \text{ m} \cong 2.5 \text{ m}$$

$$W = 2.5 \text{ m and } L = 3\text{W} = 3 \times 2.5 = 7.5 \text{ m}$$

Hence, area of the flocculator basin $7.5 \times 2.5 = 18.75 \text{ m}^2$

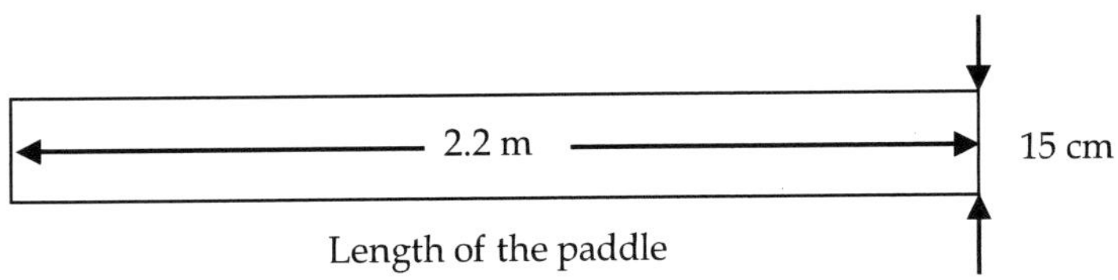

Fig. 4.37: Area of the paddle

Therefore, paddle area $A = 0.15 \text{ m} \times (2.5 \text{ m} - 0.15 \text{ m} - 0.15 \text{ m}) = 0.15 \times 2.2 = 0.33 \text{ m}^2$ (Fig. 4.37) **Ans.**

Example 4.23: Consider a flocculation tank 10 m long, 3 m wide and 3 m deep designed at a flow rate of 0.05 m^3/s. Flocculation is achieved by 3 paddle wheels each with 2 blades of 2.5 m × 0.3 m. The centre-line of the blades is 1 m from the shaft, and also at mid-depth of the tank. The paddle rotates at 3 rpm and velocity of water to be assumed 75% of paddle velocity. Calculate the power requirement for flocculator and velocity gradient?

Solution: Given that

Length, L = 10 m, width, W = 3 m, depth, H = 3 m

Paddle width = 0.3 m, Length of paddle = 2.5 m

$Q = 0.05 \text{ m}^3/\text{s}$

$N = 3$ rpm

$r = 1.0$ m

Total paddle area, $A = 3 \times 2\ (2.5 \times 0.3) = 4.5\ \text{m}^2$

The paddle velocity can be calculated as, $v_p = \dfrac{2\pi r N}{60} = \dfrac{2\pi \times 1.0 \times 3}{60} = 0.314$ m/s

and therefore,

$$v = 0.75\ v_p = 0.75 \times 0.314 = 0.235 \text{ m/s}$$

The total power required is

$P_p = \dfrac{1}{2} C_d\ \rho\ A.(v_p)^3 = \dfrac{1}{2} \times 1.8 \times 1000 \times 4.5 \times (0.235)^3 = 53$ watts

Volume of the tank $V = 10 \times 3 \times 3 = 90\ \text{m}^3$

and detention time $t = V/Q = 90/0.05 = 1800$ sec

Therefore,

$$G = \sqrt{\frac{P}{\mu . V}} = \sqrt{\frac{53}{10^{-3} \cdot 90}} = 24.27 \frac{1}{s}$$

Therefore, $G.t = 24.27 \times 1800 = 4.36 \times 10^4$ **Ans.**

Example 4.24: Design a mechanical flocculator with the following data;

$Q = 100$ MLD

G values in decreasing order = 45, 35, 20 /sec

Detention time of the tank = 25 minutes

Number of paddles in flocculator = 4

Solution: Given that $Q = 100$ MLD

G values in decreasing order = 45, 35, 20 /sec

Detention time of the tank = 25 minutes

Number of paddles in flocculator = 4

Assumptions:

$C_d = 1.8$

Two units with equal flow values

Relative velocity = 75% of paddle velocity or $k = 0.25$

Minimum spacing between two paddles = 50 cm

Assume area of paddle = 18% of cross-sectional area of the tank or compartment

Assume L = 3 W

As assumed above the mechanical flocculator can be divided in two equal units with discharge through each unit = 100/2 = 50 MLD

Flow through each unit, 50 MLD $= \dfrac{50 \times 10^6 \times 10^{-3}}{24 \times 60 \times 60} = 0.57\ \text{m}^3/\text{sec}$

Total volume of each unit of the flocculator, $V = Q \times t = 0.57 \times 25 \times 60 = 855\ m^3$

Assume three compartments in each unit

Volume of one compartment $= \frac{855}{3} = 285\ m^3$

Let the water depth in the tank = 4.5 m

Therefore, area of one compartment $= \frac{285}{4.5} = 63\ m^2$

Therefore, width $= \frac{63}{4.5} = 14\ m$

Let us assume the minimum clear distance between outer paddle and floor and water surface = 40 cm

Distance of paddle from wall = 40 cm

Therefore, $r = \frac{4.5 - 0.4 - 0.4}{2} = 1.85\ m$

Length of the paddle, $L_p = \frac{14 - 0.5 - 0.4 - 0.4}{2} = 6.35\ m$

Area of paddle = 18% cross-sectional area of the tank $= 0.18 \times 63 = 11.34\ m^2$

Number of paddles = 4

Area of one paddle $= \frac{11.34}{4} = 2.835\ m^2$

Width of one paddle, $W_p = \frac{2.835}{6.35} = 0.45\ m = 45\ cm$

However, the power delivered by one paddle can be calculated as :

$$P_p = \frac{1}{2} C_d\, \rho\, b \left(\frac{2\pi N}{60}\right)^3 (1-k)^3 (r_0^4 - r_0^4)$$

where,

$C_d = 1.8$ and $\rho = 1000\ kg/m^3$

$k = 0.25$ therefore $1\text{-}k = 0.75$

Width of the paddle = 0.45 m

$r_0 = 1.85$ m and $r_i = 1.85\ m - 0.45\ m = 1.40\ m$

Therefore, power consumed by one paddle

$$P_{p1} = \frac{1}{2} \times 1.8 \times 1000 \times 0.45 \left(\frac{2\pi N}{60}\right)^3 (0.75)^3 [(1.85)^4 - (1.4)^4]$$

$P_{p1} = 0.9 \times 1000 \times 0.45 \times 0.00115 \times 0.422 \times (11.71 - 3.84)N^3 = 1.54\ N^3$

Therefore, power imparted by 4 paddles

$P_{p1} = 4 \times 1.54\ N^3 = 6.18\ N^3$

Also, $G = \sqrt{\dfrac{P}{\mu.V}}$

and G value in first chamber $G_1 = 45$ /sec

therefore,

Compartment-I

Power required in the 1st compartment

$P_1 = G_1^2 \times \mu \times V = (45)^2 \times 10^{-3} \times 285 = 577$ watts

Hence, the power requirement can be equated with power dissipated by the paddles in the flocculator as

$$P_{p1} = 6.18\ N_1^3 = 577$$

Therefore,

$$N_1 = \sqrt[3]{\frac{577}{6.18}} = 4.47 \text{ rpm}$$

and $$v_{p1} = \frac{2\pi r N_1}{60} = \frac{2\pi \times 1.85 \times 4.47}{60} = 0.87 \text{ m/sec}$$

Compartment-II

$P_2 = G_2^2 \times \mu \times V = (35)^2 \times 10^{-3} \times 285 = 349$ watts

$P_2 = 6.18\ N_2^3 = 349$

Therefore, $N_2 = \sqrt[3]{\dfrac{349}{6.18}} = 3.79$ rpm

and $v_{p2} = \dfrac{2\pi r N_2}{60} = \dfrac{2\pi \times 1.85 \times 3.79}{60} = 0.73$ m/sec

Compartment-III

$P_3 = G_3^2 \times \mu \times V = (20)^2 \times 10^{-3} \times 285 = 114$ watts

$P_3 = 6.18\ N_2^3 = 114$

Therefore, $N_3 = \sqrt[3]{\dfrac{114}{6.18}} = 2.62$ rpm

and $v_{p2} = \dfrac{2\pi r N_2}{60} = \dfrac{2\pi \times 1.85 \times 2.62}{60} = 0.51$ m/sec

Therefore, the total power requirement of the tank can be summarized in Table 4.13 and shown in Fig. 4.38.

Table 4.13: Total power requirement of the tank

Chamber No.	G values (1/sec)	Velocity (m/sec)	N (rpm)	Power requirement, P (W)
1	45	0.87	4.47	577
2	35	0.73	3.79	349
3	20	0.51	2.62	114

Calculation of G:

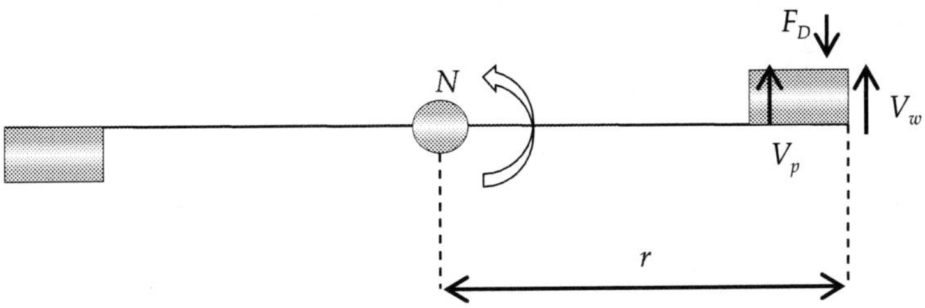

Fig. 4.38: Calculation of power requirement of tank (*P*)

$$G = \sqrt{\frac{P}{\mu v}}$$

$$P = F_D \times v_r$$

$$v_r = v_p - v_w$$

$$F_D = \frac{1}{2} C_D \rho_w A_T v_r^2$$

$$A_T = (b \times l)n$$

$$P = \frac{1}{2} C_D \rho_w A_T v_r^3$$

$$G = \sqrt{\frac{\frac{1}{2} C_D \rho_w A_T v_r^3}{\mu v}}$$

$C_D = 1.8$, for flocculation tank.

Example 4.25: A flocculation chamber 30 m long, 13 m wide & 4.5 m deep is to treat 75 MLD of water. It is equipped with 12 m long, 0.3 m wide paddles supported parallel to each other and moved by 4 horizontal shafts which rotate at the speed of 3.5 rpm. The centre-line of paddles is 1.8 m from the shaft which is situated at the mid-depth of tank. If the mean velocity of shaft is ¼ of velocity of paddle, compute (i) power consumption, (ii) time of flocculation, and (iii) value of G if v of water is 1.3×10^{-6} m^2/sec.

Solution: Size of paddle = 12 m (l) , $b = 0.3$ m , $N = 2.5$ rpm

1. Power consumption

$$P = F_D \times v_r$$

$$v_r = v_p - v_w$$

$$v_p = wr$$

$$w = \frac{2\pi N}{60}$$

$$v_p = \frac{2\pi Nr}{60} = \frac{2 \times 3.14 \times 2.5 \times 1.8}{60} = 0.47 \text{ m/sec}$$

$$v_w = \frac{v_p}{4}$$

$$v_r = \frac{3v_p}{4} = \frac{3 \times 0.47}{4} = 0.35 \text{ m/sec}$$

$$F_D = \frac{1}{2} C_D \, \rho_w A_T v_r^2$$

$$= \frac{1}{2} \times 1.8 \times 10^3 \times 0.3 \times 12 \times 8 \times 0.35^2$$

$$= 3175.2 \text{ N}$$

$$P = F_D \times v_r = 3175.2 \times 0.35 = 1.11 \text{ k watts}$$

2. Time of flocculation

$$t_D = \frac{V}{Q_D} = \frac{30 \times 13 \times 4.5 \times 24 \times 60}{75 \times 10^6 \times 10^{-3}} = 33.69 \text{ min}$$

3. Value of G if ν of water is 1.3×10^{-6} m²/sec

$$G = \sqrt{\frac{P}{\mu}}$$

$$G = \sqrt{\frac{1111}{1.3 \times 10^6 \times 10^{-3} \times 30 \times 13 \times 4.5}} = 22.06 \text{ sec}^{-1}$$

Example 4.26: Design a 3-stage horizontal flow baffled channel gravity flocculator with the following data:

Q = 100 MLD

G values in decreasing order = 70, 35, 20 / sec

Detention time of the tank = 25 minutes

Two units with equal flow values

Assume water depth = 2 m

As assumed above the gravity flocculator can be divided in two equal units with discharge through each unit = 100/2 = 50 MLD

$$\text{Flow through each unit} = 50 \text{ MLD} = \frac{50 \times 10^6 \times 10^{-3}}{24 \times 60 \times 60} = 0.57 \text{ m}^3/\text{sec}$$

Total volume of each unit of the flocculator, $V = Q \times t = 0.57 \times 25 \times 60 = 855 \text{ m}^3$

Assume total width of one unit = 15 m

Assume three compartments in each unit

$$\text{Therefore, width of one compartment} = \frac{15}{3} = 5 \text{ m}$$

$$\text{Total area of one unit} = \frac{855}{15} = 57 \text{ m}^2$$

Let the water depth in the tank = 2 m

Therefore, length of one unit $= \frac{57}{2} = 28.5$ m

Overall dimensions of the tank = 28.5 m × 15 m × 2.0 m

As we know that total width of the tank = 15 m and width of one compartment = 5 m with water depth = 2 m

$$\text{Surface area} = \frac{855}{2} = 427.5 \text{ m}^2$$

However, the number of baffles can be calculated by using equation:

$$n = \left[\frac{2\mu t}{\rho(1.44+f)}\left[\frac{HLG}{Q}\right]^2\right]^{\frac{1}{3}}$$

where, μ = Dynamic viscosity of water (Ns/m^2) = 10^{-3}

ρ = Density of water (kg/m^3) = 1000

t = Detention time per compartment (sec)

= Total detention time/no. of compartments $= \frac{25 \times 60}{3}$ =500 sec

f = Coefficient of friction = 0.3

Q = Water flow rate (m^3/sec) = 0.57 m^3/sec

L = Length of the channel = 28.5 m

G = Velocity gradient in compartment-I (1/sec) = 70

n = Number of baffles per compartment

Compartment-I

Putting the values in the above equation we get the number of baffles in compartment-I as:

$$n_1 = \left[\frac{2 \times 10^{-3} \times 500}{1000(1.44+0.3)}\left[\frac{2 \times 28.5 \times 70}{0.57}\right]^2\right]^{\frac{1}{3}} = \left[\frac{2 \times 10^{-3} \times 500}{1000(1.44+0.3)} 4.9 \times 10^7\right]^{\frac{1}{3}} = 294.4 \cong 30$$

The spacing in between each baffle can be calculated as

$$= \frac{\text{Length of the compartment}}{\text{Total number of baffles}} = \frac{28.5}{30} = 0.95 \text{ m}$$

Now head loss in chamber-I can be calculated as:

$$h_L = \frac{\mu\, t}{\rho\, g} G^2$$

where,

h_L = Head loss per chamber (m)

μ = Dynamic viscosity of water (Ns/m^2)= 10^{-3}

ρ = Density of water (kg/m^3)= 1000

t = Detention time per compartment (sec)

Total detention time/no. of compartments $= \frac{25 \times 60}{3} = 500$ sec

G = Velocity gradient (1/sec) = 70 for chamber-I

g = Acceleration due to gravity (m/sec^2) = 9.8

Putting the values in the above equation, we get

Head loss in chamber-I

$$h_{L1} = \frac{10^{-3} \times 500}{1000 \times 9.8}(70)^2 = 0.25 \text{ m}$$

The head loss increases with increasing number of baffles in the compartment. Therefore, head loss per spacing is given by equation

Head loss per baffle $= \frac{h_{L1}}{n_1}$

where, n_1 is the number of baffles in compartment-I.

Now, calculate the velocity through the chamber using the head loss equation which can be equated as:

$$\frac{h_{L1}}{n_1} = \frac{K}{2\,g}V_1^2$$

where, K = A constant with value of 1.5 (given)

g = Acceleration due to gravity (m/sec^2) = 9.8

V = Velocity through the baffles (m/sec) = ?

h_L = 0.25 m

Putting the value in the above equation the velocity in chamber-I can be calculated as

$$\frac{0.25}{30} = \frac{1.5}{2 \times 9.8}V_1^2$$

Hence, $V_1^2 = \frac{19.6}{180} = 0.108$

Therefore, $V_1 = 0.32$ m/sec

Now area of slot width in compartment-I can be calculated as:

$V = \frac{Q}{A}$ or $V_1 = \frac{Q}{A_1}$

where, $V_1 = 0.32$ m/sec

$Q = 0.57$ m^3/sec

Putting the value in the above equation, we get

$0.32 = \frac{0.57}{A_1}$

Therefore, $A = \frac{0.57}{0.32} = 1.78$ m^2

or $W \times H = 1.78$

But $H = 2$ m, therefore $W = \frac{1.78}{2} = 0.9$ m

Compartment-II

Similarly, putting the values in the above equation, we get the number of baffles in compartment-II as:

$$n_2 = \left[\frac{2\times10^{-3}\times500}{1000(1.44+0.3)}\left[\frac{2\times28.5\times35}{0.57}\right]^2\right]^{\frac{1}{3}} = \left[\frac{2\times10^{-3}\times500}{1000(1.44+0.3)}1.225\times10^7\right]^{\frac{1}{3}} = 18.44 \cong 19$$

The spacing in between each baffle can be calculated as

$$= \frac{\text{Length of the compartment}}{\text{Total number of baffles}} = \frac{28.5}{19} = 1.5 \text{ m}$$

Now head loss in chamber-II can be calculated as:

$$h_L = \frac{\mu\, t}{\rho\, g}G^2$$

where,

$t = 500$ sec

G = Velocity gradient (1/sec) = 35 for chamber-II

Other values remains the same.

Putting the values in the above equation, we get

Head loss in chamber-II

$$h_{L2} = \frac{10^{-3}\times500}{1000\times9.8}(35)^2 = 0.0625 \text{ m}$$

The head loss increases with increasing number of baffles in the compartment. Therefore, head loss per spacing is given by equation

Head loss per baffle $= \frac{h_{L2}}{n_2}$

where, n_2 is the number of baffles in compartment-II.

Now, calculate the velocity through the chamber using the head loss equation which can be equated as:

$$\frac{h_{L2}}{n_2} = \frac{K}{2\,g}V_2^2$$

where, K = A constant with value of 1.5 (given)

g = Acceleration due to gravity (m/sec^2) = 9.8

V = Velocity through the baffles (m/sec) = ?

$h_L = 0.0625$ m

Putting the value in the above equation the velocity in chamber-II can be calculated as

$$\frac{0.0625}{19} = \frac{1.5}{2 \times 9.8} V_2^2$$

Hence, $V_2^2 = \frac{19.6}{456} = 0.042$

Therefore, $V_2 = 0.20$ m/sec

Now area of slot width in compartment-II can be calculated as:

$$V_2 = \frac{Q}{A_2}$$

where, $V_2 = 0.20$ m/sec

and $Q = 0.57$ m^3/sec

Putting the value in the above equation, we get

$$0.20 = \frac{0.57}{A_2}$$

Therefore, $A_2 = \frac{0.57}{0.20} = 2.85$ m^2

or $W \times H = 2.85$

But $H = 2$ m

Therefore, $W = \frac{2.85}{2} = 1.425$ m

Compartment-III

Similarly, putting the values in the above equation, we get the number of baffles in compartment-III as:

$$n_3 = \left[\frac{2 \times 10^{-3} \times 500}{1000(1.44 + 0.3)}\left[\frac{2 \times 28.5 \times 20}{0.57}\right]^2\right]^{\frac{1}{3}} = \left[\frac{2 \times 10^{-3} \times 500}{1000(1.44 + 0.3)} 4 \times 10^6\right]^{\frac{1}{3}} = 12.88 \cong 13$$

The spacing in between each baffle can be calculated as

$$= \frac{\text{Length of the compartment}}{\text{Total number of baffles}} = \frac{28.5}{1.3} = 2.19 \text{ m}$$

Now head loss in chamber-III can be calculated as:

$$h_L = \frac{\mu\, t}{\rho\, g} G^2$$

where,

$t = 500$ sec

G = Velocity gradient (1/sec) = 20 for chamber-III

Other values remain the same

Putting the values in the above equation, we get

Head loss in chamber-III

$$h_{L3} = \frac{10^{-3} \times 500}{1000 \times 9.8}(20)^2 = 0.0204 \text{ m}$$

The head loss increases with increasing number of baffles in the compartment. Therefore, head loss per spacing is given by equation

$$\text{Head loss per baffle} = \frac{h_{L3}}{n_3}$$

where, n_3 is the number of baffles in compartment-III.

Now, calculate the velocity through the compartment using the head loss equation which can be equated as:

$$\frac{h_{L3}}{n_3} = \frac{K}{2\text{ g}} V_3^2$$

where, K = A constant with value of 1.5 (given)

g = Acceleration due to gravity (m/sec^2) = 9.8

V = Velocity through the baffles (m/sec) = ?

h_L = 0.024 m

Putting the value in the above equation the velocity in compartment-III can be calculated as

$$\frac{0.024}{13} = \frac{1.5}{2 \times 9.8} V_3^2$$

Hence, $V_3^2 = \dfrac{19.6}{812.5} = 0.024$

Therefore, V_3 = 0.16 m/sec

Now area of slot width in compartment-III can be calculated as:

$$V_3 = \frac{Q}{A_3}$$

where, V_3 = 0.16 m/sec

Q = 0.57 m^3/sec

Putting the value in the above equation, we get

$$0.16 = \frac{0.57}{A_3}$$

Therefore, $A_3 = \dfrac{0.57}{0.16} = 3.56 \text{ m}^2$

or $W \times H = 3.56$

But $H = 2$ m

Therefore, $W = \dfrac{3.56}{2} = 1.78$ m

The calculated head loss and slot width in different sections is summarized as in Table 4.14.

Table 4.14: Calculation of head loss and slot width

Chamber No.	G values (1/sec)	Number of baffles (n)	Spacing between the baffles (m)	Head loss (m)	Velocity (m/sec)	Slot width (m)
1	70	30	0.95	0.25	0.32	0.9
2	35	19	1.50	0.0625	0.20	1.4
3	20	13	2.19	0.024	0.16	1.8

Sedimentation Tank

The water after flocculation process is transferred to the sedimentation tank or settling tank that functions on the same principle as the plain sedimentation tank and designed with same approach. These sedimentation tanks are designed at lower detention time of around 2-4 hours with high value of surface loading values of 20-30 $m^3/d/m^2$ of plan area. These types of sedimentation tanks are generally designed as a combined process comprising flash mixing process, flocculation and then settling process to be carried out as a single unit as shown in Fig. 4.29. Such type of tank is generally known as sedimentation aided with coagulation tank where a plain floc chamber is provided without any mechanical devices before the water enters the settling chamber or sedimentation chamber. The floc chamber is designed at a detention period of about 15 to 40 minutes with depth half of the depth of settling chamber while the detention time of settling tank is considered in between 2 to 4 hours. The water from the mixing basin after flash mixing enters the flocculation tank, and thereafter the clarified water comes passed away from the outlet provide at the end of the tank. The tank is designed at higher depth values of 3 to 6 m as compared to plain sedimentation tank.

Example 4.27: Design a continuous flow sedimentation aided with coagulation tank for a town with 1,00,000 population with water requirement of 135 L/p/d. Assume any other required data suitably.

Solution: Given that:

Population = 1,00,000

Per capita water requirement = 135 L/p/d

Total water requirement = $1{,}00{,}000 \times 135$ L/d = 13.5×10^6 L/d = 13.500 m^3/d

(i) Design of settling chamber as shown in Fig. 4.39.

Water requirement = 13,500 m^3/d

Assume the maximum daily demand as 1.8 times the average daily demand

Therefore, total water to be treated = $1.8 \times 13{,}500$ m^3/d = 24,300 m^3/d

Assume detention time = 3 hours

$$\text{Volume of the settling tank, } V = Q \times t = \frac{24{,}300 \times 3}{24} = 3038 \text{ m}^3$$

Assume the surface overflow rate = 30 $m^3/d/m^2$ (range 24-30 $m^3/d/m^2$)

Therefore, $\frac{Q}{BL} = 30$

But $Q = 24{,}300\ m^3/d$

Putting the value of Q in the above equation, we get

$$\frac{24{,}300}{BL} = 30$$

Therefore, $\frac{24{,}300}{30}$ = B.L. = Surface area

Therefore, B.L = 810 m^2

Assume the horizontal velocity in the tank = 0.3 m/min

Therefore, length of the tank = $0.3 \times 0.3 \times 60 = 54$ m

Width of the tank $= \frac{810}{54} = 15$ m

Depth of the tank $= \frac{3038}{810} = 3.75$ m = 4.0 m

However, provide extra depth of 0.5 m for sludge storage from starting up to the end with slope (1 in 50) such that the end depth $= 4.5 + \frac{54}{50} = 4.5 + 1.08 = 5.58$

= 5.6 m

Use free board of 0.5 m at the top, therefore over all depth = 5.8 + 0.5 = 6.3 m

The dimensions of the settling tank are 54 m × 15 m × 4.0 m

(ii) Flocculation chamber

Assume the effective depth of the floc chamber

$= \frac{1}{2}$ depth of settling chamber at starting or near the floc chamber $= \frac{4.5}{2} = 2.25$ m

Assume the detention time of the flocculation chamber = 25 minutes

Volume of the flocculation chamber $= \frac{24{,}300}{24} \times \frac{25}{60} = 422\ m^3$

The cross-sectional plan area of the floc chamber $= \frac{V}{\text{depth}} = \frac{422}{2.25} = 188\ m^2$

The width of the tank = 15 m

Therefore, the length of the tank $= \frac{188}{15} = 12.5$ m

Details of the tank are given below:

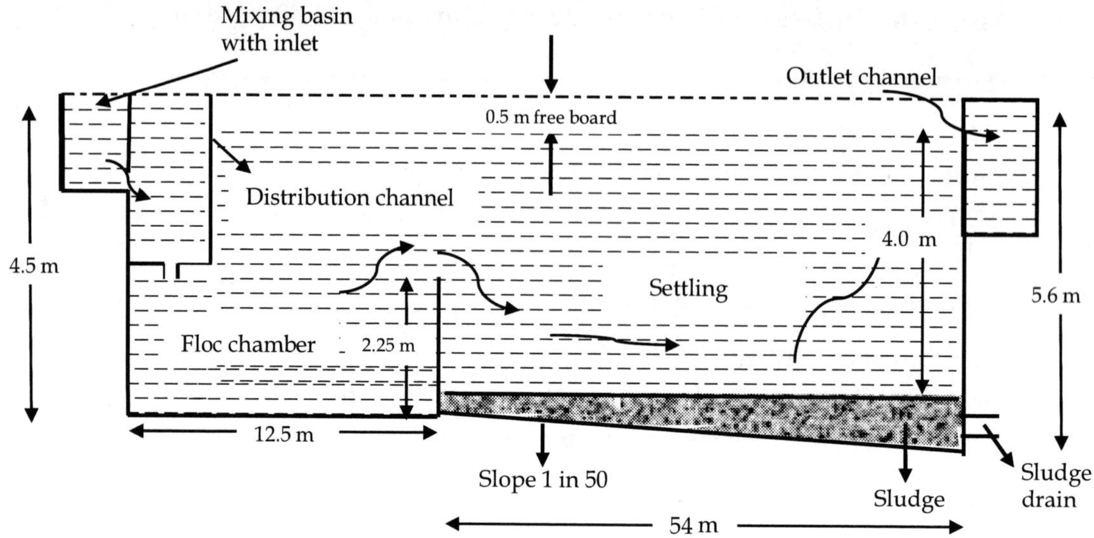

Fig. 4.39: Design of sedimentation aided with coagulation tank with computed specifications

Sludge production

The suspended solids and colloidal particles removed in the sedimentation process settle down as sludge at the bottom of the settling chamber or clarifier. This settled sludge is to be removed continuously or should be drained out using sludge pipes. The quantity of the sludge produced daily should be computed in order to provide the suitable sludge zone depth in the settling chamber. This sludge should be removed periodically and a proper disposal facility needs to be designed based on the volume of sludge produced daily. The volume of sludge produced can be calculated on the moisture content and concentration of solids in it. For example, sludge with 97% moisture content indicates 3 kg solids in 100 kg of wet sludge and therefore 1 kg of solids in 50 kg of wet sludge. However, the density of the sludge is slightly more than that of water and is generally taken as 1240 kg/m^3 for computation of volume of sludge production.

Also the sludge produced comprises coagulant added to the system and the total solids removed in the settling process. The amount of sludge produced is taken as 0.24 times the amount of alum added to the system. This can be computed using Eq.4.80 below:

$$Al_2(SO_4)_3{\cdot}18H_2O + 3Ca(HCO_3)_2 \xrightarrow{\text{yields}} 3CaSO_4 + \underset{\text{(Al hydroxide ppt.)}}{2Al(OH)_3 \downarrow} + 6CO_2 \uparrow \; + 18H_2O \tag{4.80}$$

666 g $\qquad$ $2 \times 78 = 156$ g

1 mole of alum produces = 2 mole of aluminium hydroxide as sludge

666 g of alum produces = 2×78 g of aluminium hydroxide

1 g of alum produces = 156/66 = 0.24 g of aluminium hydroxide

The other chemicals, salts and other polymers added during the coagulation process also result in production of equal amount of sludge as much amount they are added in the system. The amount of solids removed in the process is also taken equivalent of sludge produced. Therefore, the total amount of dry sludge produced can be computed using Eq. 4.81;

$$\text{Sludge produced}\left(\frac{mg}{d}\right)$$

$$= [\text{Suspended solids removed}\left(\frac{mg}{L}\right) + 0.24 \times \text{alum dose}\left(\frac{mg}{L}\right) + \text{other miscellaneous chemicals}\left(\frac{mg}{L}\right)] \tag{4.81}$$

Testing for determining optimum

It was pointed out earlier that the best coagulant dose is determined first in the laboratory and then adjusted by actual observations at the treatment plant. The common test which is performed to determine this approximate optimum quantity of coagulant is known as *jar test.*

The sample of raw water to be tested is placed in a number of jars.

Alum, which is chiefly used as a coagulant has been found to be generating the sludge equal to 0.24 times of its own weight, The addition of other miscellaneous chemicals like clay, carbon, polymers, eat. during the coagulation process may further produce sludge equal to the amounts of the substance added. The suspended solids removed from the water will also produce equal amount of sludge. These three factors, contributing to the sludge can be written down in the form of a mathematical equations 4.80 & 4.81.

Optimum coagulant dose using jar test apparatus

It was pointed out earlier that the best coagulant dose is determined first in the laboratory and then adjusted by actual observations at the treatment plant. The common test which is performed to deter mine this approximate optimum quantity of coagulant is known as jar test (Fig. 4.40)

- The jar test procedures involves the following steps:
- The jar test apparatus consists of 6 jars of 1 L capacity with flocculation device consisting of paddles. Fill all the jars with water sample for which the optimum dose is required.
- One jar will be used as a control while the coagulant dose in varying amounts is added in other 5 jars.
- The pH of the jars can be adjusted or variations of coagulant dosages can be added to determine optimum operating conditions.
- After addition of the coagulant in jar it is stirred vigorously at the rate of 100 rpm for 1 minute. The rapid mix stage helps to disperse the coagulant throughout each container.

- Thereafter reduce the stirring speed to 25-35 rpm and allow the mixing for 15 to 20 minutes. This slow mixing speed helps promote floc formation by enhancing particle collisions which lead to larger flocs.
- Turn off the mixers and allow the containers to settle for 30 to 45 minutes. Then measure the final turbidity in each container.
- Plot a graph in between residual turbidity vs. coagulant dose and determine the optimal dose of the coagulant by drawing an intercept from the maximum turbidity removal value. The values that are obtained through the experiment are correlated and adjusted in order to account for the actual treatment system.

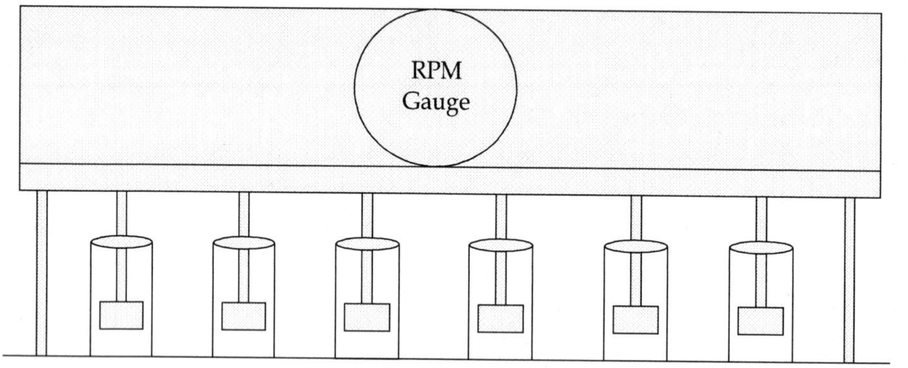

Fig. 4.40: Jar test apparatus used for the determination of optimum dose of coagulant

Type-II settling

The colloidal particles that can change their shape and size with respect to time are called flocculent particles. The process of removing such particles by addition of various chemicals as coagulants such as to form the heavier mass of flocs that can settle under mere action of gravity is termed flocculation and categorized as type-II settling or flocculant settling. This type-2 settling of flocculating particles is carried by using a dilute suspension and the behaviour of particles is entirely different from discrete particle suspensions. The Stokes' law is not applicable in this process as the particles continuously change their shape and size with respect to time. A number of factors such as large aggregate of particles collection, specific gravity because of entrapment of water in the interstitial space contribute to the flocculation process therefore no general formula has been developed for determining the settling velocities of such particles.

The settleability analysis of flocculating suspension is just similar to type-I settling but a slight modification in the column is required before the test has to be carried out. In spite of one port at the bottom of the column as in case of type-I suspension, here a number of ports at definite depth intervals are provided in the column. In the settling column the samples from each port at different depths has to be carried out at every time interval and analyzed for suspended solids concentrations. In this case instead of mass fraction remaining the mass fraction removed at each depth and for each time has to be computed by using these analysed concentrations.

$$x_{ij} = \left(1 - \frac{C_{ij}}{C_o}\right) \times 100 \tag{4.82}$$

where, x_{ij} is the removal of mass fraction in percent at the ith depth at the jth time interval as shown in Eq. 4.82. The values depicted is a graphical representation in the form of contour map indicating the isoremoval lines drawn by using graphical method. The instantaneous velocity of the fraction of particles is represented by the slope at any point on any given isoremoval line. As the slope of the isoremoval lines becomes steeper the velocity becomes higher at greater depth. The continued collision and aggregation of particles with other particles thus increases the size of these particles and subsequently increases the settling velocity with respect to time. However, at a given detention time and determining the concentration of suspended solids of samples from different ports at different depths with respect to time, the overall percentage removal can be assessed as shown in example below. However the total removal efficiency can be carried out by using the Eq. 4.83 as:

$$R = r_o + \sum \frac{\Delta r \cdot H_i}{H_o} \tag{4.83}$$

where, r_o is the percentage of solids of specific size will reach the 3-m depth in time t_o and will be removed 100 per cent. Also the term Δr is the difference between two consecutive isoremoval and H_i represents values on the y-axis obtained by drawing an intercept from the midpoints of consecutive isoremoval lines in each difference. The term H_o represents the height or depth of column used for the study as shown in Fig. 4.41.

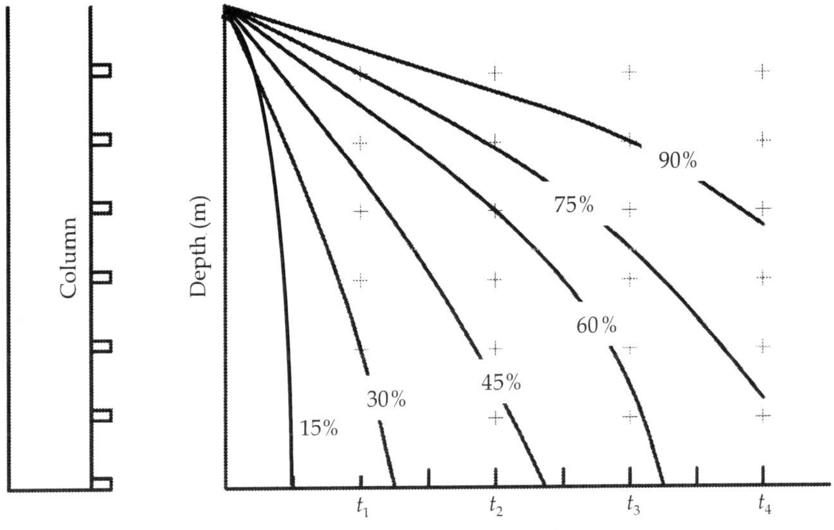

Fig. 4.41: Isoremoval lines in type-II settling analysis

Example 4.28: A settling column analysis of flocculating particles with a suspension having initial solids concentration of 300 mg/L is run in a laboratory. The obtained results are given in Table 4.15.Calculate the overall removal efficiency of a 3 m deep settling basin with a detention time of 2 hours.

Table 4.15: Concentration of sample vs Depth

Depth (m)	Sampling time (min)					
	30	60	90	120	150	180
0.5	140	85	62	45	38	25
1.0	185	135	100	78	65	48
1.5	210	160	125	102	82	65
2.0	220	175	147	122	100	82
2.5	230	195	160	135	112	91
3.0	240	205	178	145	125	102

Solution:

1. At a given suspended solids concentration C_i values with respect to initial C_o concentration, the removal rate at each depth and time can be computed using below mentioned equation:

$$x_{ij} = \left(1 - \frac{C_{ij}}{C_o}\right) \times 100$$

and tabulated below in table 4.16

Table 4.16: Removal of solids at each depth and time

Depth (m)	Sampling time (min)					
	30	60	90	120	150	180
0.5	53	72	79	85	87	92
1.0	38	55	67	74	78	84
1.5	30	47	58	66	73	78
2.0	27	42	51	59	67	73
2.5	23	35	47	55	63	70
3.0	20	32	41	52	58	66

2. Plot iso-concentration lines and construct vertical line at a detention time of 120 minutes (Fig. 4.42);

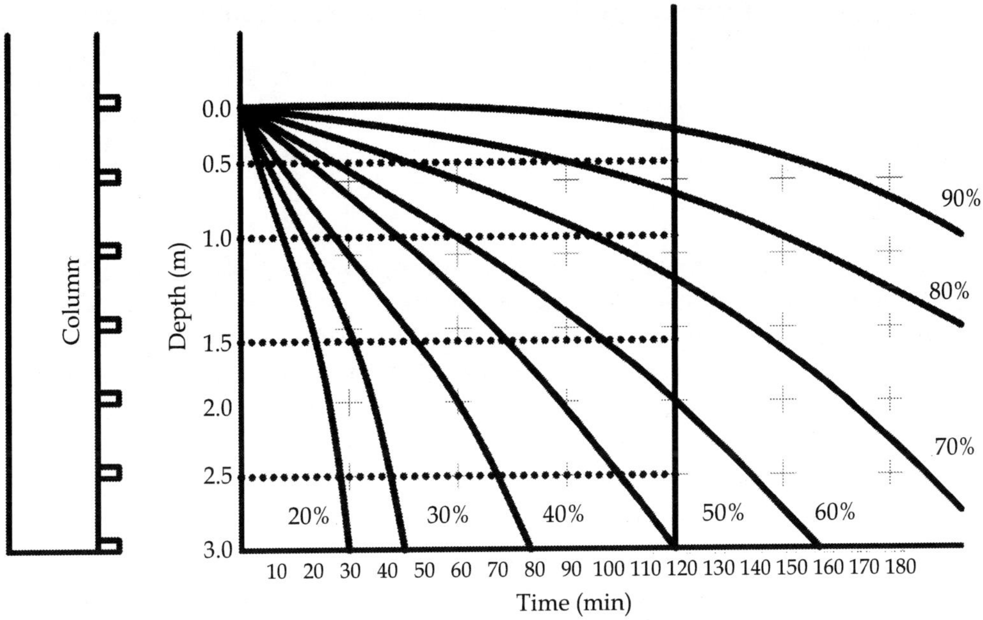

Fig. 4.42: Isoremoval lines in type-II settling analysis

3. From Fig. 4.42, it is observed that approximately 50 per cent of the solids of specific size represented as r_o will reach the 3-m depth in time t_o and will be removed 100 per cent. However, some percentage of the remaining particles will be removed and required to be computed. This can be worked out by considering the mid-point of each consecutive removal lines existing on the line of detention time and drawing intercept from these midpoints thereby fixing the mid-depth values on the *y*-axis. The difference in between two consecutive removal percentages is denoted by Δr while the respective mid-depth values of intercept on Y-axis as H_i. The multiplication of each Δr with H_i values and the summation of all the obtained values can be put up in the equation thus the total removal efficiency of the tank can be determined as shown in Table 4.17.

Table 4.17: Calculation of total efficiency of tank

S. No.	Difference in removal efficiency (Δr)	Mid-depth values on the *y*-axis (H_i)	$\Delta r \cdot H_i$
1.	0.1	2.5	0.25
2.	0.1	1.5	0.15
3.	0.1	0.8	0.08
4.	0.1	0.4	0.04
		Σ	0.52

Therefore, the overall removal efficiency of the settling basin can be determined,

$$R = r_o + \sum \frac{\Delta r \cdot H_i}{H_o}$$

$$= 0.50 + \frac{0.52}{3} = 0.50 + 0.173 = 0.673 = 67.3\%$$ **Ans.**

Tube settlers

In a conventional sedimentation process the settling efficiency of discrete particles mainly depends on the surface area of the sedimentation tank and is independent of the depth of the tank. Therefore the sedimentation tanks are constructed with higher surface area and are made as shallower as possible. But due to high cost and low availability of the land the tanks are being constructed with high volume by providing sufficient water depth. Due to such practical problems and land restrictions, the plain sedimentation tank is constructed with a depth of 3 to 5 m. In order to overcome such problems an attempt has been made by constructing the horizontal trays and placing them in the sedimentation tank.

These trays are placed such that and the tank is divided into various compartments with respect to its height in order to reduce the settling depth of the particle in the tank. This will provide the multiple surface areas while increasing the surface area of the tank and will reduce the flow rate thus enhancing the particle settlement with high efficiency in the tank. However, the use of such wide shallow trays in conventional basins directly has not been proved very much successful due to the various problems such as clogging and odours due to biological growth, build-up of oil and grease in the tank. However, very small diameter tubes having high wetted perimeter relative to wetted area, providing laminar flow conditions and low surface loading rate, have shown promising results. Such tube settling devices are called tube settlers, have been found to provide an excellent clarification with detention times of equal to or less than 10 minutes.

A group of tubes with very small diameter having high wetted perimeter relative to wetted surface area ensuring laminar flow conditions and low surface loading rates are called tube settlers. It provides a very good efficiency with detention time as low as 10 minutes. These square or circular tubes made of black colour plastic sheets with thickness around 1.5 mm are available in market. The tubes made of asbestos cement pipes have also been used many times. A large number of such tubes with size of 50 mm x 50 mm or smaller with height of 1 m or so, are arranged together side by side as a bundle called module or a tube settler module. These arrangements are made either in rectangular or circular or square in shape and are called tube settler module. The usual dimensions of such a rectangular module is about 5 m × 5 m, with height of 1 m is placed in the tank with an inclination of 60° angle for continuous gravity drainage of setlleable material.

The tubes in such a module can be placed either horizontally or steeply inclined. These tubes when placed inclined to about 60° to horizontal facilitates the continuous gravity drainage of the settleable material. In case of horizontal tubes that are normally inclined at 5° supports scouring of settled solids. However, with the increase in the angle of inclination (θ), the settlement of the tubes with the horizontal decreases as shown in Fig. 3.18, and at an inclination angle greater than 40° the efficiency decreases drastically. However, the sludge does not slide automatically with angles less than 60°. Therefore, in order to improve and enhance the performance and efficiency of the existing plants the tube settlers are inclined at an angle of 60°. However, the experimental results claim that placing 1/5th part of the outlet end of the basin with tube settlers doubles the surface area and therefore the effective surface loading on the tank is nearly halved. This indicates that the flow through the tank can be doubled without impairing the effluent quality or suspended solids removal efficiency. A number of proprietary devices such Lamella clarifier based on such principle as shown in Fig. 3.19 is available in market and are generally used for enhancing the sedimentation process of the tank.

Design of tube settlers: The efficiency of the tube settlers is generally determined by a performance parameter known by a symbol 'S' which can be calculated by using Eq. 4.84 as:

$$S = \frac{V_s}{V_o}\,[\sin\theta + L\cos\theta] \tag{4.84}$$

where,

V_s = Settling velocity of the particle vertically downward (m/sec)

V_o = Velocity of the flow along the tube (m/sec)

θ = Angle of inclination

L = Relative effective length of the settler = l/d (dimensionless)

l = Length of the tube settler (m)

d = Diameter of the tube settler (m)

Under ideal conditions it is recommended that when the S value is equal to or exceeding the S_c value for any type of particle, it is completely removed in tube settlers. However, assuming the uniform flow conditions, the S_c values under laminar flow regime conditions as recommended is prescribed to be $\frac{4}{3}, \frac{11}{8}$ and 1 for circular, square and parallel plates of the tube settler. The performance with L values up to 20 has been found to be significantly high and decreases as the value lowers from 20. Therefore, it is suggested that the tube settlers should be designed at L value in the range of 20-40 value.

However, in order to change the transition zone near the inlet to fully laminar conditions the dimensionless L value has to be increased to L' value additionally. Therefore, the equation can be further modified as Eq. 4.85 –Eq. 4.86 shown below;

$$L + L' = \frac{l}{d} \tag{4.85}$$

But,

$L' = 0.058\ R_e$ and $R_e = \frac{V_o d}{v}$

where, d = Diameter of the each tube settler

v = Kinematic viscosity of water

Therefore,

$$L' = \frac{0.058\ V_o d}{v} \tag{4.86}$$

The factor reduces the effective length of settler $(L) = \frac{l}{d} - L'$

Example 4.29: Design a tube settler module at a flow rate of 250 m^3/h with the following data:

(a) Water reduction during desludging = 3% of output required
(b) Settling velocity of the smallest particle to be removed = 125 m/d
(c) Tubes cross sections = 50 mm × 50 mm
(d) Length of the tube = 1 m
(e) Angle of inclination of the tubes = 60°
(f) Kinematic viscosity of water = $1.01 \times 10^{-6}\ m^2/sec$

Solution:

Average design flow $= \frac{250}{0.97} = 258 \frac{m^3}{h}$

Relative length of settler $= \frac{l}{d} = \frac{1\ m}{0.05\ m} = 20$

Let us increase L by L^1 where

$$L^1 = 0.058 \frac{v_o . d}{v} = \frac{0.058 \times \quad \times 0.05}{1.01 \times 10}$$

$$= 1.87 \times \frac{10^3\, v_o m}{s} = \frac{2.8 \times 10^3 . v_o}{24 \times 60 \times 60} = 0.033 v_o \left(\frac{m}{d}\right)$$

Effective relation length of settler

$$L = \frac{l}{d} - L^1 = 20 - 0.033\ V_o$$

Now using Eq. 4.84, we get

$S_c = \frac{V_s}{V_o}[\sin\theta + L\cos\theta]$, as we know that the critical value of S for square type settler is $\frac{11}{8}$ as stated earlier.

$$\text{Therefore, } \frac{11}{8} = \frac{125\left(\frac{125}{d}\right)}{V_o}[\sin 60° + (20 - 0.033\,v_o)\cos 60°]$$

or
$$\frac{11}{8} = \frac{125\left(\frac{125}{d}\right)}{V_o}[0.866 + (20 - 0.033v_o)0.5]$$

$$v_o = \frac{125 \times 8}{11}[10.866 - 0.0165\;v_o]$$

$$v_o = 1011.62 - 1.536\;v_o$$

$$v_o = \frac{1011.62}{2.536} = 399\text{ m/d}$$

Total entrance area of tubes $A = \frac{Q}{v}$

$$= \frac{258 \times 24\text{ m}^2/\text{d}}{399\text{ m/d}} = 15.52\text{ m}^2$$

Number of tubes having size 50 mm × 50 mm can be calculated as

$$n = \frac{\text{Total area}}{\text{Area of tubes}} = \frac{15.52}{0.05 \times 0.05} = 6208$$

Let us provide 79 tubes along the length and 79 tubes along the width of module.

Therefore, the total number of tubes in actual = 79 × 79 = 6241

Also, the length of the tube module can be calculated as with tube thickness assumed to be 15 mm = No. of tubes [inside dimension of square tube + 2 × thickness of each tube]

= 79[0.05 + 2 × 0.0015] = 4.18 m ≈ 4.2 m say

Also the height of the tube module for 1m length of tubes and placing them at an inclination of 60° angle

$$H = 1\sin 60° = 0.866\text{ m} = 0.87\text{ m}$$

Therefore, overall dimension of the tube module is

Over all dimension = 4.2 m × 4.2 m × 8.7 m **Ans.**

EXERCISE

1. Derive an expression for settling velocity of a spherical particle in a liquid under laminar flow condition.
2. Design coagulation-cum-sedimentation tank with continuous flow for a population of 75,000 persons with a daily per capita water allowance of 120 L. Make suitable assumptions where needed.

3. Describe the laboratory jar test method used to calculate the optimum dose of the coagulant used in water treatment.
4. A coagulation treatment plant with a flow of 0.5 m^3/sec is dosing alum at 23 mg/L. No other chemicals are added. The raw water suspended solids concentration is 37 mg/L. The effluent suspended solids concentration is measured as 12 mg/L. The sludge content is 1% and the specific gravity of the sludge solids is 3.01. What volume of sludge must be disposed of each day?
5. A WTP is being designed to process 50,000m^3/d of water with alum dosage of 40 mg/L and at a Gt value of 4×10^4 produces optimal results at water temperature 20°C. Determine (i) the monthly alum requirement, (ii) flocculation basin dimensions if three cross-flow horizontal flow paddles are to be used with flocculation tank size 12 m wide and 5 m depth, and (iii) the power requirement with paddle configuration, and (iv) paddle speed.
6. A rectangular sedimentation tank following coagulation-flocculation is to treat a flow of 3000 m^3/day with a detention time of 6 hours. It is to be hand cleaned of sludge at 6 weeks intervals. The suspended solids concentration of the water is reduced from 250 mg/L to 5 mg/L by coagulation-flocculation. The settled sludge includes 40 mg/L (based on water flow) of metallic precipitate and has a moisture content of 85% and specific gravity of 1.24. What are the basic dimensions of the tank if the water depth just before cleaning is 3 m and its length is twice its width?
7. Describe the following:
   (a) Use of copperas as coagulant with suitable equations.
   (b) Use of chlorinated copperas as coagulant with suitable equations.
8. Prove that the area and overflow rates rather than detention time govern the design of settling tank.
9. A rectangular sedimentation basin is to handle 10 MLD of raw water. A detention basin of width to length ratio of 1/3 is proposed to trap all particles larger than 0.4 mm in size. Assuming a relative density of 2.65 for the particles at average temperature of 20°C, compute the basin dimensions. If the depth of the tank is 3.5 m , calculate the detention time.
10. Draw a neat sketch of water treatment plant for a city.
11. Design criteria used for design of screens in a WTP for a city.
12. Design criteria used to design a sedimentation tank for type-I settling.
13. Design criteria used to design a sedimentation tank for type-II settling.
14. How you will calculate the power requirement for mechanical flocculator?
15. Define SOR. what values would you recommend for design of plain sedimentation tank and sedimentation aided with coagulation tank.
16. Write down the equation to check scour velocity in a sedimentation tank.

17. Describe the reaction equation when alum is used as a coagulant for water treatment.
18. How much alum sludge would be produced in a batch operation if 150 kg of alum is used for a water treatment process and no other chemical has been used?
19. Describe the reaction equation when copperas is used as a coagulant with lime for water treatment.
20. How much copperas sludge would be produced in a batch operation if 200 kg of alum is used for a water treatment process and no other chemical has been used?
21. At what pH values the chlorinated copperas or its resultant products work well?
22. What do you understand by G.t factor? What values would you recommend for flash mixing and flocculation process?
23. Describe the Hazen's equation in terms of settling velocity and temperature of water.
24. How you will calculate the displacement efficiency of a sedimentation tank?
25. On a given discharge how can the efficiency of a sedimentation tank can be increased and why so?
26. Describe a suitable equation to calculate the detention time of a circular sedimentation tank.
27. Design a 3-compartment mechanical flocculator with a configuration of paddles ⊥ to the flow with a flow rate of 15000 m^3/d and G values of 45, 30, 20 per sec in 3 successive compartments with total detention time of 30 min. Assume minimum clear spacing between outer paddle and floor and water surface to be 20 cm and the distance of the paddle from the wall to be 20 cm. The maximum depth of flocculator may be taken as 4.3 m with a Cd value of 1.8 and velocity of the liquid may be assumed as 75% of paddle speed.

REFERENCE

1. Camp, T.R. 1946, Sedimentation and the design of settling tanks. *ASCE Trans.*, p. 895.

5

CHAPTER

Filtration

The suspended solids and organic matter present in the raw water is generally removed through the screening process and large amount of it is removed during the plain sedimentation or sedimentation aided with coagulation process. However the effluent from sedimentation tank still contains fine colloidal particles, certain amount of coagulants and chemicals being added in the sedimentation and coagulation process and some amount of bacteria and microorganisms present in the water. Therefore in order to reduce or to remove the remaining impurities present in water and to make it potable and palatable, the water is allowed to pass through a bed of granular material such as sand, gravel, carbon or other such materials. The process of passing the water through a bed of such a granular materials (media) is known as filtration. The filtration process also helps in eliminating colour, odour, turbidity, and pathogenic bacteria from the water. The process of filtration not only removes 95 to 98% impurities but also is capable enough to remove bacteria and other microorganisms from water to a greater extent. Therefore, filtration is considered to be an economical, chemical-free technique implemented in the water treatment processes. Three types of filtration systems are generally used in the water treatment for public water supplies:

(a) Slow sand gravity filters

(b) Rapid sand gravity filters

(c) Pressure filters

The slow sand gravity filters (also known as slow sand filters) where the water flow takes place by gravity are capable of removing high amount of impurities and bacteria from the water as compared to rapid gravity filters. However, due to low rate of filtration in case of slow sand filters , the yield is lower than rapid gravity filters and is around 30 times less. Therefore, slow sand filters require large area and proves to be costly. However, with rapid increase in population and less availability of the land and also due to recent advancements in the area of water treatment has led to the use of rapid sand filters. Therefore, the slow sand filters are outdated and not in much use nowadays. The theory and principle behind filtration process is an important aspect that needs to be understood before designing these units.

5.1 THEORY OF FILTRATION

The filtration process can be well understood by discussing the following mechanisms that generally occur during the water treatment.

(i) **Mechanical straining:** The mechanism behind mechanical straining is that the suspended particles or impurities in water possess size greater than the voids of the filter media such as sand and therefore gets detained in these voids. Therefore the water passing through this filter media is devoid of such impurities and clear water is obtained at the effluent end of the filtration unit. This continuous detainment of impurities thus forms a layer on the top of filter bed and therefore further enhances the process of removal of impurities from water.

(ii) **Flocculation and sedimentation:** This can be well understood by coagulation and flocculation theory as the voids of the filter media acts as tiny sedimentation and flocculation tanks where the retained particles or impurities attract other impurities thus forming gelatinous flocs and the adsorption of colloidal and dissolved inorganic matter in the surface of sand grains in the form of thin film takes place thus removing other impurities from water.

(iii) **Electrolytic action:** In this mechanism the phenomena of ionization and charge neutralization take and is based on the fact that the filter media and impurities in water possess opposite charges. Therefore, when the impurities present in water come in contact with the filter media it changes the chemical characteristics of water and thus the particles or impurities are attracted to the filter media due to opposite charges in them. This electrolytic charges present on the surface of the sand particles or filter media makes it possible to bind the impurities towards them and therefore the clear water free from these impurities flows towards the effluent of the unit.

(iv) **Biological metabolism:** The microorganisms present in the water are arrested in the filter media voids or they may be initially present in the media voids itself. These microorganisms require organic matter as a food for their survival. The impurities present in the water in the form of organic matter such as algae, phytoplanktons and other materials are utilized by these microorganisms and degrade them to the harmless end products. This forms a film on the top of the media bed and is also known as Schmutzdecke or the dirty skin which also helps in arresting and removing other impurities from water. This natural biological activity due to a film of microorganisms on the top of the filter media helps in removing organic impurities from water.

5.1.1 Filter Materials

(i) **Sand:** In general, sand is a locally available filter material and is designated as fine or coarse type. The sand size is measured by the term known as effective size and is represented by the term D_{10}. This effective size D_{10} can be described as the size of sieve in mm through which ten per cent of the sand sample by mass will pass. The size of the sand is an important aspect in designing of filtration unit as low size of sand will lead to frequent clogging of the filter media and units. However, too large size of sand will allow the suspended particles and bacteria present in water without their removal. Therefore, the uniformity in the size or degree of variations in the size of in particles is denoted by a term known as *uniformity coefficient.*

The uniformity coefficient expressed in terms of $\frac{D_{60}}{D_{10}}$ may be defined as the ratio of the sieve size in mm through which 60% of the sample of the sand will pass to the effective size of sand. The various properties of the sand include that it should be free from dirt and other impurities and uniform in shape and size. The sand to be used for filtration should be hard and resistant. Also it should be such as not lose more than 5% of its weight after being placed in hydrochloric acid for 24 hours.

(ii) Gravel: The gravel layer is placed below the sand or filter media so as to support the filter media and therefore allowing the water being filtered to travel freely to the underground drains. It also allows the wash water to travel upward freely during the filter backwash and therefore helps in removing the arrested impurities present in the voids of the filter media or sand.

(iii) Other materials: Nowadays the recent advancements in various water treatment techniques and processes the sand has been replaced by certain other filter materials such as anhtrafilt, coal, activated carbons and other such materials. These media can either be used as single or in combination of the two or three types. The filter units comprising of two types of filter media are known as dual media filters or if it comprises more than two types of filter media it is known as multimedia filter units. Anthrafilt is often used as a filter media, rather than using sand. Anthrafilt is made from anthracite, a form of coal-stone that burns without flames or smoke. It is cheaper, and provides a high rate of filtration as compared to sand filters.

Classification and types of filters

Filters are classified on the basis of the filtration rate and the driving force to overcome the frictional resistance encountered by the water flowing through filter. The driving force to overcome the frictional resistance encountered by the water flowing through filter can be either the force of gravity, or the applied pressure force. As such on the basis of the driving force the fillers are classified as:

(i) Gravity filters comprising of (a) slow sand filters, and (b) rapid sand filters

(ii) Pressure filters

However, the classification based on rate of filtration is categorized as

(i) Slow sand filters
(ii) Rapid sand filters comprising of (i) Rapid gravity filters, and (ii) pressure filters

5.1.2 Capping of Sand Filters

Rapid sand filters can be converted to increase efficiency by 100 per cent for mixed media service. This conversion costs a lot less than installing additional strong sand filters. Capping requires a part of the sand being covered with anthracite. A 2-6 in, in this conversion, the 0.4-0.6 mm (0.016-0.024 inches) layer of sand is stripped from the bed surface and replaced with 4-8 inches. For anthracite of 0.9 mm (0.035 inches) when capacity improvements are needed a greater amount of sand would be replaced. Pilot tests are performed to ensure that the effluent content is not decreased by a decrease in the depth of the finer sand.

5.2 SLOW SAND FILTER

(i) The slow sand filters were first developed in England by James Simpson in 1829 working for Chelsa Water Co. Before the invention of rapid gravity filters these type of slow sand filters were used by the last decade of the 19th century. Recent advancements in the filtration systems and processes has resulted in the slow sand filters as an outdated technology. However, these types of filtration units are still in use on small scale plants at tropical and warmer climate places. The treated effluents from sedimentation tanks is taken to slow sand filter units where it is treated to maximum extent and clear water is obtained. For small towns slow sand filters are ideally suited. The sand used for filtration is determined by the effective size and uniformity coefficient. Most impurities, including bacteria and viruses, are eliminated from the raw water when they pass through the filter skin just below and the filter bed sand sheet. When the microorganisms are well developed, the filter works efficiently and generates high-quality effluent that is completely free from disease-bearing organisms and biodegradable organic matter.

5.2.1 Construction

The slow sand filter consists of a top layer of effective size 0.2-0.3 mm fine sand with uniformity coefficient in the range of 2 to 3. The thickness of the filter media layer lies in the range of 75 to 90 cm. The layer of coarse sand of size 0.2 -0.4 mm placed below the fine sand layer with voids in it does not allow the fine sand to move through it. This layer can be 90-110 cm thick. The filter media rests on the gravel layer of size 6 to 65 mm as lower layer and is around 30 to 75 cm thick. The gravel is packed in layers to the point that the smallest sizes are on top. The gravel layer is the retainer for the coarse sand layer and is spread over an open joined clay pipe or concrete pipes called under-drainage network. Water collected from underwater wastewater passes. A typical section of slow sand filter is shown in Fig. 5.1.

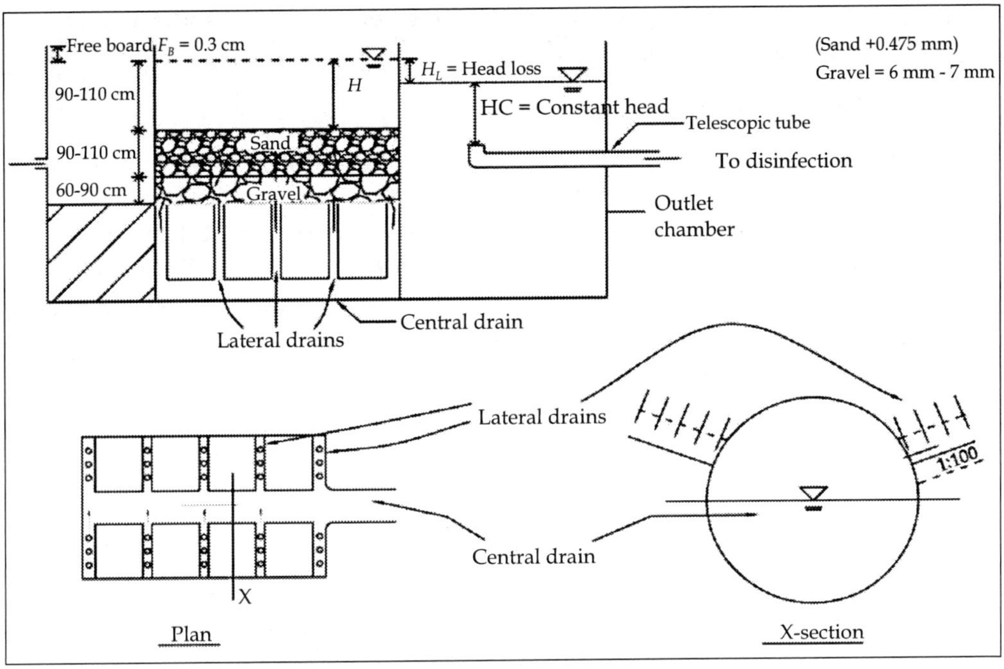

Fig. 5.1: Slow sand filter used in water purification process

(i) Enclosure tank: It consists of concrete or masonry made open water tight rectangular tank provided with bed slope of around 1 in 100 towards the central drain of collecting the treated water. The tanks are designed for a plan area in between 100-2000 m^2 or more with tank depth to be taken in between 2.5 to 3.5 m. The size may be increased or decreased as per the amount of water to be treated per day.

(ii) Filter media: The filter media consists of sand or other filtering materials such as anthrafilt, granular carbon and in case of sand it is placed in layer of 90 to 110 cm depth over the gravel support. The sand with effective sand size (D_{10}) value ranging from 0.2 to 0.4 mm and uniformity coefficient (D_{60}/D_{10}) value in the range of 1.8 to 3 is generally used for effective filtration process. The very fine uniform sand layer of around 15 cm thickness is placed at the top of the filter media. The sand of different gradation is used in the filter media with coarsest at the bottom above the gravel and finest to be placed at the top. The finer the sand better is the quality of water at the effluent.

(iii) Base material: The gravel is generally used to support the filter media in the filter unit. It is used as a base material with layer depth ranging from 30 to 75 cm comprising of diverse sizes and are placed in three to four layers with each layer of depth around 15 to 20 cm. The coarsest layer is placed at the bottom and finest gravel layer is placed at the top where the filter media rests. The bottom gravel layer consists of size 40 to 65 mm along with intermediate layer of size is 20 to 40 mm. The fine intermediate layer consists of size 6 to 20 mm and the topmost layer of size 3 to 6 mm.

5.2.2 Under-Drainage System

The under-drainage system consists of a central drainage system along with lateral drains that collect the treated water from all sides of the filter unit and the gravel is supported on the top of this under-drainage system. As shown in Fig. 5.1 these lateral drains are open joined pipe drains or some other form of porous drains sloping towards the central drain and are placed at centre-to-centre distance of 3 to 5 m apart on the bottom of the floor. However, sometimes the main drain is placed at the side of the tank with laterals sloping towards it.

(i) **Inlet and outlet arrangements:** The effluent from sedimentation tank enters the filtration unit through the inlets provided at the top of the tank that distributes the entire water evenly and uniformly on the filter bed. The outlet end is provided with filter water well that collects the filtered water from the central drain. It also maintains a constant water head and uniform discharge from the filter effluent. The inlet and outlet are provided in such a manner so as to maintain a proper head and to minimize the head loss during the operation of the filter unit. The automatic valves are provided at the inlet and outlet of the unit and the flow and other operations including backwashing is governed by these valves.

(ii) **Other appurtenances:** For smooth and efficient functioning of the filter certain appurtenances are generally provided at various desired sections and places of the unit. For proper functioning of the media layers a vertical air pipe is provided at the entire depth of the media so that the depth of water should not be varied from desired value. The head loss should also be continuously monitored by providing a gauge at the respective place and the water meter should also be installed in order to monitor the flow rate continuously. The loss of head occurs due to the resistance offered by the sand grains and void clogging with respect to time. This loss of head is known as filter head or filtering head. It also resembles the difference of water level in the filter tank and in the filtered water well. During the startup of the filtering process the filter head or head loss is small with a value of around 10-15 cm but it increases with respect to time as the filter media gets clogged and using the telescopic tube the uniform flow is maintained. However for greater head loss values in between 0.7- 1.2 m the filter unit must be backwashed or cleaned and then only be operated again.

5.2.3 Operation and Cleaning of Slow Sand Filters

Operation

The water from sedimentation tank enters the filtration unit through the inlet chamber and distributed evenly on the filter media as shown in Fig. 5.1. As soon as it passes through the filter media the impurities present in the water are arrested in the voids of the filter media. The clear water from the filter media enters the gravel bed and joins the laterals through open joints of the unit. These laterals carry this purified water to the central drain from where it is taken to the outlet in filter water well and finally to storage tanks for public supplies. During the entire process the rate of filtration is kept steady by the arrangements such as telescopic tube. It is recommended that the water passing through the filtration unit should not be treated with coagulants, because the flocs formed by coagulation process clog the filter bed and hampers its operation and working. The water depth of the filter should not be varied much and should be decided cautiously and therefore should be kept around the depth of filter media.

The maximum head loss value of 0.7 to 1.2 m is regarded as the limiting value. The limiting value is based on the consideration that when the head loss value exceeds the 0.7 to 0.8 times of depth of filter sand, the filter should be cleaned and serviced. In slow sand filter the cleaning of filter is carried out by removing the 1.5 to 3 cm top layer of sand. The top layer is scrapped out and thereafter cleaned and washed with clear water. The amount of water used for washing and cleaning of slow sand filter unit is taken as 0.2 to 0.6% of the total water filtered in that particular duration. Cleaning is carried out whenever necessary and should be continued till the sand depth is reduced to 40 cm. Thereafter, the new sand is added and the filter is made to work again. During the restart of filter unit the treated water in the effluent is not used and is allowed be discarded for 24 hours till the new layer of schmutzdecke forms on the filter media surface. The frequency of cleaning depends on the concentration of impurities in water and size of the filter media as the frequency of cleaning ranges between one to three months.

(i) **Rate of filtration or rate of loading for slow sand filters:** The rate of filtration in slow sand filter units ranges between 100-200 L/hours/m^2 of filter area. The water passes through the filter media at an average rate of 2.4 to 3.6 $m^3/m^2/$day. This filtration rate continues until the gap between the water level on the filter and in the inlet chamber is slightly lower than the water depth over the sand. The difference of water above the sand bank, and in the outlet room is the loss of head. The resistance to the passage of water and loss of head also increases during filtration as the filter media gets clogged due to the impurities in the pores.

(ii) **Efficiency and performance of slow sand filters:** The slow sand filters are highly efficient in removing the bacteria and impurities from water. The filters can remove high amount of suspended solids with more than 98-99% of bacteria removal can be achieved from water. These filters prove to be very effective in elimination of odours and tastes from water. But, such filters are less efficient in removal of colours from water. However, the slow sand filters are more effective in water with turbidity less than 50 mg/L and the effectiveness of the filters decreases for turbidities greater than 50-60 mg/L.

(iii) Uses of slow sand filters: The slow sand filters are highly effective in water having low turbidity values along with low colours and less bacterial count. However, the slow sand filters have very slow rate of filtration and therefore require large surface area. This makes them uneconomical and therefore cannot be used for treatment of water at large scale systems. Such filter units can be beneficial and best suited for smaller plants in small towns and rural areas.

5.2.4 Maintenance

The growth of algae on the overflow weir should be halted. Filtration rate should be kept constant and free from fluctuation. Head indicator filter should be in good working order. The trees around the plant should be tested to avoid dropping birds on the filter bed, no coagulant should be used before slow sand filtration as the flock easily clogs the bed.

Example 5.1: Design a six-slow sand filter beds from following data:

Population to be served = 75,000 persons

Per capita demand = 175 L/head/d

Rate of filtration = 200 L/ h/m^2

Length of each bed = Twice the breadth.

Assume max. demand as 1.8 times the average daily demand. Also assume that one unit, will be kept as stand by position (assume 8 units).

Solution: Average daily demand = Population × per capita demand

$$= 75{,}000 \times 175 \text{ L/d}$$

$$= 13.125 \times 10^6 \text{ L/d}$$

Max daily demand = 1.8 × Average daily demand

$$= 1.8 \times 13.125 \times 10^6 \text{ L/d} = 23.62 \times 10^6 \text{ L/d}$$

Rate of filtration per day = (200 × 24) L/m^2/d

$$\text{Total surface area of filters required} = \frac{\text{Max. daily demand}}{\text{Rate of filtration per day}}$$

$$= \frac{23.62 \times 10^6 \text{ L/d}}{200 \times 24 \text{ L/m}^2\text{/d}} = 4.922 \times 10^3 \text{ m}^2 = 4922 \text{ m}^2$$

Now, six units are to be used; out of them, one is to be kept as standby, hence only 5 units should provide the necessary area of filter required.

$$\text{The area of each filter unit} = \frac{1}{5} \times \text{total area required} = \frac{1}{5} \times 4922 = 984 \text{ m}^2.$$

Now, if L is length and B is the breadth of each unit, then L = 2.B (given)

$$2B \cdot B = 984 \text{ m}^2$$

$$B^2 = 492.2 \text{ m}^2$$

$$B = 22 \text{ m}$$

$$L = 2(22) = 44 \text{ m}$$

Hence, use 6 filter units with one unit as standby, each unit of size 44 m × 22 m, arranged in series with 3 units on either side. **Ans.**

5.3 RAPID SAND FILTER

Rapid sand filters, also known as gravity filters, was developed by G.W. Fuller at Louisville USA in the last decade of the nineteenth century. It utilizes the coarser sand having effective size as 0.5 mm or so. These types of filtration units provide the filtration rate of 30 times, higher as compared to slow sand filters. The water from coagulation sum sedimentation tank is taken to the filtration and the filtered water disinfected, so as to obtain potable water for public supplies.

Due to slow rate of filtration the slow sand filters purifies water at a very slow rate. However, for the treatment of high amount of water a large surface area is required and this makes the overall process uneconomical. Due to the high filtration rate ranging from 100 to 150 m^3/m^2/day and limited filter area required, rapid sand filters replace the slow sand filters. This rapid filtration rate has emerged with large amount of water treatment with lesser area and in less time. This process has led to the development of rapid sand filters also known as mechanical sand filter units. These rapid sand filters can be categorized as one which utilizes the large size sand particles or grains that, in turn, provide the higher rate of filtration and is known as rapid gravity filters.

While another types of filter units that are generally based on the development of pressure on water to be filtered during the filtration process and this ultimately increases the rate of filtration. This makes them compatible for the treatment of high amount of water with greater rate of filtration and thus requirement of surface area can be reduced to a minimum. Such filter units are known as *pressure filters*. These filters are easier to operate and are generally installed in small-scale industries, swimming pools, institutions and commercial centres. The rapid gravity filters are generally used for large scale municipal water supplies.

Construction and operation of rapid gravity filters

The cross section along with media component of a rapid gravity filter is shown through Fig. 5.2. The various components of a rapid sand filter

Filtration tank

The filtration unit comprising of water tight tank with rectangular dimensions and is constructed by masonry or concrete with depth range from 2.5 to 3.5 m. The surface of each unit is generally kept in the rage of 10-80 m^2 as this provides the uniform distribution of water over the entire area. One can estimate the number of units by using Morrell and Wallace formula as below in Eq. 5.1;

$$N = 1.22\sqrt{Q} \tag{5.1}$$

where, N = Number of filter units and Q = Flow rate or plant capacity (million litres per day)

In any case of plant capacity more than 9 million litres per day the single unit capacity should not be greater than one fourth capacity of the plant. A minimum of two filter units should be provided in a single plant.

Filter media and base material

The filtering media generally consists of sand layers supported on gravel bed. The depth of sand as a filter media is kept around 60-90 cm with effective size (D10) of sand varying from 0.35 to 0.55 mm. Uniformity coefficient $\left(\frac{D_{60}}{D_{10}}\right)$ is also considered and is kept in the range of 1.3 to 1.7. The layers of the sand should be laid as such that the finest layer should be placed at the top and coarsest at the bottom. The base material is also an important component of the rapid sand filters as it provides not only the support to the filter media or sand but it also helps in an even distribution of wash water in the filter unit during backwash period. The gravels of size range in between 60-90 cm thickness are generally used as base material of a filter unit. The total depth of 60 cm layer of gravel in four to five layers of different sizes is placed in the tank. The coarsest layer of around 40 mm in size is placed at the bottom layer and the finest size of around 3 mm is placed as the top gravel layer. Medium layers of gravel comprising of 20-40 mm are placed above the bottommost layer while the gravel size of 12 to 20 mm, 6 to 12 mm and 3-6 mm as intermediate layers are placed below the topmost layer of gravel. While placing these layers utmost care should be taken so that wash water can be distributed evenly and effectively at each and every section and corner of the tank.

Under-drainage system

In order to receive the filtered water the under-drainage system is provided at the downside of the filter unit that allows the filtered water to be collected and taken to filter well for further treatment. This under-drainage system also helps in allowing the backwash water to distribute evenly in the filter unit and thus cleansing of the filter unit efficiently.

The under-drainage system is designed both for collecting the filtered water as well as for evenly distribution of backwash water through gravel and thus cleaning of sand bed. The under-drainage system is designed at wash water upward flow rate of 300-900 L/min/m^2 of filter area. This may be due to the fact that the wash water velocity should not be higher than settling velocity of the smallest particle that lies in the range of 0.3-0.9 m/min and so is the backwash rate. This designed backwash rate is around 6-16 times the rate of filtration which is kept around 50-100 L/min/m^2. The uniform distribution of wash water is an important aspect being considered while designing the under-drainage system. The various types of under-drainage systems used in rapid gravity filters units are (i) manifold and lateral system, (ii) the wheeler bottom, and (iii) the porous plate bottom and are patented by their manufacturers. Manifold and lateral system is a form of drainage system commonly used in India. The manifold is a 40 cm diameter pipe running along the lengthwise at the centre of the filter bottom. This manifold as shown in Fig. 5.3 is connected with 10 cm diameter laterals that are placed about 15-30 cm apart and are connected with right angles.

In case of perforated pipe type system the holes are provided at the bottom side in the laterals for collection of filtered water and even distribution of water during backwashing of the tank. These holes are provided with an angle of 30° having diameter in the range of 6-13 mm and are provided with a centre-to-centre spacing of 7.5-20 cm wobbled at both sides of the pipe. In order to protect from rusting these holes are provided with brass brushings along the circumferential surface of each hole. These lateral drains are generally placed at the bottom of filter unit and are strongly firmed on the concrete blocks of 40-50 mm or so.

In case of strainer and pipe systems the hoes are not provided in the laterals attached to the manifold but strainers are provided on the lateral drains. These strainer pipes comprises perforated caps closing the top of a brass pipes and are placed on the lateral drains at a distance of about 15 cm apart. The only advantage of using pipe and strainer system is that no water is required for backwashing of the filter unit as the backwashing is carried out by introducing the compressed air in the filter unit. In the perforated pipe system unit the agitation of sand or media particles is being carried out by water jet and backwashing is carried out at high velocity with water application rate of 700-800 L/min/m^2 of filter area and is also known as high wash velocity. Also in case of pipe and strainer system the wash water is applied at the rate of 250-300 L/min/m^2 of filter area. This is because for agitating of sand particles the compressed air is being introduced in the filter unit and is known as slow wash velocity.

Design considerations in rapid sand filtration units

- The total cross-sectional area of perforation should be about 0.2 per cent of the total filter area.
- The cross-sectional area of each lateral should be about 2-4 times of the total cross-sectional area of perforations in it for diameter of perforations 13 mm and 6 mm, respectively.
- The cross-sectional area of the manifold should be about twice the cross-sectional area of the lateral drains.
- $\frac{\text{Length of each lateral}}{\text{Diameter of the lateral}} \leq 60$
- Maximum permissible velocity in the manifold in order to provide the required amount of wash water should be kept in the range of 1.8 to 2.4 m/sec.
- Other appurtenances includes: (a) wash water troughs, (b) air compressor, and (c) rate controller (d) miscellaneous accessories.

These appurtenances are summarized below:

(i) Wash water troughs: The backwash water during backwashing carries all the impurities towards the top of the filter unit that is being taken to the wash water troughs or gutters, and carried to the main gutter. These troughs or gutters may be of different shapes such as square, V-shaped, or semi-circular and may be made of cast iron, concrete, steel, or wrought iron or so. The troughs should be provided with such a height so as the backwash water with impurities will be at, or somewhere above the top of the sand rise, otherwise, sand may be washed out of the filter. Therefore, these troughs are provided with a minimum height above the sand bed by a distance of about half the depth of sand with their bottoms adjusted usually 5 cm above the top of the sand surface. This may due to the fact that sand is likely to expand by about 50% during backwashing. The flat bottom troughs are generally avoided as they can cause sand boiling during washing when the clearance above the sand is kept less than half of the depth of sand.

The spacing between the troughs should be kept around 1.5 to 2.0 m away from each other. This keeps the wash water at uniform level during filter backwashing as the spacing between these troughs restricts the horizontal distance travelled by the wash water before entering the trough to a distance of around 0.75 to 1.0 m. The trough should be laid with suitable slopes horizontally and should be large enough for working efficiently and effectively.

(ii) Air compressor: The backwashing process and agitation of filter media or sand grains in a filter unit can be carried out in different ways such as by water jet, compressed air or by mechanical rakes. In case of pipe and strainer type system the compressed air is required and therefore an air compressor of required capacity must be installed in the unit. The compressor should have the capability of supplying compressed air at the rate of about 600 to 800 L/minute/m^2 of filter area for about 4 minutes of duration or so. The compressed air can be supplied either through the laterals or through a pipe system separately. However, during the supply of compressed air in the filter unit it is ensured that it should have enough pressure so as to overcome the frictional resistances offered by the air pipes and the column of water lying above the air distribution system.

(iii) Rate controller: The water should be obtained at a uniform flow rate and therefore required to be controlled automatically flow rate irrespective of the head loss in the filter unit. This is controlled by fitting a suitable fixture at the outlet end of each filter unit. Venturi rate controller is such an equipment most commonly used controller that works on the principle of a venturimeter. The balanced disc valve is connected to a flexible that enters this valve through a venturi tube. A small pipe transmits pressure at the throat of the tube to the underside of the diaphragm. With an increase of the rate of flow ultimately results in an increase in the pressure difference between the upper and the lower sides of the diaphragm with upper side exerted upon by higher pressure because of its connection to the larger end. This results in pushing down the valve till the pressure difference decreases and becomes small enough so as to balance the pull of the lever and the weight. Clogging of filter unit with respect to time results in decrease of flow and therefore with the reverse action a constant rate of outflow is obtained.

(iv) **Miscellaneous accessories:** The miscellaneous accessories includes various appurtenances such as valves, flow control devices including other devices such as head loss indicators and meters for measuring rate of flow. The meters and gauges can be installed at each and every point of the unit where the flow rate and other parameters can be measured and the operation of filter unit can be controlled with ease. A head loss indicator is a simple mercury gauge differential equipment with one end connected to the water resting on the sand bed and the other to the effluents coming from the filter with suitable arrangements for recording the readings on a time scale.

The working of a rapid gravity filter is explained below with reference to Fig. 5.2.

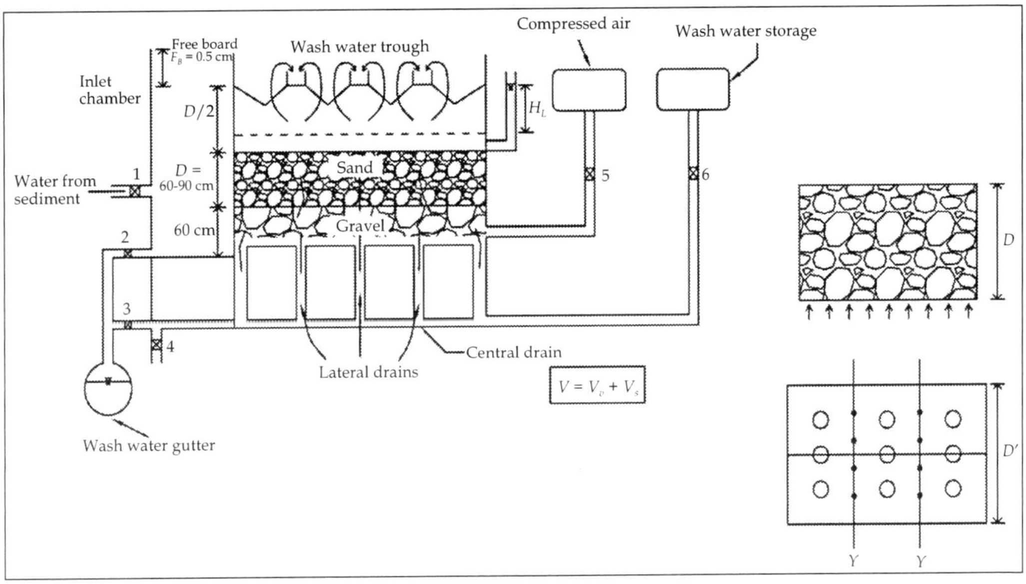

Fig. 5.2: Rapid sand filter used in water purification process at large scale

In this figure, various valves have been numbered as enumerated in Table 5.1.

Table 5.1: Valve and their functions

Valve No.	Name of valve	Function of valve
1	Inlet valve	Effluents of the coagulation sedimentation basin, to enter the inlet chamber of the filter.
2	Wastewater valve troughs	The dirty water resulting from washing, overflows into the wash water troughs, and is removed through the inlet chamber into the wash water gutter.
3	Wastewater valve	To drain water from main drain.
4	Filtered water supply valves	The valve is opened or closed as per the filtered water requirements.
5	Compressed air valve	The valve is opened for introducing compressed air at a very high rate for suitable time interval.
6	Wash water supply valve	The wash water from downside is introduced for backwashing.

The opening of valve 1 allows the water to enter the inlet chamber of the filter unit from the effluents of the coagulation sedimentation tank. The inlet water is allowed to be filtered through the filter beds and by opening of valve 4 this filtered water is taken out from the main drain to the disinfection unit. Therefore, during operation and working of filter unit valves 1 and 4 are kept open while other valves are closed for this entire duration of filtration process.

Backwashing

An excessive head loss in a filter unit indicates that the sand grains are clogged and sand has become dirty and therefore is an indicator of washing and cleansing of filter unit. For filter unit cleaning the filter unit working operation needs to be stopped by restricting the inlet raw supplies and the overall filtered supplies have to stopped by draining out all the water from the filter bed. Thereafter, the wash water is forced back upward through the filter beds and due to this forced upward movement of wash water and compressed air (if used) the filter bed is expanded with simultaneous agitation of sand particles that, in turn, removes all the suspended impurities from the filter bed. As summarized in Table 5.2 the working of these valves is as: During backwashing the valves 1 and 4 are closed while valves 5 and 6 are opened with upward forcing of the wash water and compressed air from the under-drainage through the gravel and through the sand beds. After supplying the required amount air the valve 5 is closed.

The expansion of sand bed results in removing of all the impurities arrested in the sand and is taken away by the back wash water towards upside of the tank. This dirty water with impurities after washing and cleaning of the bed overflows into the wash water troughs and simultaneously the valve 2 is opened thus removal of the dirty water through the inlet chamber into the wash water gutter. The washing process of the filters and removal of dirty water is carried out for a period of 3 to 5 minutes. After the washing process is finished the valves 2 and 6 are closed and valves 1 and 3 are opened. This reinstates the inlet water supplies through valve 1, however in the initial duration of the operation the filtered water is allowed to drain and is not collected for a few minutes through valve 3 to the gutters. This helps in removing the remains of the wash water from the voids of the filter bed and helps in the formation of a surface mat on the sand. Thereafter, valve 3 is closed and valve 4 is opened to allow the filter unit to its normal position and restoration of filtered water supplies again.

The entire duration of backwashing takes around 15 minutes and therefore while designing a filter unit the backwash time should also be considered. This backwash time not included in the entire operational time of the filter unit while designing a filter unit based on flow rate values. The amount of wash water required is taken as 2-5% of the total water filtered during that entire duration. Due to frequent clogging of rapid gravity filters the frequency of backwashing is considered to be 24-48 hours. The back wash water is applied at the rate of 15-90 cm /minute rise in upward direction. However, such backwash water with a pressure at 40 kN/m^2 at the specific rate of 45 cm/minute is quite common with a filtration rate of 500 L/minute/m^2 of bed area. The backwashing duration takes 10-15 minutes, however it takes 30 minutes for the restoration of filter unit to return back to its normal operation.

The amount of wash water used for backwashing generates high volume of wastewater after backwashing that contains low concentration of solids of 100-1000 mg/L. This wastewater contains high amount organic solids content as compared to the organic content in the sludges produced during sedimentation and coagulation process. This may be because of biological growth in such filter units. In order to reduce the moisture content and bulkiness of the produced sludge it is either filtered by filter cloths or centrifuged by using centrifugation equipment and thereafter it is disposed of at suitable lands or landfill sites. At many places the suitable disposable methods and norms are not followed as the sludge is disposed of in a very ill manner in natural drainage channels or water courses, causing atmospheric pollution and promoting pollution of rivers and streams.

The rapid gravity filters can also be washed easily and efficiently by using surface wash method in which the clean filtered water is applied to the sand bed from the top by means of using specially designed nozzles. The wash water is applied at the rate of 200 to 600 L/minute/m^2 with a water pressure of 100-200 kN/m^2 or 10 to 20 m head of water with fixed or of swiveling type of nozzles. This wash water with high pressure breaks up the filter top surface with scouring of sand particles and ultimately cleaning the top layers easily. This top layer comprising of major trapped impurities is not easily cleaned by simple backwashing. It also causes formation of mud balls which can be easily removed by using surface wash method. This improves the quality of cleaning and also increases the efficiency of the filtration.

Head loss and cleaning frequency

The raw water when enters the filter unit and percolates through the filter bed by downward movement under gravity is opposed by the impurities arrested in the sand grains and by the resistance offered of the sand grains itself. Due to this loss of energy by the percolating water it ultimately results in the loss of head which can be easily determined by the pressure or head difference in the filter and the pressure of water in the outlet pipe. This difference between the two heads is known as head loss or filter head or filtration head.

This head loss can be measured by inserting the piezometers at the stagnant water over the filter unit and providing other at the outlet pipe. In a newly established plant or newly commissioned plant the initial head loss is very small in the range of 15 to 30 cm. However, with respect to time the impurities get arrested in the sand bed and ultimately results in increase in head loss till the frictional resistance offered by the filter media exceeds the static head of water above the sand bed. This head loss is caused by top part of 10 to 15 cm sand layer that causes major resistance to the water. This ultimately creates a vacuum at the bottom where water is sucked through the filter media rather than getting filtered through it. This results in the formation of negative pressure in the centre-line of the pipe by falling of mercury level. The negative pressure releases the dissolved air of the water resulting in the formation of bubbles that stick to the sand grains and thereby affect the working of the filter unit. It is known as air binding. This air binding, restricts the functioning of the filtration process and therefore the filter efficiency gets affected and reduces considerably.

The bubble formation and air binding process result from the escaping of air into the filter media due to formation of negative head and also from the increase of water temperature when it passes through the filter bed. Some of the oxygen released by algae also accelerates the bubble formation process and is sometimes more troublesome when water becomes supersaturated with air. It leads to the hindrance in filtration process and also during backwashing process may lead to loss of media sand. This can be avoided by regularly cleansing the filter bed when the head loss reaches to an optimum value. This helps in restriction of negative head development in the filter. The head loss in any case should not be greater than 2.5-3.5 m in any case and negative head is restricted to 1.2 m. Also during the filtration process the depth of water in the filter unit should be increased by 15-20 cm from its normal depth. This, in turn, increases the static water head and ultimately reduces the negative head. This bubble formation can also be reduced by restricting the algal growth in the filter unit and preventing the water from getting warmer.

Operation and maintenance of rapid gravity filters

The other problems that generally cause the operational problems in the rapid gravity filter units are: (i) mud ball formation, and (ii) cracking of filter media

Operation

The water from the coagulation sedimentation tank reaches the filter unit through an inlet pipe and is spread evenly over the entire sand bed. Water is stored in the filtered water well after passing through the sand bed through the under-drainage method. The outlet chamber also features a filter rate controller in this system. At the start the head loss is very small. But as the bed gets clogged, the head loss increases and the filtration rate becomes very small. Therefore, the filter bed requires its washing. Design parameters of rapid gravity filters are shown in Table 5.2. and comparison of slow sand filter and rapid sand filter is tabulated in Table 5.3.

Table 5.2: Typical design parameters of rapid gravity filters

S. No.	Components	Dimensions
1.	Depth of sand	60 to 75 cm
2.	Filter gravel	2 to 50 mm size
3.	Depth of gravel	45 cm depth of water over sand
4.	During filtration	1 to 2 m overall depth of filter
5.	Including 0.5 m free board	2.6 m
6.	Area of single filter unit	100 m^2 in two parts of each 50 m^2
7.	Loss of head	Max 1.8 to 2.0 m
8.	Turbidity of filtered water	1 NTU
9.	Effective size of sand	0.45 to 0.70 mm

Table 5.3: Comparison of slow sand filter (SSF) and rapid sand filter (RSF)

S. No.	ITEM	SSF	RSF
1.	Area	Needs very large area	Needs small area
2.	Raw water turbidity	Not more than 30 NTU	Not more than 10 NTU hence needs coagulation
3.	Sand media	Effective size 0.2 to 0.3 mm uniformity coefficient 2 to 3 single layer of uniform size	Effective size 0.45 to 0.7 mm uniformity coefficient 1.3 to 1.7 multiple graded layers of sand.
4.	Rate of filtration	2.4 to 3.6 $m^3/m^2/day$	100-150 $m^3/m^2/day$
5.	Loss of head	0.6 m to 0.7 m	1.8 m to 2.0 m
6.	Supervision	No skilled supervision is required	Skilled supervision is required
7.	Cleaning of filter	Scraping of 2.5 cm thick layer washing and replacing. Cleaning interval that is replacement of sand at 1 to 2 months.	Backwash with clean water under pressure to detach the dirt on the sand. Backwashing daily or on alternate days.
8.	Efficiency	Bacterial removal, taste, odour, colour and turbidity removal	There is no removal of bacteria. Removal colour taste, odour and turbidity is good.

Rapid sand filter decreases water turbidity to 1 NTU. This filter needs continuous and professional monitoring to maintain the filter gauge, expansion gauge, flow controller rate and periodic backwash.

5.3.1 Backwashing of Rapid Sand Filter

- Before the next filter runs it must be cleaned for a filter to work efficiently. The filter runs can be very long. Some filters can operate more than one week before needed to be backwashed if the water applied to the filter is of very good quality. This is not recommended, however, as long filter runs will cause the filter media to pack up so that the bed can be difficult to expand during the backwash.
- Storage treated water is used for the backwash process. Generally, this treated water is collected from elevated storage tanks, or pumped in from the clear well.
- The backwash rate of the filter must be high enough to extend and agitate the filter media and for removal to retain the floc in the water. If the filter backwash rate is too high, however, media will be washed into the troughs and out of the filter from the filter.

The filter will be rewashed when the following conditions have been met: the head loss is so high that the filter no longer produces water at the required rate; and/or floc begins breaking through the filter and that the turbidity in the filter effluent; and/or a filter run exceeds a given operating hour..

Forms of under-drainage system

- Manifold and lateral system
- The wheeler bottom
- The porous plate bottom

Manifold and lateral systems are most widely used in India.

Manifold and lateral system:

Diameter = 40 cm manifold pipe

Lateral = 10 cm dia (placed 15 to 30 cm apart) as shown in Fig. 5.3.

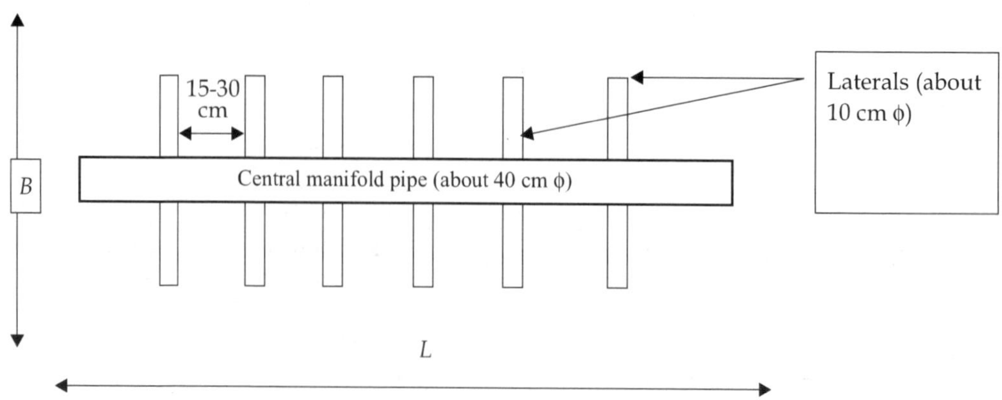

Fig. 5.3: Plan of a manifold and lateral under-drainage system for a rapid gravity filter

In perforated pipe type of this system, lateral drains with holes in bottom side:

- Dia of holes: 6 to 13 mm
- Angle of inclination: 30° with vertical
- Spacing between holes: 7.5 to 20 cm c/c

In pipe and strainer type: Strainers are placed 15 cm apart on lateral drains.

- No strainer is used, the agitation of sand particles is done by water jet, backwash of high velocity required.
- Rate of water application: 700 to 800 L/min/m^2 of filter area.
- When strainers are used, rate of water application: 250 to 300 L/min/m^2 of filter area.

5.3.2 Operational Troubles in Rapid Gravity Filters

5.3.2.1 Air Binding

- The loss of the head of water percolating through the filter is generally very small when the filter is commissioned again. The loss of head, however, tends to increase as more and more impurities get stuck in it.
- Eventually, a stage is reached when the frictional resistance provided by the filter media reaches the static water head above the bed. The top 10 to 15 cm layer of sand gives much of the resistance. The bottom sand behaves like a vacuum and instead of being filtered into it, water is sucked through the filter media.
- These established negative pressures appear to release the dissolved air and other gasses in water. Bubbles are shaped which stick to the grains of sand. This phenomenon is known as air binding because air surrounds the filter and stops working.
- To avoid these problems, the filters are cleaned as soon as the head loss reaches the maximum permissible value.

5.3.2.2 Formation of Mud Balls

The atmospheric mud usually accumulates on the surface of the sand to form a thick layer. This mud can sink down into the sand bed during insufficient washing and stick to the sand grains and other impurities that have been arrested, thus forming mud balls.

5.3.2.3 Cracking of Filters

Fine sand present in the top filter bed layers shrinks and allows shrinkage cracks to form in the sand sheet. Through the use of filters, the loss of head and, consequently, the strain on these and bed continues to increase, which further expand these cracks.

5.3.2.4 *Remedial Measures to Prevent Cracking of Filters and Formation of Mud Balls*

- Breaking the top fine mud layer with rakes and washing of the particles.
- Washing the filter with a solution of caustic soda.
- Removing, cleaning, and replacing the damaged filter sand.

5.3.3 Design Criteria Points

- Cross-sectional area of perforation = 0.2% of total filter area
- Cross-sectional area each lateral = 2 to 4 times the cross-sectional area of perforation. Diameter of perforation = 13 mm to 6 mm
- Cross-sectional area of the manifold = Twice the cross sectional area of the lateral drains.
- $\dfrac{\text{Length of each lateral}}{\text{Diamter of the lateral}} \geq 60$
- Maximum permissible velocity in the manifold required amount of wash water = 1.8 to 2.4 m/sec.

5.3.4 Other Appurtenances

- Wash water troughs
- Air compressor
- Rate controller
- Miscellaneous accessories

Wash water trough: gutters in square, V shaped, semicircular cast iron, and concrete, steel and rough iron.

- *Sand:* Expand 50 per cent at the top of the sand when the rim is over washed; change half the depth of the sand bottoms: 5 cm above the top of the sand sheet.
- Spacing b/w the tough: 1.5 m to 2 m horizontal distance travelled by the wash water before entering the trough = 0.75 to 1m.

Air compressor: During backwashing for agitation of sand grains, time of air supply: 4 min.

Rate of supply = 600 to 800 L/min/m^2 of filter area. Supplied either through the lateral or through separated pipe system.

Rate controller: Venturi rate controller, which works on the principle of venturimeter.

Miscellaneous accessories: Meters are installed for measuring discharges at entry and outlet and also at backwash.

5.3.5 Backwashing

Excessive loss of head shows makes send dirty. This forced upward movement of wash water and compressed air agitates the sand particles and thus extracts the suspended impurities from it. Close valve 1 and 4, open valve 5 and 6, close valve 5 after the necessary amount of air. Open valve to overflow from wash water troughs for the polluted water from the back wash.

Method takes time: 3 to 5 min. close valve 2 and 6, open valve 1 and 3.This restores inlet supplies through valve 1 but filtered water is not collected for a few minutes at the beginning and wastes away to gutters through valve 3. Close valve 3 and open valve 4.The amount of water needed to wash a fast gravity filter will range from 2 to 5 per cent of the total water filter. Backwashing: 24 to 48 hr.

- Rate of washing: 15 to 90 cm. rise per min.
- Normally: 45 cm per min.
- Filter rate: 500 L/$m^2$of bed area/min.
- Normally: 10 to 15 min. are used for backwash.
- Recommissioning: 30 min.
- Pressure of wash water supply = 40 kN/m^2
- Large volume of wash water (backwash low concentration of solids) = 100 to 1000 mg/L of solids. The filter back wash may contain large fraction of organic solids as sludge and therefore the sludge is desposed of after thickening.
- Surface wash from top, rate of application of wash water = 200 to 600 L/ min/ m^2
- Pressure = 10 to 20 L head of water (i.e., 100 to 200 kN/m^2)

Loss of head:

- New filter = 15 to 30 cm
- Limited value = 2.5 to 3.5 m
- Negative head = 1.2 m
- Depth of water during filtration increase by 15 to 20 cm.
- Controlling the growth of algae.
- Remediation by supersaturated with air.
- Precautions of getting water warmed.

5.3.6 Standard Design Practice of Rapid Sand Filter

- Maximum length of lateral = not less than 60 times its diameter.
- Spacing of holes = 6 mm holes at 7.5 cm c/c or 13 at 15 c/c.
- Cross-sectional area of lateral = not less than 2 times area of perforations.
- Cross-sectional area of manifold = 2 times total area of laterals.
- Maximum loss of head = 2 to 5 m.
- Spacing of laterals = 15 to 30 cm c/c.
- Pressure of wash water at perforations = not more than 1.05 kg/cm^2.

- Velocity of flow in lateral = 2 m/s. Velocity of flow in manifold = 2.25 m/s.
- Velocity of flow in manifold for wash water = 1.8 to 2.5 m/s.
- Velocity of rising wash water = 0.5 to 1.0 m/min.
- Amount of wash water = 0.2 to 0.4% of total filtered water.
- Time of backwashing = 10 to 15 min.
- Head of water over the filter = 1.5 to 2.5 m.

Example 5.2: Design a rapid sand filter unit for 5 million litres per day of supply with all the components including dimensions of the filter unit, under-drainage system and wash water troughs.

Solution: Water required per day = 5 mL.

Assuming that 4% of filtered water is required for washing of the filter, every day, we have:

$$\text{Total filtered water required per day} = \frac{5 \text{ mL.}}{0.96} = 5.20 \text{ mL/day}$$

Now, assuming that 30 minutes are required for backwashing filter every day, we have

$$\text{Filtered water required per hour} = \frac{5.20}{(24-0.5)}\frac{\text{mL}}{\text{hr}} = \frac{5.20}{23.5} = 0.22 \text{ mL/hr}$$

Now, assuming the rate of filtration to be 5500 L/h/m^2, we have

$$\text{The area of filter required} = \frac{0.22 \times 10^6}{5500}\text{ m}^2 = 40.30 \text{ m}^2$$

Now, assuming the length of filter bed (L) as 1.5 times the width of the filter bed (B), and two beds, the total area provided

$$2 \times (L \times B) = 40.30$$

or $$2 \times (1.5B)(B) = 40.30$$

or $$B^2 = \frac{40.30}{3} = 13.4$$

or $$B = 3.66 \text{ m} \approx 3.75 \text{ m}$$

So, $$L = 1.5B = 1.5 \times 3.44 = 5.6 \text{ m}$$

or use the length of the filter bed as = 5.6 m

Hence, adopt 2 filter units, each of dimensions of 5.6 m × 3.75 m. **Ans.**

Design of under-drainage system

Manifold and lateral system

Assume area of perforation = 0.2 % of filter area

$$= \frac{0.2}{100} \times 5.6 \times 3.75 = 0.042 \text{ m}^2$$

Total area of laterals = 2 × Total area of perforations

$$= 2 \times 0.042 = 0.084 \text{ m}^2$$

Area of manifold = 2 × Area of laterals

$$= 2 \times 0.084 = 0.168 \text{ m}^2$$

Therefore, diameter of manifold (d) can be calculated as

$$\frac{\pi}{4} \times d^2 = 0.168$$

$$d = 0.46 \text{ m.}$$

Hence, 46 cm diameter manifold pipe laid lengthwise along the centre of filter bottom.

Laterals emanating from the manifold at a spacing of 15 cm (c/c)

$$\text{Therefore, no. of laterals} = \frac{5.6 \times 100}{15} = 38$$

Total no. of laterals on both sides = 76

$$\text{Now, length of each lateral} = \frac{\text{Width of filter}}{2} - \frac{\text{Dia of manifold}}{2}$$

$$= \frac{3.75}{2} - \frac{0.46}{2} = \frac{3.29}{2} = 1.645 \text{ m.}$$

Now, adopt 13 mm dia of perforation in laterals,

Total area of perforations = 0.042 m^2

Assume, x = total number of perforations

$$\text{Therefore, number of perforations} = \frac{\text{Total area of perforation}}{\text{Area of one perforation}}$$

$$= \frac{420}{\frac{\pi}{4} \times (1.3)^2} = 316$$

$$\text{Therefore, number of perforations in each lateral} = \frac{316}{76} = 4.0$$

$$\text{So, area of perforations per laterals} = 4 \times \frac{\pi}{4}(1.3)^2 = 5.30 \text{ cm}^2$$

Area of each lateral = 2 × Area of perforations per lateral

$$= 2 \times 5.30 = 10.60 \text{ cm}^2$$

$$\text{Diameter of each lateral} = \sqrt{10.60 \times \frac{4}{\pi}} = 3.7 \text{ cm}$$

Hence, use 78 laterals each of 3.7 cm dia @ 15 cm c/c, having 4 perforations of 13 mm size with 46 cm dia manifold.

Check:

$$= \frac{\text{Length of each lateral}}{\text{Dia of lateral}} = \frac{1.645 \text{ m}}{3.7 \text{ cm}} = 44.45 \ (< 60 \text{ hence O.K.})$$

Now let us assume the backwash rate of filter washing to be 45 cm rise/min or 0.45 m/min

So, wash water discharge $= \dfrac{0.45 \times 5.6 \times 3.75}{60} = 0.157 \text{ m}^3/\text{s}$

Therefore, the velocity of flow in the lateral for wash water

$$= \frac{0.157}{78 \times \frac{\pi}{4} \times \left(\frac{3.7}{100}\right)^2} = \frac{0.157 \times 10{,}000}{78 \times 10.75} = 1.87 \text{ m/s}$$

Similarly, the velocity of flow in manifold $= \dfrac{\text{Total discharge}}{\text{Area}}$

$$= \frac{0.157}{\frac{\pi}{4} \times (0.46)^2} = \frac{0.157}{0.166} = 0.94 \frac{\text{m}}{\text{s}}$$

$= (< 1.8 - 2.4 \text{ m/s})$, hence O.K.

Design of wash water troughs

Wash water troughs = 1.5 to 2 m apart with total length of the filter bed to be 5.6 m

Therefore, spacing between the troughs along the filter length @ 3 troughs $= \dfrac{5.6}{3} =$ 1.87 m

Total wash water discharge of 0.157 m^3/s enters in 3 troughs.

Therefore, the discharge in each trough $= \dfrac{0.157}{3} = 0.052 \text{ m}^3/\text{s}$

The flat bottom trough can be designed by suitable formula as :

$Q = 1.376\, b \cdot y^{3/2}$

where, Q = Discharge m^3/s

b = Width of the trough = 0.2 m (assumed)

y = Water depth in trough (m)

Putting the value in the above equation the dimensions of each trough can be calculated as :

$0.052 = 1.376 \times 0.2 \times y^{3/2}$

$y = 0.32 \text{ m} = 32 \text{ cm}$

Keep 5 cm freeboard the depth of trough = 32 + 5 = 37 cm

Provide 3 wash water troughs of size 37cm × 20 cm at c/c distance of 1.87 m

Example 5.3: In a treatment plant, two tanks of rapid sand filtration are proposed after sedimentation. Each filter bed has a surface area of 3 m × 2 m. The design flow rate to each bank of filters is 0.044 m^3/sec. The design loading rate to each bank of filters is 150 $\text{m}^3/\text{day}/\text{m}^2$. Determine the number of filter beds in each bank of filters. Determine the loading rate when one filter is out of service. The maximum acceptable loading rate is 235 $\text{m}^3/\text{day}/\text{m}^2$. (ESE Mains, Civil Engg. 2015)

Solution: Area of each filter bed = $3 \times 2 = 6$ m^2

Design flow rate to each bank of filter = 0.044 m^3/s

Design loading rate to each bank of filter = 150 m^3/day/m^2

$$\text{Total area of filter beds required} = \frac{\text{Design flow rate}}{\text{Design loading rate}} = \frac{0.044 \times 3600 \times 24}{150}$$

$$= 25.344 \text{ m}^2$$

$$\text{Number of filter beds required} = \frac{\text{Total area of filter beds required}}{\text{Area of one filter bed}} = \frac{25.344}{6} = 4.224$$

Provide 5 filter beds.

When one gets out of service,

$$\text{Area of filter beds} = \frac{0.044 \times 360 \times 24}{24} = 158.4 \text{ m}^3/\text{day}/\text{m}^2$$

Hence ,design is safe even if one filter bed gets out of service.

Example 5.4: A square filter box is to be designed for a filtration rate of 2.8l/m^{2}s. What are the required surface area and side dimension of the unit if the flow rate is 6 ML/d? If the filter is backwashed once a day for 12 minutes at a rate of 10 L/m^{2}s, what percentage of the total flow is used for cleaning the filter?

(ESE Mains, Civil Engg. 2014)

Solution: Filtration rate = 2.8 L/m^{2}s

Flow rate, Q = 6 MLD

Filtered water is used for backwashing.

Filtered water required per day + Water required for backwashing = Water filtered at a rate of 2.8 L/m^{2}s in a day.

Let the dimension of filter be 'B'.

$$6 \times 10^6 + 10 \times B^2 \times 12 \times 60 = 2.8 \times B^2 \times 24 \times 60 \times 60$$

$$= 6 \times 10^6 = [2.8 \times 86{,}400 - 7{,}200] \times B^2$$

$$= B^2 = \frac{6 \times 10^6}{2{,}34{,}720} = 25.56 \text{ m}^2$$

$$B = 5.06 \text{ m}$$

The percentage of the flow rate used for cleaning the filter.

$$= \frac{10 \times B^2 \times 12 \times 60}{6 \times 10^6 + 1 \times B^2 \times 12 \times 60} \times 100 = \frac{184049.1}{6184049.1} \times 100 = 2.976\%$$

Design of filter media for rapid gravity filters

The most conventional and commonly used filter media is silica sand and is used in granular filter units. However, advancements in filtration process with research and modern applications make use of various types of filter media such as anthracite, anthrafilt coal and garnet sand with silica sand. The filtration process generally depends on the media properties such as size of the media, density and its distribution in the filter bed. The smaller size media particles are good enough to remove maximum impurities from water and thus provide better filtration efficiency in terms of solids and bacteria removal. The small size media possesses a drawback of high head loss due to frequent clogging of the media due to arrested impurities in them. However, in case of large size media particles the head loss is less with adequate flow rate but results in lower filtration efficiency.

It is very difficult to procure a uniform size media for the filter units and may also prove to be expensive to obtain a filter media of narrow range diameter. This non-uniform media is an important aspect of the filter unit as during the backwashing process the granular bed expands with rising of smaller particles towards the top of the filter while the larger size particles do not move significantly. This creates stratification in the filter bed as the backwash cycle ends and therefore the finer particles settle at the top resulting in small pore openings at the top of filter. This inefficient arrangement results in high head loss frequently thus affects the filtration process at the surface. The density of the filter bed material is also an important parameter that needs to be considered while placing the two different types of the media in a single unit. The media particles when used with diverse specific gravity values will have different settling rates after backwashing. Therefore, the lighter material should be placed over denser material of the same size.

The effective size is an important parameter used for the design of filter bed as it is represented by the term D_{10} size and is defined as the sieve size in mm through which 10% of sand sample by mass will pass. This size is an important factor as described earlier as too fine size leads to frequent clogging and too coarse size leads to escape of impurities and bacteria present in the water. The uniformity coefficient has also been described earlier as the ratio of D_{60} by D_{10} value and is represented by U. It has been recommended that the effective size (D_{10}) for silica sand to be used for rapid gravity filters should lie in the range of 0.35-0.55 mm and must not be more than 1 mm in any case. The recommended UC value should lie in the range of 1.3-1.7. It also indicates that the higher the value of UC the lower will be the uniformity grain size distribution. Therefore, due to practical constraints and from economical point of view the UC value should not be less than 1 and should not be greater than 1.7 in any case. Therefore, in order to obtain an economical and low cost media the locally available sand is used. The characteristics and important properties of this stock sand can be determined by carrying out a sieve analysis and plotting the grain size distribution curve. The values obtained from the grain size distribution curve helps to achieve the optimum characteristics as the coarser sand particles are removed or screened out while the finer sand particles are washed out. The D_{10} and D_{60} values are utilized for the removal of coarser and finer particles and can be calculated as: the sand particles that are too fine and too coarse do not meet the requirements of D_{10} and D_{60} size may be called P_f means that too fine and P_c means that too coarse. However, the useful portion may be regarded as P_{use} as indicated in Eq. 5.2,

$$P_{use} + P_f + P_c = 100 \quad (5.2)$$

where, P_{use} is the usable portion that lies in the range of $D_{10} - D_{60}$

Let us assume that with respect to D_{10} size the percentage finer than P of total stock sand may be termed $P_{st(10)}$ while $P_{st(60)}$ representing the percentage of stock sand finer the P and represents the specific size dia D_{60}. The values can be determined of the grain size distribution curve plotted after sieve analysis.

$$P_{st(10)\text{ - finer than } P \text{ of stock sand or}}\ D_{10}$$

$$P_{st(60)\text{ - coarser than } P \text{ of stock sand or}}\ D_{60}$$

It is also assumed that the sand that lies in between the range of D_{10} and D_{60} comprises 50% of the specified sand and therefore the usable sand can be calculated as shown in Eq. 5.3:

$$P_{use} = 2[P_{st(60)} - P_{st(10)}] \quad (5.3)$$

However it is also assumed that the ten per cent of usable sand can be below the specified D_{10} size and this 10% of usable sand below D_{10} size can be specified as 0.1 P_{use}. Therefore, the percentage of stock sand that is too fine as in Eq. 5.4 - Eq. 5.5:

$$P_f = P_{st(10)} - 0.1\ P_{use} \quad (5.4)$$

or $$P_f = P_{st(10)} - 0.2\ [P_{st(60)} - P_{st(10)}] \quad (5.5)$$

It is desirable to remove the finer particles that are smaller than specified D_{10} size and retaining of large size of particles is advantageous.

Now the percentage of particles that are too coarse is represented as P_c and therefore can be calculated by using the Eq .5.6-Eq. 5.8:

$$P_c = 100 - P_f - P_{use} \quad (5.6)$$

$$P_c = 100 - [P_{st(10)} - 0.2\ \{P_{st(60)} - P_{st(10)}\}] - 2\ [P_{st(60)} - P_{st(60)}] \quad (5.7)$$

$$P_c = [100 - P_{st(10)} - 1.8\ \{P_{st(60)} - P_{st(10)}\}] \quad (5.8)$$

The sand with close value of $P_{st(60)}$ size should be retained as much as possible, while the size P_c sand consists of large size particles and therefore should be removed from the media to be used for rapid gravity filter unit. The too fine sand grain particles can be removed in rapid gravity filters during backwashing as the stratification takes place and the unwanted portion at the top can be removed easily.

Example 5.5: Filter sand is to be prepared for a rapid sand filter from the stock sand for which the details of sieve analysis are given below Table 5.4. The required effective size and the uniformity coefficient of the filter media is 0.4 mm and 2.0, respectively. Determine the amount of usable sand and percentage and size above and below which the sand is too coarse or too fine.

Table 5.4: Sieve analysis details

Sieve size mm	Mass retained (g)	Sieve size mm	Mass retained (g)
2.0	0	0.50	165
1.5	95	0.40	162
1.2	120	0.30	160
1.0	155	0.20	100
0.9	215	0.10	120
0.8	110	0.08	47
0.75	95	0.06	37
0.70	175	Finer	04
0.60	240		

Solution: Required sand for filter, $D_{10} = 0.4$ mm is given

$D_{60} = U \times D_{10} = 2.0 \times 0.4 = 0.80$ mm.

Calculate the total mass: The sieve analysis being carried out is reported and the data is analyzed in terms of percentage P each of given size (d) with P_{st} representing the stock sand values along with sieve analysis study as summarized in Table 5.5 below.

Table 5.5: Percentage mass retained and cumulative % finer than *d* size

Sieve size *d* (mm)	Mass retained (g)	% mass retained $\frac{Col.(2)}{2000} \times 100$ $\frac{Col.(2)}{20}$	Cumulative % mass retained	Cumulative % finer than *d* size (P_{st}) = 100 – col. (4)
Column (1)	Column (2)	Column (3)	Column (4)	Column (5)
2.0	0	0	0	100
1.5	95	4.75	4.75	95.25
1.2	120	6.00	10.75	89.25
1.0	155	7.75	18.50	81.50
0.9	215	10.75	29.25	70.75
0.8	**110**	**5.50**	**34.75**	**65.25**
0.75	95	4.75	39.50	60.50
0.70	175	8.75	48.25	51.75
0.60	240	12.00	60.25	39.75
0.50	165	8.25	68.50	31.50
0.40	**162**	**8.10**	**76.60**	**23.40**
0.30	160	8.00	84.60	15.40
0.20	100	5.00	89.60	10.40
0.10	120	6.00	95.60	4.40
0.08	47	2.35	97.95	2.05
0.06	35	1.75	99.70	0.30
Finer	06	0.30	100.00	0
Σ	2000	100.00	–	–

From Table 5.5 the specified D_{10} and D_{60} values can be find out and the values can be calculated as below:

$P_{st(10)}$ = % age finer than 0.4 mm size = 23.40%

$P_{st(60)}$ = % age finer than 0.80 mm size = 65.25%

$$P_{use} = 2[P_{st(60)} - P_{st(10)}]$$

And therefore $P_{use} = 2(65.25 - 23.40) = 83.70\%$

Therefore, 83.7% of stock sand can be utilized however the remaining 16.3% of the sand will either be screened or washed away.

Also too fine (P_f) for removal can be computed as

$$P_f = P_{st(60)} - 0.1\ P_{use} = 23.40 - 0.1 \times 83.7 = 15.03\%$$

Hence, 15.03% of finest sand will have to be washed out.

However, too coarse sand to be screened can be calculated as:

$$P_c = 100 - P_f - P_{use}$$

or
$$P_c + P_f + P_{use} = 100$$
$$P_c + 15.03 + 83.7 = 100$$
$$P_c = 100 - 98.73 = 1.27\%$$

Hence, 1.27% of the coarse sand will have to be screened. The grain size distribution curve is plotted between the values of particle size as (d) on x-axis and percentage finer than P_{st} on y-axis on a semilog graph paper as shown in Fig. 5.4. The P_{st} values can be read out on corresponding D_{10} = 0.4 mm and D_{60} = 0.8 mm values. To determine the size of 1.27% coarsest sand, the particle size (d) can be read out corresponding to P_{st} = 100 – 1.27 = 98.73%.

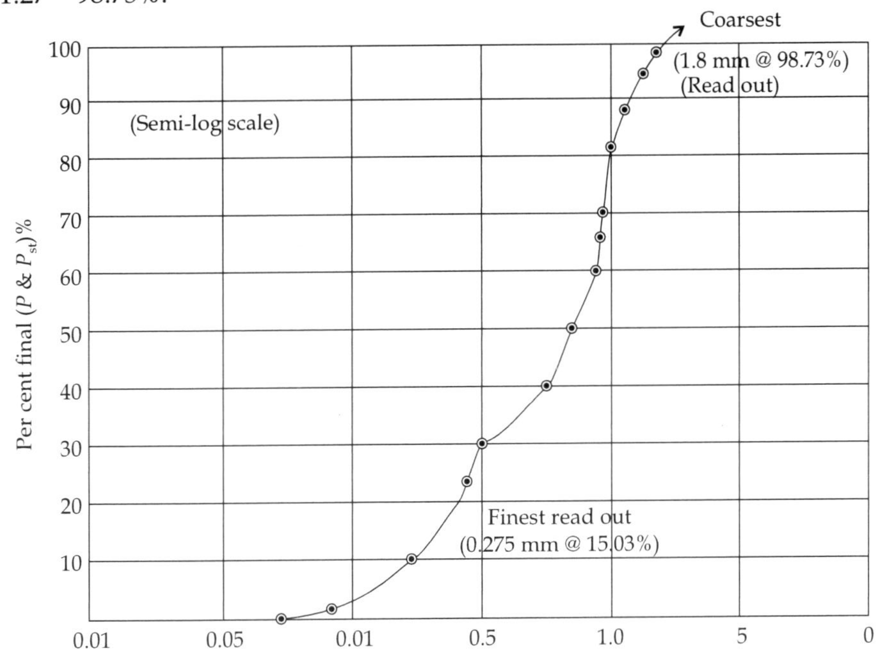

Fig. 5.4: Grain size distribution curve

Also the size (d) for finest sand at 15.03% can be assessed and read out from the graph at intercept P_{st} = 15.03%.

5.4 HYDRAULICS OF GRAVITY FILTERS AND FILTER HEAD LOSS

The filter hydraulics comprising of two separate design components that include the filtration process and another important aspect is backwashing process that is generally used for regular cleansing of the filter bed after saturation with arrested impurities. Both the operations are very important during the working of filter. The design of filter media is based on head loss interval as well as frequency of the filter cleaning. The head loss in a filter bed is caused by resistance offered by filter media and the arrested impurities in the media to the gravitational movement of water through the filter bed. However, in case of newly commissioned plant the initial head loss is caused by the filter media or the sand grains. This initial head loss in the filter unit depends on the filter media characteristics such as porosity (n) and diameter of the sand grains. The other important parameters that govern the overall filtration process are filtration velocity (v), depth (D), drag coefficient (C_D), shape factor (ϕ) and acceleration due to gravity (g). The initial head loss can be computed by using the Rose equation and also by Carmen-Kozeny as both the equations can be used to calculate the head loss in a clean filter. The equations are described below:

5.4.1 Rose Equation for Head Loss Determination

The Rose equation can be used for determination of initial head loss in a clean stratified filter unit as shown in Eq. 5.9 and reduced form as indicated in Eq. 5.10.

$$h_L = \frac{1.067\, v^2 \cdot D}{\phi \cdot g \cdot n^4} \sum \frac{C_D \cdot (f)}{d} \tag{5.9}$$

where,

h_L = Frictional head loss through filter (m)

v = Apporach velocity or filtration velocity (m/s)

D = Depth of filter (m)

ϕ = Shape factor (non-spherical particles)

d = Diameter of sand particles (m)

g = Acceleration due to gravity (m/s^2)

n = Porosity

C_D = Newton's drag coefficient

f = Mass fraction of sand particles of diameter (d)

The drag coefficient (CD) can be calculated by using the Reynolds number (Re) values with diameter (d) of the non-spherical sand grains multiplied by shape factor (ϕ) and therefore the diameter (d) is represented as ($\phi.d$) for non-spherical particles.

The term $\sum \frac{C_D \cdot (f)}{d}$ can be calculated by utilizing the sieve analysis values being obtained in the experiment. Each value is computed and summarized as $\sum \frac{C_D \cdot (f)}{d}$.

Also if the sand with particle diameter (*d*) is uniform in nature the term $\sum \frac{C_D \cdot (f)}{d}$ is reduced to $\frac{C_D}{d}$ only.

The rose equation can be reduced as:

$$h_L = \frac{1.067\, v^2 \cdot D}{\phi \cdot g \cdot n^4} \sum \frac{C_D}{d} \tag{5.10}$$

The equation corresponds to the new filter unit with clean bed and does not provide the effect of sand grains on the head loss values during filtration process. However, it is suggested that an initial head loss greater than 0.6 m in the filter indicates that either the media or sand grains are too fine or the hydraulic loading rate is too high. Therefore, it is recommended that the filter box must be designed at high head loss value of minimum 3 m or so.

5.4.2 Head Loss and Filter Backwash

The head loss in a filter unit increases with time and reaches to the maximum value where the filter bed is required to be cleaned in order to get it prepared for operation. This may be due to the fact that when solids begin to accumulate in the media or sand grains the porosity of the bed changes and with decrease in porosity there is an increase in head loss in the filter.. This process of filter bed cleaning is carried out by backwashing process. Therefore, in order to clean the filter bed and to remove the arrested impurities in the media or sand grains it is necessary to expand the filter bed such that the filter media granules are not in contact with each other and all the surface is exposed for cleansing. This rate of solids accumulation in the filter corresponds to change in head loss and generally depends on the factors such as nature of the suspension, the media characteristics and the operating conditions of the filter.

This expansion of the porous filter bed represented by h_{Le} is the uplift during backwashing and must be equal to the total buoyant weight of the filter bed. It can be represented as shown in Eq. 5.11 followed by Eq. 5.12-Eq. 5.14:

$$h_{Le} \cdot \gamma_w = D \cdot \gamma_{sub} \tag{5.11}$$

where,

h_{Le} = Head loss through filter bed to initiate expansion (m)

γ_w = Unit weight of water (kN/m^3)

D = Depth of the filter bed (m)

γ_{sub} = Submerged unit weight of sand in bed at depth D (kN/m^3)

and

$$\gamma_{sub} = \frac{\gamma_w(G-1)}{1+e} \tag{5.12}$$

where, G = Specific gravity of sand grains, e = Void ratio

$$\text{Porosity } (n) = \frac{e}{1+e} \tag{5.13}$$

$$1 - n = 1 - \frac{e}{1+e} = \frac{1}{1+e}$$

$$\gamma_{sub} = \gamma_w(G - 1)(1 - n)$$

$$h_{Le}.\gamma_w = D \cdot \gamma_w(G - 1)(1 - n)$$

or

$$h_{Le} = D(1 - n)(G - 1)$$

$$\frac{h_{Le}}{D} = (1 - n)(G - 1) \tag{5.14}$$

It is assumed that during the expansion of bed depth D_e the head loss through expanded bed do not changes as the total buoyant weight of the bed remains constant. The weight of expanded bed or the fluidized bed is expressed as in Eq. 5.15.

Weight of fluidized/ expanded bed = $D_e(1 - n_e)(G - 1)$

and therefore $h_{Le} = D_e(1 - n_e)(G - 1)$ (5.15)

where,

D_e = Depth of expanded bed (m)

n_e = Porosity of expanded bed

The equation can be further modified as shown in Eq. 5.16:

$D(1 - n)(G - 1) = D_e(1 - n_e)(G - 1)$

therefore,

$$\boxed{D_e = \frac{(1-n)D}{(1-n_e)}} \tag{5.16}$$

The calculation is also helpful in draining out the wash water carrying the impurities through backwash troughs provided at the top and the height of these troughs should be kept a minimum of 0.15 m above the filter bed during its expansion or fluidization so as to avoid the loss of bed material.

However, when the sand media of a filter bed material is not of uniform size the above equation can be further modified in order to evaluate the value of expanded bed depth D_e as shown in Eq. 5.17 :

Therefore, the D_e value can be computed when sand is not uniform:

$$D_e = (1-n)D \cdot \sum \frac{f}{1-n_e} \tag{5.17}$$

where, f = Mass fraction of sand of different size particles obtained after sieve analysis

The term n_e is a function of terminal velocity of the particles and the backwash velocity. The higher the backwash velocity, the greater will be the filter bed expansion. The backwash velocity and bed expansion can be expressed by Eq. 5.18 as:

$$n_e = \left(\frac{v_b}{v_s}\right)^{0.22*} \tag{5.18}$$

where, n_e = Porosity of expanded bed

v_b = Backwash velocity (m/sec) $= \frac{\text{Backwash flow}}{\text{Total filter area}}$

v_s = Settling velocity of size d (m/s)

0.22* = The value for sand but more correct value as per Amirtharrajah (1978) can be taken as 0.4-0.5 for the sand filters.

The expanded bed is a function of settling velocity (v_s) which can be determined by using the equation.

Therefore, settling velocity (v_s) (is calculated as shown in Eq. 5.19:

$$v_s = \sqrt{\frac{\frac{4}{3}gd(G-1)}{C_D}} \tag{5.19}$$

where,

d = Diameter of the particle (m)

g = Acceleration due to gravity

G = Specific gravity of the particle

C_D = Coefficient of drag depends on Reynolds number (R_e) which can be calculated as

$$R_e = \frac{v_s d\phi}{v}$$

v_s = Settling velocity of size (d) (m/s)

ϕ = Shape factor

ν = Kinematic viscosity

Therefore, after determination of diameter (d) and specific gravity (G) of the sand grains and calculating the Reynolds number, the C_D and v_s can be calculated. The v_s value can also be fixed after number of iterations with respect to Reynolds number as the calculations are repeated till the final solution is obtained.

For backwashing process the wash water should be applied to the filter at the rate of 15 cm/min to 90 cm/min with specific value of 45 cm/min rise. The backwash rate is recommended to be decided on the basis of settling velocity of the smallest particle of the filter media or sand grains.

Example 5.6: In a water treatment plant raw water is passed through a filter bed of uniform sand at a velocity of 5 m/h the filter is made of sand grains of diameter is 0.4

mm, shape factor is 0.85 and specific gravity is 2.65. The depth of bed is 0.67 m and the porosity is 0.4. Determine the head loss through bed. Density of water is 1000 kg/ m^3 the dynamic viscosity is 1×10^{-3} kg/ms.

Solution: Using Rose equation

$$h_L = \frac{1.067\, v^2 D}{\phi \cdot g \cdot n^4} \sum \frac{C_D}{d}$$

$$R_e = \frac{v \cdot d\phi}{v}$$

$$V = 5 \text{ m/h} = \frac{5}{60 \times 60} = 1.389 \times 10^{-3} \text{ m/s}$$

$$d = 0.4 \times 10^{-3} \text{ m}$$

$$\phi d = 0.85 \times 0.4 \times 10^{-3} = 0.34 \times 10^{-3} \text{ m/s}$$

$$\text{v} = \frac{\mu}{\rho_w} = \frac{1 \times 10^{-3}}{1000} = 1 \times 10^{-6} \text{ m}^2\text{/s}$$

$$R_e = \frac{1.389 \times 10^{-3} \times 0.34 \times 10^{-3}}{1 \times 10^{-6}}$$

$$R_e = 0.471 < 0.5 \text{ (laminar flow condition)}$$

$$C_D = \frac{24}{R_e} = \frac{24}{0.471} = 50.92$$

$$h_L = \frac{1.067 \times (1.389 \times 10^{-3})^2 \times 0.67 \times 50.92}{0.85 \times 9.81 (0.4)^4 \times 0.4 \times 10^{-3}} = 0.882 \text{ m.}$$ **Ans.**

Example 5.7: Design a rapid sand filter for a city with population of 2,40,000 with per capita water demand of 180 L/d. The rate of filtration may be taken as 17 m^3/m^2/h. The backwashing period of 30 minutes may be considered and for backwash of 5% of water is to be stored. Determine the number of filters required and one standby unit. The surface area of the filter unit may be 8 m $\times$ 3.5 m. If the undisturbed original depth of the filter is 0.6 m calculate the up-flow velocity and head loss value if the bed is expanded to 0.66 m from its original depth. The porosity of the bed and specific gravity may be taken as 0.50 and 2.5, respectively. The average particle diameter may be taken as 0.5 mm with a drag coefficient value of 5.0. Assume transitional flow conditions.

Solution: The daily water requirement can be calculated as = Population $\times$ Per capita demand

= 2,40,000 $\times$ 180 L/P/d = 43.2 MLD

The amount of water used for backwashing = 5% of total water

= 5% $\times$ 43.2 = 2.16 MLD

Total amount of water to be filtered daily = 43.2 + 2.16 = 45.36 MLD

The total time required for backwashing as given = 30 minutes = 0.5 h

Therefore, the net working time of the filter in a day = 24 – 0.5 = 23.5 h

Therefore, filtered water required per hour $= \frac{45.36}{23.5} = 1.93$ ML/h

The filtration rate as given = 17 $m^3/m^2/h$

Therefore, the area of the filter unit $= \frac{Q}{v} = \frac{\text{Total flow rate}}{\text{Velocity or filtration rate}} = \frac{1930}{17} = 114\ m^2$

Given size of the filter unit = 8 m × 3.5 m = 28 m^2

Therefore, the number of units $= \frac{114}{28} = 4$

One unit is taken as standby unit

Therefore, the total number of units required = 4 + 1 = 5

For settling velocity determination the equation is:

$$v_s = \sqrt{\frac{\frac{4}{3} gd(G-1)}{C_D}}$$

Given that:

$g = 9.8\ m/sec^2$

$d = 0.5\ mm = 0.5 \times 10^{-3}\ m$

$G = 2.5$

$C_D = 5.0$

Putting the values in the above equation, we get

$$v_s = \sqrt{\frac{\frac{4}{3} \times 9.8 \times 0.5 \times 10^{-3}(2.5-1)}{5}} = 0.044\ m/sec$$

Calculate the porosity of expanded bed by the following equation as

$$D_e = \frac{(1-n)D}{(1-n_e)}$$

where,

$D_e = 0.66$

$D = 0.6$

$n = 0.5$

Putting the values in equation, we get

$$0.66 = \frac{(1-0.5)6}{(1-n_e)}$$

$$n_e = 0.545$$

The backwash velocity now can be calculated as

$$n_e = \left(\frac{v_b}{v_s}\right)^{0.22}$$

$$0.545 = \left(\frac{v_b}{0.044}\right)^{0.22}$$

$v_b = 2.79 \times 10^{-3}$ m/sec = 0.279 m/sec = 16.8 cm/min.

The backwash rate is 16.8 cm/min

Head loss through the expanded bed can be calculated as:

$h_{Le} = D_e(1 - n_e)(G - 1)$

therefore,

$h_{Le} = 0.66(1 - 0.545)(2.5 - 1) = 0.455 \times 0.66 \times 1.5 = 0.45$ m **Ans.**

Head Loss and Backwashing by Carmen-Kozeny Equation

The determination of head loss in a filter unit with respect to actual filtration process and the backwashing is also governed by the Carmen-Kozeny equation. Carman-Kozeny equation is used to calculate head loss when the filter bed is of uniform size of particles as indicated in Eq. 5.20:

$$h_f = \frac{f'L(1-e)V_S^2}{e^3 g d_p} \qquad (5.20)$$

where,

h_f = Friction loss through the bed of particles of uniform size, d_p(m)

f' = Friction factor

e = Porosity of the bed

L = Depth of filter bed or layer (m)

d_p = Grain size diameter (m)

V_s = Superficial (approach) filtration velocity (m/s)

g = Acceleration due to gravity, 9.81 (m/s^2)

However, the friction factor f' depends on the type of flow and drag coefficient that ultimately depends upon the Reynolds number. In case of filter velocities and under laminar flow conditions the value of f' can be calculated as indicated in Eq. 5.21:

$$f' = 150\frac{(1-e)}{R_e} + 1.75 \qquad (5.21)$$

where,

R_e = Reynolds number

μ = Dynamic viscosity, N-s/m^2

ρ_w = Density of water kg/m^3

where, R_e value can be calculated as shown in Eq. 5.22:

$$R_e = \frac{\phi \rho_w V_s d}{\mu} \qquad (5.22)$$

For most of the filter bed material the shape factor (ϕ) value can be taken in between 0.75 and 0.85.

However, in case of non-uniform bed material, the sieve analysis needs to be carried out and the weight fraction (x_{ij}) between the consecutive sieve sizes is to be determined. The average size of the particle is denoted as d_{ij} and is assumed to be halfway between the two consecutive sieve sizes. The depth of the particle can be determined between the adjacent sieve sizes by using the term $x_{ij.}L$ and can be expressed as in Eq. 5.23;

$$h_f = \frac{L(1-e)V_S^2}{e^3 g}\sum\frac{f_{ij}\,x_{ij}}{d_{ij}} \tag{5.23}$$

where,

h_f = Friction loss through bed of particles of uniform size, d_p, (m)

f' = Friction factor

e = Porosity of the bed

ϕ = Particle shape factor (1.0 for spheres, 0.82 for rounded sand, 0.75 for average sand, 0.73 for crushed coal and angular sand)

L = Depth of filter bed or layer (m)

d_p = Grain size diameter (m)

V_s = Superficial (approach) filtration velocity (m/s)

g = Acceleration due to gravity, 9.81 (m/s^2)

R_e = Reynolds number

μ = Dynamic viscosity (N-s/m^2)

ρ_w = Density of water (kg/m^3)

Example 5.8: Calculate the head loss in a filter bed of uniform size particles of diameter 0.5 mm with a shape factor of 0.80 having specific gravity of 2.5. The water having temperature of 20°C passing through the filter with a filtration velocity of 5.5 m/h. The bed depth under normal conditions is 0.66 m with porosity of 0.4.

Solution: Given that

h_f = ?

f' = ?

e = 0.4

ϕ = 0.8

G = 2.5

L = 0.66

$d_p = 0.5 \times 10^{-3}$ m

V_s = 5.5 (m/h) = 1.5×10^{-3} m/s

g = Acceleration due to gravity, 9.81 (m/s^2)

R_e = ?

$\mu = 1.002 \times 10^{-3}$ (kg/m.s) at 20°C

$\rho_w = 998.2(\text{kg/m}^3)$

In order to assess the flow type conditions the Reynolds number is required to be calculated as:

$$\text{Therefore, } R_e = \frac{\phi \rho_w V_s d}{\mu} = \frac{0.8 \times 998.2 \times 1.5 \times 10^{-3} \times 0.5 \times 10^{-3}}{1.002 \times 10^{-3}}$$

$$= 0.59 < 1.0 \text{ (Laminar flow)}$$

Therefore,

$$f' = 150\frac{(1-e)}{R_e} + 1.75 = 150\frac{(1-0.4)}{0.59} + 1.75 = 154.29$$

The head loss can be determined as:

$$f' = \frac{f'L(1-e)V_S^2}{e^3 g d_p} = \frac{154.29 \times 0.66(1-0.4)(1.5 \times 10^{-3})^2}{(0.4)^3 \times 9.8 \times 0.5 \times 10^{-3}} = 0.44 \text{ m} \qquad \textbf{Ans.}$$

Example 5.9: Calculate the head loss in a filter bed of non-uniform size particles with a shape factor of 0.80 having specific gravity of 2.5. The water with 20°C temperature is passing through the filter with a filtration velocity of 4.5 m/h. The bed depth composed of non-uniform sand under normal conditions is 0.70 m with porosity of 0.4. The size distribution of the granules is given below in the Table 5.6.

Table 5.6: Size distribution of granules

U.S. Sieve no		Particle size range (mm)		Average size d_{ij}, mm	Mass fraction in size, x_{ij}
Passing	Retained	Passing	Retained		
	14		1.45	1.45	0.013
14	20	1.45	0.88	1.17	0.017
20	25	0.88	0.75	0.82	0.22
25	30	0.75	0.65	0.70	0.33
30	35	0.65	0.55	0.6	0.23
35	40	0.55	0.45	0.5	0.15
40		0.45		0.45	0.04

Solution: Given that

$h_f = ?$

$f' = ?$

$e = 0.4$

$G = 2.5$

$\phi = 0.8$

$L = 0.70$

$d_p = d_{ij}$ m

$V_s = 4.5$ (m/h) $= 1.25 \times 10^{-3}$ m/s

g = acceleration due to gravity, 9.81 (m/s^2)

$R_e = ?$

$\mu = 1.002 \times 10^{-3}$ (kg/m.s) at 20°C

$\rho_w = 998.2$ (kg/m^3)

Therefore, Reynolds number can be computed as :

$$R_e = \frac{\phi \rho_w V_s d}{\mu} = \frac{0.8 \times 998.2 \times 1.25 \times 10^{-3} \times d_{ij}}{1.002 \times 10^{-3}} = 979\ d_{ij}$$

Also f' can be calculated as:

$$f' = 150\frac{(1-e)}{R_e} + 1.75 = 150\frac{(1-0.4)}{979\ d_y} + 1.75 = \frac{0.092}{d_{ij}} + 1.75$$

Now determine the $\sum f_{ij}\frac{x_{ij}}{d_{ij}}$ as summarized in Table 5.7 below:

Table 5.7: Calculation of mass fraction to average size

Particle size, d_{ij}(m) × 10^{-3}	x_{ij}	f_{ij}	$f_{ij}\frac{x_{ij}}{d_{ij}}$
1.45	0.013	65.20	584.54
1.17	0.017	80.38	1167.95
0.82	0.22	113.95	30570.64
0.70	0.33	133.18	62784.18
0.6	0.23	155.08	59448.61
0.5	0.15	185.75	55725.00
0.45	0.04	206.19	18328.40
	$\sum f_{ij}\frac{x_{ij}}{d_{ij}}$		228609.32

Putting the above computed values in the table, the head loss can be calculated as below:

$$h_f = \frac{L(1-e)V_S^2}{e^3 g}\sum\frac{f_{ij}\,x_{ij}}{d_{ij}} = \frac{0.70(1-0.4)(1.25\times10^{-3})^2}{(0.4)^3 \times 9.8} \times 228609.32 = 0.57 \text{ m}$$ **Ans.**

The backwashing of the filter bed is carried out after the head loss reaches to the maximum limiting value and the filter cleaning is required. The water with high up-flow velocity enters the filter bed from downside therefore expands the filter bed. This allows the granules no longer in contact with each other and the arrested impurities in the sand grains are washed away towards the top side thus cleaning the filter bed. Therefore, during the hydraulic expansion of porous bed it is assumed that the total head loss must be equal to buoyant weight of the particles in the bed.

Therefore, considering the filter having unit area the head loss can be expressed as in Eq. 5.24:

$$h_{fb} = L(1-e)\frac{\rho_m - \rho_w}{\rho_w} \tag{5.24}$$

where,

h_{fb} = Head loss at initial expansion of the bed (m)

L = Bed depth (m)

$1 - e$ = Fraction of the packed bed composed medium

ρ_m = Density of the medium (kg/m^3)

ρ_w = Density of the water (kg/m^3)

However, the head loss through the expansion of the bed also remains unchanged. This may be due to the buoyant weight of the bed is constant.

Therefore,

Weight of packed bed = Weight of the fluidized bed (Eq. 5.25)

$$L(1-e)\frac{\rho_m - \rho_w}{\rho_w} = L_{fb}(1-e_{fb})\frac{\rho_m - \rho_w}{\rho_w}$$

or

$$L_{fb} = L\frac{(1-e)}{(1-e_{fb})} \tag{5.25}$$

where,

L_{fb} = Depth of fluidized bed

e_{fb} = Porosity of fluidized bed

This porosity of fluidized bed (e_{fb}) is a function of settling velocity and ultimately the backwash velocity. The higher bed expansion is due to the high backwash velocity and the relationship between the two can be determined by Eq. 5.26.

$$e_{fb} = \left(\frac{V_B}{v_t}\right)^{0.22*} \tag{5.26}$$

where, e_{fb} = Porosity of expanded bed

V_B = Backwash velocity (m/sec) = $\dfrac{\text{Backwash flow}}{\text{Total filter area}}$

v_t = Terminal settling velocity of size d (m/s)

0.22* = The value for sand but more correct value as per Amirtharrajah (1978) can be taken as 0.4-0.5 for the sand filters.

$$L_{fb} = L\frac{(1-e)}{1-\left(\frac{V_B}{v_t}\right)^{0.22}} \tag{5.27}$$

For a stratified bed of non-uniform particles layer the equation can be further modified as:

$$x_{ij} = \frac{L_{ij}}{L}$$

where, x_{ij} is weight fraction of the consecutive sieve size with L_{ij} as the depth of the bed layer of media by assuming the uniform bed porosity. The bed during expansion can be represented as in Eq. 5.28;

$$L_{fb,ij} = x_{ij} \frac{L(1-e)}{1-\left(\frac{V_B}{v_{t,ij}}\right)^{0.22}} \tag{5.28}$$

The sum of individual layers provides the total expansion by suing the following Eq. 5.29:

$$L_{fb} = L(1-e)\sum \frac{x_{ij}}{1-\left(\frac{V_B}{v_{t,ij}}\right)^{0.22}} \tag{5.29}$$

The total expanded depth is reported to be 120 to 155 per cent as compared with normal depth. Amirtharajah (1978) reported the expanded bed porosities in the range of 0.65 to 0.70.

Example 5.10: A filter bed of uniform size particles of diameter 0.5 mm with a shape factor of 0.80 having specific gravity of 2.5 is treating the water having temperature of 20°C passing through the filter with a filtration velocity of 5.5 m/h. The bed depth under normal conditions is 0.66 m with porosity of 0.4. The expanded bed porosity has been determined to be 0.65 during backwashing. Determine the expanded bed depth and backwash velocity.

Solution: Given that:

$e = 0.4$

$e_{fb} = 0.65$

$\phi = 0.8$

$G = 2.5$

$L = 0.66$

$d_p = 0.5 \times 10^{-3}$ m

$V_s = 5.5$ (m/h) $= 1.5 \times 10^{-3}$ m/s

g = Acceleration due to gravity, 9.81 (m/s^2)

$R_e = ?$

$\mu = 1.002 \times 10^{-3}$ (kg/m.s) at 20°C

$\rho_w = 998.2$ (kg/m^3)

Settling velocity of the medium can be calculated by using Stokes' law:

$$v_t = \frac{g\rho_w(G-1)d^2}{18\mu} = \frac{9.8 \times 998.2(2.5-1)(0.5 \times 10^{-3})^2}{18 \times 1.002 \times 10^{-3}} = 0.20 \text{ m/s}$$

The Reynolds number needs to be calculated as

$$R_e = \frac{\phi \rho_w V_t d}{\mu} = \frac{0.8 \times 998.2 \times 0.2 \times 0.5 \times 10^{-3}}{1.002 \times 10^{-3}} = 80 > 1.0 \text{ (transitional flow)}$$

Therefore,

$$C_D = \frac{24}{R_e} + \frac{3}{R_e^{0.5}} + 0.34 = \frac{24}{80} + \frac{3}{80^{0.5}} + 0.34 = 0.98$$

$$v_t = \sqrt{\frac{\frac{4}{3} gd(G-1)}{C_D}} = \sqrt{\frac{\frac{4}{3} \times 9.8 \times 0.5 \times 10^{-3}(2.5-1)}{0.98}} = 0.1 \text{ m/s}$$

Therefore putting the value of $v_t$in above equations and by repeating the above steps the values can be calculated as :

$$R_e = \frac{\phi \rho_w V_t d}{\mu} = \frac{0.8 \times 998.2 \times 0.1 \times 0.5 \times 10^{-3}}{1.002 \times 10^{-3}} = 40 > 1.0 \text{ (transitional flow)}$$

$$C_D = \frac{24}{R_e} + \frac{3}{R_e^{0.5}} + 0.34 = \frac{24}{40} + \frac{3}{40^{0.5}} + 0.34 = 1.41$$

$$v_t = \sqrt{\frac{\frac{4}{3} gd(G-1)}{C_D}} = \sqrt{\frac{\frac{4}{3} \times 9.8 \times 0.5 \times 10^{-3}(2.5-1)}{1.41}} = 0.08 \text{ m/s}$$

The backwash velocity can be calculated as :

$$0.7 = \left(\frac{V_B}{0.08}\right)^{0.22}$$

$$V_B = 0.65^{4.55} \times 0.08 = 0.011 \text{ m/s}$$

The expanded bed length can be calculated as:

$$L_{fb} = L\frac{(1-e)}{(1-e_{fb})} = 0.66\frac{(1-0.4)}{(1-0.7)} = 1.32 \text{ m}$$ **Ans.**

Example 5.11: Calculate the head loss in a filter bed of non-uniform size particles with a shape factor of 0.80 having specific gravity of 2.5. The water with 20°C temperature is passing through the filter with a filtration velocity of 4.5 m/h. The backwash velocity can be taken as 0.015 m/s. The bed depth composed of non-uniform sand under normal conditions is 0.70 m with porosity of 0.4. The size distribution of the granules is given in Table 5.8.

Table 5.8: Size distribution of granules

U.S. Sieve no.		Particle size range (mm)		Average size d_{ij}, mm	Mass fraction in size x_{ij}
Passing	Retained	Passing	Retained		
	14		1.45	1.45	0.013
14	20	1.45	0.88	1.17	0.017
20	25	0.88	0.75	0.82	0.22

25	30	0.75	0.65	0.70	0.33
30	35	0.65	0.55	0.6	0.23
35	40	0.55	0.45	0.5	0.15
40		0.45		0.45	0.04

Solution: Given that

$e = 0.4$

$G = 2.5$

$\phi = 0.8$

$L = 0.70$

$d_p = d_{ij}$ m

$V_s = 4.5\ (\text{m/h}) = 1.25 \times 10^{-3}\text{m/s}$

$V_B = 0.015\ \text{m/s}$

g = Acceleration due to gravity, 9.81 (m/s^2)

$R_e = ?$

$\mu = 1.002 \times 10^{-3}\ (\text{kg/m.s})$ at 20°C

$\rho_w = 998.2(\text{kg/m}^3)$

For bottom layer d_{ij}= 1.45 and x_{ij} = 0.013

The settling velocity can be calculated as:

$$v_t = \sqrt{\frac{\frac{4}{3}gd(G-1)}{C_D}} = \sqrt{\frac{\frac{4}{3}\times 9.8 \times d_{ij} \times 10^{-3}(2.5-1)}{0.4}} = \sqrt{0.049 \times d_{ij}}$$

$$= \sqrt{0.049 \times 1.45} = 0.27\ \text{m/s}$$

Now Reynolds number can be calculated as below:

$$R_e = \frac{\phi \rho_w V_t d}{\mu} = \frac{0.8 \times 998.2 \times 0.27 \times 1.45 \times 10^{-3}}{1.002 \times 10^{-3}} = 322 > 1.0 \text{ (transitional flow)}$$

$$C_D = \frac{24}{R_e} + \frac{3}{R_e^{0.5}} + 0.34 = \frac{24}{322} + \frac{3}{322^{0.5}} + 0.34 = 0.58$$

Putting the C_D value in settling velocity determination equation

$$v_t = \sqrt{\frac{\frac{4}{3}gd(G-1)}{C_D}} = \sqrt{\frac{\frac{4}{3}\times 9.8 \times d_{ij} \times 10^{-3}(2.5-1)}{0.58}} = \sqrt{0.033 \times d_{ij}}$$

$$= \sqrt{0.033 \times 1.45} = 0.22\ \text{m/s}$$

Repeat the above steps and calculate the Reynolds number and C_D based on v_t values:

$$R_e = \frac{\phi \rho_w V_t d}{\mu} = \frac{0.8 \times 998.2 \times 0.22 \times 1.45 \times 10^{-3}}{1.002 \times 10^{-3}} = 254$$

$$C_D = \frac{24}{254} + \frac{3}{254^{0.5}} + 0.34 = 0.62$$

The expanded porosity can be determined as:

$$e_{fb} = \left(\frac{V_B}{v_{t,ij}}\right)^{0.22} = \left(\frac{0.015}{0.22}\right)^{0.22} = 0.55$$

Therefore,

$$\text{Determine} = \frac{x_{ij}}{1 - e_{fb}} = \frac{x_{ij}}{1 - \left(\frac{V_B}{v_{t,ij}}\right)^{0.22}} = \frac{0.013}{1 - 0.55} = 0.028$$

Repeat the steps again and calculate for all the x_{ij} and $v_{t,ij}$ values as summarized in Table 5.9 below:

Table 5.9: Calculation of settling velocity

Average particle (mm)	$v_{t,ij}$	$\left(\frac{V_B}{v_{t,ij}}\right)^{0.22}$	$1-\left(\frac{V_B}{v_{t,ij}}\right)^{0.22}$	x_{ij}	$\frac{x_{ij}}{1-\left(\frac{V_B}{v_{t,ij}}\right)^{0.22}}$
1.45	0.22	0.55	0.45	0.013	0.03
1.17	0.20	0.57	0.43	0.017	0.04
0.82	0.16	0.59	0.41	0.22	0.54
0.70	0.15	0.60	0.40	0.33	0.83
0.6	0.14	0.61	0.39	0.23	0.59
0.5	0.13	0.62	0.38	0.15	0.40
0.45	0.12	0.63	0.37	0.04	0.11
$\sum \frac{x_{ij}}{1-\left(\frac{V_B}{v_{t,ij}}\right)^{0.22}}$					2.53

The expanded bed depth now can be calculated as:

$$L_{fb} = L(1-e)\sum \frac{x_{ij}}{1 - \left(\frac{V_B}{v_{t,ij}}\right)^{0.22}} = 0.7\ (1 - 0.4)\ 2.53 = 1.063 \text{ m}$$

The expanded bed depth in term of percentage of original can be calculated as:

$$= \frac{1.063}{0.7} \times 100 = 152\%$$

Ans.

5.5 PRESSURE FILTERS

Hot process softeners usually use pressure filters to allow high-temperature operation and to prevent heat loss. The application of pressure filters removes the need to repump filtered water. Pressure filters are similar to gravity filters as they include filter media, bed support, under-drain system, and control device; however, there are no wash water troughs in the filter shell. Pressure filters have cylindrical steel shells and dished heads, which are constructed vertically or horizontally. To allow individual backwashing these filters are divided into compartments as shown in Fig. 5.5.

Do not use cold water to backwash a hot process filter. It will cause the metallurgy system to expand and contract, which would contribute to metal fatigue. The cold water laden with oxygen will also accelerate corrosion.

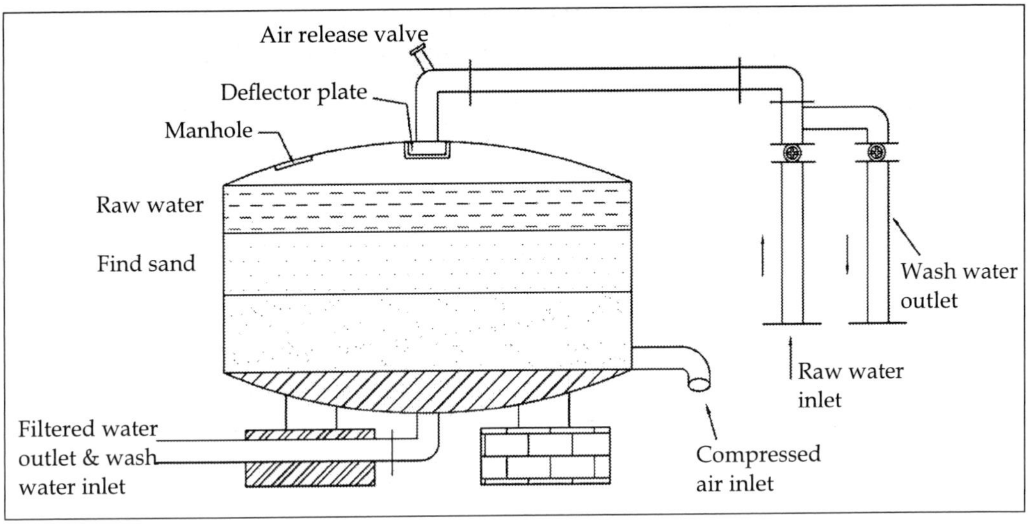

Fig. 5.5: Pressure filter

- Diameter: 1.5 to 3.0 m
- Height: 3.5 to 8.0 m

5.5.1 Working Operation

Inside these filters alum is commonly used with flocculation and coagulation. Rewashing is performed as per fast gravity filters.

Rate of filtration of pressure filters

- 2 to 5 times than RGF
- Rate of filtration = 6000 to 15000 L/h/m^2 of filter area

Efficiency

More effective than RGF in the reduction of bacteria and turbidity. They can be built for home, business, swimming pools, train stations, etc.

5.5.2 Advantages and Disadvantages of Pressure Filters

Advantages

- Compact and automatic operation.
- Ideal for small estates and small waterworks.
- Require small area for installation.
- Small number of fittings are required.

- Filtered water comes out under pressure, no further pumping is required.
- 6.No sedimentation and coagulant tanks are required.

Disadvantages

- Due to heavy cost on treatment, they cannot be used for large quantity of water at waterworks.
- Proper quality control and inspection is not possible because of closed tank.
- The efficiency of removal of bacteria and turbidity is poor.
- Change of filter media, gravel and repair of drainage system is difficult.

5.6 UPFLOW FILTERS

Upflow units produce medium graded sand with a single filter. At the top of the bed is the finest sand with the coarsest sand below. Gravel is retained at the bottom of the unit that grids in a fixed position. Gravel purpose is to ensure proper distribution of the water during the service period. Another layer over the graded sand avoids media fluidization. Air injection during cleaning (not called backwash as the flow path is the same as when in-service) helps to eliminate the solids and reclassify the filter material. The bigger, coarse solids are removed at the bottom of the bed during service, while smaller solid particles are allowed to penetrate further into the media. Typical service flow rates are 510 gpm/ft². An example of this unit is shown in Fig. 5.6.

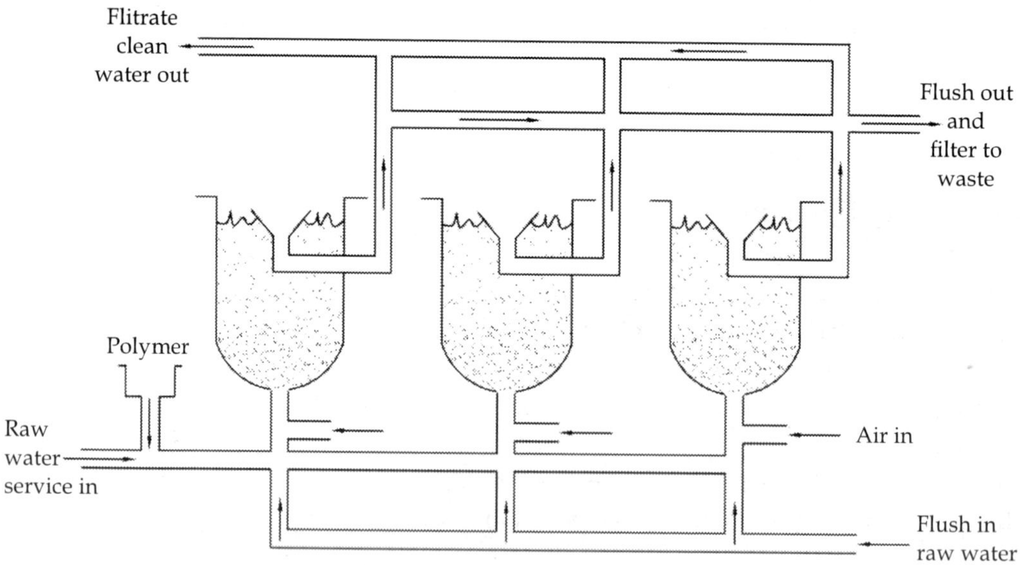

Fig. 5.6: Upflow filters used in water purification process (solutions)

5.7 GRAVITY FILTERS

Gravity filters are open vessels that depend on system gravity head for operation. Apart from the filter media, the essential components of a gravity filter include the following.

- The filter shell may be square, rectangular, or circular, either concrete or steel. Rectangular reinforced concrete structures are used most frequently.
- The support bed, which prevents fine sand or anthracite from being lost via the under-drain network. The support bed, usually 1-2 ft deep, often distributes water from the backwash.
- An under-drain network that ensures uniform filtered water collection and uniform backwash water distribution. The device may consist of a header and laterals, with suitably spaced perforations or strainers. Falsified tank bottoms with suitably spaced strainers are often used for underwater systems.
- Wash water troughs, wide enough to store water from the backwash without flooding. The troughs are spaced so that the backwash water's horizontal movement doesn't reach 3-3 ft. Wash troughs are mounted roughly 2 ft above the filter surface in standard sand bed systems. Enough freeboard must be given during service to prevent loss of a portion of the filter media at maximum backwash levels.
- Control devices optimize the efficiency of filter operation. Flow rate controls powered by effluent line venturi tubes automatically keep filtered water supply uniform. Controllers are also used for backwash flow volume. Flow rate and head loss gauges are essential to successful service.

5.7.1 Automatic Gravity Filters

Several manufacturers have developed gravity filters, which are automatically backwashed at a preset head loss. Head loss (water level above the media) works on a siphon backwash and pulls water from storage through the bed and out to waste through the siphon pump. The siphon breaks by a low level in the backwash storage area, and the filter returns to operation. Automatic gravity filters with diameters of up to 15 feet are available. A single large-diameter unit can filter as much as 1,000 gpm when fitted with a high rate, multilayer screen.. An example is shown in Fig. 5.7.

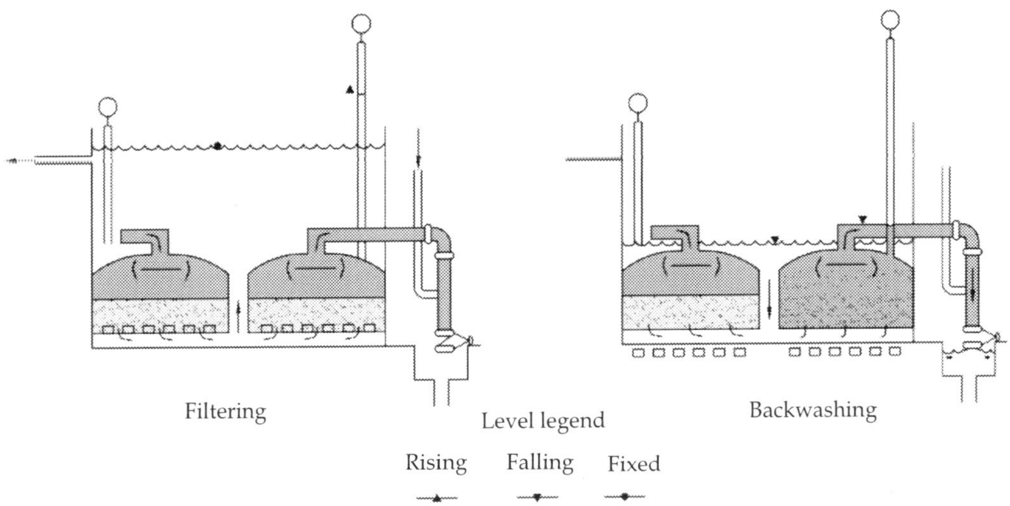

Fig. 5.7: Mono valve gravity filters used in water purification process (solutions)

5.8 CONTINUOUS CLEANING FILTERS

Continuous filtering systems remove off-line backwash cycles by constantly rewashing parts of the filter or portions of the filter media. Various designs have been introduced. An example is shown in Fig. 5.8.

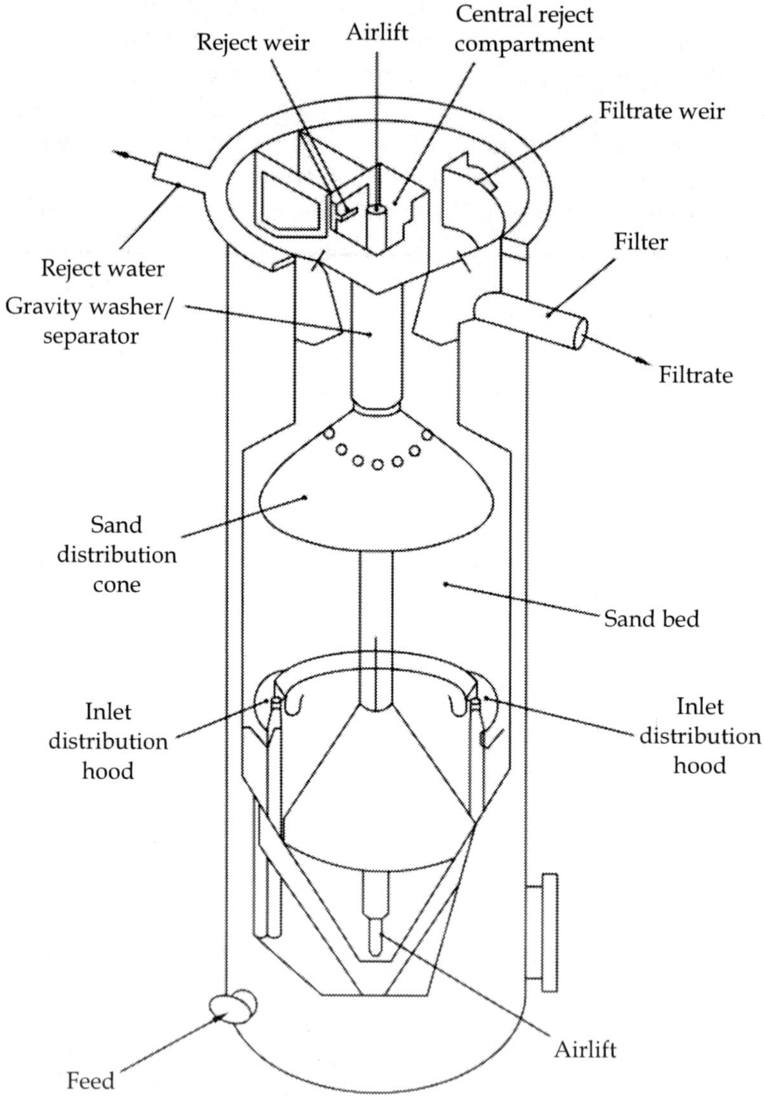

Fig. 5.8: Dyna sand continuous cleaning filter (solutions)

5.9 IN-LINE CLARIFICATION

In-line clarification is the removal of suspended solids by in-line coagulant addition accompanied by rapid filtration. Often, this method is called in-line filtration, or touch filtration. Suspended solids are extracted without the use of sedimentation basins. Coagulation may be achieved in in-line clarification by either of two methods:

- An inorganic aluminum or iron salt used alone or with a high molecular weight polymeric coagulant.
- A strongly cationic organic polyelectrolyte.

Since metal hydroxides form precipitates, the inorganic coagulant programmes can only use dual-media filters. Floc particles have to be treated with coarse-to-fine graded media in filters to avoid rapid filter blindness and to remove backwashing difficulties. Where a high polymeric coagulant of molecular weight is used, feed levels of less than 0.1 ppm optimize the removal of solids by increasing floc size and facilitating particle absorption inside the filter. This filtration technique readily delivers effluent turbidities below 0.5 NTU. The second coagulant pretreatment process requires the use of one particular compound, a strongly charged cationic polyelectrolyte. This treatment does not create floc particles from precipitation, and usually no floc formation is evident in the powerful filter. Within the bed, solids are collected by adsorption and by flocculating colloidal matter directly onto the sand or anthracite media surface. The cycle can be visualized with positive cationic charges as seeding of the filter bed surfaces to create a strong pull on the negatively charged particles. Since precipitates of gelatinous hydroxide are not present in this process, single-media or upflow filters are suitable for poly-electrolyte clarification as shown in Fig. 5.9.

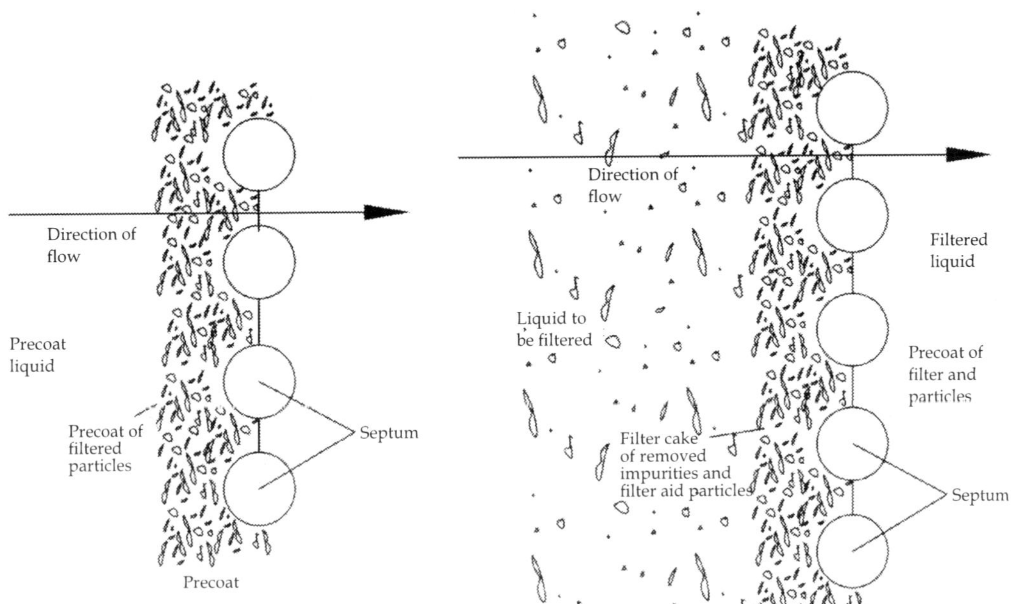

Fig. 5.9: Principles of diatomite filtration (solutions)

In-line clarification is an excellent way of enhancing the performance of extracting solids from turbid surface waters. With this form, effluent turbidity levels of under 1 NTU are normal.

5.10 PRECOAT FILTRATION

Precoat filtration is used to remove very small particulate matter, oil particulate matter and even water bacteria. This approach is only feasible for relatively small volumes of water that contain low contaminant concentrations.

Following traditional clarification methods, precoat filtration can be used to generate water with very low content of suspended solids for different application requirements. Precoat filters, for example, are often used to extract the oil from polluted condensate. The precoat media, usually diatomaceous earth, serves as the filter media in precoat filtration, and forms a cake on a permeable base or septum. The base must stop the passage of the precoat media without limiting the flow of filtered water and must be able to withstand the differentials in high pressure. Base materials such as filter cloths, porous stone tubes, porous paper, wire screens, and wire wound tubes are used. The base material supporting it is first coated with a slurry of precoat media. During the filter run additional slurry (body feed) is normally applied. When the accumulation of material removed by filtration causes a decrease in high pressure through the filter, the filter surface becomes sloughed off by backwashing. Afterwards the filter bed is precoated and returned to service. Chemical coagulants are typically not required, but were used where an ultrapure effluent is required.(S. Elaiyarasu)

5.11 SOFTENING

Softening methods of precipitation are used to reduce hardness of the raw water, alkalinity, silica, and other constituents. It helps to prepare water for direct use as a cooling tower make-up or as a first-stage treatment followed by an exchange of ions for boiler making or process use. The water is treated with lime or a mixture of soda and lime (carbonate ion) ash. Such chemicals react to the water's hardness and natural alkalinity, forming insoluble compounds. The compounds are precipitated by sedimentation and usually filtration, and are separated from water. Waters with low to high concentrations of hardness and alkalinity (150-500 ppm as $CaCO_3$) are mostly viewed that way.

5.12 OTHER TREATMENT PROCESSES

The above treatment method is all engineered by removing microorganisms and suspended solids to make drinking water safe. Drinking water, however, may also contain chemical pollutants, especially from groundwater sources, which must be removed. The removal of chemicals from water is usually more complex and costlier than removing microbiological or physical pollutants. The basic techniques of filtration and coagulation are usually not effective for most chemicals.

Since there are several different chemicals that may be dealt with, there will be a few important examples. Iron may be a major component of both land and surface waters (where bacteria and algae are generally associated with this). Although iron is not a health risk, it creates water acceptability problems as many consumers find the color off-putting and it stains clothes. The key way of extracting iron from the water is by aerating or oxidizing the Fe^{2+} to the Fe^{3+} material. This can be easily accomplished by flowing the water over a simple waterfall and then sedimenting. Note aeration is often used in waters considered to be anoxic or deficient in oxygen. A number of methods for extracting organic and inorganic pollutants, including ion exchange and precipitation are used. Fluoride, for example, may be extracted using calcinated burnt bone or trigger alumina by coagulation with lime or by ion exchange. Granulated activated carbon (GAC) is widely used as an adsorptant to remove pesticides. This is costly but sadly no other method appears to work effectively, so GAC remains the only alternative.(S.Elaiyarasu)

5.13 SELECTING TECHNOLOGY

When choosing treatment equipment and systems it is important that a image of the source water quality as thoroughly as possible is available. Until attempting to develop appropriate treatment systems it is necessary to know what is in the water. It is equally necessary to maintain a comprehensive monitoring system across the plant to ensure that each treatment stage is operating efficiently and effectively. Before they are fit for human consumption, both waters will need treatment. Surface water is more susceptible to pollution than groundwater. In addition, all water supplied by distribution systems should be disinfected to provide a residual disinfectant that helps to protect against bacterial growth and survival.

5.14 WATER TREATMENT PLANT

The need to handle all surface waters and any groundwater contamination could be microbiological, chemical or physical. Microbiological pollution is of utmost importance because it causes highly infectious diseases. Short-term effects of chemical pollution appear to have long-term effects on the survival of microbial suspended solids and the acceptability of water also disinfects the source of water. The need for more than one procedure during treatment avoids breakdown in one cycle leading to complete source of treatment failure must be well protected.

Many processes available, the suitability of each depends on the nature of the source, the capability of the operator and the selection of financial resources technology must be made on the basis of the above to ensure sustainability. Prefiltration is a physical process that eliminates prefilters of suspended solids that can be horizontal, vertical upflow or vertical upflow-downflow. The key benefit is small working parts do not use chemical drawbacks and doesn't include poor ability to remove fine material, microbial removal poor and may need frequent cleaning. Sedimentation is accomplished by settling particles in slow moving water. Simple sedimenters do not use chemical coagulants and do not extract fine material effectively (S.Elaiyarasu).

5.14.1 Section Key Points OHP

Through adding coagulants, settling is improved to shape larger aggregates that accelerate settling and eliminate fine material modular and plate settlers improve settling performance, whereas other aggregates include polyelectrolytes and ferric salts such as sulphate and chloride.

Advantages include removal of fine particles, removal of some viruses, quick, compact disadvantages include expense, need for good monitoring capacity, and trained operators

- Sand filtration can be rapid or slow
- Slow sand filtration is a biological process and rapid sand filtration is a physical process
- Slow sand filters, a biologically active top layer called the schumtzdecke which is composed of predatory bacteria. Schumtzdecke kills bacteria and viruses.
- Require cleaning @ every 2 months, takes 3-4 days to recover.
- Rapid sand filters work at much faster rates and remove suspended solids.
- dvantages of slow sand filtration include production of good quality water, relatively simple to operate.

Disadvantages include large land requirement, labour intensive, requires low turbidity water.

Advantages of rapid sand filtration include small land requirement.

5.14.2 Treatment Plant Assessments

Assessments of treatment plants may be carried out for a number of different reasons

- Routine appraisals are frequently performed by water suppliers to ensure reliable and optimized production.
- Assessments should also be performed when there is a water quality failure or a failure to generate water of sufficient quality.
- Assessments include assessing each unit process to ensure it operates efficiently and to identify any process deficiencies and causes of deficiencies.
- Assessment may also evaluate the suitability of technology combinations.

5.15 WATER PURIFICATION

Water purification is the mechanism by which unnecessary chemicals, materials and biological pollutants are extracted from polluted water. The goal is to generate water that is fit for a specific purpose. Some water is filtered for human use (drinking water), but water purification can also be planned for a range of other uses, including satisfying medical, pharmacological, chemical and industrial requirements. The methods used typically include physical processes such as filtration and sedimentation, biological processes such as slow sand filters or activated sludge, chemical processes such as flocculation and chlorination and the use of electromagnetic radiation such as ultraviolet light. Water purification method may minimize the concentration of particulate matter including suspended particles, insects, bacteria, algae, viruses, fungi; and a variety of dissolved and particulate matter resulting from the surfaces with which water may have come into contact after falling as rain. The drinking water quality specifications are usually set by states or international guidelines. Usually, these requirements would set minimum and maximum concentrations of pollutants for the water usage to be made. (S. Elaiyarasu)

It is not possible to tell, by visual inspection, whether water is of an acceptable quality. Simple procedures such as boiling or using a household activated carbon filter are not adequate to remove all potential pollutants from an unknown source which may be present in water. Even natural spring water, which in the 19th century was considered healthy for all practical purposes, must now be checked before deciding what kind of treatment, if any, is required. Although costly, chemical analysis is the only way to get the details needed to decide on the correct method of purification. According to a 2007 study from the World Health Organization, 1.1 billion people lack access to sufficient drinking water supply, 88 per cent of the 4 billion annual cases of diarrhoeal disease are due to contaminated water and poor sanitation and hygiene, and 1.8 million people die each year from diarrhoeal diseases. The WHO reports that 94% of such diarrhoeal cases may be avoided through environmental changes, including exposure to clean water

- Effective methods for cleaning water at home, such as chlorination, filters and solar disinfection, and storing it in healthy containers, may save an immense number of lives each year.
- Reducing water-borne disease deaths is a major health concern.

EXERCISE

1. A filter unit is 4.5 m by 9.0 m. After filtering 10,000 m^3/day in 24 hours, the filter is backwashed at a rate of 10 L/m^2/sec for 15 minutes. Compute the average filtration rate, quantity and percentage of treated water used in washing and the rate of washwater flow in each trough. Assume 4 troughs.
2. Define filtration. Classify various types of filters. Distinguish between various types of filters in terms of (i) rate of filtration, (ii) quantity of washwater, (iii) method of cleaning, (iv) effective size of sand and UC, (v) depth of bed for gravel and sand, and (vi) backwash rate

3. A rapid sand filter is to be provided for a town with population of 3,00,000 and the water demand is 150 L/c/d. The rate of filtration 15 m^3/m^2/hour and backwash allowance can be assumed 5% with a backwashing period of 30 min. Determine the number of filters required with one standby unit.
4. Distinguish between slow sand and rapid sand filters with reference to (a) rate of filtration, (b) quantity of washwater, (c) method of cleaning, (d) effective size of sand, and (e) depth of bed for gravel and sand.
5. Design a rapid sand filter unit for 9 MLD supply with all its principal components including under-drainage system and washwater troughs?
6. A rapid sand filter is to be provided for a town with population of 7,00,000 and the water demand is 250 L/c/d. The rate of filtration 15 m^3/m^2/hour and backwash allowance can be assumed 5% with a backwashing period of 30 min. Determine the number of filters required with one standby unit. The available surface area of filter unit is 10,000 mm × 4000 mm. Also compute the up-flow velocity and head loss to expand the bed to 0.66 m from its original undisturbed bed depth of 0.6 m. The porosity of the bed is 0.45 and specific gravity is 2.5. The average particle size is 0.6 mm. The coefficient of drag is 5.02.
7. Design criteria used for slow sand filters in terms of filter material, filtration rate and backwash velocity for WTP.
8. Design criteria used for RGF in terms of filter material, filtration rate and backwash velocity for WTP.
9. Define head loss in a filter unit and how it is calculated along with suitable equation.
10. Design criteria used for design of under drainage system for a RGF.
11. What range of values of backwash velocity is kept in filtration units.
12. Name different theories behind filtration.
13. Major difference between slow sand filters and rapid gravity filters.

REFERENCES

1. S. Elaiyarasu. (N.D.). *A Course Material on Environmental Engineering I.* Vijayamangalam: Department of Civil Engineering, Sasurie College of Engineering.
2. Solutions, S. W. (N.D.). *Filtration.* Retrieved From Handbook on Industrial Water Treatment: https://www.Suezwatertechnologies.com/Handbook/Chapter-06-Filtration#In Line%20clarification

6

CHAPTER

Disinfection

Disinfection is a process of killing the infectious bacteria present in the water during its distribution thus making the water safe for the end users. During water treatment process the water coming out after filtration may contain some disease causing bacteria. These microorganisms are required to be killed so that water can be safe for drinking and can be supplied to the public. The chemicals or substances used in disinfection process to destroy such bacteria or microorganisms are called disinfectants. In water treatment process the disinfection process is very important as water containing bacteria or microorganisms may lead to spread of various diseases and sometimes epidemics caused by these microorganisms may be disastrous. The turbidity, colour and various minerals in water may not be as much dangerous as compared to the presence of these bacteria or viruses. Therefore, disinfection is an important aspect of the water treatment process that needs to be carried out before the water is supplied to the public.

The significance of the process can be well understood by considering the fact that in case of emergencies or wartime, when the treatment of water is not possible by using processes such sedimentation and filtration, the only process of water disinfection is generally carried out. The disinfectant not only removes bacteria from the treated water but also ensures the entire safety of water in the distribution system. Therefore, the chemical that can provide residual sterilizing effect for a longer period with protection against recontamination may be recommended. The disinfectant should be economical and can be easily measurable by simple test procedures. It should destroy 100% harmful pathogens in a short duration and make it safe for use. It should be readily available and should not require high skills for its applications. The water does not become poisonous and offensive against the consumer after treatment. The concentration of such disinfectant can be calculated easily and expeditiously. Chlorine is considered to be an ideal disinfectant as it possesses all the above-mentioned properties. Due to its versatile properties as a disinfectant the chlorines is accepted universally. Other than chlorine there are minor methods of disinfection of water. These methods are discussed below.

6.1 MINOR METHODS OF DISINFECTION

The various minor methods of disinfection of water are given below.

6.1.1 Boiling of Water

Boiling of water for a long time destroys the bacteria present in water. This is the most effective form of water disinfection. However, this approach cannot, for economic reasons, be used to disinfect large-scale of water for public water sources. In fact, it can only destroy the current bacteria and can't take care of any potential contaminations. However, boiling may be adopted by consumers individually for disinfecting water before use for drinking and domestic purposes in emergencies such as accidental contamination of public water supply.

6.1.2 Treatment Using Excess Lime

It has been found that *E. coli* and other bacteria present in water cannot survive when the pH value of the water is greater than 9.5. Therefore, when ample lime is applied to water to raise its pH value to 9.5 or more, most of the bacteria are destroyed. In this scenario, the elimination of bacteria leads to 99.9% or even 100%. The necessary dose of lime required for disinfection of water ranges between 10 and 20 mg/L. Moreover, after disinfection, the excess lime should be extracted from the water by an effective re-carbonization process before providing it to customers. It is also recommended that most water-borne bacteria cannot thrive if the water is highly acidic with a pH value below 3. Therefore, water can be disinfected even by using acid bath and then neutralize it.

6.1.3 Iodine Treatment

The iodine (I_2) is also used in water purification process for water disinfection. Iodine is a bluish black solid that forms hypoiodic acid (HIO) when applied to the water and dissociates the hypoiodic acid resulting in the formation of hypoiodite ion (IO). At a pH value of 7 the percentage of iodine (I_2), hypoiodous acid (HIO) and hypoiodite ion (IO) for a total iodine residual of 0.5 mg/L has been observed to be 52, 48 and 0%, respectively. All iodine and hypoiodine acid are similarly effective disinfectants. Iodine is less reactive to organic matter than chlorine, and is relatively soluble in water. Additional iodine does not react to iodamine formation with ammonia, but it oxidizes ammonia. It oxidises phenols, too. Because of these factors it takes less iodine to acquire free residual iodine.

The maximum iodine dose of around 8 mg/L for 5 minutes contact time with water is quite effective. The iodine is also essential for disinfection, in the form of pellets or small tablets. Iodine is less dependent than chlorine on the pH value and water temperature, contact time, and nitrogen impurities. It can also destroy amoebic cysts that chlorine does not do. Compared to chlorine iodine, it also offers longer-lasting protection against bacteria and decreases unpleasant tastes and smells. Owing to these benefits over chlorine, iodine is better for post-disinfection compared to chlorine. However, iodine is more expensive than chlorine, which has restricted the usage of iodine for disinfection to just small amounts of water, such as swimming pools, field army troops, etc.

6.1.4 Bromine Treatment

Bromine (Br) is also used for water disinfection process. It is a thick, dark reddish-brown element that forms hypobromous acid (HOBr) when applied to the water and dissociates the hypobromous acid resulting in hypobromite ion (OBr) formation. The bromine's bactericidal activity almost resembles that of chlorine. More bromine also reacts to monobromanine and dibromanine formation with ammonia. In comparison to monochloramines, monobromanin is a potent bactericide and free bromine. However, bromine is not used for the disinfection of water in large water sources due to higher cost and poor quality. It is limited to be used to clean swimming pool waters.

6.1.5 Ozone Treatment

Ozone (O_3) is used as a disinfectant in water purification systems. It is a pungent odorous, weakly blue gas. It is an unstable gas that appears to break down to normal oxygen (O_2) and nascent oxygen (O) (which contains single oxygen atoms).The nascent oxygen as (O) is very effective in oxidizing the organic matter and very potent in destroying bacteria. Due to its high instability it cannot be stored and therefore must be generated at the point of usage.

Ozone is created by moving dry air or oxygen between two high-potential electrodes. The mixture of air or oxygen and ozone created by the ozone generator is injected into the water to be treated by injecting it into a mixing chamber or diffusing it. The ozone concentration is approximately 2 to 3 ppm to obtain 10 ppm residual ozone with a contact time of approximately 10 minutes. During ozone application in water to be treated, it reacts first with the chemical impurities present in the water. Thereafter, it destroys chemical impurities present in water the remaining free residual ozone destroys the microorganisms present in it and thus requisite water disinfection is achieved. The advantages and disadvantages of ozone treatment for disinfection of water are as indicated below.

Advantages of ozone treatment

- Ozone is a disinfectant more active than chlorine. Ozone is effective in destroying pathogens such as cysts and other viruses which chlorine cannot destroy.
- The efficiency of ozone disinfection remains uninfluenced by water temperature and pH across a wide range.
- The bactericidal action with ozone is faster than with chlorine, and thus short contact time is required.
- Ozone is highly efficient in extracting water from tastes, odours, colour, iron and manganese.

- Unlike chlorine, ozone does not transmit unpleasant tastes and odours to water nor typically does it contain harmful substances such as chlorinated hydrocarbons.
- There is no chance of over-treatment because ozone decomposes to oxygen.
- Organic impurities found in water are rapidly oxidized by ozone.
- There are no transport and storage problems because ozone is produced at the point of usage. This also does not involve storing hazardous chemicals.

Disadvantages of ozone treatment

- Ozone is more expensive than chlorine.
- Because ozone is fairly unstable, residual protection against recontamination is not given.
- Ozone is expected to be generated as needed and energy is expected for its production. Therefore, it may be difficult to adapt the treatment to differences in load or changes in water quality with regard to ozone demand.
- Because ozone cannot be stored and supplied in cylinders, it is important to install complicated ozone manufacturing equipment called ozonizer at the treatment plant.

While ozone treatment has many advantages, it was not widely used for disinfecting water in our country's wide public water sources due to its high cost. Ozone has, however, been widely used in European countries for water disinfection for large public water sources. Further ozone treatment was used for small scale disinfection of water such as swimming pool waters.

6.1.6 Potassium Permanganate Treatment

Potassium permanganate ($KMnO_4$) is another important chemical compound used for the disinfection of water. Potassium permanganate acts as a strong oxidizing agent, and is found to be effective in destroying bacteria from cholera. It is less effective at destroying bacteria from other water-borne diseases, however. This disinfection method is not used for large public water sources but is widely used in rural areas where water supplies are primarily from wells that typically contain less bacteria. The standard potassium permanganate dosage ranges from 1 to 2 mg / L for a contact time period of 4 to 6 hours. Further, the well water should not be used to apply potassium permanganate to it for at least 48 hours. The addition of potassium permanganate in water gives it pink colour appearance. Furthermore, with the passing of time, the water treated with potassium permanganate creates a dark brown coating on porcelain vessels and this coating is difficult to remove even with scratches or rubbing.

6.1.7 Silver Treatment

Silver also has an important property of disinfecting water and is quite effective in killing the bacterial spores and algae present in fairly clear water. Gold may be inserted into water either in the form of a silver salt or by immersing silver or silver coated electrodes in water and adding about 100 volts of electrical potential. The later method is called as Electro-Katadyn method. The silver dose range of 0.05 to 1ppm with a contact period of 15 minute to 3 hour is highly effective in killing the microorganisms in water. The silver treatment produces no taste and odour in water and does not have any adverse effect on the human body. The disinfectant in water treatment using silver proves to be a costly and expensive method. Therefore, silver as a disinfectant is limited to private individual houses only.

6.1.8 Ultraviolet Ray Treatment

The ultraviolet rays provide an important tool for water disinfection. The ultraviolet rays are the invisible rays of light. Essentially, these rays are produced at sunlight. As such, water exposure to sunlight contributes to microorganism damage that is mainly due to ultraviolet rays. Sunlight treatment also involves a wide area of exposure and a long period. Ultraviolet rays are therefore artificially produced by a system consisting of a mercury-vapor lamp enclosed in a quartz globe. These lamps are commonly supplied with 220-volt DC. The ultraviolet rays kill both active bacteria and spores, which otherwise are hard to destroy. The destructive force of these rays starts with a wavelength of 0.490 μ in the blue-green area of the spectrum and raises its effectiveness up to 0.149 μ.

For efficient disinfection by ultraviolet rays the following conditions are necessary

- Water should be free of suspended and turbid-causing colloidal compounds. Water turbidity does not exceed 15 to 20 ppm.
- Water does not contain light absorbing substances such as phenols, alkyl benzene-sulphonate (ABS) and other atomic compounds.
- Water should flow in thin films or sheets not exceeding 10 cm thick, which should be close to ultraviolet rays.
- Water should be stirred well but not mixed with air.
- Apply sufficient duration and time of ultraviolet ray exposure.

Ultraviolet rays treatment poses advantages that exposure to it for short periods, actually ingests no foreign matter and creates no taste and odor with no adverse effects of exposure. The drawback of this procedure is as no residual effect is available to protect against recontamination, and therefore a rapid field test is required in order to determine the efficacy of the procedure. In addition, ultraviolet treatment is expensive and is inadequate for large scale disinfection of water for municipal water sources. However, it may be implemented for water supply systems of private homes, private institutions, office buildings, and swimming pools in particular where, if chemicals are used for disinfection, the skins of people taking bath are likely to experience harmful effects.

6.1.9 Chlorination

The process of applying chlorine to destroy the microorganisms and organic matter present in water is known as chlorination. The process is widely practised in all the developed countries for public water supply systems. Chlorine is available in all the three forms as a liquid (liquid Cl_2), solid (bleaching powder) and gaseous form (chlorine gas).The major advantage of chlorine is that it can be manufactured easily by using electrolytes of common salts (NaCl). It is also a potent oxidant, which can penetrate the organism's cell wall and its contents. The chlorine dosage can be precisely regulated and can be easily detected by using basic orthotolidine test. It does not form harmful constituents upon inorganic reaction in water. It also leaves required residue in water to protect the water from re-contamination. The major disadvantage of using chlorine is that it is extremely corrosive when used as chlorine gas for water disinfection and lethal to inhalation. Hence, it must be carefully placed at a distance in sealed jar. Also, the chlorine when added to water containing phenolic compounds reacts with them resulting in cancer causing substances.

6.1.9.1 Residual Chlorine and Chlorine Demand

When chlorine is added in water, a part of it is consumed to destroy the bacteria in water while some part is used to oxidize the organic matter present in the water. The remaining part of chlorine left during the application process is perceived as residual chlorine. Therefore, its requirement may be calculated as the difference between the amount of chlorine added per litre of water to be treated and the residual chlorine after consumption in the process. Polluted waters exert a greater demand for chlorine. Chlorine demand may be decreased if water is pretreated by sedimentation and aeration. For water distribution system residual chlorine of 0.2 mg/L is the recommended value for drinking water.

6.1.9.2 Chlorine as a disinfectant in water

When chlorine is dissolved in water it forms hypochlorous acid or hypochlorite ions with hydrochloric acid as indicated in Eq. 6.1.

$$Cl_2 + H_2O \xrightarrow{pH>5} HOCl + HCl \tag{6.1}$$

Thereafter, the unstable hydochlorous acid further ionizes and dissociates as follows in Eq. 6.2;

$$HOCl \overset{pH>8}{\Leftrightarrow} H^+ + OCl^- \tag{6.2}$$

The two prevailing species hypochlorous acid (HOCl) and hypochlorite ions (OCl^-) are called free available chlorine and are responsible for the disinfection of water. Chlorine reacts with ammonia in water to form monochloramines (NH_2Cl), dichloramines ($NHCl_2$) and trichloramines (NCl_3) released and their distribution depends on the pH-value of water. At pH value greater than 10, only OCL ions exist, while pH value less than 7 and more than 5, HOCl ions exist without any dissociation. However, at pH value less than 5 the chlorine does not react and remains as elemental chlorine. Of all the available forms of chlorine the hypochlorous acid is most destructive and is 80 times more effective than hypochlorite ions. Therefore, the pH value of water is kept slightly below 7 during the chlorination process.

When chlorine is added to a water containing ammonia, the chlorine instantly reacts with this available ammonia and different forms of chloramines are produced as shown in Eq. 6.3-Eq. 6.5;

$$\underset{\text{Monochloramines}}{NH_3} + HOCL \xrightarrow{\text{yields}} NH_2Cl + H_2O \tag{6.3}$$

$$\underset{\text{Dichloramines}}{NH_2Cl} + HOCL \xrightarrow{\text{yields}} NHCl_2 + H_2O \tag{6.4}$$

$$\underset{\text{Trichloramines}}{NHCl_2} + HOCL \xrightarrow{\text{yields}} NCl_3 + H_2O \tag{6.5}$$

In water to be treated, at pH value greater than 7.5 the mono-chloramines (NH_2Cl) predominates and at pH value in between 5-6.5 the di-chloramines ($NHCl_2$) dominates. However, at pH value less than 4.4 tri-chloramines (NCl_3) are maximum. These formed chloramines possess disinfecting properties in water. Therefore, the chlorine when combines with ammonia in water and forms chloramines is known as combined chlorine and is around 25 times less effective than normal chlorine.

6.1.9.3 Dosage of Chlorine

When chlorine is added to water it immediately reacts with inorganic impurities of the water such as S^-, Fe^{++}, Mn^{++}, NO_2^-, etc., that convert this chlorine into chlorides having no oxidizing power. Thereafter, the chlorine reacts with ammonia present in the water to from chloramines and at the same time the chlorine reacts with organic matter present in the water. The total amount of chlorine being utilized in all the above processes is known as chlorine demand of water. After all the above consumption of the total chlorine applied to water, the remaining chlorine is known as *free chlorine*. Free chlorine and combined chlorine cause the germicidal action on bacteria and pathogens present in water. This free available chlorine will kill the pathogens while the combined chlorine has long-term germicidal effect to protect water for long time. It is recommended that the free chlorine in water should be 0.2 mg/L after 10 minutes of its application.

As the concentration of free chlorine greater than 0.2 mg/L makes bad taste in water and even higher concentration is toxic to human beings. However, the free chlorine concentration of less than 0.2 mg/L can re-contaminate the water after some time. Therefore, the optimum dosage of chlorine requirement in particular water before its distribution is required to be assessed and is added accordingly. During rainy season and epidemics high dose of chlorine should be added to water as compared to normal dosage. Also the presence of different types of viruses may sometimes requires different dosage of chlorine in order to kill each type of virus individually. The poliomyelitis and hepatitis viruses require chlorine dose of 0.1 and 0.4 mg/L. However, the cysts of histolytica organisms that cause amoebic dysentery and tuberculosis organisms require chlorine dose of 3 mg/L. The coxsaickie virus requires a chlorine dose of 21-138 mg/L. The contact time of 30 minutes at pH value less than 8 may be considered after application of chlorine in such type of waters.

6.1.9.4 Forms of Chlorine Application

Plain chlorination: It is a method of chlorine addition when the surface water needs no further treatment. The lake and spring water is clean and can be used after simple chlorination. The water having turbidity in between 20 and 30 mg/L can be easily treated and bacteria, organic matter and clour present in the water is removed comfortably. The standard dose is a rate of 0.8 mg /L/ hour at 15 N/cm^2 pressure to sustain in residual chlorine concentration of 0.2 mg/L.

Pre-chlorination: The application of chlorine before sedimentation and filtration is known as pre-chlorination. It is required when the water is highly polluted and to minimize the taste, odour, color and growth of algae in the water treatment units. Pre-chlorination improves coagulation and can reduce post-chlorination dosages as the load on filters decreases. A normal chlorine dose of 5–10 mg/L is required with residual chlorine dose of 0.1–0.5 mg/L. Thereafter, post-chlorination is must in order to ensure the safe water being obtained in the treatment process.

Post-chlorination: The normal method of chlorine application after complete treatment of water is known as post-chlorination. This is a method generally employed after filtration process before water is to be distributed for general public supplies. The residual chorine of about 0.1–0.2 mg/L after a contact period of 20 minutes ensures no further contamination in water distribution system.

Double chlorination: The above two processes of pre- and post-chlorination are combinedly known as double chlorination. The process of post-chlorination is normally used in all water supply systems.

Break point chlorination: It may be defined as the amount of chlorine required to kill the microorganisms and oxidize the organic matter and ammonia present in the water. The remaining chlorine after complete utilization acts as the residual chlorine as shown in Fig. 6.1. When chlorine is added to water it instantly reacts with ammonia present in it thus forming chloramines. As the free chlorine content increases it interacts with the chloramines and reduces the chlorine available. Some of the added chlorine is consumed in killing of microorganisms of water and thus the amount of chlorine added is less than the chlorine added as depicted through curve AB in Fig. 6.1. The fall down of curve BC indicates the chlorine consumption after its further addition in oxidizing the organic matter present in the water. The residual chlorine at point C is minimum and beyond this point any further addition of chlorine will act as free chlorine in water to be treated. Therefore, the point C is called *break point* as any further addition of chlorine beyond this point will act as residual chorine. The further addition of chlorine beyond this point is called *break point chlorination*. A residual chlorine dose of 0.2-0.3 mg/L is quite effective in controlling the potential re-contamination of water in distribution systems, if sunlight is avoided.

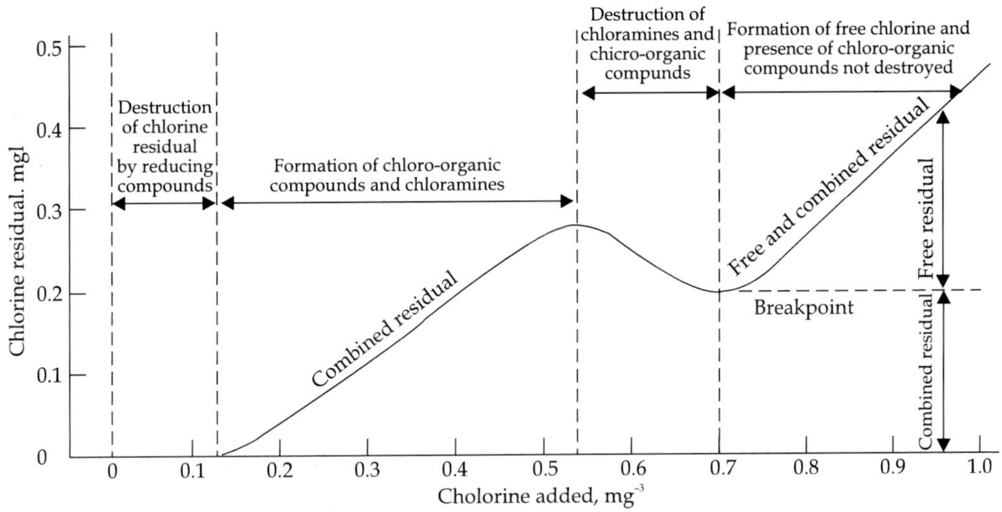

Break pint chlorination by the formation of chloramins. The free chlorine residuals first rises then falls until the reaction with ammonia have been completed. As additional chlorine is applied and ammonia is consumed, the chlorine residual again rises.

Fig. 6.1: Break point chlorination

Super chlorination: Super chlorination is characterized as the administration of a dose considerably exceeding that needed for the adequate purification of water by bacteria. In situations such as when epidemic outbreak the breakout water is to be dechlorinated before supply to the distribution system. An approximately dose of 10 to 15 mg/L is applied with a contact time of 10 to 30 minutes.

Dechlorination: Removal of excess chlorine in whole or in part from water is called *dechlorination*. In water, excess chlorine gives the pipe edges a pungent smell and corrodes. Excess chlorine therefore needs to be extracted before delivery. Physical methods such as aeration, heating, and absorption can be implemented on charcoal. Chemical approaches such as sulphur dioxide (SO_2) gas with dose of 0.3-0.6 ppm for 10-15 minutes, sodium bi-sulphite ($NaHSO_3$), sodium thiosulphate ($Na_2S_2O_3$), sodium metabisulphate ($Na_2S_2O_5$), activated carbon and ammonia (NH_4OH) are used to remove chlorine from such waters. However, it is ensured that the residual chlorine dose of 0.2–.03 mg/L should remain in the system.

Re-chlorination: Chlorine residual can fall in long distribution systems resulting in unsafe water tendering. Application of excess chlorine to compensate for this will lead to unpleasant smell to customers at the points closer to the treatment point. In these cases chlorine is again added and is rechlorinated at intermediate points usually at service reservoirs and booster pumping stations.

Kinetics of disinfection: A disinfectant is considered to be good if it destroys the microorganisms of water at concentrations well below the thresholds to humans and animals. It should kill the microorganisms at a very fast rate and should remain for a long time to prevent the re-contamination in water distribution system. The killing rate of microorganisms can be described by first order reaction as in Eq. 6.6;

$$\frac{dN}{dt} = -k.N \quad \text{or} \quad \ln\frac{N_t}{N_o} = -k.t$$

$$\boxed{N_t = N_o e^{-k.t}} \tag{6.6}$$

where,

N_o = number of microorganisms present initially

N_t = Number of microorganisms remaining at time t

k = Killing rate constant per unit of time

The factors that generally hamper the process of disinfection are turbidity and resistant organisms. Turbidity generates colloids that provide shelter to the microorganisms thus protects them from killing and therefore makes the disinfection process less effective. Also the disinfectant can adsorb on the surface of particulate matter present in the water. As compared to bacteria the viruses, cysts and ova present in water are more resistant to disinfection process. Therefore, such microorganisms require high dose of disinfectant with longer contact time.

The relationship between efficiency of chlorine removal and various species of free chlorine is pH dependent. Empirically, it is recommended that the chlorine dosage should be increased to compensate the higher pH values. The concentration of chlorine with respect to contact time can be mathematically calculated using Eq. 6.7 and shown in Fig. 6.1.

$$C^n t_p = \text{k} \tag{6.7}$$

where,

C = Concentration chlorine (mg/L)

t_p = Time required to kill the % of microorganisms (minutes)

n, k = Experimental derived constants

6.2 WATER SOFTENING

The process of removing hardness and stiffness from water is known as water softening. Hardness is characterized as the capacity of the water to absorb soap. Besides making water more comfortable for washing purposes, water softening may also be useful in preventing encrustation and scaling within boilers, water heaters, hot water lines and some industrial processes. The advantage of water softening is mainly useful in reduction of soap consumption, low cost of maintenance of water and fixtures and also improves the taste of food preparations. Therefore, hardness of water should be removed by using low cost treatment methods. Typically, hardness is represented in terms of "ppm as $CaCO_3$," or mg/L as $CaCO_3$. Typically, the home water softener industry tests hardness in the form of grains per gallon (gpg), where 17 ppm hardness is equal to 1 gpg whereas one degree of hardness is equivalent to 14.25 mg/L. In public water supplies the permissible hardness as prescribed in India standard is 75-115 mg/L.

Calcium (Ca) and magnesium (Mg), in particular calcium, are the primary constituents in water which cause hardness. Iron (Fe) and manganese (Mn) can also enhance water hardness, but to a much lower degree. Calcium and manganese hardness is usually considered a carbonate hardness since calcium and manganese are found in water in the form of calcium bicarbonate, $Ca(HCO_3)_2$, and $Mg(HCO_3)_2$. Typically, this type of hardness is called carbonate hardness, or temporary hardness. On the other hand, sulphates, chlorides, and nitrate salts of calcium and magnesium are generally referred to as permanent hardness, because they cannot be precipitated easily.

Calcium carbonate is slightly soluble in water resulting in calcium carbonate compound that dissolves in natural water containing carbon dioxide. When the water containing calcium carbonate is heated the carbon dioxide escapes forming a precipitation of calcium carbonate that settles and is separated from the water. The same can be represented by Eq. 6.8:

$$Ca(HCO_3)_2 + \text{Heat} \rightarrow CaCO_3\downarrow + CO_2\uparrow + H_2O \tag{6.8}$$

where, $CaCO_3$ is calcium carbonate, white, crusty matter, precipitates out of the water. Water heaters suffer from hard water as the insoluble carbonate form of calcium would be precipitated when water containing calcium bicarbonate is heated. It is this white, crusty matter that builds up in water heaters, hot water lines, and on the kitchen tap as well as in the sink.

Water hardness varies with the source water from which it came. The mineral content and water hardness usually reflect the existence of geological formations from which the water comes into contact, e.g., calcareous. Soil-water sources are usually much tougher than surface waters. The groundwater with a hardness of 300 ppm is not uncommon. No "fully cooled" drinking water done. A hardness of about 80-90 ppm is normally quite enough. Water softening science and chemistry can be challenging and frustrating for water operators, but it is also a fascinating process that exemplifies the wonders of science and the natural cycle.

6.2.1 Lime/Soda Ash Method

Lime softening, or softening of lime/soda ash, is the most popular water softening process used in water plants. The explanation for this is that lime and soda ash is readily available and cost-effectively softens the vast quantities of water that are pumped into a water plant. Hydrated lime is fed at an appropriate dose in the simplest sense to allow the hardness of the carbonate to precipitate, settle and be extracted from the water.

Quicklime (CaO) when added to water generally forms hydrated lime [$Ca(OH)_2$] is added to the water. The carbonates and bicarbonates of magnesium and bicarbonates of calcium present in water react with this hydrated lime and the following reactions (Eq. 6.9-Eq. 6.11) take place.

$$\underset{\substack{\text{Mag. Carbonate}\\\text{(soluble)}}}{MgCO_3} + \underset{\text{Hydrated lime}}{Ca(OH)_2} \xrightarrow{\text{yields}} \underset{\substack{\text{Mag. Hydroxide}\\\text{(insoluble)}}}{Mg(OH)_2 \downarrow} + \underset{\substack{\text{Calcium carbonate}\\\text{(insoluble)}}}{CaCO_3 \downarrow} \quad (6.9)$$

$$\underset{\substack{\text{Mag. Carbonate}\\\text{(soluble)}}}{Mg(HCO_3)_2} + \underset{\text{Hydrated lime}}{Ca(OH)_2} \xrightarrow{\text{yields}} \underset{\substack{\text{Mag. Hydroxide}\\\text{(insoluble)}}}{Mg(OH)_2 \downarrow} + Ca(HCO_3)_2 \quad (6.10)$$

$$\underset{\text{Cal. bicarbonate}}{Ca(HCO_3)_2} + Ca(OH)_2 \xrightarrow{\text{yields}} 2CaCO_3 \downarrow + 2H_2O \quad (6.11)$$

The precipitates obtained after the overall reaction occur in form of magnesium hydroxide and carbonates that settles down by sedimentation process and thus can be removed easily from the tank. The process is generally used for removal of temporary hardness in water. However for the removal of permanent hardness special procedures of water softening are generally adapted.

6.2.2 Permanent Hardness Removal Methods

The permanent hardness can be removed by special methods of water softening. The various methods of water softening used in water treatment process are:

(i) Lime-soda process

(ii) Base exchange process known as Zeolite process

(iii) Demineralization process

6.2.2.1 Lime

In this method of softening hydrated lime as $Ca(OH)_2$ and soda ash as Na_2CO_3 are added to the water to be softened. These chemicals when reacts salts of calcium and magnesium thus forming the precipitates of calcium carbonate and magnesium hydroxide with the following reactions Eq. 6.12-Eq. 6.20.

$$\underset{\substack{\text{Mag. Carbonate}\\\text{(soluble)}}}{MgCO_3} + \underset{\text{Hydrated lime}}{Ca(OH)_2} \xrightarrow{\text{yields}} \underset{\substack{\text{Mag. Hydroxide}\\\text{(insoluble)}}}{Mg(OH)_2 \downarrow} + \underset{\substack{\text{Calcium carbonate}\\\text{(insoluble)}}}{CaCO_3 \uparrow} \quad (6.12)$$

$$\underset{\substack{\text{Mag. Carbonate}\\\text{(soluble)}}}{Mg(HCO_3)_2} + \underset{\text{Hydrated lime}}{Ca(OH)_2} \xrightarrow{\text{yields}} \underset{\substack{\text{Mag. Hydroxide}\\\text{(insoluble)}}}{Mg(OH)_2 \downarrow} + Ca(HCO_3)_2 \quad (6.13)$$

$$\underset{\text{Cal. Bicarbonate}}{Ca(HCO_3)_2} + Ca(OH)_2 \xrightarrow{\text{yields}} 2CaCO_3 \downarrow + 2H_2O \quad (6.14)$$

$$\underset{\substack{\text{Non-carbonate}\\\text{Hardness of Mg}}}{MgCl_2} + \underset{\text{Hydrated lime}}{Ca(OH)_2} \xrightarrow{\text{yields}} \underset{\text{Magnesium Hydroxide}}{Mg(OH)_2\downarrow} + CaCl_2 \tag{6.15}$$

$$\underset{\substack{\text{Non-arbonate}\\\text{Hardness of Mg}}}{MgSO_4} + \underset{\text{Hydrated lime}}{Ca(OH)_2} \xrightarrow{\text{yields}} \underset{\text{Magnesium Hydroxide}}{Mg(OH)_2\downarrow} + \underset{\substack{\text{Non-carbonate}\\\text{hardness of Ca}}}{CaSO_4} \tag{6.16}$$

$$\underset{\text{Dissolved } CO_2}{CO_2} + \underset{\text{Hydrated lime}}{Ca(OH)_2} \xrightarrow{\text{yields}} \underset{\text{Calcium carbonate}}{CaCO_3\downarrow} + H_2O \tag{6.17}$$

$$\underset{\substack{\text{Non-carbonate}\\\text{Hardness of Ca}}}{CaCl_2} + \underset{\text{Soda}}{Na_2CO_3} \xrightarrow{\text{yields}} CaCO_3\downarrow + 2NaCl \tag{6.18}$$

$$\underset{\substack{\text{Non-carbonate}\\\text{Hardness of Ca}}}{CaSO_4} + \underset{\text{Soda}}{Na_2CO_3} \xrightarrow{\text{yields}} CaCO_3\downarrow + Na_2SO_4 \tag{6.19}$$

$$Mg(No_3^-)_2 + Ca(OH)_2 \xrightarrow{\text{yields}} Mg(OH)_2\downarrow + Ca(NO_3^-)_2 \tag{6.20}$$

The precipitates obtained after the overall reaction occur in the form of magnesium hydroxide and carbonates that settle down by sedimentation process and thus can be removed easily from the tank. Therefore, from the above reaction equation it is seen that lime facilitates in removing the carbonate hardness caused by Ca and Mg. From Eq. 6.13 it can be inferred that lime also reacts with non-carbonate hardness of Mg and converts it into non-carbonate hardness of Ca. Thereafter, the remaining non-carbonate hardness of Ca is finally eliminated by soda. Lime is also helpful in removal of dissolved carbon dioxide from water (Eq. 6.14).

The formation of precipitation of $CaCO_3$ and $Mg(OH)_2$ is generally governed by the pH of water. The optimum pH value for $CaCO_3$ precipitation after addition of lime should be in the range of 9.0 to 9.5. However, the pH value for precipitation of $Mg(OH)_2$ is recommended to be 11.0. As most of the waters to be treated possess the pH value less than these values therefore the pH of the water is to be raised for removal of hardness from water effectively. Also the CO_2 concentration of 10 mg/L or greater in water should be removed prior softening confirms to be economical and valuable.

6.2.2.2 Caustic Soda

The different forms of hardness can also be removed by use of caustic soda (NaOH), the reaction equations for calculation may be described as shown in Eq. 6.21-Eq. 6.24.

$$CO_2 + 2NaOH \xrightarrow{\text{yields}} 2Na^+ + CO_3^{2-} + H_2O \tag{6.21}$$

$$Ca^{2+} + 2(HCO_3)^- + 2\,NaOH \longrightarrow CaCO_3\downarrow + 2\,Na^+ + CO_3^{2-} + 2\,H_2O \tag{6.22}$$

$$Mg^{2+} + 2(HCO_3)^- + 4\,NaOH \longrightarrow Mg(OH)_2\downarrow + 4\,Na^+ + 2CO_3^{2-} + 2\,H_2O \tag{6.23}$$

$$Mg^{2+} + SO_4^{2-} + 2\,NaOH \longrightarrow Mg(OH)_2\downarrow + 2\,Na^+ + SO_4^{2-} \tag{6.24}$$

In order to precipitate magnesium hydroxide the pH value of 11 should be kept during soda ash process of water softening. Therefore, in order to achieve this, an excess of 1.25 mequiv/L of sodium hydroxide is added in water.

6.2.2.3 Stabilization

During the water softening process the chemical precipitation occurs as the hardness is removed from water. But still some of the hardness remains in water and therefore should be removed. It has been reported that in treated water the concentration of up-to 40 mg/L of $CaCO_3$ and 10 mg/L of $Mg(OH)_2$ still persists. However, in softening process the finally produced soluble sodium salts are unobjectionable in water. During the entire process of water softening still the finely divided particles of remaining precipitates of magnesium hydroxide and carbonates slowly form sand grains and are deposited as incrustation of filter media or pipes of the distribution system. Therefore, stabilization of water is necessary by dissolution of supersaturated $CaCO_3$ back into Ca_2^+ and HCO_3^-. It can be achieved by the addition of any one of the several acids as shown in (Eq. 6.25-Eq. 6.26);

$$2\,CaCO_3 + H_2SO_4 \rightarrow 2Ca^{2+} + 2(HCO_3)^- + SO_4^{2-} \quad (6.25)$$

$$Mg(OH)_2 + H_2SO_4 \rightarrow Mg^{2+} + SO_4^{2-} + 2\,H_2O \quad (6.26)$$

This can also be removed by recarbonation process of passing carbon dioxide through the water. The carbon dioxide gas can be produced by burning of coke, oil or gas or by liquid CO_2 under pressure. The recarbonation process can be simply understood by the following equations as (Eq. 6.27-6.28)

$$\underset{\text{Insoluble Ca Carbonate}}{CaCO_3\downarrow} + CO_2 + H_2O \rightarrow Ca^{2+} + 2(HCO_3)^- \quad (6.27)$$

$$Mg(OH)_2 + 2CO_2 \xrightarrow{\text{yields}} Mg^{2+} + 2(HCO_3)^- \quad (6.28)$$

When the pH of water is increased in order to enhance the magnesium precipitation, the excess of hydroxyl ions produced should be neutralized before the stabilization process. Therefore, the two stage process is required and can be represented by following reaction equations as:

When sulphuric acid is used as shown in Eq. 6.29-Eq. 6.30;

$$Ca^{2+} + 2OH^- + H_2SO_4 \rightarrow Ca^{2+} + SO_4^{2-} + 2\,H_2O \quad (6.29)$$

$$2Na^+ + 2OH^- + H_2SO_4 \rightarrow 2Na^+ + SO_4^{2-} + 2\,H_2O \quad (6.30)$$

When carbon dioxide is used as shown in Eq. 6.31-Eq. 6.32;

$$Ca^{2+} + 2OH^- + 2CO_2 \rightarrow CaCO_3 + H_2O \quad (6.31)$$

$$2Na^+ + 2OH^- + CO_2 \rightarrow 2Na^+ + CO_3^{2-} + H_2O \quad (6.32)$$

It is recommended that before achieving the considerable stabilization to occur, the pH of water should be lowered to 9.5 value. In a lime soda process of water softening, the requirement of lime and soda generally depend on the type and degree of hardness in the water to be softened. The softening is generally carried out as per the type of hardness present in water to be treated as lime or soda or both can be added accordingly.

6.2.2.4 Role of Alkalinity in Hardness Removal

It has already been discussed that alkalinity can be considered to be the sum of carbonate and bicarbonate ions present in the water as Ca, Mg and Na ions. It is the fact that if sodium carbonate or bicarbonate is not present in the water, then the entire alkalinity will be caused by the presence of carbonates and bicarbonates of Ca and Mg or simply by the carbonate hardness of water. Also there would be no sodium alkalinity in water if the sulphates or chlorides of calcium or magnesium are present in water as these salts will react with sodium alkalinity and covert it into calcium or magnesium alkalinity. Therefore, it can be inferred that the presence of such a non-carbonate hardness in water will indicate the absence of sodium alkalinity. It is therefore, recommended that the water with zero sodium alkalinity and with non-carbonate hardness will have carbonate hardness equal to its alkalinity value.

However, for a water containing sodium alkalinity and not having any non-carbonate hardness, the alkalinity will be greater than carbonate hardness. Therefore, the carbonate hardness will be equal total hardness and therefore, carbonate hardness may be considered equal to the alkalinity or the total hardness, whichever is less.

A large quantity of sludge in the form of insoluble precipitates of calcium carbonate and magnesium hydroxide requires to be disposed of appropriately. However, from the chemical equations it is evident that for each mg/L of hardness to be removed in form of $CaCO_3$ 1 mg/L of sludge is produced as $CaCO_3$. Also for each mg/L of magnesium hardness being removed produces 0.58 mg/L of sludge in the form of $Mg(OH)_2$ with each mg/L of lime added in the system will produce 1 mg/L $CaCO_3$ as sludge. The sludge produced during the process of water softening can be calculated as shown in Eq. 6.33;

$$\text{Sludge produced (mg/L)} = C_{ah} + M_{gh} + L_{iR} \quad (6.33)$$

where,

C_{ah} = Calcium hardness removed (mg/L as $CaCO_3$)

M_{gh} = Magnesium hardness removed (mg/L as $CaCO_3$)

L_{iR} = Lime added (mg/L as $CaCO_3$)

Softening the lime is sort of like battling fire with stone. Feeding large doses of hydrated lime is not unusual, which is why a lime softener plant uses a grain-bin lime feeder and not a small dry chemical feeder using 25 kg bags of hydrated lime. Complete removal of hardness cannot be accomplished by chemical precipitation Now, if non-carbonate hardness adds to the water's hardness, soda ash, Na_2CO_3, may be fed to eliminate non-carbonate hardness, in addition to the hydrated lime. The liquid caustic soda, NaOH, can also be used instead of soda ash. The advantages of taking caustic soda are as follows:

- Simple to manage and to feed.
- Lack of deterioration in storage.
- It makes less sludge of calcium carbonate.
- Can precipitate hardness of both carbonate and non-carbonate, so the caustic soda will actually substitute some of the hydrated lime required.

- Whether using soda ash or caustic soda depends on chemical costs and characteristics of the raw water.

6.2.2.5 Chemical Requirement Calculations

The concentration and quantity of chemicals required in water softening process can be calculated by using appropriate formulae. The same can be stated as below with following example.

Single Stage softening:

Example 6.1: A water with ionic characteristics as shown through bar diagram in Fig. 6.2 needs to be soften to a minimum Ca hardness by lime and soda ash process. It is not mandatory to remove magnesium.

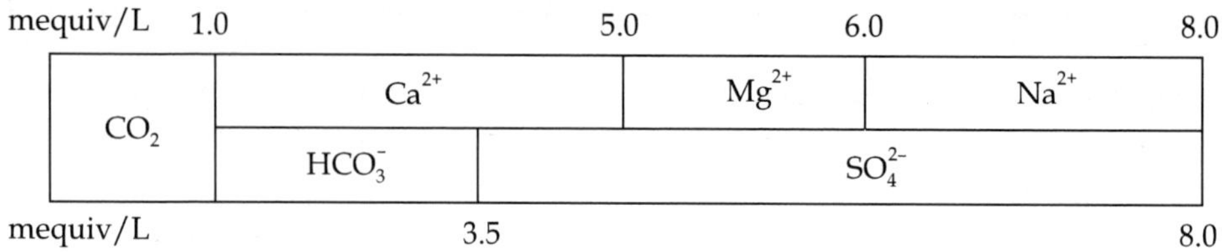

Fig. 6.2: Concentration of ions (mequiv/L)

- Calculate the chemical requirement and the solids produced in milli equivalents per litre.
- Draw the diagram of finished water.
- For a flow rate of 25,000 m^3/day calculate the dose of chemicals required daily and the mass of solid produced. Assume the lime and soda ash to be 90% and 85% pure respectively.

Solution: The chemical equation for the lime addition is as follows:

$$1.0\ CO_2 + 1.0\ CaO \xrightarrow{\text{yields}} 1.0\ CaCO_3 \downarrow$$

$$2.5\ (Ca^{2+} + 2HCO_3^-) + 2.5\ CaO \longrightarrow 5.0\ CaCO_3 \downarrow + 5.0\ H_2O$$

$$1.5\ (Ca^{2+} + SO_4^{2-}) + 1.5\ (Na_2CO_3) \xrightarrow{\text{yields}} 1.5\ CaCO_3 \downarrow + 1.5\ (2\ Na^+ + SO_4^{2-})$$

Second stage recarbonation will be required to stabilize the water. Assuming a $CaCO_3$ concentration of 40 mg/L in effluent from the settling basin, out of which 25 mg/L should be converted to reach the equilibrium of 15 mg/L of $CaCO_3$.Concentration of finished water (mequiv/L) is shown in Fig. 6.3.

$$0.5CaCO_3 + 0.5\ CO_2 + 0.5H_2O \rightarrow 0.5\ Ca(HCO_3)_2$$

Therefore, total chemical requirements are:

Lime = 1 + 2.5 = 3.5 mequiv/L

Soda ash = 1.5 = 1.5 mequiv/L

CO_2 = 0.5 mequiv/L

Therefore, solids produced as $CaCO_3$ = 1.0 + 5.0 + 1.5 – 0.8 = 6.7 mg/L

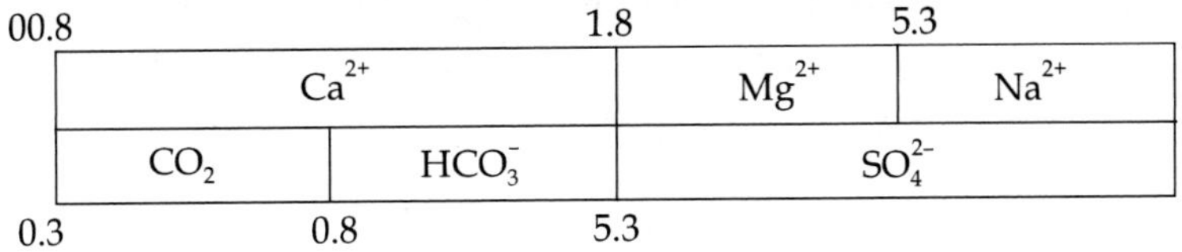

Fig. 6.3: Concentration of finished water (mequiv/L)

The equivalent mass of lime and soda can be calculated as:

$$\text{Lime} = \frac{40+16}{2} = 28 \text{ mg/milliequiv}$$

$$\text{Soda as} = \frac{2(23)+12+3(16)}{2} = 53 \text{ mg/milliequiv}$$

$$\text{Carbon dioxide} = \frac{12+2(16)}{2} = 22 \text{ mg/milliequiv}$$

The daily requirements can be calculated as :

$$\frac{1}{1.9}\times 28\frac{\text{mg}}{\text{milliequiv}}\times 3.5\frac{\text{milliequiv}}{\text{L}}\times 25\times 10^6\,\frac{L}{d}\times\frac{1 \text{ kg}}{10^6 \text{ mg}} = 2722 \text{ kg/d}$$

$$\frac{1}{0.85}\times 53\frac{\text{mg}}{\text{milliequiv}}\times 1.5\frac{\text{milliequiv}}{\text{L}}\times 25\times 10^6\,\frac{L}{d}\times\frac{1 \text{ kg}}{10^6 \text{ mg}} = 2338 \text{ kg/d}$$

$$22\frac{\text{mg}}{\text{milliequiv}}\times 0.5\frac{\text{milliequiv}}{\text{L}}\times 25\times 10^6\,\frac{L}{d}\times\frac{1 \text{ kg}}{10^6 \text{ mg}} = 275 \text{ kg/d}$$

The mass of dry solids produced per day is

$$50\frac{\text{mg}}{\text{milliequiv}}\times 6.7\frac{\text{milliequiv}}{\text{L}}\times 25\times 10^6\,\frac{L}{d}\times\frac{1 \text{ kg}}{10^6 \text{ mg}} = 8375 \text{ kg/d}$$

Example 6.2: Water with ionic characteristics shown below in Fig. 6.4 is to be softened to the minimum possible hardness by the lime soda ash excess lime process. Calculate the amount of chemicals required in milliequivalents per litre. Draw a bar diagram of finished water.

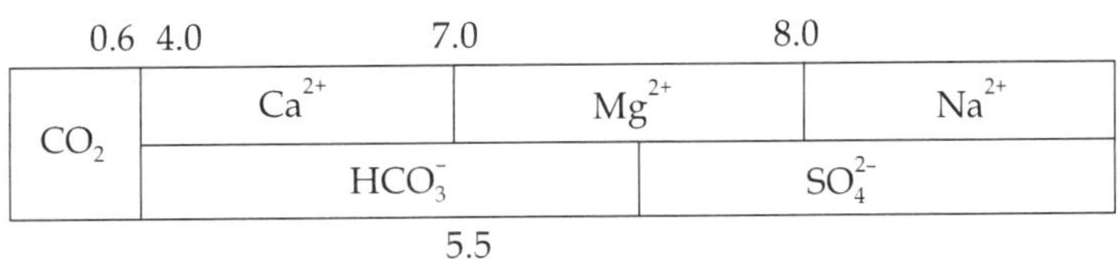

Fig. 6.4: Concentration of ions in water (mequiv/L)

The chemical requirement can be calculated using following chemical equations as:

$$0.6\ CO_2 + 0.6\ CaO \xrightarrow{\text{yields}} 0.6\ CaCO_3\downarrow$$

$3.4\,(Ca^{2+} + 2HCO_3^-) + 3.4\,CaO \rightarrow 6.8\,CaCO_3\downarrow + 6.8\,H_2O$

$1.5(Mg^{2+} + 2HCO_3^-) + 3CaO \rightarrow 1.5\,Mg(OH)_2\downarrow + 3\,CaCO_3\downarrow$

$1.5(Mg^{2+} + SO_4^{2-} + 1.5CaO + 1.5H_2O \rightarrow 1.5\,Mg(OH)_2\downarrow + 1.5(Ca^{2+} + SO_4^{2-})$

$1.5(Ca^{2+} + SO_4^{2-}) + 1.5\,Na_2\,CO_3 \rightarrow 1.5\,CaCO_3\downarrow + 1.5(2Na^+ + SO_4^{2-})$

Excess lime = 1.25 milliequiv/L

For first stage recarbonation, use of CO_2 to neutralize excess lime

$1.25(Ca^{2+} + 2OH^-) + 1.25CO_2 \rightarrow 1.25\,CaCO_3\downarrow + 1.25\,H_2O$

Assuming 40 mg/L $CaCO_3$ and 10 mg/L $Mg(OH)_2$ remaining in solution after second stage settling

$0.2\,Mg(OH)_2 + 0.4CO_2 \rightarrow 0.2\,Mg(HCO_3)_2$

$0.5CaCO_3 + 0.5CO_2 + 0.5H_2O \rightarrow 0.5Ca(HCO_3)_2$

Total quantity of chemical required are:

Lime = 0.6 + 3.4 + 3.0 + 1.5 + 1.25 = 9.75

Soda ash = 1.5

CO_2 = 1.25 + 0.4 + 0.5 = 2.15

Bar diagram of final water is shown below in Fig. 6.5.

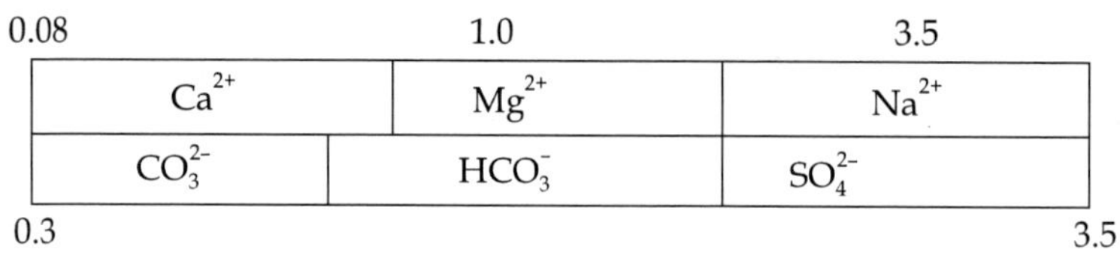

Fig. 6.5: Bar diagram of final water

6.2.3 Softening Operation

It consists of several steps may be carried out in one or two stages including mixing of chemicals, flocculation process to enhance the precipitation, settling of precipitates and stabilization. The systems of operation are similar like removing of turbidity using coagulation process. The design criteria are summarized in Table 6.1.

Table 6.1: Design criteria for softening operations

Parameter	Mixer	Flocculator	Settling basin	Solids contact basin
Detention time (min)	5	30-50	120-240	60-240
Velocity gradient (s^{-1})	700	10-100	NA	NA
Flow through velocity (m/s)	NA	0.045-0.138	0.045-0.138	NA
Overflow rate (L/min/m^2)	NA	NA	0.577-1.16	2.9

The water containing high magnesium concentration is treated by split treatment process. The first stage unit in softening is avoided from incoming water. Thereafter, excess lime is added to remove the magnesium in the first stage and in spite of neutralizing it the same is used to precipitate the calcium hardness in the bypassed water in the second stage. Therefore, as no magnesium is removed in the process the initial magnesium hardness and allowable magnesium hardness in the finished water will decide the overall quality of water that needs to be bypassed as shown in Eq. 6.34;

$$Q_x = \frac{M_{gf} - M_{g1}}{M_{gr} - M_{g1}} \tag{6.34}$$

where,

Q_x = Fraction of total flow bypassed

M_{gf} = Magnesium concentration in finished water (40-50 mg/L as $CaCO_3$, usually acceptable)

M_{gr} = Magnesium concentration in raw water (mg/L as $CaCO_3$)

M_{g1} = Magnesium concentration remaining in the fraction of water receiving first stage treatment (limit of 10 mg/L as $CaCO_3$)

Example 6.3: Use split treatment to soften the water with ionic strength as indicated in Fig. 6.6. Assume that a final hardness of less than 100 mg/L is acceptable, provided the magnesium concentration to be less than 45 mg/L. Calculate the chemical requirement and draw the final bar diagram for the same?

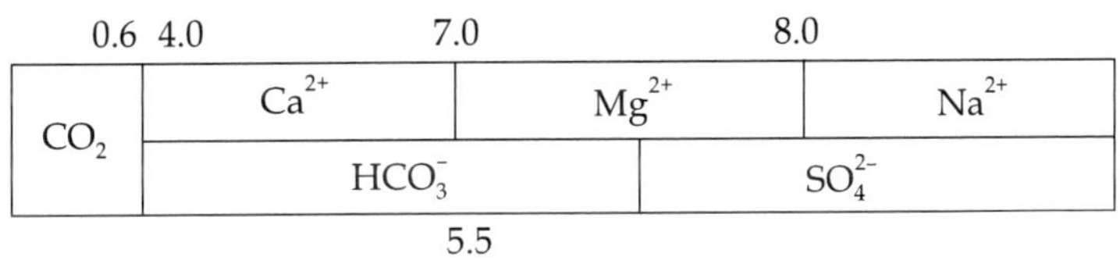

Fig. 6.6: Bar diagram of ionic strength of water (mequiv/L)

Solution: For the softening of water, calculate the bypass fraction required:

$$Q_x = \frac{M_{gf} - M_{g1}}{M_{gr} - M_{g1}} = \frac{0.9 - 0.2}{3.0 - 0.2} = 0.25$$

The quantity of chemicals required can be calculated as:

$0.6\ CO_2 + 0.6\ CaO \xrightarrow{\text{yields}} 0.6\ CaCO_3 \downarrow$

$3.4\ (Ca^{2+} + 2HCO_3^-) + 3.4\ CaO \rightarrow 6.8\ CaCO_3\downarrow + 6.8\ H_2O$

$(1.0\text{-}0.25)1.5(Mg^{2+} + 2(HCO_3)^- + (1.0\text{-}0.25)3CaO \rightarrow 1.13\ Mg(OH)_2\downarrow + 2.25\ CaCO_3\downarrow$

$(1.0\text{-}0.25)\ 1.5(Mg^{2+} + SO_4^{2-}) + (1.0\text{-}0.25)1.5CaO + (1.0\text{-}0.25)1.5H_2O \rightarrow$

$1.13Mg(OH)_2\downarrow + 1.13(Ca^{2+} + SO_4^{2-})$

$(1.13(Ca^{2+} + SO_4^{2-}) + 1.13Na_2CO_3 \rightarrow 1.13\ CaCO_3\downarrow + 1.13(2Na^+ + SO_4^{2-}))$

The mount of lime required $= \frac{(0.6 + 3.4)0.25}{0.75} = 1.33 > 1.25 \frac{\text{mequiv}}{L}$ (hence acceptable)

However, for second stage recarbonation:

$$0.5CaCO_3 + 0.5CO_2 + 0.5H_2O \rightarrow 0.5Ca(HCO_3)_2$$

$(1.0\text{-}0.25)\ 0.2\ Mg(OH)_2 + (1.0\text{-}0.25)0.4CO_2 \rightarrow 0.15\ Mg(HCO_3)_2$

Total quantity of chemicals required are:

Lime = 0.6 + 0.34 + (1.0-0.25) (3.0 + 1.5) = 7.38 mequiv/L

Soda ash = 1.13 mequiv/L

CO_2 = 0.3 + 0.5 = 0.8 mequiv/L

The ionic strength of finished water can be calculated as:

Ca^{2+} = 0.8

Mg^{2+} = 0.75 × 0.2 (I Stage) + 0.25 × 3.0 (in by pass) = 0.9

Na^{+} = 1.0 + 1.13 = 2.13

CO_3^{2-} = 0.3

HCO_3^{-} = 0.5 (conversion of $CaCO_3$) + 0.15 (conversion of $Mg(OH)_2$) + 0.25 × 0.15 (passed Mg) = 1.03

SO_4^{2-} = 2.5

Bar diagram of final water is shown below in Fig. 6.7.

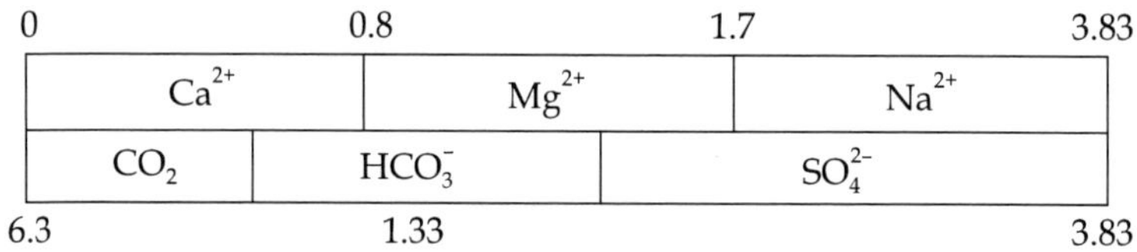

Fig. 6.7: Bar diagram of final water (mequiv/L)

6.2.4 Ion-exchange Method

The purification of water carried out using solid polymeric resins that exchange ions between two electrolytes in water is known as ion exchange process. Other than water the process is also used extensively in nuclear facilities, industrial processes and medical and pharmaceutical operations to control the purity. Typical ion exchangers are ion-exchange resins (functionalized porous or gel polymer), zeolites, montmorillonite, clay, and soil humus. Ion exchangers are either cation exchangers, which exchange positively charged ions (cations), or anion exchangers, which exchange negatively charged ions (anions). There are also amphoteric exchangers that are able to exchange both cations and anions simultaneously. However, the simultaneous exchange of cations and anions can be more efficiently performed in mixed beds, which contain a mixture of anion- and cation-exchange resins, or by passing the treated solution through several different ion-exchange materials. Ion exchange operations are usually conducted in enclosed structures containing the bed with resin medium. Water is forced through the resin bed of media under pressure up to 0.4 $m^3/min.m^2$.

Ion exchange is not nearly as cost-efficient as lime softening when producing large amounts of water. Therefore, water plants typically do not use ion exchange to soften the finished product water. However, some water plants do need to soften smaller amounts of water used in some treatment process, for example fluoridation, which can be sufficiently handled with ion-exchange methods. Most water softeners at home are likewise ion-exchange devices. Ions may exist on a positive charge or on a negative charge. An ion with a positive charge is called a cation. A negative ion of the charge is considered an anion. Usually, positive charged cations are calcium (Ca) and magnesium (Mg), the principal elements that induce hardness. The basic principle of which ion-exchange softening uses is therefore to substitute one ion for another. Anion-exchange water softener has three main components:

- A resin tank containing small beads of synthetic resin
- A brine tank
- The control valve

The resin from brine tank is treated with sodium (Na). Sodium is also a positive charged cation, with plus one charge, hence, Na+. The sodium ions are combined with calcium and magnesium ions as water moves through the resin tank, since calcium and magnesium ions have a higher positive charge than sodium ions, Ca^{++} and Mg^{+}. The resin therefore acts as a negative charge. Since calcium and magnesium ions have a greater positive charge than sodium ions, the resin "exchanges" sodium for calcium and magnesium so there is more attraction between resin and calcium and magnesium than between sodium and resin.

This process of ion exchange continues until all available exchange sites are filled, at which point the resin is depleted, and the sodium (salt bags) from the brine tank must regenerate. The biggest downside, as you can see, of ion-exchange softeners is that they place sodium in the water. This can pose health problems for individuals who do not consume excessive quantities of sodium. That is why I suggest that home water softener installations be installed on the water line that goes to the hot water heater. The benefits of the hot water pump, the advantages of the hot water pipes, the soap use is minimized so you don't have to drink it from the cold water tap. A benefit of ion-exchange softeners is that they can remove iron (Fe) to some extent because iron is also a positive charge cation.

6.2.4.1 Zeolite Softening

Zeolite is a micro-porus mineral used as catalyst in many industrial processes of water and air purification. The zeolites are hydrated alumina silicates with general composition of $Al_xSi_yO_{2(x+y)}$ without the molecule of water. Zeolites may be natural and synthetic or artificial. The natural zeolites used for water softening are gluconites or greensand. However permutit a natural zeolite with chemical formula $Na_2Al_2Si_2O_8xH_2O$ or $Na_2OAl_2O_3ySiO_2xH_2O$ with x and y values ranging from 2 to 10 is mostly used in water softening process. It is more porous, glassy and possesses greater softening capacity as compared to greensand. The characteristics of these zeolites are soft to moderately hard with light density and insoluble in water. They acts as base exchangers when comes in contact with cations present in water. Therefore, Ca^{2+} and Mg^{2+} ions are removed easily when water containing hardness passes through the bed of these zeolites as the sodium ions are replaced by Ca^{2+} and Mg^{2+} ions.

Natural kaolin is the most common mineral of a group of hydrousaluminum silicate, which results from the breaking of aluminium-rich silicate rock, such as feldspar and nephalinsyenite, either through weathering or hydrothermal activity. Kaolin when pure occurs in white clay-like masses and has been identified as a very important group of clay mineral, which is copiously utilized as an industrial mineral commodity. The various physical attributes of kaolin mineral have made it useful in various ways industrially and these include paints, ceramics, and rubber, paper, and petroleum and glass industries.

The water softening ion-exchange process has been widely used in smaller water systems and individual homes. This is based on the ability of zeolite, the ion-exchange resin, to exchange one ion from the water being treated to another ion in the resin. Zeolite resin trades sodium ions for water-borne ions such as calcium and magnesium that induce hardness. Sodium is one part of the softening salt and the other is chlorine. Zeolites of tening units is shown in Fig. 6.8.

6.2.4.2 Chemistry of the Ion-Exchange Reaction

The following chemical reactions show the exchange process. The X represents the zeolite, the exchange material (Eq. 6.35-6.40).

Removal of carbonate hardness

$$\underset{\text{Ca salt}}{Ca(HCO_3)_2} + \underset{\text{Sodium zeolite}}{Na_2X} \rightarrow \underset{\text{used Ca zeolite}}{CaX} + \underset{\text{Non-hardness Na salt}}{2NaHCO_3} \quad (6.35)$$

$$\underset{\text{Mg salt}}{Mg(HCO_3)_2} + \underset{\text{Sodium zeolite}}{Na_2X} \rightarrow \underset{\text{used Mg zeolite}}{MgX} + \underset{\text{Non-hardness Na salt}}{2NaHCO_3} \quad (6.36)$$

Removal of non-carbonate hardness

$$\underset{\text{Ca salt}}{CaSO_4} + \underset{\text{Sodium zeolite}}{Na_2X} \rightarrow \underset{\text{used Ca zeolite}}{CaX} + \underset{\text{Non-hardness Na salt}}{Na_2S} \quad (6.37)$$

$$\underset{\text{Ca salt}}{CaCl_2} + \underset{\text{Sodium zeolite}}{Na_2X} \rightarrow \underset{\text{used Ca zeolite}}{CaX} + \underset{\text{Non-hardness Ca salt}}{CaCl_2} \quad (6.38)$$

$$\underset{\text{Mg salt}}{MgSO_4} + \underset{\text{Sodium zeolite}}{Na_2X} \rightarrow \underset{\text{used Mg zeolite}}{MgX} + \underset{\text{Non-hardness Na salt}}{Na_2SO_4} \quad (6.39)$$

$$\underset{\text{Mg salt}}{MgCl_2} + \underset{\text{Sodium zeolite}}{Na_2X} \rightarrow \underset{\text{used Mg zeolite}}{MgX} + \underset{\text{Non-hardness Na salt}}{2NaCl} \quad (6.40)$$

The calcium and magnesium zeolites when used for longer period get saturated and therefore need to be regenerated. The regeneration of these used zeolites can be carried out by treating them with 10% of NaCl solution and reconverting them into sodium zeolite. The regeneration reaction equation can be represented as (Eq. 6.41-6.42):

Regeneration of Zeolite

$$\underset{\text{Ca used zeolite}}{CaX} + \underset{\text{Na salt solution}}{2\,NaCl} \rightarrow \underset{\text{regenerated Na zeolite}}{Na_2X} + CaCl_2 \quad (6.41)$$

$$\underset{\text{Mg used zeolite}}{MgX} + \underset{\text{Na salt solution}}{2\,NaCl} \rightarrow \underset{\text{regenerated Na zeolite}}{Na_2X} + MgCl_2 \quad (6.42)$$

These reaction equations generally reflect the exchange of cations or positive ions. After saturation of these zeolites the units must be regenerated with water containing 80-160 kg of sodium chloride per cubic metre of resin with 5-20 % of salt brine solution at a flow rate of 40 $L/min.m^2$ to replenish the sodium ions. The zeolites can be regenerated and may be reused many times. The cycle of ion-exchange softening does not alter the pH or alkalinity of the gas. However, due to the removal of calcium and magnesium an increase in dissolved solids, water equilibrium is altered. The total dissolved solids increase by 0.15 ppm for each ppm of calcium extracted and replaced with sodium. The total dissolved solids will rise by 0.88 ppm for each ppm of magnesium extracted and replaced by sodium. In the ion-exchange method, hardness is expressed as grains per gallon instead of mg/L of calcium carbonate; with a value of 1 grain is equivalent to 17.12 mg/L.

6.2.4.3 Advantages of the Ion-Exchange Process

Zeolites have huge potential as a cost-effective, environmental-friendly solution that can improve the efficiency in various applications. The various advantages of ion-exchange process in water softening as compared to lime-soda ash process is its compactness and lower cost. The chemicals can be handled safely by the operator and thus makes it much easier to operate and also can be virtually completely automated. Since the resins have the potential to extract all the hardness from the surface, the processed surface must be combined with water that has been bypassed through the softener to achieve the degree of hardness as required. At present, synthetic zeolites are used commercially more often than natural zeolites due to the purity of crystalline products and the uniformity of particle sizes.

Many utilities have considered the ion-exchange cycle to be the most cost efficient way for their customers to obtain clean water. When the zeolite units are used to soften surface water, the usual surface water treatment units of strong mixing, flocculation, sedimentation, and filtration have to be followed. This is necessary to remove any substance otherwise capable of fouling the resin. So the lime-soda cycle is more likely than the ion-exchange one to be the more cost-effective approach for surface-water softening.

6.2.4.4 Components of Ion-Exchange Process

There is a range of materials which can serve as cation exchangers for softening, including some forms of soils. Natural green sand called glauconite has very strong potential exchange capabilities, and it was once commonly used. The most widely used ones now are synthetic zeolites, known as polystyrene resins. Their cost is fair, and the consistency of the resin is easy to monitor. They also have ion-exchange capacities much greater than the natural material. The resin's ability to reduce water hardness is related to the resin volume in the tank. The zeolite resin can contain about 50,000 hardness grains per cubic foot of resin. The resin holds the hardness ions until they are regenerated with a salt brine solution. The hardness ions are exchanged for the sodium ions in the salt brine. The capacity of an ion exchange materials ranges from 2-10 mequiv/g or about 15-100 kg/m^3.

Salt storage: The salt is processed as brine and is able to be used for resin regeneration. The amount of salt required for every 1000 grains of hardness extracted varies from 0.25 to 0.45 pounds. To avoid corrosion of the tank walls, the tank should be covered with a salt-resistant coating. The salt used should follow the American Water Works Association (AWWA) sodium chloride requirements. Rock or pellet salt is best for brine preparation. Rock salt is not suitable because of the soil it contains, salt-storage tanks should be sealed in order to avoid salt pollution. To avoid contamination by floodwater or rainwater adding of salt in the form of brine with a raised curb should be given at each access door. The brine is thicker than water and sinks down to the tank floor and is usually pumped from the tank to the units for ion exchange. The introduction of water when producing brine must be through an air gap due to the risk of back siphoning of the brine to the water network.

Brine feeding equipment: Concentrated brine feeding equipment contains approximately 25 per cent salt. During recovery, the brine should be diluted to around 10 per cent before being applied to the tank. It is normally pumped by the use of a venturi or by the use metering pump. Salt solubility decreases as temperature rises which forces the salt out of solution. The water that remains out of the solution after the salt has separated is subject to freezing. The brine tubing should therefore be safe from cold temperatures.

Devices for blending: A properly controlled ion-exchange unit produces zero-hardness water with high corrosivity. Since the most acceptable total hardness of 85 to 100 mg/L is, the effluent from the ion-exchange unit is usually combined with source water to increase the hardness in the finished water. Mixing is usually achieved by metering both the softener effluent and the added raw water. In both lines meters are mounted so the operator can change the blend and track it.

Operation of units: The basic steps in the operation of an ion-exchange softening unit are: softening cycle, backwash, regeneration, slow rinse and fast rinse.

Softening cycle: The softening process involves feeding water into the unit before the hardness emerges in the effluent. The process ends with detection of 1 to 5 mg/L hardness in the effluent. Nearly all softening systems have warning on the water meter to show when the exchange unit has passed through a certain volume of water. Loading concentrations for synthetic resins should be lower than the media surface area of 10 to 15 gpm / square feet (gallons per minute /square feet).

Backwash cycle: The softener unit would need to be regenerated until the hardness breaks through. First, the resin must be rewashed in downflow systems in order to release the water and any extracted substance carried by the resin. The rate of backwash is usually 6 to 8 gpm / square feet of zeolite bed surface area. The operator must add ample backwash water to expand the resin bed by about 50%. Usually, the backwash water is discharged into a box containing orifice plates which set and calculate the flow rate. At the top of the building, distributors arrange for clear water delivery and uniform selection of wash water. Under-drains support help the uniform distribution of the backwash water on the bottom of the resin.

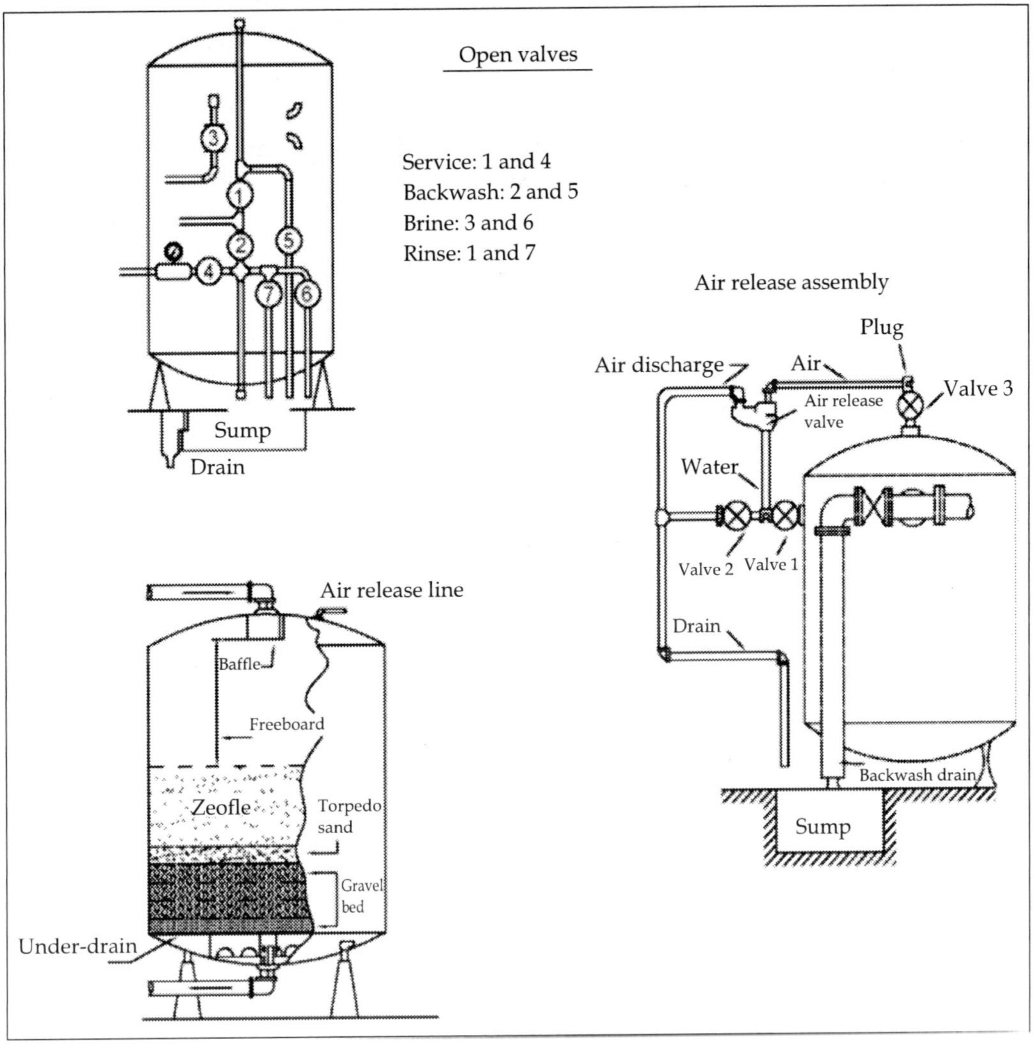

Fig. 6.8: Zeolite softening unit

Regeneration: Concentrated brine is pumped from the storage tank to the unit to regenerate an ion-exchange unit. The brine is diluted to a solution containing about 10 % salt through the injector before flowing through the resin. The regeneration time needed is about 20 to 35 minutes. The brine flow rate through the resin is expressed in gallons per minute per cubic foot of media (gpm/ft^3).The brine needs to be in contact with the resin long enough to allow maximum exchange with sodium ions in the brine of the hardness ions in the resin. Due to time constraints in the system when the resin is not completely recharged, the softening duration required in next batch reduces accordingly.

Rinse cycle: A rinse cycle obeys the regeneration process, in order to remove the remaining brine from the tank. The total water needed for rinsing is 20 to 35 gallons per cubic foot of resin. The rinse is started at a slow rate of 2 gpm/square foot of surface area and continued until the effluent chloride concentration, which should be regularly tracked, is quite low. The overall amount of the brine used during a regeneration process, along with the resulting rinse water, ranges from 1.5 to 7% of the unit's softened water content. In this wastewater the chloride concentration may be as high as 35,000 to 45,000 mg/L. This amount of chlorides will upset a wastewater treatment plant, so the brine usually will have to be disposed of by lagooning method.

Example 6.4: The hardness of water with the following data is to be removed using ion exchange process with a flow rate of 20,000 m^3/day as shown in Fig. 6.9. The medium has an adsorption capacity of 100 kg/m^3 at a flow rate of 0.5 m^3/min.m^2.

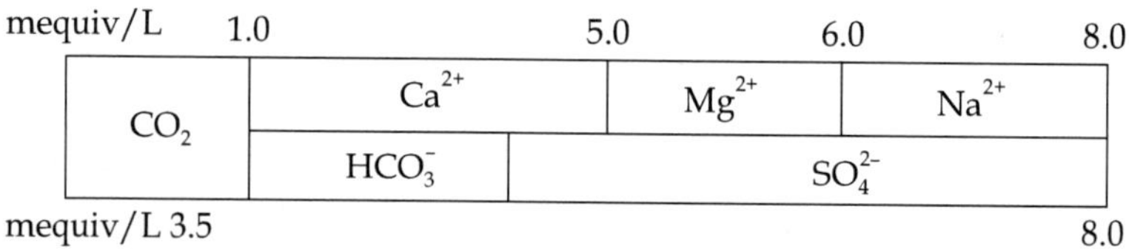

Fig. 6.9: Bar diagram showing ionic strength of water (mequiv/L)

The regeneration is carried out by using 140 kg of NaCl per cubic metre of resin in 12 % brine solution. Design the system with volume of media required with physical arrangements of continuous fixed bed system. Determine the total dosage of chemical required and the regeneration time of this system.

Solution: Volume of the media can be calculated as :

(i) Total hardness = 6 mequiv/L × 50 mg/mequiv = 250 mg/L.

Assume 80 mg/L as acceptable limit of hardness in treated water. Therefore, the percentage of water to be bypassed = 80/250 = 0.32 or 32% of the inflow.

Therefore water to be treated = (1-0.32) × 20,000 = 13,600 m^3/d.

(ii) Hardness to be removed = 5 equiv/m^3 × 13,600 m^3/d × 50 g/equiv × 1 kg/1000 g = 3400 kg/d

(iii) Volume of media per day required = 3400 kg/d × 1 m^3/100 kg = 34 m^3

(iv) Surface area and height of the media required

= 13,600 m^3/d × 1 d/1440 min = 9.45 m^3/min

Therefore, area required = 9.45 m^3/min × 1 min/0.5 = 18.9 m^2

(v) Using tank of 2 m diameter we can calculate the area of one tank

$= \pi d^2/4 = \pi \times (2)^2/4 = 3.14$ m^2

Therefore, tanks required are = 18.9/3.14 = 6

Use 6# of tanks and the height of the tank

= Total volume/Total area = 34/18.9 = 1.8 m ≈ 2 m

(vi) Assume 2 tanks required for carrying out the regeneration process , therefore the total volume of exchange resins as

= Number of tank × Area × Height = (6 + 2) × 3.14 × 2 = 31.4 m

(vii) Chemical dosage requirement for regeneration can be calculated as:

(a) Volume of single unit = 3.14 × 2 = 6.28 m^3

(b) Salt requirement = 140 × 6.28 = 879 kg

The regeneration of 6 units will require = 6 × 879 = 5,274 kg/d of NaCl

(c) Using 12 % brine solution the volume of liquid during regeneration is calculated as = 879 kg/0.12 = 7325 kg = around 7 m^3 for each unit

(d) Assuming the standard loading rate of 0.04 m^3/m^2.min, the regeneration time (t) can be calculated as $t = \dfrac{7\ m^3}{0.04\dfrac{m^3}{m^2.min} \times 3.14\ m^2} = 55.7$ mniutes

Therefore, assuming the regeneration time to be around one hour all the 8 units will require 8 hours of the workday.

6.2.4.5 Common Ion Exchange System Problems

Like with every complex operating system that combines electrical and mechanical equipment and chemical processes, the ion exchange systems have issues. The problems typically leading to low effluent efficiency, shortened service run lengths or increased regenerant consumption. Should be addressed to keep the ion exchange system operating efficiently and reliably.

Operational problems: Changes in the quality of the raw water have a major effect on both the run duration and the consistency of the effluent created by an ion exchanger. While most of the well waters are of high consistency, most compositions of surface water differ greatly over time. A rise in water hardness by 10 per cent to a sodium zeolite softener induces a decrease in duty run length by 10 per cent. An increased sodium-to-total cation ratio triggers increased sodium leakage from a demineraliser system. To identify these differences, routine chemical analysis should be performed of the powerful water to ion exchangers. Other causes of ion exchange operational problems include the following:

- Improper regenerations caused by inappropriate flows, times or concentrations of regenerants. Recommendations from manufacturers should be followed when ion exchange resins are regenerated.
- Channelling, whether due to high or low flow levels, increased loading of suspended solids or inadequate backwashing. This triggers premature fatigue even in a regenerated state where most of the bed is in.

- Fouling or deterioration of resin, caused by poor quality of regenerant.
- Lack of silica removal from resin that can result from low regenerant caustic temperature. This might contribute to increased leakage of silica and to short runs of service.
- Residual pollutants in the resin due to accumulated fatigue loads from previous service. Since the resin is filled with more pollutants than a standard regeneration is designed to extract, a double regeneration after an extended service run is needed.

Mechanical problems

- Valves leaking, causing low effluent consistency and lengthy rinses.
- Processor broken or clogged, leading to channelling.
- Resin loss due to inadequate under-drain screening or support media backwashing or failure.
- Cation resin in the anion unit, which triggers prolonged rinse times and sodium leakage in demineralized water.
- Instrumentation issues, such as defective totalizers or conductivity meters, which may suggest a problem when there is none, or may introduce water of poor quality for operation. Instrumentation should be tested regularly in the demineraliser region.

Resin fouling and degradation: Fouling and degradation are the two most frequent cause of ion exchange device. Resin can be fouled with pollutants that interfere with the process during the plant operation. Chemicals that can cause irreversible damage, may also strike the resin. Some products, such as natural organics, initially foul resins and the degraded resin over time can cause fouling problems.

- **Manganese and iron:** Iron can exist in water as an inorganic ferrous or ferric salt, or as an organic sequester complex. Exchange of ferrous iron in water produces ferric iron which is insoluble. Therefore, cation resin forms ferric iron plates and creates operational problems. This iron can be extracted by using some strong acid solutions that act as a strong reducing agent. Organically bound iron passes through a unit of cations and fouls the resin with anions. Along with the raw content it must be discarded. Manganese, present in the water of certain wells, fouls the resin in the same way as iron
- **Aluminum:** Aluminum is normally present as aluminium hydroxide, arising from the use of aluminum or sodium aluminate in clarification or softening of precipitation. Aluminum flocs coat the resin in a sodium zeolite softener, if brought through filters. It is extracted by either acid or caustic cleaning. Aluminium is typically not a foulant in a demineralization process because it is extracted from the resin during natural regeneration. After filtration by post-precipitation, hardness precipitates bring through a filter from a precipitation

softener. These precipitate foul resins which are used to soften the sodium zeolite. They are washed off with acid.

- **Precipitation by sulphate:** Precipitation of calcium sulphate may occur in a strong acid cation unit that is operating in the hydrogen cycle. The surface of the resin bed at the end of a maintenance period is high in calcium. If sulphuric acid is used as the regenerant, whether it is applied at too high a concentration or too low a flow rate, calcium sulphate precipitation occurs, thus fouling the resin. It is very difficult to redissolve it after the formation of calcium sulphate. Therefore, resin fouled by calcium sulphate is usually discarded. Mild cases of calcium sulphate fouling may be reversed with a prolonged soak in hydrochloric acid.
- **Barium sulphate and fouling oil:** It is even less soluble than calcium sulphate. Recovery of hydrochloric acid should be addressed where a source of water contains measurable concentrations of barium. Oil coats oil, which prevents the passage of ions to and from sites of trade. To extract oil a surfactant can be used. To pick a surfactant that doesn't foul resin, caution must be exercised. Oil-fouled anion resins can only be treated with nonionic surfactants.
- **Fouling microbiology:** Microbiological fouling can occur in resin beds, particularly in beds where no service flow is allowed to sit. Microbiological fouling can lead to extreme resin bed plugging, and even mechanical damage due to an excessive drop in pressure through the fouled resin. If microbiological fouling is an issue in standby systems, the problem can be reduced by using a steady flow of recirculating water. Severe conditions can require the application of appropriate agents and surfactants for sterilization.
- **Fouling on silica:** If the regenerant temperature is too small, or if the effluent caustic from the SBA unit used to regenerate the weak base unit contains too much silica, a silica fouling can occur in strong base anion resins. Polymerization of silica can occur at low pH levels in a weak base resin. It can also be a problem in an exhausted strong base anion resin. Silica fouling is removed by a prolonged soak in warm (120°F) caustic soda.

Causes of irreversible resin degradation

Oxidation: Oxidizing agents, such as chlorine, destroys resins in both cations and anions. Oxidants damage the cross-links of divinylbenzene in a cation resin, thereby reducing the overall resin bed power. The cation resin begins losing its spherical form and rigidity as the attack continues, causing it to compact during operation. This compaction increases the pressure drop across the resin bed and contributes to channeling which decreases the unit's effective power. The anion resin is not directly affected in the case of raw water chlorine since the chlorine is absorbed by the cation resin. However, some degradation products from oxidized cation resin are fouled off downstream solid base anion resins. If chlorine is present in raw water it should be extracted with activated carbon filtration or sodium sulphite before ion exchange. The intake of 1 ppm of chlorine requires about 1.8 ppm of sodium sulphite. Oxygen-saturated water, such as that found after a forced decarbonization draft, accelerates the destruction of strong base exchange sites which occur naturally over time. It also speeds up degradation caused by organic fouling.

Thermal degradation: Thermal degradation happens when the anion resin becomes overheated during the regeneration or duty process. This is especially true for acrylic resins with temperature limits as low as 100°F and strong base anion resins of Type II which have a temperature limit of 105°F when in the form of hydroxide. Organic fouling is the most popular and costly method of fouling and degrading resin. Typically, only small amounts of organic materials are present in well waters. Nevertheless, hundreds of parts per million of natural and man-made organic matter can be contained in surface waters. Natural organics are derived from the vegetation that decays. In fact, they are aromatic and acidic, and can contain complex heavy metals, including iron. Tannins, tannic acid, humic acid, are among such toxins.

Organics initially block the solid base sites on a resin. This blockage triggers long final rinses and reduces the capacity to break the salt. As the foulant stays on the resin it starts degrading the strong base sites, reducing the resin's salt splitting ability. The site's functionality transitions from a strong base to a weak base, and in the end to a non-active site. Even a resin in the early stages of degradation exhibits high overall ability, but decreased potential for salt splitting. Cleaning up the resin at this point can still restore some, though not all, of the lost operating power. The reduction in the capacity to break salts limits the resin's ability to absorb silica and carbonic acid.

Organic fouling of anion resin is indicated by the effluent colour from the anion regeneration unit, varying from tea coloured to dark brown. The treated water has higher conductivity during service, and a lower pH. The following methods are used, either alone or in combination, to reduce organic fouling:

(a) Clarification and prechlorination: At the source water is prechlorinated and then clarified by an organic removal aid.

(b) Filtration by activated carbon: It should be noted that a carbon filter has a finite ability to remove organic material, and regularly track the carbon removal efficiency.

(c) Macroporous: A low base resin with strong base resin in a head. The poor base or macroporous resin absorbs the organic material and during regeneration is eluted.

(d) Fabric resins: It possesses acrylic and other specialty resins which are less prone to organic fouling.

Cleaning and testing: A regimen of routine inspection and cleaning of the ion exchange device, as well as other preventive procedures, helps preserve the life of anion resin. Most of the cleaning procedures use one of the following.

(a) Brine and caustic water at 49°C: To boost cleaning it can be applied which mild oxidants or solubilizing agents.

(b) Chlorochlorinated acid: Hydrochloric acids are used because the resins are not fouled with large quantities of carbon.

(c) 0.25-0.5 per cent of sodium hypochlorite solutions: This process kills the organic material, but degrades the resin significantly too. Purification of hypochlorite is the last resort.

It is necessary to clean an organically fouled resin before the strong base sites undergo excessive permanent degradation. Cleaning after irreversible deterioration takes away large quantities of organic material but does not increase efficiency of the device. Resin quality should be closely tracked to assess the optimal cleaning schedule.

Resin testing and analysis: The resin should be regularly tested and analyzed for physical stability, foulant levels, and the ability to perform the necessary ion exchange to monitor the condition of ion exchange resin and decide the best time to clean it. Samples should be descriptive of the entire bed of resins. Samples should either be collected at various rates within the bed, or a grain thief or hollow pipe should be used to get a "core" sample. The inlet and regenerant distributor should be checked during sampling, and the condition at the top of the resin bed should be noted. Excessive hills or valleys in the resin bed are indicative of issues with the flow distribution.

Signs of fouling and damaged or broken beads should be investigated microscopically with the resin sample. Physical properties such as density and moisture content should also be checked .The amount of organic and inorganic foulants in the resin should be determined and compared with the resin's previous state and established requirements. Finally, for evaluating the rate of degradation or organic fouling, the salt splitting and total efficiency should be measured on anion resin samples.

6.3 DEMINERALIZATION

For most high-pressure boiler feedwaters and many process streams, especially those used in the manufacture of electronic equipment, softening by itself is insufficient. Such processes also involve removal of all dissolved solids, such as sodium, silica, alkalinity, and mineral anions (Cl^-, SO_4^{2-}, NO_3^-), in addition to removing hardness. Water demineralization is a process of removing all minerals or inorganic salts from water through ion exchange. It is a process in which the minerals are eliminated from water to a desired content and proves to be very suitable for various industrial applications. Demineralization produces water close to distillation efficiency at a lower cost. The complete demineralization process is carried out by passing the water through a media bed of cation exchange resins and thereafter through a bed of anion exchange resins. In this process, strong hydrogen shaped acid cation resin transforms dissolved salts into their corresponding acids, and strong hydroxide shaped base anion resin eliminates those acids. The cation exchange resins are products produced from phenol aldehyde condensation which during the sulphonation process generates base exchange properties resins.

6.3.1 Operation

The system used for demineralization of cation-anions is similar to that used in zeolite softening. The main difference is that the tubes, valves and pipes must be made of (or filled with) materials which are resistant to corrosion. Rubber and polyvinyl chloride (PVC) are widely used for vessel linings for the ion exchange. The controls and regenerating mechanisms for demineralisers are more complex, allowing for such changes as stepwise acidic and warm caustic regenerations. The demineralisers are similar to zeolite softeners in service. For a demineralizer the operation flow rate recommendations vary from 6 to 10 gpm per square foot of resin. Flow levels of more than 10 gpm per square foot of resin cause increased sodium content and silica leakage with certain waters. Anion resin is slightly lighter than cation resin.

Backwash flow rates for anion exchange resins are also much lower than those for cation resins, and expansion of anion resin is more influenced by the water temperature than expansion of cation resins. To avoid accumulation of hardness salts in the alkaline anion resin bed, the water used for each stage of anion resin regeneration should be free of hardness. Continuous conductivity devices and silica analyzers are widely used to track water quality of the anion effluent and identify the need for regeneration. For certain situations, conductivity tests are placed above the under-drain collectors in the resin bed to detect resin exhaustion.

6.3.2 Principles of Demineralization

A demineralizer system consists of one or more ion exchange resin columns, which include a strong acid cation unit and a strong base anion unit. The cation resin exchanges hydrogen for the raw water cations as shown by the Eq.6.43;

$$\begin{bmatrix} Ca \\ Mg \\ 2Na \end{bmatrix} . \begin{bmatrix} 2HCO_3 \\ SO_4 \\ 2Cl \\ 2NO_3 \end{bmatrix} + 2\,Z.H \rightarrow 2Z. \begin{bmatrix} Ca \\ Mg \\ 2Na \end{bmatrix} . \begin{bmatrix} 2HCO_3 \\ SO_4 \\ 2Cl \\ 2NO_3 \end{bmatrix} \tag{6.43}$$

This obtained water produced during the above cation exchange process conatins diluted carbonic acid, hydrochloric acid, sulphuric acid and others. These acids can now be eliminated by passing the water through a bed of anion exchange resins. The anion exchange resins are manufactured and produced by the condensation of amines with formaldehyde that possess a property of replacing the anions with their hydroxyl ions. The final treated water is thus free from minerals and acids. This can be represented by the following chemical reaction equation (Eq. 6.44-Eq. 6.45) as :

$$\underset{\text{Fresh anion exchange resin}}{ROH} + HCl \xrightarrow{\text{yields}} \underset{\text{Exhausted resins}}{RCl} + H_2O \tag{6.44}$$

$$\underset{\text{Fresh anion exchange resin}}{2ROH} + H_2SO_4 \xrightarrow{\text{yields}} \underset{\text{Exhausted resins}}{R_2SO_4} + H_2O \tag{6.45}$$

The free mineral acidity (FMA) is one indicator of the overall concentration of the strong acids in the cation effluent. During a typical service run, the FMA material, as shown in Fig. 6.10, is stable much of the time. If cation exchanges were 100% efficient, the exchanger FMA would be equal to the waters theoretical mineral acidity (TMA). The FMA is typically significantly lower than the TMA, since the cation exchanger leaks a small amount of sodium. The amount of sodium leakage is dependent on the degree of regeneration, the rate of flow and the proportion of sodium in the raw water to the other cations. Sodium leakage typically decreases as the sodium- to-total-cations ratio decreases.

When an exchange unit for cations nears saturation, FMA falls dramatically in the effluent, signalling that the exchanger should be withdrawn from service. The resin can be regenerated at this time with an acid solution, which returns the hydrogen source to the exchange sites. Despite its low cost and availability, sulphuric acid is commonly used. Improper use of sulphuric acid, however, can cause irreversible fouling of calcium sulphate resin. Typically, the sulphuric acid is applied at a high flow rate of 1 gpm per square foot of resin (gpm/ft^2) and an initial concentration of 2% or less to avoid this phenomenon. To complete regeneration, the acid concentration is slowly increased to 6-8%.

Many systems regenerate with hydrochloric acid. It includes the use of special construction materials inside the regenerant method. As with a sodium zeolite unit, it needs an excess of regenerant (sulphuric or hydrochloric acid) up to three times the theoretical dosage. Water from the cation unit is passed through a strong base anion exchange resin in the form of hydroxide, to complete the demineralization cycle. The resin exchanges hydrogen ions for both highly ionized mineral ions and the more weakly ionized carbonic and silicic acids, as shown in Eq. 6.46;

$$\begin{bmatrix} 2H_2CO_3 \\ H_2SO_4 \\ 2HCl \\ 2H_2SiO_3 \\ 2HNO_3 \end{bmatrix} + 2Z.\,OH \rightarrow 2Z. \begin{bmatrix} 2HCO_3 \\ SO_4 \\ 2Cl \\ 2\,HSiO_3 \\ 2NO_3 \end{bmatrix} + 2H_2O \tag{6.46}$$

The above reactions suggest that demineralization eliminates the cations and anions from the water entirely. In fact, some leakage occurs because ion exchange reactions are equilibrium reactions. Most cation-unit leakage is sodium. In the anion units this sodium leakage is converted into sodium hydroxide. Thus, the effluent pH of a demineralizer device with two beds cation-anion is slightly alkaline. The caustic formed in the anions is causing a small amount of leakage of silica. The degree of the anion leakage depends on the composition of the processing water and the use of the regenerant dosage. Demineralization by using heavy anion resins removes silica and other dissolved solids. Effluent silica and conductivity are critical criteria for control during a service run of the demineraliser. At the end of the fast rinse both silica and conductivity are small, as shown in Fig. 6.10.

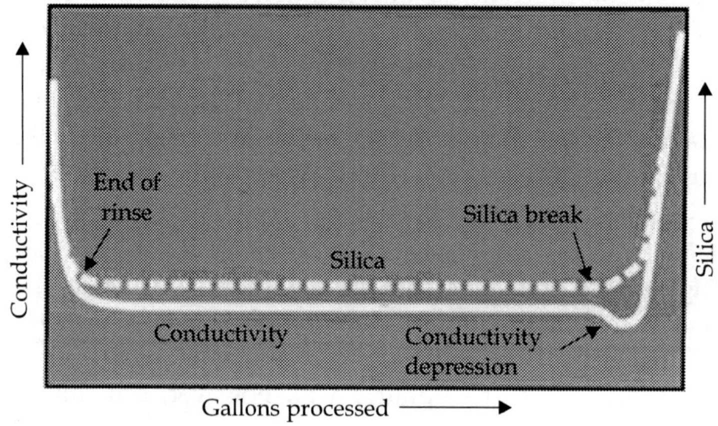

Fig. 6.10: Conductivity/silica profile for strong base anion exchanger
(*Source*: Suez Water Technologies)

When the silica breakthrough occurs at the end of a service cycle, the level of silica treated water rises dramatically. Sometimes, the water's conductivity decreases momentarily, and then increases. This temporary decrease in conductivity is explained quite easily. During normal operation, the majority of the effluent conductivity is due to the low sodium hydroxide level provided by the anion exchanger. The hydroxide is no longer usable when a silica breakthrough happens, so the sodium from the cation exchanger is converted to sodium silicate which is much less conductive than sodium hydroxide.

As saturation of anion resin progresses, the more conductive mineral ions break through, causing subsequent conductivity increases. When a demineralizer run is detected to end, the device must be removed immediately from service. If the demineralizer is allowed to remain in operation beyond the breakpoint, the silica level in the treated water that rise above that of the influential water due to the silica concentration that occurs in the anion resin during the operation run. With a 4 per cent sodium hydroxide solution, strong base anion exchangers are regenerated. As with cation regeneration, the regeneration process is powered by the relatively high hydroxide concentration. The regenerant caustic is usually heated to 120°F or to the temperature specified by the resin manufacturer to improve the removal of silica from the resin surface. The removal of silica is also facilitated by a preheat phase on the resin bed before the addition of warm caustic.

6.3.3 Advantages and Limitations

Demineralisers can provide high purity water for almost any use. Demineralized water is commonly used in feedwater for high pressure boilers and in other process waters. The water quality produced is comparable with the purified water, usually at a fraction of the cost. Demineralisers come in a number of different sizes. Systems range from laboratory columns generating only a couple of gallons an hour to systems generating thousands of gallons in one minute. Like other ion exchange systems, demineralisers need filtered water for efficient operation. Prior to demineralisation, resin foulants and corrosion agents, such as iron and chlorine, should be avoided or withdrawn. For many surface water sources anion resins are very susceptible to fouling and attacking from the organic materials present. A demineraliser does not eliminate all types of silica, known as colloidal, or non-reactive. Warm, alkaline boiler water dissolves the colloidal content, forming simple silicates in a soluble shape similar to those that join the boiler. As such, they can form deposits and volatilize into the steam on tube surfaces.

6.4 DEALKALIZATION

Boiler or process operating conditions frequently require eliminating hardness and reducing alkalinity while not eliminating the other solids. Softening of zeolite does not minimize alkalinity, and demineralization is too costly. A method of dealkalization is used for these circumstances. The most commonly used processes are sodium zeolite/ hydrogen zeolite (split stream), chloride-anion dealkalization, and low acid cation dealkalization.

6.4.1 Sodium Zeolite/Hydrogen Zeolite (Split Stream) Dealkalization

A part of the raw water flows through a sodium zeolite softener inside a split stream dealkalizer. The remainder flows into a solid acid cation unit (zeolite of hydrogen) produced by hydrogen. The sodium zeolite effluent is mixed with the zeolite effluent for hydrogen. The hydrogen zeolite unit effluent includes carbonic acid, derived from alkalinity of the raw water, and free mineral acids. As the two sources are mixed, free mineral acidity of the hydrogen zeolite effluent transforms sodium carbonate and bicarbonate alkalinity into carbonic acid in the sodium zeolite effluent as shown below in Eq. 6.47.

$$2NaHCO_3 + \begin{bmatrix} 2HCl \\ H_2SO_4 \\ 2HNO_3 \end{bmatrix} \rightarrow \begin{bmatrix} 2NaCl \\ Na_2SO_4 \\ 2NaNO_3 \end{bmatrix} + 2H_2CO_3 \tag{6.47}$$

Carbonic acid in water is unstable. This shapes gas and water by carbon dioxide. The mixed effluents are sent to a decarbonator or degasser, where a countercurrent stream of air removes the carbon dioxide from the water. Figure 6.11 demonstrates a standard dealkalization device for split streams.

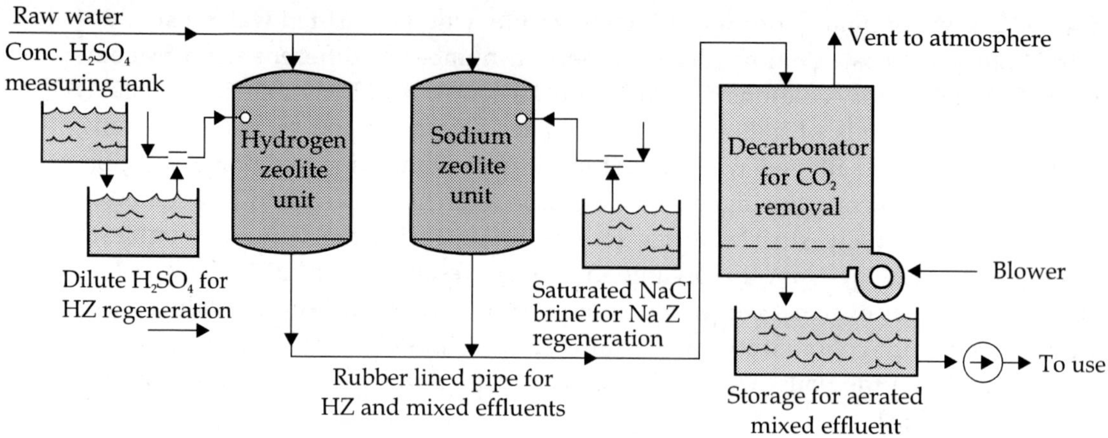

Fig. 6.11: Sodium zeolite/hydrogen zeolite split stream softener

The optimal level of alkalinity of the blended water can be preserved by regulating the ratio of sodium zeolite and zeolite water in the mixture. A higher percentage of sodium zeolite water results in increased alkalinity, and a higher percentage of zeolite hydrogen water decreases alkalinity. A split stream dealkalizer decreases the total dissolved solids of the water, in addition to reducing alkalinity. It is critical in high alkalinity waters, because the conductivity of these waters affects the process and can restrict concentration boiler cycles.

6.4.2 Sodium Zeolite / Chloride Anion Dealkalization

Heavy chloride-shaped base anion resin can be used to reduce the alkalinity of water. Water flows through a zeolite softener and then into anion network, replacing carbonate, bicarbonate, sulphate and nitrate ions with chloride ions as seen in Eq. 6.48:

$$\begin{bmatrix} Na_2SO_4 \\ 2NaHCO_3 \\ 2NaCO_3 \\ 2NaNO_3 \end{bmatrix} + 2Z.\,Cl \rightarrow 2NaCl + 2Z.\begin{bmatrix} SO_4 \\ 2HCO_3 \\ 2NO_3 \end{bmatrix} \tag{6.48}$$

where, Z represents the resin. The carbonic acid (H_2CO_3) formed is removed by a decarbonator or degasser as in a split stream system

A split stream dealkalizer lowers the total dissolved water solids, in addition to reducing alkalinity. This is critical in high-alkalinity waters, as the conductivity of these waters affects the process and can restrict concentration boiler cycles. The optimal effect for a weak acid cation system has a degree of hardness equal to that of alkalinity (both expressed as $CaCO_3$ in ppm). The alkalinity is not reduced to its lowest level in waters which are higher in alkalinity than in hardness. Some hardness remains after treatment in waters which contain more hardness than alkalinity. Usually, to eliminate hardness these waters have to be polished by a sodium zeolite softener.

Many cations associated with mineral anion exchange occur during the initial portion of a weak acid cation supply (the first 40-60 per cent), creating small quantities of mineral acids in the effluent. As the operation period continues, the effluent exhibits alkalinity. When the effluent alkalinity reaches 10 per cent of the powerful alkalinity, the unit is withdrawn from operation and regenerated with a solution of 0.5 per cent sulphuric acid.

Regenerant acid concentration should be kept below 0.5-0.7 per cent to avoid accumulation of calcium sulphate in the resin. Weak exchange of resin cation acids is very efficient. The amount of acid produced therefore is nearly equal (chemically) to the quantity of cations extracted during the service period. If the building materials for the downstream machinery or the overall process cannot withstand the mineral acidity present during the initial portions of the service cycle, a brine solution shall pass through the regenerated weak acid resin prior to final rinsing. This solution extracts the mineral acidity without impacting the consistency or duration of the subsequent run significantly. Equipment used for a weak acid cation dealkalizer, with the exception of resin, is similar to that used for a strong acid cation exchanger. One common design variation uses a layer of weak acid resin on top of a heavy acid cation resin. The weak acid resin stays on top, since it is lighter. The sulphuric acid and then the sodium chloride brine regenerate the porous resin network. The brine solution transforms the strong acid resin into a type of sodium. The resin then serves as a softener for polishing.

6.4.2.1 Direct Acid Injection

Acid is used to transform alkalinity to carbonic acid in the process of direct acid injection and decarbonisation. The carbonic acid dissociates to form carbon dioxide and water, and a decarbonator extracts the carbon dioxide. The use of an acid injection system should be treated with caution, because an acid overfeed or a breakdown in the pH control system can produce acidic feedwater which corrodes the iron surfaces of feedwater systems and boilers. Proper pH monitoring and regulated caustic feed are required after decarbonisation.

6.4.3 Advantages and Limitations of Dealkalization Systems

Ion exchange handling systems deliver hardness-free, low alkalinity water at fair cost, and with a high degree of reliability. They are well suited for processing feedwater for medium-pressure boilers, and beverage industry process water. Split stream and low cation systems reduce the overall dissolved solids as well. In addition to these advantages it is important to consider the following disadvantages:

- Dealklizers do not eliminate all alkalinity and do not influence the silica content of water.
- Dealkaliser agents require the same powerful purity as other methods of ion exchange; filtered water low in potential foulants must be used.
- Water produced by the dealing device using a forced draft decarbonator is oxygen-saturated and therefore potentially corrosive.

6.4.4 Other Demineralization Processes

For several systems the traditional cation-anion cycle was changed to reduce the use of costly regenerants and waste generation. Modifications include the use of decarbonators and degassers, weak acid and low base resins, high base anion caustic waste (to regenerate weak base anion exchangers), and reclamation of a portion of spent caustic for subsequent regeneration cycles.

6.4.4.1 Decarbonators and Degassers

Decarbonators and degassers help many demineralization schemes economically, as they reduce the amount of caustic regeneration. A cation exchanger's water is divided into tiny droplets by sprays and trays, or packing in a decarbonator. The water then flows in the opposite direction through a stream of liquid. Carbonic acid dissociates in carbon dioxide and water within the effluent of the cation. The carbon dioxide is separated by air from the water, thereby raising the anion exchanger load. Typical forced decarbonators will eliminate carbon dioxide by as much as 10-15 ppm. Water effluent from a decarbonator is therefore oxygen-saturated.

Water droplets are inserted in a vacuum degasser into a packed column which is worked under a vacuum. Because of their reduced partial pressure in a vacuum, carbon dioxide is separated from the atmosphere. A degassing vacuum normally lowers carbon dioxide to less than 2 ppm and often eliminates much of the oxygen from the water. However, vacuum degassers are more expensive to purchase and operate than forced draft decarbonators.

6.4.4.2 Weak Acid and Weak Base Resins

Weak functionality resins have a much higher regeneration efficiency than their strong functionality counterparts. Weak acid cation resins, as described in the dealkalization section, exchange with cations associated with alkalinity. Soft base resins are mixed in a heavy acid solution with the mineral acid anions (SO_4^{2-} Cl^-, NO_3^-). Poor resin regeneration efficiency is essentially stoichiometric, removing 1 kg of ions (as $CaCO_3$) requires just slightly more than 1 kg of the regenerant ion (as $CaCO_3$) is required.

With the same removal of contaminants strong resins need three to four times the regenerant. Weak base resins are so successful that regeneration of a weak base exchanger with a portion of the "spent" caustic from the regeneration of the strong base anion resin is standard practice. The first fraction of the caustic from the solid base unit is sent to waste to prevent the weak base resin from fouling with silica. The residual caustic is used to regenerate the damaged resin foundation. An additional attribute of weak base resins is their ability to retain natural organic materials that foul and release strong base resins throughout the regeneration process. Because of this ability, weak base resins are widely used against harmful organic fouling to protect strong base resins.

Regenerant reuse: Because of the high cost of caustic soda and the growing waste disposal problems, many demineralization systems are now fitted with a caustic reclaiming feature. At the start of the next regeneration process the reclaim device uses a portion of the accumulated caustic from the previous regeneration. To complete the regeneration, the reused caustic is replaced by fresh caustic. In the next regeneration the existing caustic is then reused for use. Sulphuric acid is not usually reclaimed, as it is lower in cost, and precipitation with calcium sulphate is a potential issue.

6.4.4.3 Condensate Polishing

The use of ion exchange is not limited to processing and boiler making steam. Ion exchange may be used to purify or polish the condensate returned, eliminating products of corrosion that may cause hazardous deposits in boilers. The particulate iron and copper are usually the pollutants in the condensate system. Low levels of other pollutants that reach the system through seal leaks from condensers and pumps, or the transfer of boiler water into the steam. Condensate polishers filter the particulates and exchange ions eliminate soluble pollutants.

Most paper mill condensate polishers work at temperatures close to 93 °C, which prevents the use of anion resin. In such applications, cation resin, which is stable until temperatures above 132°C, is used for deep bed condensate polishing. As in a zeolite softener, the resin is regenerated with the sodium chloride brine. In cases where sodium leakage from the polisher adversely affects the internal chemical system for boiler water or steam attempting to purify water, the resin can be regenerated with an ionized amine solution to avoid such problems. Compared to a traditional softener, the service flow rate for a deep bed polisher (20-50 gpm per square foot of resin surface area) is very high. High flow levels are reasonable because the condensate level of soluble ions can normally be very low. Particulate iron and copper are extracted by filtration, while dissolved contaminants are reduced through the sodium or amine exchange in the resin.

In a manner similar to that used for traditional sodium zeolite regeneration, the deep bed cation resin condensate polisher is regenerated with 15 pound of sodium chloride per cubic foot of resin. A solubilizing or reducing agent is widely used to help extract iron. A supplementary backwash header is often located just below the resin bed level. Used prior to backwashing, this subsurface distributor uses water to break down the crust that forms during regeneration on the resin surface. A significant factor is the option of a condensate polishing resin. Because high pressure drops are caused by high duty flow rates and particulate charge, and since many systems work at high temperatures, the resin structure is put under tremendous stress.

A quality cellular or macroreticular resin can be used for deep-bed polishing applications with condensate. A mixed bed condensate polisher may be used in systems requiring complete dissolved solids and removal of particulate matter. The condensate temperature should be below 60 °C, which reflects the average continuous operating temperature for the anion resin. The flow through the unit is also generally reduced to approximately 20 gpm/ft^2.Ion exchange resins are also used for polishing of condensate as part of a precoat filtration method, as shown in Fig. 6.12.

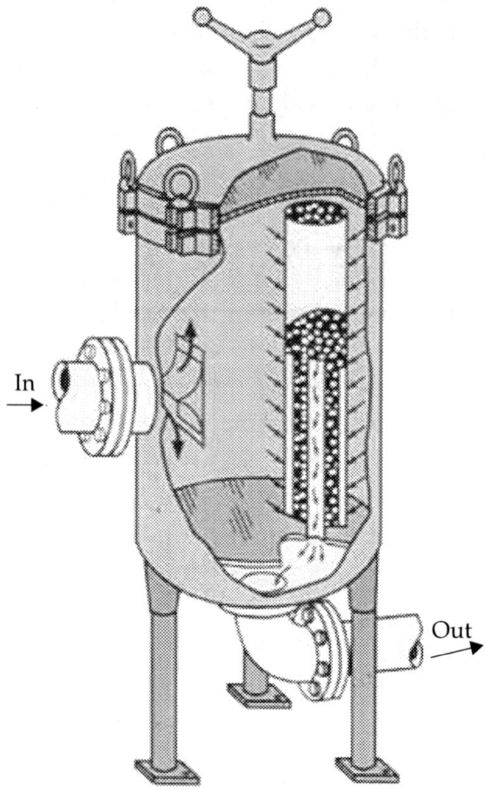

Fig. 6.12: Powdered resin condensate polisher

The resin is compressed and mixed into a slurry which is used in a filter vessel for coating individual septums. The powdered resin is a very fine filtering medium that collects particulate matter and transfers ions to eliminate certain soluble contaminants. The precoat content is disposed of when the filter media is clogged, and the septums are coated with a fresh slurry of powdered resin.

6.5 MEMBRANE PROCESSES

The use of membrane technologies in wastewater treatment was earlier mainly limited to reverse osmosis. In the last few years, their use has been extended by the development of a variety of membranes due to advancement in polymer chemistry and technology as well as by employing the membranes for newer applications such as membrane bioreactors Thus, membrane technologies are now beginning to play an increasing role in water and wastewater treatment, worldwide. In the US, more stringent public water supply quality requirements make its use increasingly necessary, while in countries like India, their use is becoming new as freshwater resources are being polluted requiring better and more advanced treatment before public and industrial supplies can be made. In both the US and India, membrane technologies are being increasingly considered where reuse of the treated wastewater is envisaged.

Membranes are produced from a wide variety of materials such as cellulose acetate, polyamide polysulfones polypropylene, nylon, polyacrylonitrile, polycarbonate, polyvinyl alcohol poly-tetrafluoroethylene, ceramic and metal composites, etc. They are manufactured to remove the smallest desired material which is normally stated as molecular weight cut-of (generally expressed in daltons). The four most common configurations are: tubular, plate and frame spiral wound and hollow fibre. Of these, the hollow fibre membranes are the most commonly used in wastewater treatment and reuse because they have the highest membrane surface area for a given footprint. Membrane replacement is generally required every 3-5 years. Membranes are semipermeable materials designed to separate particulate, colloidal and dissolved substances from liquid solutes Essentially, they allow substances smaller than the membrane pores to flow through the system, while holding back substances larger than the pores. When a membrane is used to separate water from dissolved impurities, the water from the dilute solution gradually passes through the membrane towards the concentrated one till "osmotic equilibrium" is attained between the two liquids. If pressure is applied and increased gradually on the concentrated side, the flow of water continues to reduce till the applied pressure is equal to the osmotic pressure. Any pressure in excess of the osmotic pressure reverses the direction of flow of water and water from the concentrated side enters the dilute side (reverse osmosis). Thus, the concentrated side becomes more concentrated until at some point it becomes more economical to reject it. The influent is called feed stream, The portion which passes through the membrane is called the permeate while the portion retained is called concentrate. The rate at which the permeate flows through is called the flux, kg/m-d or L/m d.

Popular membrane processes are ultrafiltration (UF), reverse osmosis (RO), electrodialysis (ED), and reversal of the electrodialysis (EDR). Both processes (except UF) eliminate most ions; RO and UF systems also effectively reduce non-ionized organics and particulates. Since UF membrane porosity is too high for ion shedding, the UF process is used to minimize pollutants such as oil, grease and suspended solids.

6.5.1 Reverse Osmosis

Osmosis is the movement of water across a semi-permeable membrane, from a concentrated solution to a dilute solution. This flow is the product of driving force created by the pressure difference between the two solutions. Osmotic pressure is the pressure that must be applied to the side of the concentrated solution to avoid the flow of the solvent through the membrane. Reverse osmosis is the method of reversing the flow, moving water from a concentrated solution to a dilute solution via a membrane to obtain filtered water. In RO systems, the membranes have pores < 2 nm and have the lowest mol.wt. cut-off. They are, therefore used in desalination operations to remove ionic species from solution. They also remove sodium, nitrates, sulphates, heavy metals, TDS and other molecules and, therefore, have applications in treating feedwaters for high pressure boilers in industries. Other use is in further treating waters pre-treated by MF and UF to produce waters of high quality for indirect potable reuse applications. They require a relatively high operating pressure >200 lbs/sq in (>1400 kPa). Flux rates vary from 300 - 500 L / m^2/day. Reverse osmosis is produced by adding sufficient pressure to the concentrated solution to overcome the osmotic pressure. Feedwater pumps provide the energy. From the RO membrane's high-pressure side, accumulated pollutants (brine) are decreased, and from the low-pressure side, filtered water (permeate). Figure 6.13 is a simplified RO cycle schematic.

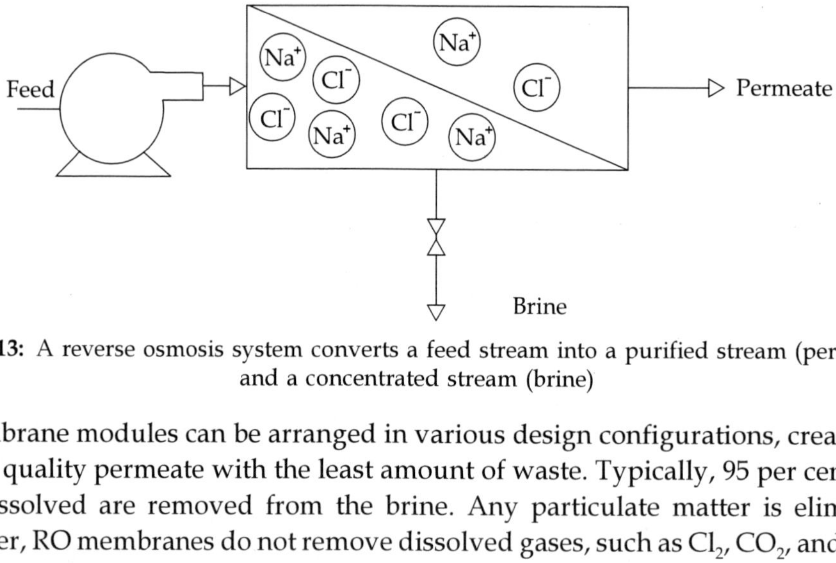

Fig. 6.13: A reverse osmosis system converts a feed stream into a purified stream (permeate) and a concentrated stream (brine)

Membrane modules can be arranged in various design configurations, creating the highest quality permeate with the least amount of waste. Typically, 95 per cent of the salts dissolved are removed from the brine. Any particulate matter is eliminated. However, RO membranes do not remove dissolved gases, such as Cl_2, CO_2, and O_2 due to their molecular porosity.

6.5.2 RO Membranes

Cellulose acetate (CA), and polyamide (PA) composite are the two most common RO membranes used in industrial water treatment. Most membranes are currently spiral wound; however, configurations of hollow fibre are possible. In the spiral wound configuration, the permeate collection tube is wound around a flat sheet membrane and spacers to create flow channels for permeate and feed water. This design maximizes flow while reducing the dimensions of the membrane assembly. Hollow fibre systems are bundles of tiny membrane hairs like tubes. Ions are rejected as the feedwater permeates these tubes walls, and permeate is obtained through the fibres' hollow centre.

6.5.3 Electrodialysis

Electrodialysis (ED) conducts the transfer of ions of dissolved salts through membranes, leaving behind filtered water. Ion motion is caused by electrical fields with direct current. A negative electrode (cathode) attracts cations, and attracts anions by a positive electrode (anode). For stacks, structures are compartmentalized by alternating membranes for cation and anion transfer. Concentrated brine and filtered permeate bear alternating compartments. 40-60 per cent of the dissolved ions are usually lost or refused. Staging (operation of stacks in series) enables further increase in water quality. ED procedures, such as silica, do not eliminate particulate contaminants or weakly ionized contaminants. Figure 6.14 is a simplified schematic of an ED process.

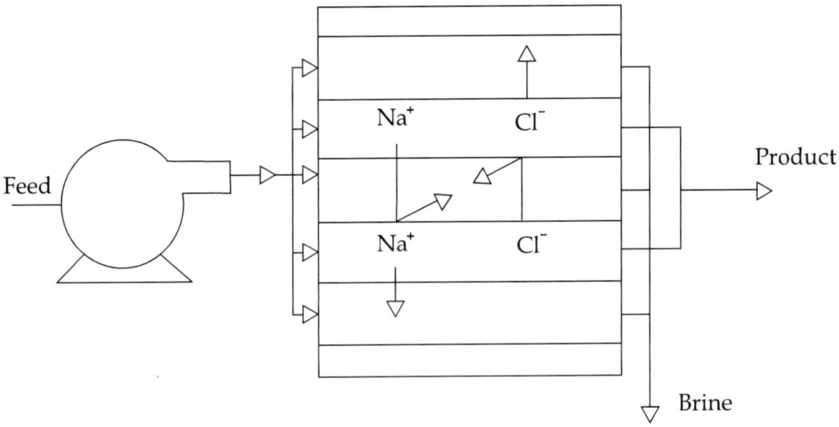

Fig. 6.14: Electrodialysis reversal ions pass through alternating cation and anion transfer membranes

6.5.4 Electrodialysis Reversal

Electrodialysis reversal (EDR) processes work on the same principles as ED; however, operation with EDR reverses the polarity of the device (usually 3-4 times an hour). This reversal prevents the aggregation of condensed membrane solutions and thereby decreases the accumulation of inorganic and organic contamination on the surface of the membrane. EDR systems are similar to ED systems which are designed to collect both liquid water and brine with sufficient chamber space. EDR provides water of similar quality to that of ED.

6.5.5 Microfiltration

MF membranes (pores > 50 nm) are the least expensive membranes and have been used in wastewater treatment for turbidity removal, solids separation after biological treatment, as in membrane bioreactors (MBR), removal of helminth ova, other organisms, etc. Operating pressures are generally below 50 lbs/sq in (< 350 kPa). Their flux rates average between 400-1600 L/m^2-d. They are often used in MBRs for producing reclaimed water for non-potable purposes and for pretreatment of waters/ wastewaters before NF and RO. Ultrafiltration: UF membranes (pore sizes 2-50 nm) have been used in wastewater treatment for many of the same applications as MF membranes except that UF systems give a better separation of finer colloids, bacteria, viruses, etc. They are also used in MBRs to separate biosolids after activated sludge. Operating pressures vary from 50-100 lbs/sq in (350-690 kPa). Flux rates vary from 400-800 L /m-d. Besides their use in MBRS, they are also used to pretreat waters/ wastewaters before using NF and RO systems.

6.5.6 Nanofiltration

In NF membranes the pores are actually similar to those in RO systems (< 2 nm) but operating pressures are lower. Pressures vary between 75-200 lbs/sq in (520-1400 kPa) and flux rates vary from 200-800 L/m^2/day. They are commonly used in water softening to remove divalent cations and in water treatment to remove colour, humic organic matter and other low molecular weight compounds. They can also remove viruses. They are often used to treat waters that are pre-treated by microfiltration or ultrafiltration to produce water for indirect potable reuse applications such as groundwater injection.

6.5.7 Ultrafiltration

Reduction of dissolved ions is not needed in many processes and wastewater applications but is successful for reduction of inorganic or organic colloidal molecules. Ultrafiltration (UF) membrane configurations and device designs are identical to the one-stage RO-process. Since the large UF-removed molecules show negligible osmotic strain, operating pressures are typically much lower than in RO systems. Figure 6.15 demonstrates how ultrafiltration membranes work.

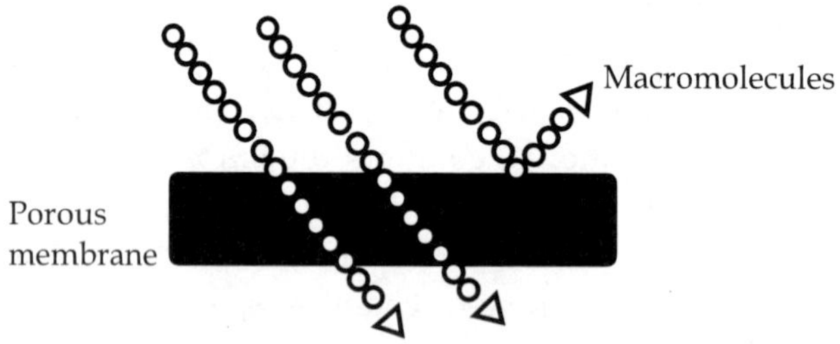

Fig. 6.15: Ultrafiltration membranes pass inorganic ions but reject large organic molecules and colloidal particles

Typical uses include oil and grease reduction and remediation of essential pollutants in plant waste streams

Pretreatment: Processes which rely on microporous membranes have to be protected against fouling. Fouling the membrane causes a loss of water supply (flux), reduced permeate strength, and increased drop in transmembrane pressure. Membrane fouling is usually caused by deposition of inorganic salts, metal oxide particulates, colloidal silt, and membrane surface accumulation or development of microbiological species. These fouling problems can cause serious damage and require more frequent membrane replacement.

Solids reduction: Feedwater for the membranes should be completely free of colloidal particulates. The silt, iron oxides, and manganese oxides are the most common particulates present in industrial membrane systems. Testing of the silt density index (SDI) should be used to validate adequate water quality for the specific membrane device used. The SDI assesses feedwater capacity for fouling a 0.45 μm filter. Unacceptable SDI measurements can be generated even though, according to most industrial water treatment requirements, water quality is relative high. Where pretreatment is insufficient or ineffective, chemical dispersants can be used to enable activity at SDI values which are higher than recommended. RO systems are highly susceptible to particulate fouling, ED and EDR systems are more forgiving, and UF systems are designed to handle dirty waters.

6.6 DESIGN OF MEMBRANE SYSTEMS

There are three forms of membrane system: (a) small cells, (b) bench units or (c) pilot plants. Description of the mass transport of water and solutes through membrane is helpful for membrane studies. Product flux and product purity characterize the performance of a given membrane. Flux is a rate of flow per unit area of membrane and depends on membrane thickness chemical composition of feed, membrane porosity, time of operation, pressure across membrane, and feedwater temperature. However, product purity is a function of the rejection ability of the particular membrane. The useful equations are Eq. 6.49-Eq. 6.55.

$$F_w = K_w(\Delta_P - \Delta\pi) = \frac{Q_p}{A} \tag{6.49}$$

$$F_s = K_s \Delta C = \frac{Q_p C_p}{A} \tag{6.50}$$

$$R = \frac{Q_p}{Q_r} \tag{6.51}$$

$$Q_f = Q_c + Q_p \tag{6.52}$$

$$Q_f C_f = Q_c C_c + Q_p C_p \tag{6.53}$$

$$\frac{C_s - C_p}{C_f - C_p} = e^{\frac{F_w}{k}} \tag{6.54}$$

$$r = \frac{Q_p}{Q_r} \tag{6.55}$$

ΔC = Concentration gradient (M/L³), $\left(\frac{C_f + C_c}{2 - C_p}\right)$

C_f = Feed stream solute concentration (M/L³)

C_c = Concentrate stream solute concentration (M/L³)

C_p = Permeate stream solute concentration (M/L³)

K = Diffusion coefficient from the surface to the bulk (L³/L² t)

ΔP = Pressure gradient (L), $\left(\frac{P_f + P_c}{2 - P_p}\right)$

$\Delta\pi$ = Osmotic pressure (L), $\frac{\pi_f + \pi_c}{2 - \Delta_P}$

K_w = Solvent mass transfer coefficient (L² t)/M)

F_w = Water flux (L³/L² t)

F_s = Solute flux (M/L² t)

K_s = Solute mass transfer coefficient (L² t)/M)

Q_f = Feed stream flow (L³/(t)

Q_p = Concentrate stream flow (L³/(t)

Q_P = Permeate stream flow(L³/(t)

R = Recovery

A = Membrane area (L²)

r = Recycle ratio

The cost and performance of membrane system are also important factors. So, the models predicting the cost and performance of membrane system are:

$$C_p = \frac{K_s C_f e^{\frac{F_w}{k}}}{F_w\left(\frac{(1-r)(2-r)}{2+2r-R}\right) + K_s e^{\frac{F_w}{k}}} \tag{6.56}$$

Equation 6.56 illustrates diffusion controlled mass transfer including alkalinity, hardness, TDS, sodium, chlorides;

$$C_p = \phi C_v \tag{6.57}$$

Equation 6.57 is used to explain sieving controlled mass transfer.

The silt density index(SDI), modified fouling index (MFI) and the mini-plugging factor index (MPFI) are the three fouling indices. These indices have been shown to not statistically correlate to flux or mass transfer decline during membrane operation but are used as guides for membrane pretreatment. Collection of index data is useful to estimate pretreatment, can be easily done and could provide data that releases pretreatment requirements of varying water qualities. The equations for fouling indices are (Eq. 6.58-Eq. 6.60) shown below.

$$\text{SDI} = \frac{100\left(1 - \frac{t_i}{t_f}\right)}{T} \tag{6.58}$$

$$\text{MFI} = (Q.V)^{-1} \tag{6.59}$$

$$\text{MPFI} = \frac{Q}{t} \tag{6.60}$$

where,

t_i = Time to collect initial 500 mL

t_f = Time to collect final 500 mL

T = Time between t_i and t_f (15 min)

Q = Flow

V = Volume

Solvent mass transfer coefficient and the rate of decline of productivity measure, which is essential for successful membrane applications. The correlation is shown in Eq. 6.61;

$$K_w = \frac{F_w}{\Delta P - \Delta\pi} \tag{6.61}$$

$$K_w = K_{w_0}^{-mt}$$

Monitoring K_w over time normalizes flux based on pressure and provides truer assessment of productivity.

6.7 ISSUES IN OPERATION AND MAINTENANCE OF MEMBRANE UNITS

The design of the membrane process and fouling affect productivity which is essential to water treatment facility. Treatment characteristics specify membrane and operating conditions for a particular membrane process. Thus, fouling is the prime contemplation affecting productivity. The causes of fouling are particle deposits on the membrane surface, macro-molecules adsorbing onto the surface or into the bulk membrane material, or pore blocking.

Pretreatment

Requirements to avoid fouling are:

1. Scaling can be controlled in either surface or groundwaters for all RO/NF membrane systems by addition of acid and/or anti scalent.
2. Control of plugging in either surface or groundwaters for all RO/NF membrane systems is attained by feed water turbidities and SDU's of less than 0.2 NTU and 2, respectively.
3. Bio-fouling occurring in aerobic surface or ground waters is controlled by NH_4CI or addition of other bactericidal agents.

4. The control of organic fouling in surface water systems with TOC > 3-6 mg/L is achieved by coagulation, sedimentation and filtration. Another problems caused due to membrane fouling are poor salt rejection characteristics and increasing pressure differential across the membranes. As membrane fouling is the most common reason for performance problems, monitoring of pre-treatment plant and biofouling control can reduce fouling.

Scale control: Membrane processes create a gradient of concentration of the dissolved salts that enter the membrane surfaces. Membrane concentration can exceed the solubility limits of some organisms. Typical precipitates formed are calcium carbonate ($CaCO_3$), and calcium sulphate ($CaSO_4$). Membrane deposits most also classify silica, barium, and strontium salts. Because of their low solubility, very low barium or strontium levels of feedwater can cause fouling of the membranes. To avoid calcium carbonate fouling, various saturation indexes, such as the Stiff-Davis and Langelier, should be kept below precipitating values in the brine (through pH control or deposition control agents).

Microbiological fouling: Microbiological activity can degrade the cellulose acetate membranes. Proper preservation of the chlorine residuals may prevent these membranes from being damaged microbiologically. Polyacrylamide membranes are microbiologically resistant to degradation, but are vulnerable to chemical oxidation. Hence, chlorination is not an appropriate therapy. Microbiological fouling can become a problem when inoculation occurs. If there is considerable potential for microbiological fouling, use should be made of non-oxidizing antimicrobials and bio-dispersants.

6.8 SOME PHYSICO-CHEMICAL METHODS AND MEMBRANE TECHNOLOGIES

Cleaning of membranes is vital for production of high quality hygienic potable water and the functionality and capacity of the membrane plant. The good cleaner for cleaning of membrane could be optimized by removing the soil and does not adversely influence the membrane characteristics. Also, physical aspects of membrane cleaning and chemical activities play important role in membrane cleaning processes. In addition to this, the water for membrane cleaning should be of best quality.

First use of a combination of nanofiltration (NF) and reverse osmosis (RO) was reportedly made in India in 1975. Presently, over 5000 installations exist, mostly for producing required quality feed water for high pressure boilers while a few installations also exist for reuse of water, especially where so-called "zero" discharge is required (Chilekar, 2005). NF and RO are generally required where MF and UF are not adequate for the purpose. In fact, ME and UF are often used to give pretreatment to wastewaters to prepare them for application to NF and RO systems. Both NF and RO can give adequate treatment to pretreated wastewater to make them fit for reuse in high-pressure boilers and for various non-potable uses/reuses in industry and for groundwater recharge. RO units are really required where dissolved solids (TDS) are to be reduced. Their use in treating brackish waters and seawaters is well known.

A term often used in deciding on need and extent of pretreatment by conventional methods or by MF or UF is the SDI value or its silt density index. The value of SDI has to be less than 3.0 for successful use of RO systems with reasonably long filter runs. Higher values would need unduly frequent cleanings. Parameters to be satisfied after pretreatment by UF are suggested by Chilekaras shown in Table 6.2;

Table 6.2: Parameters to be satisfied after pre-treatment by ultra-filtration

S. No.	Parameter	Concentration
1.	Turbidity	< 0.3 NTU
2.	SDI	< 3.0
3.	Temperature	< 45°C
4.	Oil & grease	Absent
5.	Free chlorine	< 0.1 ppm
6.	Organics	Free from harmful organics
7.	pH	3 to 10

Cases

The membranes are made from cellulose acetate or polyamides or other materials and have flux rates and operating pressures as mentioned earlier. Moreover, owing to the higher pressures they have to withstand their manufacture and assembly are done in different specialized ways by the manufacturers, all of which are generally more expensive. It is also necessary to take the manufacturers help in ascertaining the extent of pretreatment required depending upon the quality of the feed water. Heater power requirement is also much higher (5 kWh/m′ for nanofiltration and 10 kWh m or RO plants) than for MF and UF systems.

6.9 MISCELLANEOUS TREATMENT METHODS OF WATER PURIFICATION

The various conventional water treatment processes as discussed earlier can treat the water up to certain extent but still the water contains some impurities and minerals that may harm the people during general water supplies. The groundwater may sometimes contain high concentration of dissolved solids. However, the surface water may be contaminated with industrial effluents and other human activities and may contain naturally occurring organics, colours or products of algal blooms. All such contaminants can only be removed by providing additional treatment with a series of additional steps to further reduce organics, turbidity, dissolved solids, hardness, nitrogen, phosphorus, metals, and pathogens. Most processes involve advanced treatment options such as activated carbon adsorption of organics, reverse osmosis, ultrafiltration, demineralization and additional disinfection and for specific pollutant removal, disinfection, etc. The advanced oxidation processes using ozone, UV radiation and chlorination are also the treatment techniques for water containing such impurities. These treatment options remove a variety of pollutants but have certain limitations related to cost and significant water rejects as in case of reverse osmosis. The various techniques being employed in removing inorganic and organic impurities present in water make it potable for use.

6.9.1 Taste and Odour

The taste and odour in water generally occurs due to the presence of organic matter or inorganic salts. Taste and odour may generally occur due to the presence of dissolved gasses such as hydrogen sulphide (H_2S), methane (CH_4), ammonia (NH_3) or carbon dioxide (CO_2) and many other odorous gases. The presence of H_2S an end product of anaerobic decomposition imparts rotten egg smell in water. The taste and odour in water may also be due to the presence of mineral substances and salts such as NaCl, various compounds of iron, carbonates, sulphates, hydroxides, carbonates, bicarbonates, organic compounds such as phenols, organic compounds or oily compounds and matter. The presence of certain species of algae adds oily matter in water. The bitter taste in water is due to the presence of alkaline material while metallic salts. The drinking water should not contain taste and odour as the consumer is pleased when the water is free from these contaminants. Methane gas is a major byproduct of anaerobic decomposition of organic matter tastes like garlic and highly explosive and risky and can be removed by aeration process. Hydrogen sulphide and ammonia can also be removed from water by aeration. Adsorption of carbon is also efficient but costlier.

Surface water and groundwater are the major sources of taste and odour problems as pollutants such as gasoline, chemical solvents, or a wide range of volatile organics contaminate the surface water or groundwater. The removal of these pollutants from surface water is expensive as the use of aeration and powdered activated carbon, requires huge energy. Therefore, in order to control the taste and odour in water the processes such as oxidation, aeration and adsorption may be efficient and also enhance the coagulation and filtration process.

6.9.2 Odour Measurements

The taste in water can be measured directly if the causative agents are identified. However, various methods such as liquid chromatography and quantitative method are used in determination of taste and odour in water. The threshold odour test is one of the most common quantitative methods used for measuring odour in water. The water sample to be tested is diluted in series of different dilutions with odour free distilled water to a final amount in a flask up to 200 mL of each dilution. The prepared samples are arranged with increased dilution concentrations and smelled by an observer. The sample in which no odour is detected or observed is the odour value of that sample and is denoted in terms of threshold odour number (TON). The dilution ration and threshold odour number can be calculated as shown in Eq. 6.62 and Eq. 6.63;

$$\text{Dilution ratio} = \frac{\text{Final volume}}{\text{Original volume}} \tag{6.62}$$

$$\text{Threshold odour number} = \frac{A + B}{A} \tag{6.63}$$

where,

A = Volume of odourous water (mL)

B = Volume of odour-free water (mL) required to produce a 200 mL mix water sample. The Indian standard prescribes the odour limit to be 1 TON and permissible limit to be 3 TON for domestic water to be used for drinking purpose and other usages.

During the test, the observer is provided a known odour-free blank for comparison. Individuals vary in their reactions to other types of odors. A stimulus to the odour that is pleasing to one can be irritating to another. The attempts to predict the odour strength of the mixtures exacerbate these variations.

6.10 TASTE AND ODOUR REMOVAL PROCESSES

6.10.1 Oxidation

Oxidation is a chemical process of adding such an oxidant in water that can oxidize the organic matter present in water. It is one of the safest and easiest methods to eliminate the taste and odour from water without any harm. Of all the chemicals used for oxidation process, potassium permanganate is the most commonly used and accepted oxidant. As per California Department of Health Services Water Treatment Plant Activity a dose concentration of potassium permanganate between 0.1 and 0.5 mg/L is quite effective in controlling the taste and odour in water. However, ozone with a dose of 2 to 5 mg/L is quite effective and an efficient oxidizer that can be used for removal of taste and odour in water. Several pilot scale studies indicate the combined effect of ozone and hydrogen peroxide to be very effective in eliminating taste and odours from water than ozone alone. Chlorine dioxide or chorine is another easily available and economical chemical being used to control taste and odour in water. However, for removal of organics from water proper use of such chemicals must be carefully assessed due to the formation of trihalomethanes (THMs) and chlorophenol in water.

6.10.2 Aeration

The addition of oxygen in water naturally or mechanically is called aeration and is a simple but sometimes energy-oriented technique (when used mechanically). It is generally applied for the control of taste and odour in water. It is mostly used when volatile compounds and gases such as hydrogen sulphide, ammonia are present in the water. However, it cannot be used when taste and odour is caused by the algae. (see aeration chapter)

6.10.3 Adsorption

The process of accumulating the solid substances or gases as contaminants present in water is an interface between two phases. The process is quite useful for removing the pollutants from water or wastewater. It is carried out by providing or formulating such an interface that can adsorb the contaminant from liquid to the formulated solid surface. The contaminant removed from water is known as adsorbate while the solid surface on which the contaminant adsorbs is known as adsorbent. The basis for separation by adsorption technology is the development of an adsorbed phase with a composition different from that of the bulk fluid phase. In a bulk substance, all the bonding properties of the constituent atoms of the substance (whether ionic, covalent, or metallic) are filled in by other atoms in the sample. Atoms on the adsorbent surface, however, are not completely surrounded by other adsorbent atoms, and can therefore attract adsorbents. The exact nature of the bonding depends on the species details involved, but the process of adsorption is generally classified as physicsorption (characteristic of the weak forces of van der Waals) or chemisorption (characteristic of covalent bonding).

Adsorbents of different forms are categorized into natural adsorbents and synthetic adsorbents. The natural adsorbents include iron, clays, minerals of clay, zeolites, and ores. In many cases these natural materials are relatively inexpensive, plentiful in supply and have considerable potential for modification and eventually enhancement of their ability to adsorb. Synthetic adsorbents consist of adsorbents extracted from agricultural products and waste, household waste, industrial waste, sewage sludge and polymer adsorbents. Adsorbent has its own characteristics of its adsorbing surfaces, such as porosity, pore structure and composition. Many waste materials used include fruit waste, coconut shell, scrap tyres, bark and other tannin-rich products, sawdust, rice husk, petroleum waste, fertilizer waste, fly ash, sugar industry waste blast furnace slag, chitosan and seafood production waste, seaweed and algae, peat moss, clays, red mud, zeolites, sand and soil, mineral ore, etc.

Activated carbon is specially treated carbon possessing attracting properties is the most commonly adsorbent used in wastewater treatment worldwide. It has excellent properties of adsorbing contaminants present in gases, liquids and finely divided solids and therefore easily removes tastes and odors from water for public supplies.

It can be easily manufactured by burning carbonaceous material such as wood, coal, saw dust and other petroleum residues at high temperature of 500 °C in the absence of air in a closed vessel. This char is further activated at high temperature conditions at 800 °C to create porous structure after removal of hydrocarbons and is also known as activated carbon. This activation creates pores in the solid mass resulting in increase in surface area per unit volume or per unit of mass. The surface area ranging from 500 to 1500 m^2/g is reported in the literature thus proves to be very useful in removing various contaminants and impurities from solid, liquid and gases. The activated carbon exists under different patent names such as Darco, Nuchar and others and added to water in form of granules ranging from 0.1 to 2 mm diameter or pulverized powdered very fine powder before or after coagulation but before filtration.

Adding powdered activated carbon to water or using granular activated carbon (GAC) can remove taste and odour in the water filter. Powdered activated carbon (PAC) is the preferred method for moderate and infrequent taste and odour. The dosage of powdered activated carbon can vary between 1 and 50 mg/L as many industry experts suggests a dose of 25 mg/L as the optimum value. Two basic types of PAC feed systems are dry storage and dry feeding. If the hourly feed rate is under150 pounds, a solution tank should be used for the feed system. If the hourly feed rate is greater than 150 pounds, consideration should be given to a slurry device.

The adsorption properties and mechanisms of using granular activated carbon and powdered activated carbon are same but application techniques are generally dissimilar. Granular activated carbon (GAC) systems used for contaminants removal consists of a cylindrical tank that holds a bed of the GAC material .The water to be treated is passed through the GAC bed with adequate retention time for achieving the adsorption process satisfactorily. Such systems are provided with a fixed-bed or a moving-bed GAC material. However, the fixed beds are operated in batch processes and the beds are made offline when gets saturated with contaminants and the adsorbing capacity weakens. Therefore, such beds of GAC are cleaned by passing a superheated steam through it. This regeneration process resembles the fresh activation process only as the organics are burnt at 800°C in the absence of air followed by high temperature steam to remove the remaining organics.

However, in a moving bed system the exhausted carbon is removed continuously from the bed by replacing it with fresh regenerated carbon at the top. Nowadays in advanced moving bed systems water is forcibly introduced from the downside of the system where the GAC bed expands and the arrested impurities present in the bed are generally taken away by the moving water. The major problem linked with granular activated carbon contact system is clogging of the bed by suspended solids present in the water. Provisions of backwashing the bed in a fashion to similar filter unit has been made in the vessel that helps to fluidized the bed during its backwashing. It provides a sufficient upflow velocity in the filter bed for 10 per cent expansion at all times so that suspended solids in the influent can be removed easily. This operation mode is advantageous as it increases the density of the carbon in the adsorbed organics and the spent carbon migrates to the bottom of the fluidized bed for removal to the regeneration process.

The granular activated carbon systems are generally designed on the basis of flow rates and contact times. The volume and cross section of the tank is commonly designed on flow rate value of 0.08 to 0.4 m^3/m^2.min with contact time ranging from of 10 to 50 min based on empty tank conditions. The plant should have a downtime of up to 40 per cent capacity with 5 to 10 per cent makeup carbon being provided after each regeneration cycle. For increase in capacity the carbon columns can be arranged in parallel while increasing the contact time the columns should be arranged in series. However, in a fixed-bed arrangement the powdered activated carbon (PAC) cannot be used in due to such a small size thus results a high head loss that will restrict the passing of water through it. The PAC is generally added in open vessels containing water and is allowed to remain suspended in water to provide necessary contact time for the solids to adsorb on it and thus can be eliminated from water successfully.

Mostly fluidized bed systems are generally adapted where an arrangement in such a manner where carbon in suspension is held up by a mixture of steam and other hot gases and the regeneration processes is carried out. However, in some cases the sand is used with carbon so that during fluidization it could facilitate to retain the heat in the system.

In wastewater treatment during biological process for removal of refractory and biodegradable organics, the powdered activated carbon can be added to the aeration basin and thereafter removed as sludge with the biological solids in the secondary clarifier. The biomass growth on the carbon surface utilizes the biodegradable fraction and enhances the biodegradable organic matter removal efficiency of the process. However, the sludge produced during the treatment process contains mixed biological solids and carbon and which cannot be separated. Regeneration of the carbon using thermal process results in destruction of the biosolids. This will eliminate the need of sludge processing and disposal techniques but will increase capacity of carbon regeneration system.

In current scenario, activated powdered carbon is mostly used for the treatment of potable water use. The moving bed, granular activated carbon systems are generally used in advanced wastewater systems. Better regeneration procedures would greatly enhance the use of powdered activated carbon in wastewater treatment, particularly in the secondary processes. Activated carbon may be applied either in a dry form or as a suspension or as a slurry, However, its use in powdered form may create problems, as in such cases being blown off due to winds. In rapid gravity filters or in pressure filter used for water treatment the activated carbon can be used in the granular form as a filter media in place of sand. The use of activated carbon is advantageous as it helps in coagulation process when used in powdered form before coagulation. It also reduces the chlorine demand of treated water and removes the organic matter present in water and its overdose is not harmful. The tastes, odours and colours caused by iron, manganese, phenols, excess chlorine and hydrogen sulphide in water can be removed easily and effectively.

6.10.3.1 Sorption Isotherms

Freundlich model: It assumes that sorption of nutrient ion occurs on heterogeneous sorbent surface as shown in Eq. 6.64.

$$\log C_s = \log k_f + \frac{1}{n}\log C_e \tag{6.64}$$

C_s = Sorbed concentration of NH_4^+ (mmol/g)

C_e = Equilibrium concentration in aqueous solution (mmol/L)

K_f = Freundlich parameter $\left(\frac{\text{mmol/g}}{\text{mmol/L}}\right)^{-n}$

$\frac{1}{n}$ = Indicator of energetics of sorption

Langmuir model: It assumes that sorption of two nutrients occurred on a homogenous sorbent surface by monolayer sorption with no interaction between the sorbed ions as shown in Eq. 6.65.

$$\frac{C_e}{Q_e} = \frac{1}{bQ_o} + \frac{C_e}{Q_o} \tag{6.65}$$

Q_e = Amount of nutrients sorbed at equilibrium (mmol/g)

Q_0 = Monolayer capacity of sorbent (mmol/g)

Biochar: Biochar prevents nutrients from leaching during runoff and make them available to plants through adsorption. It inhibits nitrification process of ammonia fertilizers. Biochar can increase the crop yield while some time reducing the use of fertilisers.

Gap identification

- Maximum adsorption and retention capacity of various nutrients is necessary to ensure efficient use of biochar with different properties.
- Bioavailability of digested metabolized novel path of adsorbed nutrients.
- Date of nutrient release upon soil application of biochar.
- Effect of storage on nutrient enrichment in long-term stability of biochar.
- Develop effective and economical system that can be integrated into the existing structures.
- To quantify the parameters for designing these systems.

Sorption Kinetics is illustrated in Eq. 6.66-Eq. 6.67.

$$\log(Q_e - Q_t) = \log Q_e - \frac{k_1}{2.303}t \tag{6.66}$$

$$\frac{t}{Q_t} = \frac{1}{k_2 Q_e^{\,2}} + \frac{t}{Q_e} \tag{6.67}$$

Q_e, Q_t = Amount of nutrients sorbed $\left(\frac{\text{mmol}}{\text{g}}\right)$ at equilibrium

k_1, k_2 = Rate constants

6.10.4 Treatment Using Copper Sulphate

Copper sulphate with chemical formula $CuSO_47H_2O$ is usually available in powder or crystal form. It is quite effective in removing colours, tastes and odours from water and therefore may also be added in the distribution pipes of water supply systems. However, it is more economical and advantageous when adding it to the open reservoirs or lakes as it kills the algae or to rather check the growth of algae, even before its production in any such waterbody. The water containing high amount of algae may possess the disagreeable taste in it. This may be due to the fact that the end of algae lifecycle results in oily matter as such decomposed end products impart highly disagreeable tastes and odours in water. A normal dosage of copper sulphate in between 0.5 to 0.65 mg/L is quite effective in removing such taste and odours from water as such a range of concentration is not harmful in treated public water supplies. The presence of copper sulphate is having harmful effects on the fish and other aquatic life present in lakes and reservoirs. Therefore, a dose requirement of copper sulphate needs to be optimized before applying it for controlling the algal growth in lakes and reservoirs.

It can either be added directly to the distribution pipes or in open reservoirs. In the former case, copper sulphate solution of necessary strength is prepared and fed to the water just before the distribution system. In case of reservoirs or lakes the copper sulphate crystals are grounded in powder form and sprinkled by boat on the water surface. Alternatively, it can be applied by means of crystal bags, which are pulled by boats across mud. The algae killing develops taste and odours in water. Hence, doses in smaller amounts are often successful to prevent troubles.

6.11 FLUORIDATION AND DEFLUORIDATION IN WATER

As discussed earlier the concentration of fluorides in drinking water is generally limited to a very narrow range of 1-1.5 mg/L. However, in general case an optimum value of 1 mg/L should be present in drinking water. Therefore, when the concentration of fluoride in drinking water is less than 1 mg/L the fluoride is added to the water and the process is known as fluoridation. However, when the concentration of fluoride in water is greater than the limiting value 1.5 mg/L the fluoride is removed from water and the process is known as defluoridation.

6.11.1 Fluoridation

The fluoride concentration of less than 1 mg/L in drinking water is considered harmful as it may cause the dental caries in children thus resulting in weak tooth enamel leading to early tooth decay. The presence of fluoride can be advantageous for older people in reduction of arteries hardening and proves to be helpful in treatment of osteoporosis. Therefore, fluoride should be added with extra dose when water is deficient of fluoride in public supplies. The extra fluoride addition to public water supplies is called fluoridation. The process is largely adopted in western countries and in some parts of the U.S.A. Fluoride deficient water used for drinking purpose is augmented with extra dosage of fluoride. The compounds used for fluoridation in water are sodium fluoride (NaF), sodium silico fluoride (Na Si Fe) and hydro-fluosilicic acid (H,Si Fe) of which most commonly used compound for fluoridation is sodium fluoride which is fed under pressure or as a solution, through various types of chemical feed equipment. The fluoride dosage greater than 1.5 mg/L is also harmful as it causes the disease such as spotting and discolouration (mottling) of teeth ultimately causes permanent grey to black discolouration of the enamel (dental fluorosis). The fluoride concentration in drinking water greater than 5 mg/L develops severe pitting of the enamel. However the drinking water with high fluoride concentration of 30 to 50 mg/L can cause diseases such as deformation of bones, skeletal problems, skin irritation and gastroenteritis.

6.11.2 Defluoridation

Fluoride during water consumption when enters the human body combines with bones due to its affinity for calcium phosphate in the bones. The water containing high concentration of fluoride causes skeletal fluorosis disease that causes permanent bone and joint deformation. Another disease caused by excessive fluoride intake is dental fluorosis results in discolour, blackened, mottled, or chalky white teeth. High intake of fluoride in water also causes gastrointestinal problems and neurological disorders and can also damage a foetus, even IQ of children is highly affected. Excessive fluoride can also cause diarrhoea, abdominal pain, constipation, neurological manifestations include nervousness, excessive thirst, and tendency to urinate frequently as the patients get confused with arthritis, spondylitis or non-specific backache.

Therefore, it is very much essential to regulate the fluoride concentration in drinking water in the range of 1-5 mg/L to circumvent such diseases. Therefore, presence of fluoride concentration greater than prescribed limits of 1.5 mg/L in water should be removed and the process of removing excessive fluoride is known as defluoridation.

6.11.3 Methods of Defluoridation

The following technologies are generally used for removing fluoride from water:

(1) Absorption by activated alumina (AA), commonly known as Prasanti Technology.

(2) Ion exchange adsorption method.

(3) Nalgonda technique

(4) Reverse osmosis process.

The methods are described below.

Prashanti Technique using Adsorption by Activated Alumina (AA)

It is an adsorption process of removing fluoride from water. The water with high fluoride concentration is allowed to pass through the granular bed containing adsorbents such like activated alumina (AA) or activated carbon, or serpentinite, or activated bauxite, where the fluoride from water get adsorbed and treated water passes away. Activated alumina, though costly, is an excellent adsorbent for the removal of excess fluoride from water. The fluoride absorption capacity of media decreases with the presence of bicarbonates in water and the presence of silica in water creates interference in the fluoride adsorption capacity on an adsorbent. At pH value in the range value of 5-7, the adsorption process proves to be effective. Due to continuous usage the adsorbent bed of activated alumina becomes saturated and exhaustive after a certain period of time and therefore needs to be cleaned periodically. The regeneration of activated alumina can be carried out by backwashing of the bed with 1% NaOH solution.

Ion Exchange Adsorption Method

The process uses anion exchange resin (zeolite) as a strong base in the form of chloride. The water when passed through such a resin bed the fluorides along with other anions like arsenic, nitrates, present in the water are exchanged with the chloride ions of the resin. This results in releasing of chlorides into water and ultimately adsorbing the fluoride ions into the resin with the removal of arsenic and nitrate ions from the water. Due to continuous exchange of anions like fluoride, nitrate, arsenic the resin bed gets saturated and exhausted thus decreasing the fluoride removal efficiency. Therefore, the resin bed needs to be regenerated with 5-10% sodium chloride solution also known as brine solution. During regeneration, the reverse exchange process is carried out with the adsorption of anions on the resin with replacing by chloride ions as the water in effluent is discharged with excess chloride ions. The plant capacity of ion exchange process ranges from 500 L/h to 5000 L/h. The method proves to be very costly due to regular resin replacement using large amount of salt (NaCl) for regeneration process. Due to this the method is not suitable for rural areas in villages. The wastewater produced during regeneration contains high concentrations of toxic fluoride, nitrates and arsenic ions therefore the disposal of such wastewater poses a serious environmental threat to the waterbodies.

Nalgonda Technique

Another method for removal of excess fluoride from groundwater is generally used in rural areas and is known by the name Nalgonda technique. In this technique no regeneration of any resin is required, therefore it is comparatively simpler and economical than fixed bed ion-exchange processes. The technique is quite easy to operate and maintenance as it employs readily available chemicals such as aluminium salt (alum) and uses local skills. However, other than fluoride removal the technique also helps in removing colour, odour, turbidity, bacteria and organic contaminants from water.

The requisite amount of lime (CaO) or sodium carbonate (Na_2CO_3) is added and mixed in raw water adequately and thereafter alum solution is added thoroughly with stirring of water slowly for 10 minutes. The precipitates formed in the water are then allowed to settle for nearly one hour and the sludge formed is discarded. The treated water as the clear supernatant containing allowable concentration of fluoride is withdrawn for usage. For effective hydrolysis of aluminium salts in order to ensure the elimination of the residual aluminium in the treated water lime or sodium carbonate is added that helps to maintain the required alkalinity. Sometimes before the lime addition bleaching powder is added with lime to carry out the disinfection process and removal of organic matter from water.

Reverse Osmosis Process

In this method, a semipermeable membrane barrier is used through which the raw water is passed as it permits the flow of clear water through it and restricts the flow of salts including fluorides. This method is generally adopted for desalination for removing salt from water. This method is generally not encouraged in India due to its high cost maintenance although the method is quite effective in fluoride removal without use of any chemicals.

The technology and processes to be used generally depend on the concentration of fluoride in water and the amount of water to be treated. The various merits and demerits of all the used methods have been discussed in above section. Due to high cost and low availability of service providers the process such as ion exchange and reverse osmosis are not encouraged in rural areas. However, Nalgonda technique of fluoride removal from water is quite popular in rural India. But due to improper management and scarcity of funds to meet the running operation and maintenance costs (RMO) it sometimes proves to be failure. An example of it is the two such plants for fluoride removal based on Nalgonda technique were installed in Gujarat in Tawadiya village and Thakarasen village of Patan district in 1988 and 1992, respectively. The Thakarasen plant of 60 m^3 capacity having annual RMO cost of Rs 1.83 lakh was installed with a cost of around Rs 15 lakh and the treated water costing around Rs 7.69 per kL. While, the plant at Tawadiya with capacity of 10 m^3 was set up with a total cost of around Rs 3.5 lakh. The annual RMO cost of this plant has been found to be around Rs 1.81 lakh with treated water costing around Rs 8.26 per kL.

The Government and the local people could not manage to meet such huge expenses on treating the water and use to lack of sufficient funds both the plants were shut down after a few months of their commissioning. Due to similar reason another plant based on Nalgonda technique in Mehsana district was shut down by the Gujarat Water Supply and Sewerage Board (GWSSB). Such plant failures are not failures of Nalgonda technology, but insufficient funds for its RMO and also due to the negligence of the local government and municipalities. It is a fact that the plants are quite successful at places where local people and NGOs are running them with their mutual involvement on sharing basis. The such example is Muniwar Abad Charitable Trust, an NGO, is running such a plant for 4 years successfully by involving the local community to share the RMO cost before being shut down in the year 2002, when municipal piped water supply through Dharoi surface water

6.12 THE FLUORIDE CONTENTS OF INDIAN GROUNDWATERS

The overexploitation and withdrawal of groundwater in India has tremendously increased in the past few decades through excavation of deep boreholes that has resulted in increase in fluoride concentration in the groundwater. The municipal corporation in Indian towns and villages have generally drilled deep tubewells for the supply of drinking water in order to withdraw the bacteria-free water. However, due to the development of proper drainage systems and increase in high percentage of paved surface areas has reduced the groundwater recharge drastically as during rainy season the surface run-off drains away without reaching the groundwater table. This development of surface paving and proper drainage system has led to the fall of groundwater-table continuously almost most of the places in the country. This fall in water table and decrease in volume of water in aquifers resulted in excessive enrichment of fluorides in deep waters. This may be due to the fact that as per the geological strata in Indian context the fluoride bearing minerals are generally present at deeper depths.

It has been reported that a total of 59,111 villages in India has been identified with fluoride concentration greater than 1.5 mg/L. The large rural population in India is affected by fluorosis due to consumption of high fluoride in water. In India about 65 million people are reported affected from fluorosis problem. The places with high concentration of fluoride in India are 42.5 mg/L in Muktsar district of Punjab, 32.5 mg/L in Delhi region, 23.4 mg/L in Karbi Anglong district in Assam, 17.3 mg/L in district Bhiwani of Haryana, 14.47 mg/Lin district Birbhum of West Bengal, 12.9 mg/Lin Mehsana district of Gujarat, 9.2 mg/L of Koraput district in Odisha, 8.12 mg/Lin Gaya district of Bihar, 7.79 mg/L in Kolar district of Karnataka, 7.67 mg/L in district Kannauj of U.P., 7.1 mg/L in Krishna district of Andhra Pradesh, 7 mg/Lin Erode district of Tamil Nadu.

The maximum population of the country is dependent on groundwater and using it for drinking and domestic purposes without any water quality testing. No such protocol of fluoride testing has been prepared or implemented by the government of India to test the water being extracted daily from more than 40 lakh hand pumps and nor such types of specific laboratories so far has been established. Therefore, there is a need of developing such a proper measurement system to be installed in most of the treatment plants being used for the treatment of groundwater used for public supplies. Also the government should strictly enforce the implementation groundwater recharge method to the public and private organizations and individuals by constructing rainwater harvesting structures in their premises. Such welfare schemes by the government will definitely help to improve the groundwater quality on a large scale. Moreover, the groundwater resources should be protected with proper management by making awareness in the society and among the public for efficient use of water and consider it an important resource.

EXERCISE

1. Describe the following: (a) disinfection of water, (b) superchlorination, (c) pre-chlorination, (d) breakpoint chlorination, and (e) dichlorination.
2. The analysis of hard water shows the following compositions:

   Free carbon dioxide = 3 mg/L

   Alkalinity = 68 mg/L

   Non-carbonate hardness = 92 mg/L

   Total magnesium = 15 mg/L

   Assume that it is possible to remove all but 35 mg/L of carbonate hardness with lime and that the treated water is to have a total hardness of 80 mg/L. Determine the amount of hydrated lime and soda required for treating per million litres of water.
3. The analysis of hard water shows the following compositions:

   Free carbon dioxide = 5 mg/L

   Alkalinity = 150 mg/L

   Non-carbonate hardness = 100 mg/L

   Total magnesium = 25 mg/L

   Assume that it is possible to remove all but 50 mg/L of carbonate hardness with lime and that the treated water is to have a total hardness of 100 mg/L. Determine the amount of hydrated lime and soda required for treating per million litres of water.
4. Write brief notes with neat sketches on the following:
   (a) Breakpoint chlorination
   (b) Disinfection of water
   (c) Ozonation
   (d) Fluoridation and defluoridation
   (e) Demineralization and ion exchange process
   (f) Electrodialysis
   (g) Ultra filtration
   (h) Reverse osmosis
5. Describe briefly the salient points on base exchange process and lime soda process in water softening? What are the different methods of removing colours, odours and tastes from waters?
6. Explain in detail by using a figure or flow circuit (a) direct filtration and cross filtration, (b) different types of pressure drive membrane processes, and (c) the permeate flux limiting phenomenon.

7. An unltrafiltration membrane has a pure water flux of 250 L/m^2.h at 3 bars. When an oil-water emulsion is concentrated at 4.0 bars the flux reduces to 30 L/m^2.h due to a build-up of a cake layer on membrane surface. The specific resistance of this cake is $r_c = 1.8 \times 108$ m^{-2}. Calculate the thickness of the cake. The viscosity of the emulsion is similar to that of the water.
8. Describe various minor methods of disinfection. Derive an equation for kinetics of disinfection.
9. Explain the lime and soda ash method used for water treatment along with stochiometric equations involved in water treatment.
10. What are the different permanent hardness methods used in water treatment along with all the reaction equations?
11. Define the term water stabilization and also describe the role of alkalinity in hardness removal from water.
12. Define the term zeolite softening. What are the various advantages of ion exchange process?
13. The hardness of water with the following data is to be removed using ion exchange process with a flow rate of 15,000 m^3/day. The medium has an adsorption capacity of 150 kg/m^3 at a flow rate of 0.75 m^3/min.m^2.

mequiv/L 1.0 5.0 6.0 8.0

<table>
<tr><td rowspan="2">CO_2</td><td colspan="2">Ca^{2+}</td><td>Mg^{2+}</td><td>Na^+</td></tr>
<tr><td>HCO_3^-</td><td colspan="3">SO_4^{2-}</td></tr>
</table>

mequiv/L3.5 8.0

14. What are the common problems that exist in an ion exchange process system?
15. Define membrane process and explain the various membrane processes used in water treatment?
16. Differentiate between Freundlich model and Langmuir model used as sorption isotherms in adsorption of pollutants during water treatment process.

REFERENCES

1. Technologies, S. W. (n.d.). *Membrane systems.* Retrieved from Handbook on Industrial Water Treatment: https://https://www.suezwatertechnologies.com/handbook/chapter-05-clarification
2. Suezwatertechnologies.Membranesystems.https://www.suezwatertechnologies.com /handbook/chapter-09-membrane-systems)
3. Chilekar, S. (2005), Overview of Membrane Applications and Basics of Membrane Processes, Workshop on Membrane Technologies for Recycle and Reuse, Ind. Env. Assoc., Mumbai, July.

7

CHAPTER

Water Distribution and Supply

The water after thorough treatment in water treatment plant (WTP) becomes wholesome and safe. Therefore, from the treatment plant it needs to be supplied to the consumers through the roads and streets in the city, and finally to the individual houses. To deliver this safe water to individual consumers with suitable quality, quantity and pressure requires broad system of pipelines, pipes, storage reservoirs, pumps and its related appurtenances. The system comprising of all such units and equipment's is termed as distribution system. The distribution system consists of pipe lines of diverse sizes and lengths that conveyance the water from storage reservoirs to streets through pipes fitted with different types of flow controlling valves. The connections of fire hydrants are also provided with water mains in order to release water during occurring of fires.

The amount of water being transmitted to the users is generally measured by fitting the water meters on the delivery pipes, service connections ending at the individual homes. The water measuring meters are also provided at the pumps utilized for lifting and transmitting the water into the distribution pipes, distribution reservoirs for storing the treated water to be fed into the distribution pipes. The water supply is provided to the consumers either continuously for 24 hours or it may be supplied certain for fixed hours of the day intermittently. The water is stored in a distribution reservoir and then transmitted through the distribution pipes to the consumers or it may be pumped directly into the distribution pipes thus depending upon the situation required.

Requirements of a water distribution system

Water supply system is a network of hydrological and hydraulic engineered components utilized for the supply of safe and wholesome water to the public. A comprehensive water supply system should be consisting of the following:

- A drainage basin.
- A natural water collecting point available in below or above the ground such as a lake, river, or underground water aquifers.

- Entire water treatment facilities to make it safe and wholesome for the public usage.
- Pumping station consisting of various types of pumps for water extracting and transmitting it to the WTP when gravitation flow is not available.
- Transfers of treated water to underground or overhead water reservoirs or water towers through water pipes or conduits by pumping or gravity flow.
- The water may be collected in cisterns or pressure vessels by smaller water systems.
- A water distribution pipe network for water supply to end users and other points of use (such as fire hydrants). The customers may be private houses or manufacturing units, commercial or educational establishments.
- The water from reservoirs through pipe network is supplied to high-rise buildings also and requires enough pressure to reach up to the topmost floor of the building. Therefore, additional water-pressurizing components such as pumping stations are required to be located either at the underground outlets or above the ground reservoirs (if gravity flow is impractical).
- The system includes the main, sub-mains, branches, and intervals (called feeder), which carry the water to the streets and to individual houses or property by laterals.

7.1 DISTRIBUTION SYSTEM REQUIREMENTS

The miscellaneous requirements of a good water distribution system to work efficiently are given below.

- The designed water distribution network system should maintain enough pressure head proficient enough of supplying water at all the proposed places within the city.
- It should possess accommodating capacity of requisite amount of water supply for firefighting throughout the 24 hours during requirements.
- The network should be configured in such a way that the supply can satisfy maximum demand per hour. For the towns with high population a peak factor of 0.5 is suggested. A factor of 2.0 would be sufficient for larger population.
- The distribution system comprises about 70% of the total cost of the entire water supply system.
- The water supply system should be economical in construction, operation and maintenance cost. It should be simple in operation and easy in repair, therefore leading to RMO cost to be minimized.
- The water pipe lines should always be placed away from the sewerage and drainage lines by adequate distances so that the water to be supplied may not get contaminated or mix with sewage.

- The water supply pipelines should be ensured to be fairly watertight so that the water may not interact with sewerage lines or leakage losses may be minimized to possible extent.

7.1.1 Distribution Pipes and Other Appurtenances

The entire distribution system consisting of pipelines of various shapes and sizes are categorized as supply mains distributing to sub-mains further distributing it to streets with branches and ultimately leading to our houses through service connections known as laterals. These mains and sub-mains of cast iron pipes are joined by means of spigot and socket joints. The laterals are made of galvanized cast iron pipes or PVC pipes. In order to utilize the available head to a maximum and to counterbalance the head loss during water transmission through pipes the water mains and sub-mains are usually laid with sloping from higher level to the lower level areas.

The water flowing in the different sections of the distribution system is controlled by providing the sluice valves along the length of the pipes at suitable intervals. Along with all the junctions and branching off points. In order to drain the water during repair works, the drain valves are placed at all the low points in the distribution system pipelines. Also in order to avoid contamination of water in pipe line by back flow the drain valves are connected properly through check valves. Air valves are also provided at all the high points in order to remove air from the pipe during the filling operations and emptying of the pipe. All different types of valves are discussed in chapter 6.

The water distribution in smaller cities and towns is carried out through the distribution mains directly from the treatment plant itself, and distribute the water into different branches and laterals. The water distribution in large cities is generally carried out by dividing the large city into different zones. A separate distribution reservoir is installed in each zone of the city connected with distribution mains that carry the water to the different localities of the zone. In order to avoid any future contamination of water through the leakages from sewage, the laying out of these distribution pipes is in such a way that it is laid about 2 m vertically above and 3 m away in horizontally from the sewers. Also the laying should also be such as laid on one side of the roads and the streets, usually below the footpaths. In a street, if space is not a constraint the water pipes should be laid on the opposite side of sewer lines.

7.2 WATER DISTRIBUTION SYSTEM

7.2.1 Drainage Basin

A drainage basin is a region or area of land where rainwater, ice or snow melting water as surface run-off meets at a single point is known as basin, that further meets another water source, such as a rivers, lakes, reservoirs, estuaries, wetlands and ultimately meets the sea, or ocean. When water converges to a single point within the reservoir in a closed drainage basin is termed as pool. This pool may be a permanent lake, dry lake, or a point through which surface water is lost underground. The drainage basin consists of both the streams and rivers that carry water, as well as the land surfaces through which water flows into it and is divided by a drainage from its neighboring basins.

The drainage basin caters to all the draining water through funnelling effect and channelizes it to a single point within the area covered by the basin. The drainage basin is topographically isolated by geographical barriers such as a cliffs, hills or mountains from neighbouring basins. The drainage basin is the chief body that helps in the collection of run-off water to a single point since the bulk of the water that discharges from the outlet of the basin originated as precipitation falls on the basin. A portion of the water entering the groundwater system beneath the drainage basin can also flow to the outlet of another drainage basin. This may be due to the direction of groundwater flow beneath the earth is not always same as its preceding drainage. In order to determine the discharge of water from a lake, a stream gauge can be fixed at the outlet of it.

7.2.2 Distribution System

The water after thorough treatment can be directly supplied to the consumers or should be stored in the reservoirs so that it can be distributed to the public effectively. The main objective of a distribution system is to supply water to every customer via the pipeline and maintain adequate water pressure at all the points of water supply system. The treated water to be supplied is involved with cost and economy and therefore such potable water should reach to all the consumers without any leakage. The water system performance relies on careful preparation, implementation, and maintenance. The distribution of water generally depends on the type of city and its local conditions along with topography of the area. Therefore, the distribution system should be designed on the level of water source and its transmission for the public usage in an area. The various systems for the transmission of water with adequate pressure can be described as:

- Gravitational system
- Pumping system
- Combined gravity and pumping system

7.2.2.1 Gravitational System

In order to maintain adequate pressure up to the end users, the supply system should be mounted at a sufficient elevation or height above the consumer and the water is distributed under gravity without any pumping. In order to be more effective and for proper functioning of the system, the difference of head available between the source and the consumer must be sufficient enough after allowing the frictional and other losses in the pipes. Therefore, the method proves to be more economical as no pumping is required at any stage. The method can be implemented to any type of source such as a lake or a reservoir. An example of such a system exists in Mumbai city where the reservoir is situated at the hills and water is distributed to consumers through gravity. Therefore, the method as shown in Fig. 7.1, can be best implemented in cities situated at the foothills and the source of supply is available somewhere in the hills so as to have sufficient elevation to distribute water to the city.

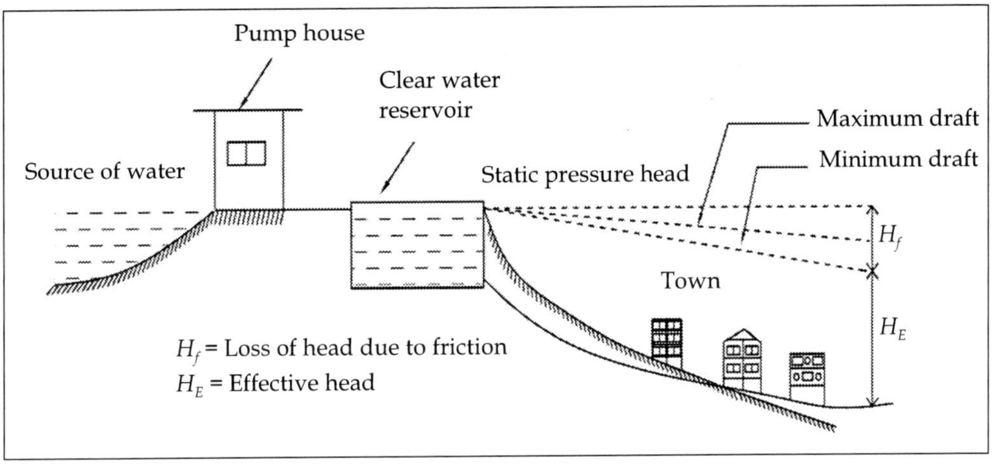

Fig. 7.1: Water distribution gravity system for a city

However, when the treatment plant (WTP) is also situated almost at the same level on the hill itself and the sources such as lake is also situated at a hill, therefore makes easy conveyance of water from lakes to WTP by using low lift pumping. Thereafter, the treated water can be distributed to the consumers situated at the foothill by gravity without pumping. However, sometimes to maintain adequate pressure during fires the use of motor pumps is required.

7.2.2.2 Pumping System

The process of supplying the treated water directly into the distribution mains using high lift pumps without storing it anywhere is called a pumping system and is also known as pumping without storage system. The pumps are operated at variable speeds in order to meet the fluctuations of water demand. Therefore, the system requires continuous monitoring and attention at the pumping station for ensuring the desired flow in the distribution system. Also in case of power failures the water stops completely and may be disastrous in case of fire breaks out in a city at that particular interval of time. Therefore, the method is generally not used so as to avoid any such calamity and disaster. The system possesses the only advantage of forcing large amount of water under high pressure in the required direction during break out of fires. The system showing maximum and minimum demand along with hydraulic gradient is shown through Fig. 7.2.

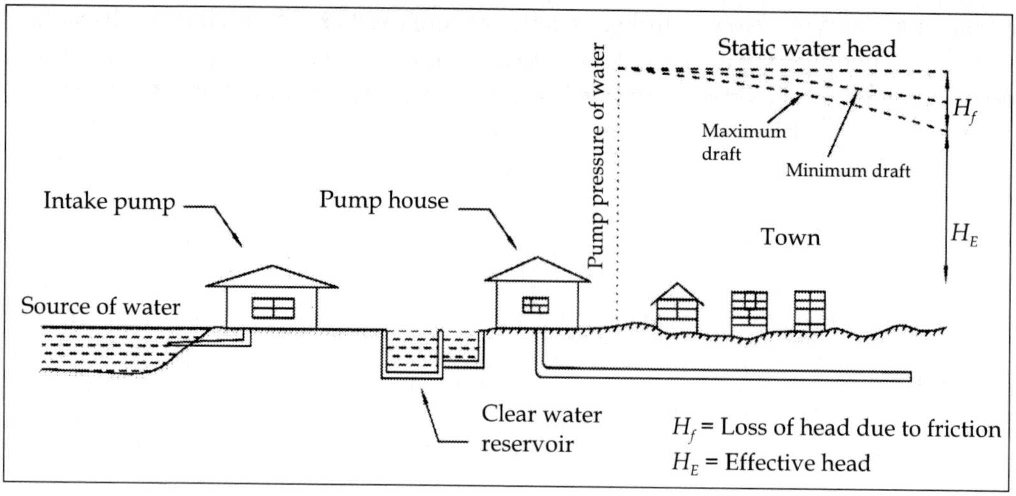

Fig. 7.2: Pumping system for water distribution in a city

7.2.2.3 Combined Gravity and Pumping System

This is a dual system consisting of pumps and storage reservoir. The treated water at a steady rate is pumped and stored in a reservoir at the required elevation and thereafter distributed to the consumers through gravity. Sometimes, the entire water is first of all pumped into the distribution reservoir and stored initially during the time of minimum water demand but when the demand increases water is distributed in the system – both by pumping and by the elevated reservoir as shown in Fig. 7.3. This method is sometimes called pumping with storage system. This is considered to be more efficient, cost-effective and efficient method as the pumping rates are standardized in order to meet both low and maximum demands. The water stored in the elevated reservoir meets all types of demands and requirements during pump breakdown and firefighting.

This system helps in reducing the wear and tear of pumps and increases their efficiency by operating them at constant speed at their rated capacities. The system is usually adopted and accepted generally due to the following advantages:

- In such a system the necessary water pressure during fire occurrence can be maintained for firefighting. This can be accomplished by closing down the supply of neighbouring areas or by utilizing the motor pumps and modern firefighting squads being provided at intended pace of the distribution network.
- The efficiency of the pumps can be increased by operating them at uniform flow rate with their rated capacities and ultimately reducing the wearing and tearing of the pumps. This system does not also require much supervision and maintenance as compared in the case of direct pumping method.
- In such a system pumping is carried out for 8 to 16 hours of the day in small towns and cities.

- The method has advantage as the stored water in a reservoir can be supplied even during power failures also.
- The method of water distribution is cheap, economical and very efficient.

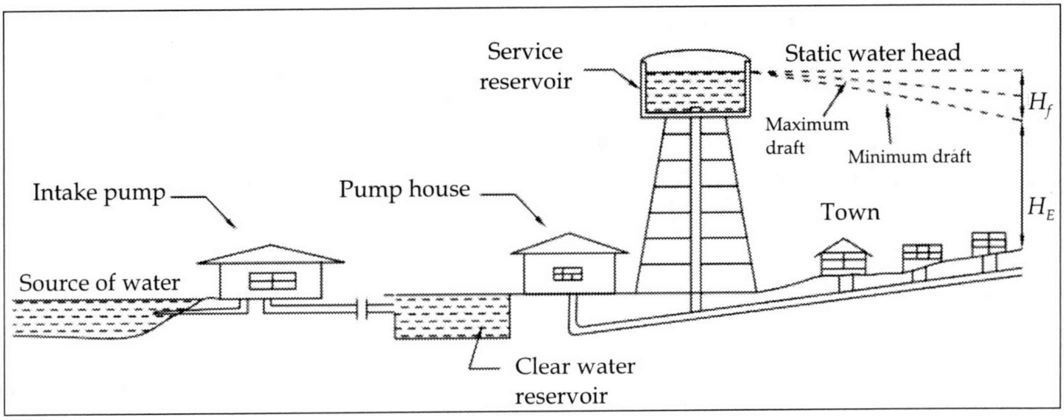

Fig. 7.3: Combined gravity and pumping system for water distribution in a city

7.2.3 Pressure Requirement in the Distribution System

The water pressure in a water distribution process during its conveyance from storage reservoirs through the distribution mains is lost constantly in overcoming friction of the pipes, during contraction and expansion of pipes in water bends, fitted valves and meters installed in the pipelines till it reaches to the end users. Therefore, the total head is reduced due to these losses as it reaches the consumers. The available head at this particular point is known as available head or effective head (H). From the pressure point of view it is the most important head as it makes the water to rise even up to the top of high-rise buildings. The more the height of the building the greater is pressure required in the system which allows the water to rise. For an example, a building having top floor height of 12 m requires a pressure of 1.2 kg/cm^2(120 kN/m^2). The system should be designed in such a manner so as to keep the losses in the distribution pipes and fittings to be optimum and minimum.

In order to maintain high pressure in the distribution system it requires high quality pumps, better quality, thicker and stronger material pipes. This increases the installation and operational cost of the system and therefore proves to be uneconomical. Also high pressure in pipes will have high leakage of water thus leading to increase in wastage of water. The greater the design pressure the more costly it is but proves to be beneficial for the consumers. Therefore, such design systems need to be optimized and thus choosing of a proper value for the design pressure to be maintained in the distribution pipes. However, in large cities the firefighting requirement is the governing factor as the firefighting squads are generally equipped with motor pumps for lifting the water and the design pressure will be maintained accordingly. In small towns the system is not equipped with such firefighting squads and motor pumps, therefore requires high pressure to be maintained in the distribution pipes so as to reach the water up to 3, 4 storeys easily. The water meters being installed with individual house connections cause a significant head loss in the system and therefore ensure a high pressure requirement.

In India the water supplies by municipal corporations and jal boards is in extremely poor conditions except in some areas. However, during the maximum demand of the day the water supply only reaches to ground floor and never reaches the first or second floor of the building. Therefore, the consumers residing on first or second floor install their pumps and boosters privately.

In U.S.A. the available normal minimum pressure head being provided is around 20 t0 30 m which is very high as compared to 8-15 m of maximum pressure head available in India. In India it is enough to design a distribution system with minimum pressure head of 14 m, (140 kPa) under ordinary circumstances. Therefore, high-rise buildings have their own boosters and elevated storage tanks. Under ideal conditions a pressure head of at least 28 metres (280 kPa) is required when all the houses are metered along with the firefighting squads with no motor pumps. A high pressure value greater than 70 m in the distribution system is generally not encouraged and is undesirable as the maximum allowable pressure of the order of 100 m, generally equals the design pressure for plumbing fixtures, like pipe fittings, etc.

The towns and cities located on the hilly areas are divided into different zones with different elevations. This proves to be economical and desirable so as to feed each zone as per the elevation by a separate distribution reservoir. The system should be equipped with pressure relief valves at all the places in the pipes passing through low level areas. This ensures safety of the pipelines against bursting due to high pressure.

7.3 SYSTEMS OF SUPPLY

The water supplied may be done either continuously for all the 24 hours of the day or may be supplied intermittently only during the peak duration of morning and evening hours. The intermittent supply system possess an advantage as it may lead to less water consumption and less losses that can save some saving in due to losses occurring for lesser time and more vigilant use of water by consumers. This method, however, can be implemented at places where there is a shortage of water at the source. However, the intermittent system has certain disadvantages due to certain limitations and should not be continued for long term and be replaced by the continuous system whenever required. The disadvantages and limitations are discussed below as:

- During non-supply periods the fire requirement is not considered in intermittent supply of system as the fire during this duration may sometimes be disastrous in terms of damaging the life and property.

- The collection and receiving of water during restoration of supply after some obstruction due to unavoidable circumstances may cause some major inconveniences to the end users.
- The system proves to be costlier as it necessitates the extra vigilance and requires more staff due to daily opening and closing of various air valves and sluice valves fitted in the distribution system.
- While emptying the entire pipe during stoppage of water supply a partial vacuum develops in the pipe thus creating suction of the surrounding dirt and contaminants near the pipe through the leaky joints. When the supply is restored this contamination mixes with the fresh supplied water.
- However, the water demand of the whole day in this system is to be supplied within 6 to 8 hours of the day. This obligates the use of larger diameter size distribution mains for water supply during the entire town or area at the same time.

The system is quite effective in water supply with adequate pressure to the areas located in high altitude of the city without considering the water saving practice in the system. For maintaining a better pressure during water supply in different hours of the day, the entire city is divided into a number of zones. However, having so much disadvantages and limitations the intermittent supply system is very much common in India. Also during non-supply hours the repair works in the system can be carried out easily. The capital of India, Delhi, follows the intermittent supply system of distribution. The lift of water can be increased even at same horsepower of the pumps with less discharge thus results in maintaining the improved pressures.

7.3.1 Distribution Reservoirs

The treated water from water treatment plant is stored in storage reservoirs for supplying it during peak and high demand hours of the day, fire, breakdown and repairs, and such storage reservoirs are known as distribution reservoirs. The important functions of the distribution reservoirs are discussed below.

Distribution reservoirs and their utilities

- In order to improve the efficiency of the system the water treatment units and pumps are generally operated at constant rate. This ultimately reduces the hourly variation in demand and thus reduces the RMO costs.
- During the peak hours or high demand period the pressure generally falls down thus helps in maintaining constant pressure in the distribution mains.
- In this system around 8-16 hours are sufficient in order to meet the various types of water demands in spite of whole day pumping and therefore the pumping of water in shifts is possible without affecting the supply.
- During various emergencies such as fire, breakdown of pumps the stored water in the reservoir can be utilized easily.

- The optimum sizes of pipelines and low capacity of pumps and treatment units makes the system overall economical.

7.3.1.1 Types of Distribution Reservoirs

In order to store high amount of water load and to protect the structure from seepage or any leakage, the distribution reservoirs are generally made of steel, R.C.C. or masonry. However as per the construction of these reservoirs depending upon their elevation with respect to the ground, may be classified in two types and are discussed below.

- Surface reservoirs
- Elevated reservoirs

Surface reservoirs: The reservoirs, circular or rectangular tanks, constructed at ground level or below the ground are called ground reservoirs. In order to make the gravity flow easy and economical the tanks are generally constructed in the city at the different points of higher elevation wherever required. For proper and effective distribution of water to the consumers in different zones of the city the reservoirs are constructed zone-wise separately. The treated water stored in the ground service reservoir is fed to the distribution system directly with gravitational type of flow. However, in case of combined gravity and pumping system the treated water stored in ground reservoir is pumped into an elevated service reservoir by special pumps from where it is fed to the distribution mains for supplying water to the consumers.

A typical section of a surface reservoir is shown in Fig. 7.4. Construction of a surface reservoir in two compartments connected by shut-off valve or sluice valves is a

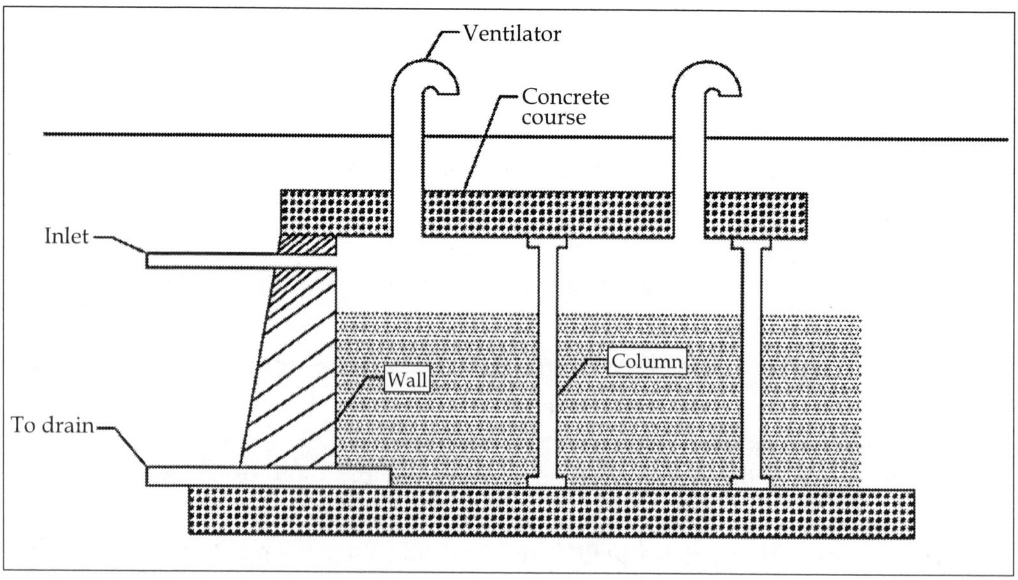

Fig. 7.4: Underground reservoir

standard practice, so that one compartment can be used while the other is cleaned or repaired. In order to maintain a constant water level in the tank the overflow pipes are provided at full supply level. The roof slabs are equipped with ventilator for the free circulation of air. As the treated water is stored in the tank some minute material settles down as sludge and should be removed periodically through the washout pipes installed at proper intervals. These outlet pipes are laid on the cement concrete floor with sloping towards the central washout pipe of the tank.

Elevated reservoirs: The overhead tanks supported on towers constructed in rectangular, circular, or elliptical shape at suitable elevation above the ground level are known as elevated reservoirs. The reservoirs are constructed at significant elevation above the ground surface when the use of stand pipes becomes impracticable. This fulfills the requirement of necessary pressure in the system, and constructed in the combined gravity and pumping system type of distribution. Water from the treatment plant unit or from the surface reservoir is pumped into the elevated reservoir and then fed to the distribution system to supply water to the consumers.

In order to suit the architectural requirements and advancement in structural engineering, the tanks can be given any shape and are generally made of R.C.C., steel, or prestressed concrete. Due to lower cost and maintenance with major resistant against corrosion the R.C.C. tanks are preferred nowadays to the steel tanks. The R.C.C. tanks (Fig. 7.5) are structurally more strong and economical and therefore preferred. Also use of prestressed concrete is preferred due to its high strength and economy. The elevated tanks constructed by using prestress concrete can save up to 60% of steel and 40% of concrete.

Due to the high cost and possible location in the heart of the city these elevated reservoirs are designed for capacity of not more than 6 to 8 hours of the average daily supply of the city. In order to avoid the contamination of water due to dirt, birds, insects or other external interactions the top of these tanks are covered with roof slabs provided with ventilators, so as to allow free circulation of air. The different components and appurtenances installed in such a reservoir are:

- Inlet pipe for entry of water and for the exit of water the outlet pipe connected to the distribution mains.
- For maintaining constant level of water in the tank an overflow pipe discharging into drain gutters should be installed.
- The depth of water can be read out from outside itself as the tank is provided with a float gauge or an indicator and having automatic devices to stop pumping when the tank is full.
- For draining out the water during cleaning process a washout pipe (or drain pipe) should be placed along with ladders to be provide around the tank surface and inside the tank in order to reach the top and then up to the bottom of the reservoir for inspection.

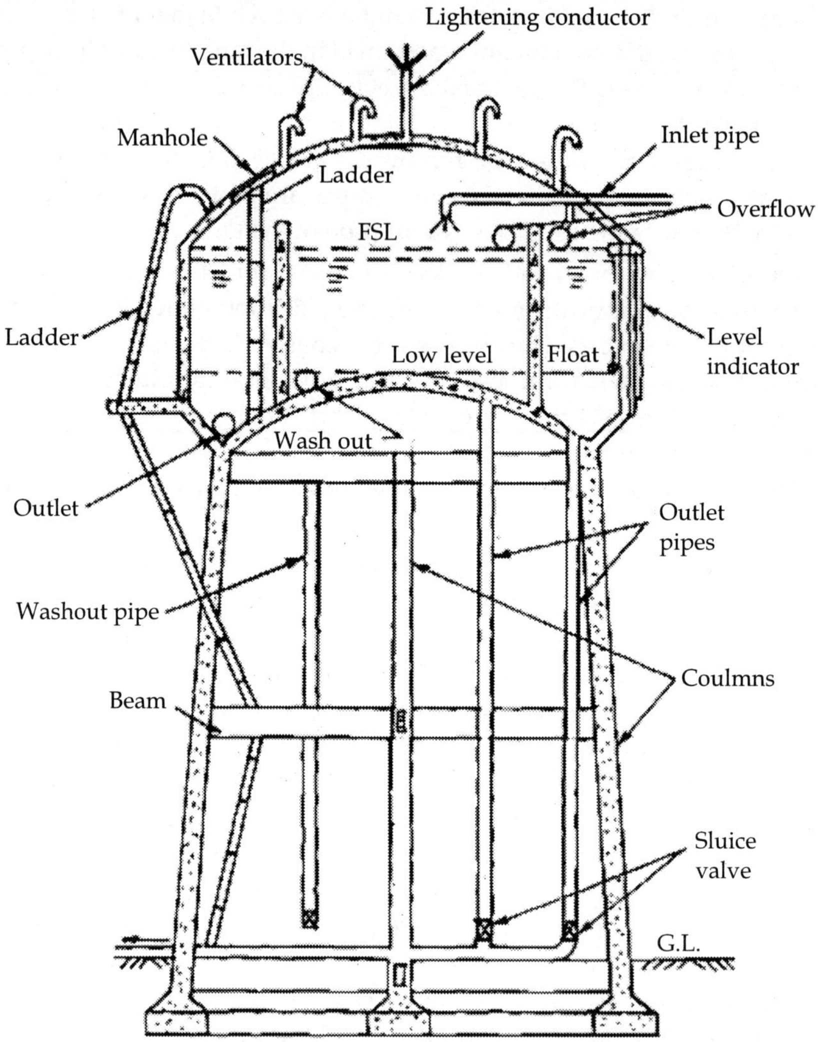

Fig. 7.5: Elevated rectangular type reservoir

- For inspection suitable manholes should be provided into the entry of the tank for inspection purposes along with ventilator near the rooftop for fresh air circulation. All such components and accessories are shown in Figs. 7.8 and 7.9.

Stand pipes: Stand pipes are tall cylindrical shells (Fig. 7.6) are type of elevated tanks constructed without any erected towers for resting the tank body and are therefore directly rests on the ground. The stand pipes are generally made of R.C.C. or steel containing 02% copper with diameter of 10 to 15 m and may be 15 to 30 m high. The use of 2% copper restricts the corrosion in steel pipes. The R.C.C. stand pipes are not preferred and encouraged to be used for greater than 15 m heights. This is due to the fact that at large heights, the hoop tension developed in the bottom portion of such tanks is very high as such high stresses are unbearable. Therefore, the stand pipes made of R.C.C. is unable to remain water-tight under greater than 15 m heads.

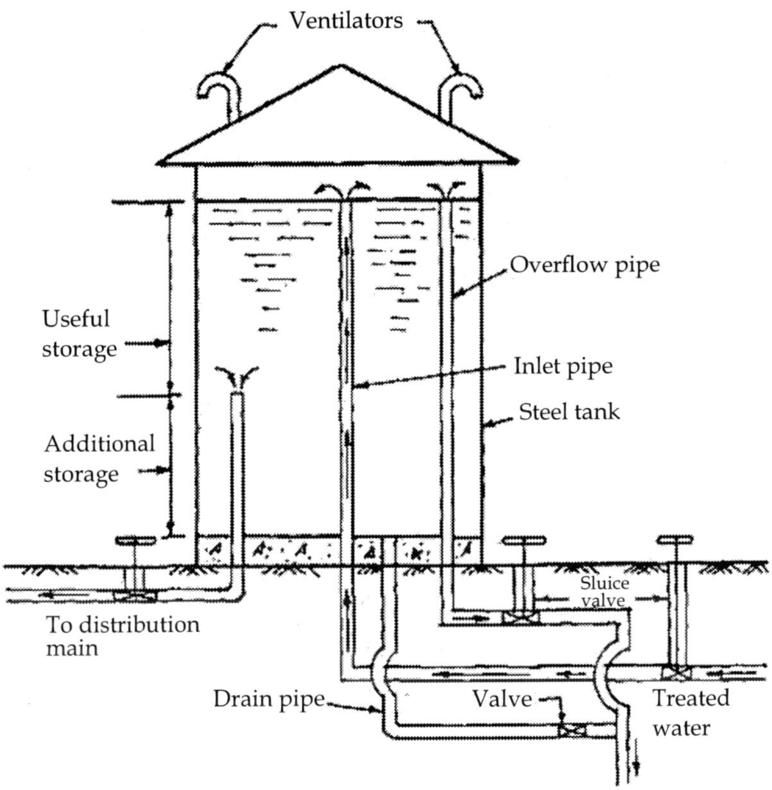

Fig. 7.6: Cross section of stand pipe

The volume above the elevation provides the advantage to the storage capacity of a stand pipe as it can provide the requisite pressure in the distribution system. However, using the booster pumps the stored water below the head can also be used during emergency such as firefighting. These stand pipes are also equipped with inlet pipe, outlet pipe, drain or washout pipe, overflow pipe along with manholes and ladders for inspection, repairs and maintenance. On the other side, the lower portion of the storage situated below the outlet pipe entrance cannot usually be used and it acts only as a support for usable storage and is thus called a storage support.

The economic height limit for stand pipes is reached when the supporting framework for an elevated reservoir is less costly than the lower ineffective portion of the stand pipe. Distribution reservoirs should be situated centrally or as close as possible to the areas they are to represent. A central position of a distribution reservoir would reduce friction losses in the distribution pipes as the length of the pipes is reduced. Further, in this case the pressure will be uniform over the entire distribution area during periods of high and low demands. On the other hand, if a distribution reservoir is not centrally located, large head loss will occur, and by the time water reaches the tail end of the area served by it, the pressure will be too low for satisfactory water supply to the consumers at the tail end.

7.4 DISTRIBUTION NETWORK SYSTEM

The distribution pipes are generally laid below the road pavements in the direction as per the road exists. Based on different road classification network system these pipes are also laid accordingly. It is advantageous of laying these pipes below the footpath instead of laying them below the paved roadway as it makes easy during excavations while repairs. The four different types of pipe network system either singly or in combinations and also depending upon the local conditions and orientation of roads are generally used for the water distribution in a city. The distribution network systems can be classified as:

- Dead end or tree system
- Grid iron system
- Circular or ring system
- Radial system

These systems are discussed below.

7.4.1 Dead End or Tree System

In a dead end system, sometimes also called tree system, a number of sub-main pipes are connected in right angles and originate from one main supply pipe. Each sub-main is connected with a number of branch pipes also known as laterals. These laterals further end up with the consumer's property by connecting them with service pipes as shown in Fig. 7.7.

These type of layouts are generally found in old towns and cities having unplanned roads with growth and development in haphazard pattern. The water supply mains have to be laid as per the existing roads and conditions. However, the taken off branches from the mains are provided wherever the streets and roads are leading. This ultimately results in creation large number of dead ends. Therefore, the system is best suited in zones of towns and cities with irregular growth and expansion without any planning. The pipes are thus laid in such an irregular random pattern of road network system.

It possess the following advantages:

- The discharge and pressures at different points in the system can be computed easily and helps in solving of the distribution network easily and effectively.
- The system requires less number of sluice valves used as cut-off valves and pipes shorter size makes the laying of the pipes simple and easier.
- The easy laying and expansion of the system makes it simple, cheap and economical.

However the system has certain disadvantages.

- Any damage or repair work in the pipeline at a particular point completely stops the water supply of the whole area fed by that pipe. This results in major hassle to the consumers as the considerable areas may get affected due to water supply stoppage till supply resumes.
- The water after supply got traps and remains stagnant at the various dead ends in the system. Due to lack of free circulation of water the quality of this stagnant water starts deteriorating. This stale water therefore, should be removed from time to time with installation of scour valves at each dead end. This requires additional vigilance on the installed valves with wastage of water and proves to be uneconomical sometimes.
- Also during firefighting the pressure cannot be increased as the water supply is from one direction only.
- Hence, it can give only limited supplies and may sometimes prove to be a serious handicap. This system is suitable for irregular towns or cities. In this system water flows in one direction only into sub-mains and branches. The diameter of pipe decreases at every tree branch as shown in Fig. 7.4 and the termination or end point of the pipe is known as a dead end.

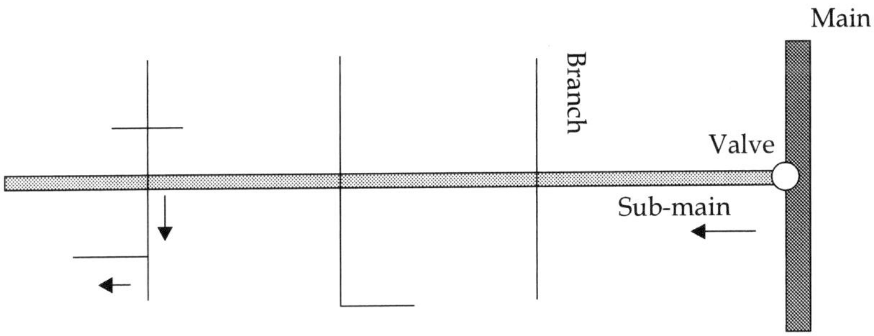

Fig. 7.7: Dead end or tree system

7.4.2 Grid Iron System

In this system, all the mains, sub-mains and branches are organized in such a manner that they remain interconnected with each other and also known as interlaced system or reticulation system as shown in Fig. 7.8. Such types of systems are generally developed in well planned city or town consisting of grid-iron pattern road network and the pipelines in such a system follows the same road network. The suitable example of such a well planned system is Chandigarh city in India. The water from the mains enters in all the branches at all junctions of equal diameter connected sub-mains. Also from two directions of interconnected network of pipes the pressure at any point can be balanced easily and effectively.

The advantages of this system are:

- The friction loss, and the size of the pipe is reduced with flow being carried by each pipe and the water reaches at different places through more than one route.
- During the maintenance and repairs only small area is affected as the supply to other points and places is not affected and reaches from some other route as each point is supplied from two directions and with higher pressure.
- The interconnections of all the pipes keeps the water in continuous circulation and abolishes the dead ends completely therefore water quality is not affected as in case of dead end system which is liable to pollution due to stagnation.
- During the occurrence of fire the various cut-off valves of area can be closed in order to divert more water towards the place of fire occurrence.

However the system has certain disadvantage.

- A larger number of cut-off valves as sluice valves are required with requirement of more length of pipe lines in the system thus makes it costlier.
- The design is difficult as the calculations for determining the pipe size and pressure accurately at different points is not an easy task and requires design experts and software thus make it costly.

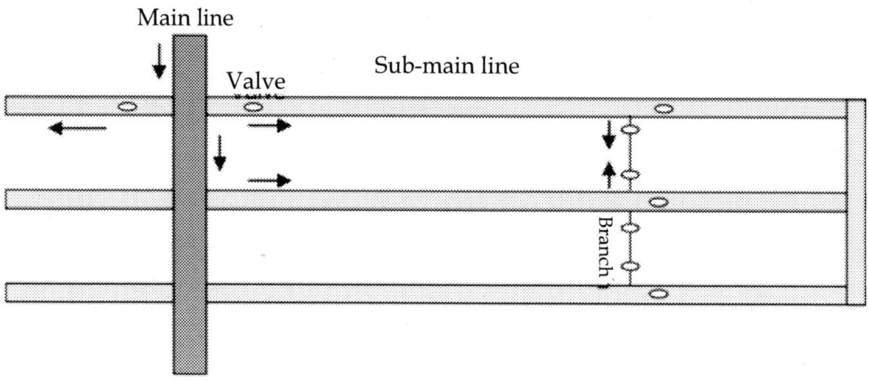

Fig. 7.8: Grid iron system

7.4.3 Circular or Ring System

The system comprising of main pipes in a closed ring forming circular or rectangular shape around the area to be served and sometimes also called circular system as shown in Fig. 7.9. The main water pipes are laid along with the sub-mains on the periphery of the rectangular or circular blocks formed in the distribution area. The system is used as a "looped feeder placed centrally around a high demand area" and is very suitable for towns and cities having well planned roads in circular or ring system. The system has the advantages of adequate pressures at various points and also enhances the capacity of the grid iron system. It has the same pressure as of having pipes with smaller diameters. The advantages and disadvantages of this system are the same as that of the grid iron system.

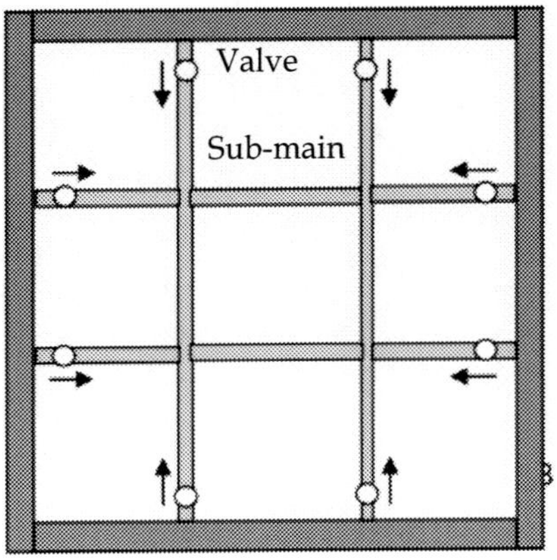

Fig. 7.9: Ring system

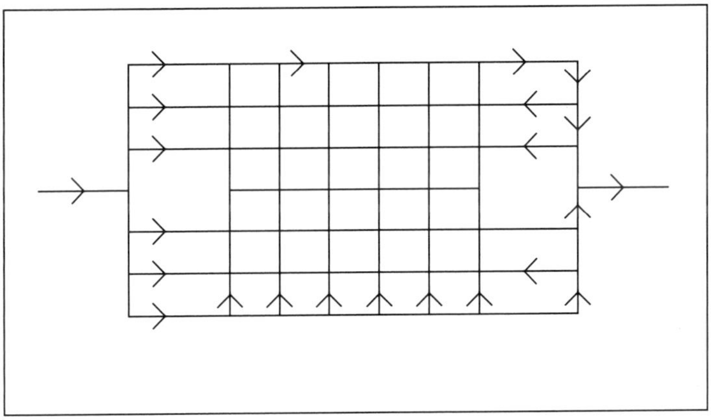

Fig. 7.10: Circular or ring system

7.4.4 Radial System

This system is quite popular in cities or towns having a system of radial roads and joins the various city centres. In the system the pipelines are generally laid radially and placing the distribution reservoirs at these centres. The water is taken from the water mains, and pumped into the distribution reservoirs placed at different centres, as shown in Figs. 7.10 and 7.11. This helps in easy distribution of water through such radial load distribution pipes. The method is quite effective in maintaining adequate high pressure along with distribution of the water efficiently and effectively. Road configurations need to be radial to prevent head loss in bends. This method is also most economical if combined pumping and gravity flow are adopted. This makes the design calculations for pipe sizes and lengths easy and simple.

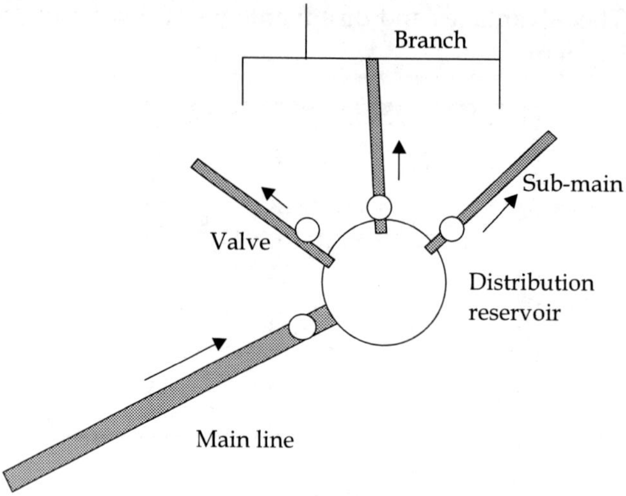

Fig. 7.11: Radial system

7.4.5 Branching vs. Grid Systems

- A grid system is generally preferable to a branching system because it can provide a supply from at least two directions to any point. The branching system has dead ends, so it does not require supply from more than one direction. Wherever possible, should be avoided.
- In situations where there are drastic differences in topography (hilly or mountainous regions), the division of the delivery network into two or more service areas is a standard practice.
- Pressure should be sufficiently large to meet consumers and firefighting needs adequately.
- No undue pressure: cost consideration, leakage and maintenance increase
- Capacity is calculated on the basis of local water needs plus the need for firefighting.
- Pipe size should be chosen to avoid high speeds: Pipe size should be chosen based on flow speeds of 3-5 fps – where firefighting is needed, minimum pipe diameter is 6 in.

7.5 DISTRIBUTION RESERVOIRS

In a distribution network, distribution or service reservoirs are used to provide storage to meet fluctuations in water demand, to provide storage for firefighting and emergencies such as breakdowns, repairs, etc., and to manage distribution system pressures. These reservoirs may be made of brick masonry, stone masonry, and concrete-plain cement, reinforced or pre-stressed and steel. To avoid pollution and to prevent algal growth these reservoirs are sealed. Additional acceptable provisions for manholes, mosquito-proof ventilation, access ladders, scour and overflow arrangements, and water level indicator are made.

The various purposes served by service or distribution reservoirs are as follows:

- The availability of such reservoirs allows the pumps to run at a constant pace if pumps are required.
- In the case of a water supply gravity scheme, the installation of such reservoirs results in a decrease in the scale of the distribution mains.
- These reservoirs have water storage facilities to accommodate the variability in the hourly water demand.
- They help to maintain constant pressure on distribution mains. The pressure in distribution mains will decrease as water demand rises in their absence.
- The construction of these reservoirs contributes to a decrease in the total size of the pumps, pipes and treatment systems. The method of production is economical.
- These reservoirs serve as storage for emergencies such as pump breakdowns, power outages, heavy fire demand, power outages, temporary floods, etc.

7.5.1 Storage Capacity of Distribution Reservoir

The storage capacity of a distribution reservoir is based on the following requirements:

7.5.1.1 Balancing Storage (or Equalizing Storage or Operating Storage)

The key function of a distribution reservoir is to provide storage of water from the water treatment plant with a constant rate of water pumping into the reservoir in order to meet the fluctuating water demand. The quantity of water needed to be retained in the reservoir to offset or equalize this variable water demand against the constant pumping rate is known as storage offset or balancing reserve or the storage capacity of the balancing reservoir.

The balancing storage of a distribution reservoir can be determined by the following two methods.

(a) **Hydrograph or mass curve method:** In this method the hourly demands of water are plotted against the respective hours of the day for a typical maximum day, and a hydrograph is obtained as shown in Fig. 7.12. The pumping rate for a standard 24-hour pumping would be equal to the mean hourly water demand, and the same is plotted as shown in Fig. 7.12 by line *PQ*. The storage required for the distribution reservoir is then obtained by determining the area (shaded shown) between the *BCDE* curve and the *PQ* axis. When converted into volume units, this area gives the storage required for the reservoir distribution. From the hydrograph, it is clear that a very high pumping rate would be required in the absence of a distribution reservoir, but by providing a distribution reservoir with the appropriate storage capacity, the required pumping rate would be reduced by nearly 30 to 40 per cent.

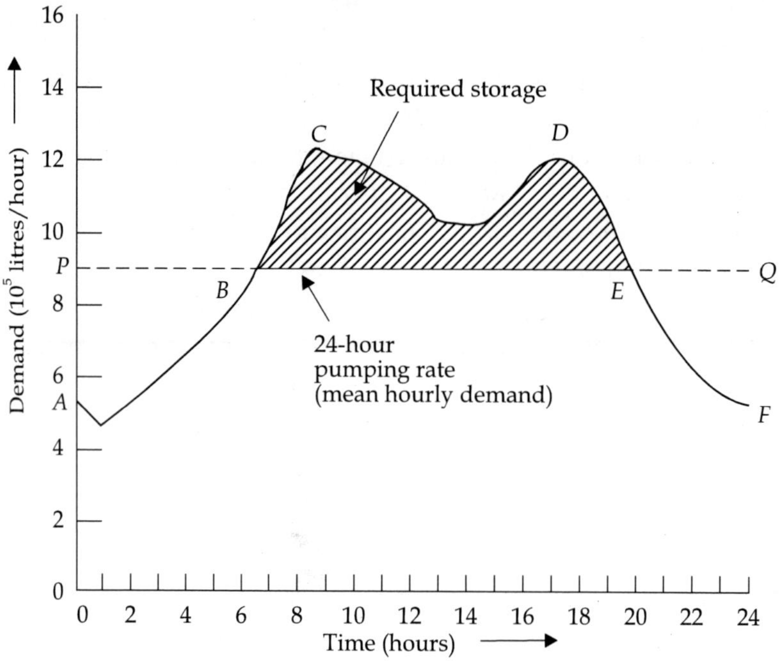

Fig. 7.12: Hydrograph for determining storage required for distribution reservoir

The storage needed for a reservoir distribution can also be analytically calculated. In addition, since pumping is done for only limited hours of the day, the analytical approach is more convenient to calculate the storage needed for a reservoir distribution.

(b) Mass curve method: It is a plot of cumulative supply is considered as inflow and cumulative demand as outflow with respect to time. The mass curve line known as supply line is drawn first and thereafter it is overlaid by demand curve line. By continuously adding the hourly water demand for a typical maximum day, values of cumulative water demand are obtained at successive hours, and the same is plotted against the corresponding hours to obtain the mass demand curve as shown in Fig. 7.13. The mass market curve increases continuously. The steepness of the mass demand curve indicates a higher demand rate, whereas flat part indicates a lower demand rate. If line *AB* is connected to the two ends *A* and *B* of the demand mass curve, then line *AB* represents the pumping mass curve into the delivery reservoir if the pumping takes place at a consistent rate for 24 hours, and line *AB* slope represents this consistent pumping rate. Draw tangents through the lowest point *C* and highest point *D*, parallel to line *AB*, to evaluate the storage needed for distribution reservoirs.

The maximum vertical distance *EF* between the two tangents gives the reservoir storage needed for the distribution. This is because at *C* there is maximum surplus supply equal to *CC′* that should be stored, and at *D* there is maximum *DD′* shortfall that must be removed from the stock and replenished. Thus, the storage (*S*) needed for the distribution reservoir will be the amount of the maximum excess supply through E_p pumping i.e., *CC′* and maximum excess demand (or maximum deficiency) E_d, i.e., *DD′*.

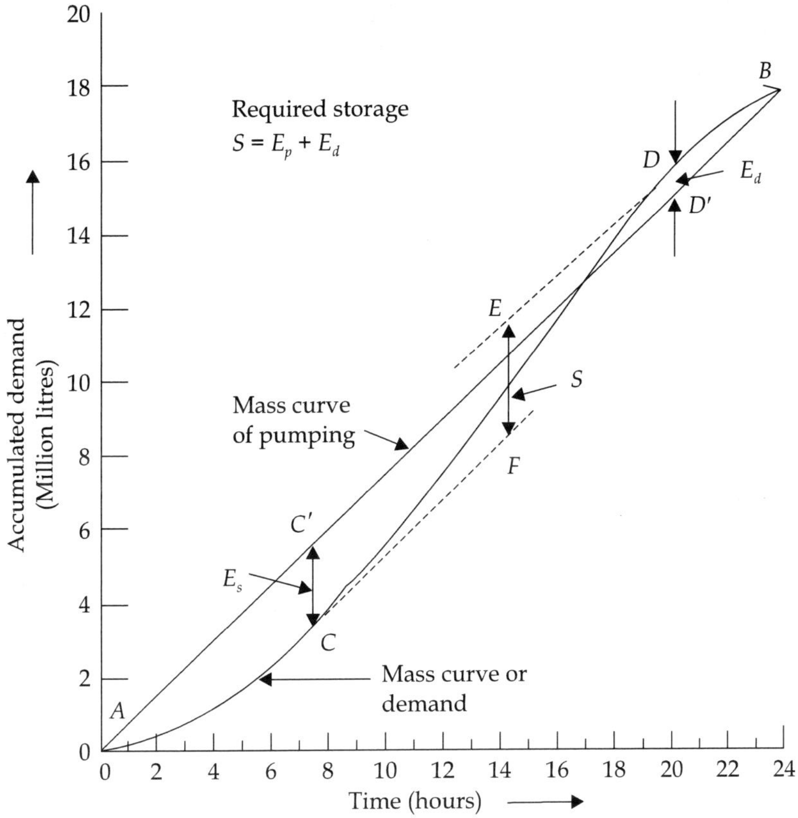

Fig. 7.13: Hydrograph for determining storage required for distribution reservoir

If pumping is conducted at a constant rate but only for a limited time then on the demand mass curve, the pumping mass curve defined by line *CD* is drawn from the beginning of the pumping cycle to its end as shown in Fig. 7.14. In this case, *CC* represents the maximum excess demand (or maximum deficiency) that E_d and *DD′* serves by pumping the maximum excess supply. The storage (*S*) required for distribution reservoir is again given by Eq. 7.1 as:

$$S = E_p + E_d \tag{7.1}$$

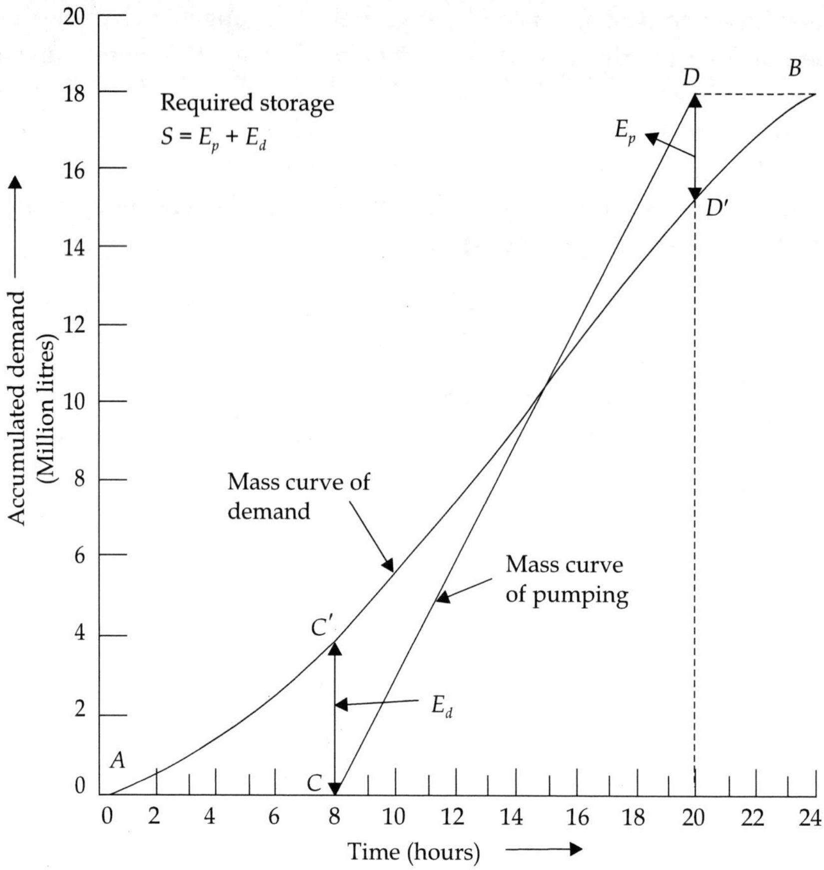

Fig. 7.14: Mass curve for determining storage required for distribution reservoir when pumping is done/or limited period

If pumping is continued for all the 24 hours of the day, the supply line will be a straight line extending diagonally across the graph of the whole day.

(c) **Analytical solution:** In this method, the cumulative hourly demand and cumulative hourly supplies for all the 24 hours duration are tabulated. The cumulative hourly excess of demand as well as the cumulative hourly excess of supply is then computed. The maximum value of the excess of demand and the maximum value of excess of supply will be selected and the summation of both will provide the required storage capacity of a reservoir.

Example 7.1: A city supplied with water at 200 L/person/day having population of 1.5 lakh with probable hourly variations in the rate of the demand is summarized in Table 7.1. Determine the capacity of the balancing reservoir to be provided for meeting the variable demands against a constant pumping rate as (i) if the pumping is carried out for 24 hours, and (ii) if the pumping is done from 6 a.m. to 12 noon and 3 p.m. to 9 p.m. Determine the rate of pumping in both the cases. Solve the problem by using (a) mass curve method, and (ii) analytical method.

Table 7.1: Percentage of average hourly flow expected

S. No.	Period of the day(hours) (1)	Percentage of average hourly flow expected (2)
1.	0-1	12
2.	1-2	15
3.	2-3	18
4.	3-4	22
5.	4-5	28
6.	5-6	42
7.	6-7	78
8.	7-8	120
9.	8-9	180
10.	9-10	200
11.	10-11	220
12.	11-12	170
13.	12-13	110
14.	13-14	70
15.	14-15	70
16.	15-16	100
17.	16-17	150
18.	17-18	180
19.	18-19	170
20.	19-20	170
21.	20-21	130
22.	21-22	90
23.	22-23	35
24.	23-24	25

Solution: Average daily supply = 200 × 1,50,000 L = 30 million litres

$$\text{Average hourly demand} = \frac{30 \text{ ML}}{24} = 1.25 \text{ ML},$$

The given demand in Table 7.1 can be converted to cumulative demand as below in Table 7.2:

Table 7.2: Cumulative demand (million litres)

S. No.	Period of the day(hours)	Percentage of average hourly flow expected	Period (hours) (1)	Demand units (avg hourly demand) (2)	Cumulative demand (avg hourly demand) (3)	Cumulative demand (million litres) (4) = col. 3 × 1.25
1.	0-1	12	0	-	-	
2.	1-2	15	1	0.12	0.12	0.15
3.	2-3	18	2	0.15	0.27	0.34
4.	3-4	22	3	0.18	0.45	0.56
5.	4-5	28	4	0.22	0.67	0.84
6.	5-6	42	5	0.28	0.95	1.19
7.	6-7	78	6	0.42	1.37	1.71
8.	7-8	120	7	0.78	2.15	2.69
9.	8-9	180	8	1.20	3.35	4.19
10.	9-10	200	9	1.80	5.15	6.44
11.	10-11	220	10	2.00	7.15	8.94
12.	11-12	170	11	2.20	9.35	11.69
13.	12-13	110	12	1.70	11.05	13.81
14.	13-14	70	13	1.10	12.15	15.19
15.	14-15	70	14	0.70	12.85	16.06
16.	15-16	100	15	0.70	13.55	16.94
17.	16-17	150	16	1.00	14.55	18.19
18.	17-18	180	17	1.50	16.05	20.06
19.	18-19	170	18	1.80	17.85	22.31
20.	19-20	170	19	1.70	19.55	24.44
21.	20-21	130	20	1.70	21.25	26.56
22.	21-22	90	21	1.30	22.55	28.19
23.	22-23	35	22	0.90	23.45	29.31
24.	23-24	20	23	0.35	23.8	29.75
			24	0.20	24	30.00

Case-I with analytical method: The analytical solution for the above-mentioned problem can be calculated with all the computations are shown and discussed below as:

$$\text{Rate of supply when pumping is done for 24 hours} = \frac{30\text{ ML}}{24} = 1.25\frac{\text{ML}}{\text{h}}$$

From the above table it can be analyzed that the maximum excess of demand as calculated = 6.06 ML (Table 7.3)

The maximum of excess of supply as calculated = 1.94 ML (Table 7.3)

Therefore, the total capacity of the storage reservoir as required = 6.06 + 1.94 = **8.0 million litres**

Table 7.3: Maximum of excess of supply and demand(million litres)

When the pumping is done for 24 hours of the day (hours) (1)	Cumulative demand calculated in col.4 above (million litres) (2)	Constant pumping rate (million litres/hour) (3)	Cumulative pumping (million litres) (4)	Excess of demand (million litres) (5) = col.2–col.4	Excess of supply (million litres) (6) = col.4–col.2
1	0.15	1.25	1.25	-	1.10
2	0.34	1.25	2.5	-	2.16
3	0.56	1.25	3.75	-	3.19
4	0.84	1.25	5	-	4.16
5	1.19	1.25	6.25	-	5.06
6	1.71	1.25	7.5	-	5.79
7	2.69	1.25	8.75	-	6.06
8	4.19	1.25	10	-	5.81
9	6.44	1.25	11.25	-	4.81
10	8.94	1.25	12.5	-	3.56
11	11.69	1.25	13.75	-	2.06
12	13.81	1.25	15	-	1.19
13	15.19	1.25	16.25	-	1.06
14	16.06	1.25	17.5	--	1.44
15	16.94	1.25	18.75	-	1.81
16	18.19	1.25	20	-	1.81
17	20.06	1.25	21.25	-	1.19
18	22.31	1.25	22.5	-	0.19
19	24.44	1.25	23.75	0.69	-
20	26.56	1.25	25	1.56	-
21	28.19	1.25	26.25	1.94	-
22	29.31	1.25	27.5	1.81	-
23	29.75	1.25	28.75	1.00	-
24	30.00	1.25	30	0.00	-

Case-I using mass curve method as shown in Fig. 7.15.

Average daily supply = 200 × 1,50,000 L = 30 million litres

Rate of supply when pumping is done for 24 hours $= \frac{30 \text{ ML}}{24} = 1.25 \frac{\text{ML}}{\text{h}}$

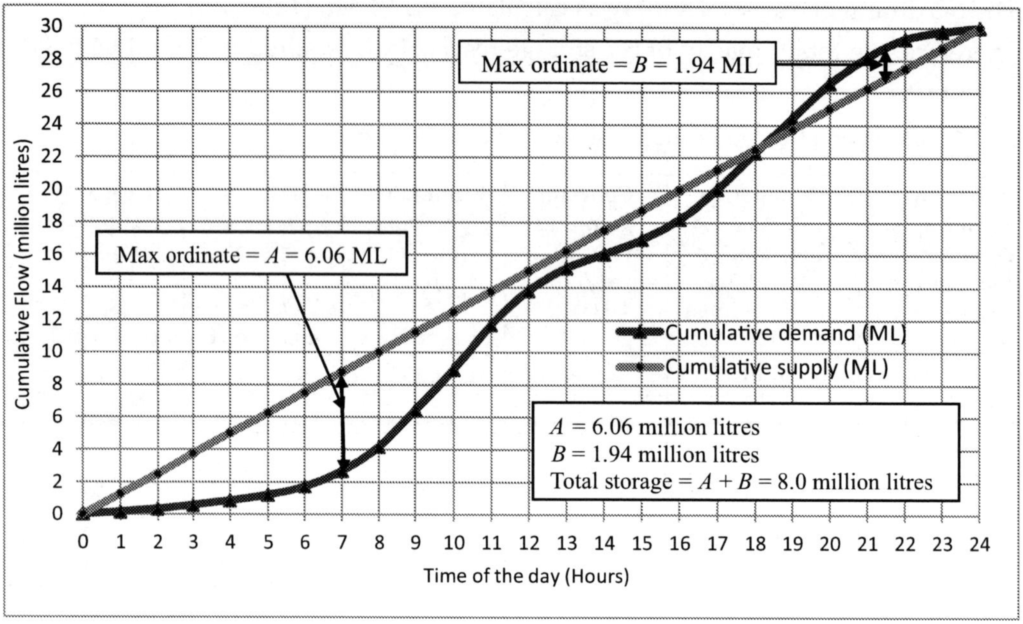

Fig. 7.15: Determination of storage capacity using mass curve method (Case I)

Case-II Using Analytical method:

However, when the pumping is carried out for a limited period of time, the analytical solution in the case can be calculated as shown in Table 7.4.

The intermittent pumping is to be carried out from 6 a.m. till 12 noon = 6 hours

Also the pumping is to be carried out from 3 p.m. till 9 p.m. = 6 hours

The pumping hours required = 12 hours

Therefore, the rate of supply $= \dfrac{30\text{ ML}}{12} = 2.5\dfrac{\text{ML}}{\text{h}}$

Therefore, based on the pumping rate of 2.5 ML/h the storage capacity of the reservoir can be calculated as below in Table 7.4.

Table 7.4: Analytical solution when pumping is resorted only for limited period

Time of the day (hours) (1)	Cumulative demand calculated in col. 4 above (million litres) (2)	Constant pumping rate (million litres/ hour) (3)	Cumulative pumping (million litres) (4)	Excess of demand (million litres) (5) = col. 2-col. 4	Excess of supply (million litres) (6) = col. 4-col. 2
1	0.15	-	-	0.15	-
2	0.34	-	-	0.34	-
3	0.56	-	-	0.56	-
4	0.84	-	-	0.84	-

5	1.19	-	-	1.19	-
6	1.71	-	-	1.71	-
7	2.69	2.5	2.5	0.19	-
8	4.19	2.5	5	-	0.81
9	6.44	2.5	7.5	-	1.06
10	8.94	2.5	10	-	1.06
11	11.69	2.5	12.5	-	0.81
12	13.81	2.5	15	-	1.19
13	15.19	-	15	0.19	-
14	16.06	-	15	1.06	-
15	16.94	-	15	1.94	-
16	18.19	2.5	17.5	0.69	-
17	20.06	2.5	20	0.06	-
18	22.31	2.5	22.5	-	0.19
19	24.44	2.5	25	-	0.56
20	26.56	2.5	27.5	-	0.94
21	28.19	2.5	30	-	1.81
22	29.31	-	30	-	0.69
23	29.75	-	30	-	0.25
24	30.00	-	30	-	-

From Table 7.4 it can be assessed that:

The maximum excess of demand as calculated = 1.94 ML

The maximum of excess of supply as calculated = 1.81 ML

Therefore, the total capacity of the storage reservoir as required = 1.94 + 1.81 = **3.75 million litres**

Case-II using mass curve method as shown in Fig. 7.16.

However, using mass curve method when pumping is done for limited hours of the day

The intermittent pumping is to be carried out from 6 a.m. till 12 noon = 6 hours

Also the pumping is to be carried out from 3 p.m. till 9 p.m. = 6 hours

The pumping hours required = 12 hours

Therefore the rate of supply $= \dfrac{30 \text{ ML}}{12} = 2.5 \dfrac{\text{ML}}{\text{h}}$

7.5.1.2 Break Down Storage or Emergency Storage

During the emergencies caused by electricity failure, pump failure or any such malfunctioning in the system the breakdown storage, or often called, emergency storage is preserved in order to cope up under such situations. The exact amount of

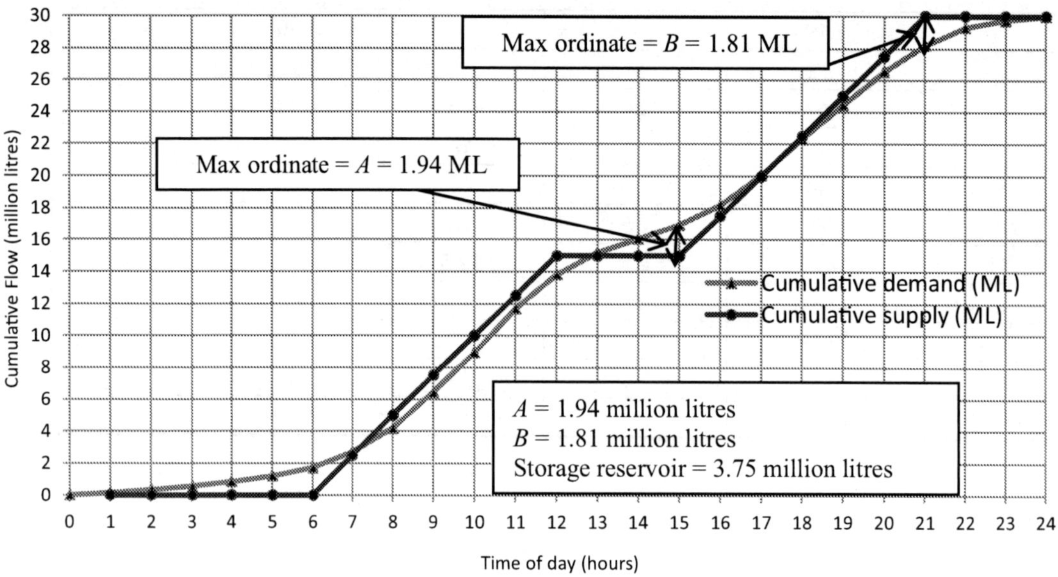

Fig. 7.16: Determination of storage capacity using mass curve method (Case II)

provision under such circumstances is difficult to predict as it depends upon the frequency and its occurrence. It also depends upon the severity of the failure being caused and the time required for carrying out its repairs. Therefore, a provision of lump sum amount is normally taken for this storage. However, under all circumstances an extra storage value of around 25% of the total storage capacity of the reservoir, or about 1.5 to 2 times of the average hourly supply may be considered. However, during extreme conditions of long power cuts or for longer repair durations higher allowances should be considered.

7.5.1.3 Fire Storage

The fire storage is much important and third major component to be considered in assessing the total reservoir storage capacity. The requirement of water for extinguishing fires is prerequisite as during electricity failures at the waterworks and occurrence of fire simultaneously, the storage reservoir should have sufficient amount of water at that time for throwing it over the fire at that particular instant of time. The firefighting requirements have already been discussed in Chapter 2 in section 2.1. It is stated that the total volume of water required for firefighting is generally in the range of 1 to 5 L per day per person. It is recommended that for extinguishing a fire, the three jet streams each with a discharge of about 1100 L/minute must be thrown with one at the fire and the other two on either side of the property where fire has occurred. Therefore with desired provision of 10 hours of fire fighting per day, the volume of water for storage is:

$$= (3 \times 1100) \times (10 \times 60) \text{ litres/day} = 198 \times 10 \text{ litres/day} \approx 2 \text{ million litres/day}$$

The storage for firefighting is based on the fact that the chances of fire break outs in the city. For towns with population more than 6000 and up to 2 lakhs the National Board of Underwriters of America recommends the provision for extinguishing a serious fire of 10 hours at a rate of about 10,000 to 50,000 L/minute. However, for smaller communities with population of 4000, 2000 and 1000, the provisions for extinguishing a serious fire may be taken as 8, 6 and 4 hours, respectively. In a developing country like India the provisions are difficult to implement and are considered to be too high. As in India depending upon the type of the city under normal conditions the water is stored with a provision of about 1 to 4 litres per person per day as the essential fire storage. However, the total reservoir storage capacity is computed by adding these three storages (i) balancing storage, (ii) breakdown or emergency storage, and (iii) fire storage.

7.5.2 Location and Height of Distribution Reservoir

The height and location of a reservoir is most important and critical part of the distribution system and therefore utmost care is important before deciding the location and height of the distribution reservoirs. The following points need to be understood:

- In order to command and cater the maximum area they should be located in the heart or centre of the city and should be ensured closer to the point of heaviest demand. This will allow the water to travel small distances in reaching the different points from the distribution reservoir and will result in lesser head loss due to friction.
- In order to maintain sufficient pressure at all the time in the distribution system the reservoirs should be ensured at high elevations or otherwise should be located at the top of some natural ridges if it exists.
- Based on the distribution area location and the arrangement of the pumping stations the reservoirs may be positioned in two different ways (a) they can be placed between the pumping station and the distribution area; or (b) they can be placed at the farther end of the distribution area. In the first case due to high elevation of the reservoir the water to be supplied will firstly always pass through the reservoir. However, in second case the reservoir elevation can be lower and during high demand period the supply can be made available both directly from the pump as well as from the reservoir. In the first case the length of the rising main carrying the water from the pumping station of waterworks to the reservoir will be less and vice versa. Also in the first case due to use of lesser length of the pipe the head loss would also be less and vice versa. However, in both the cases the total head against which the pumping has to be done will remain the same.

7.6 WATER WASTAGE IN DISTRIBUTION SYSTEM

In a water distribution network system due to a number of unauthorized water connections and various leakages and other issues a substantial amount of water vanishes and is wasted. The amount of losses with respect to total consumption can vary from 15% in an efficient managed system to 40% in such a distribution system having various flaws. This wastage of water increases the production cost incurred in producing per unit of treated water to be supplied to the public. Therefore, a proper management system will help to identify such areas of leakages and unauthorized connections so that a proper action against such issues will rectify the huge wastage of water. This will help to save our valuable resource and also the money that incurs on water treatment.

The various methods that are generally used for detecting water leakage in the pipe joints and in other pipe junctions, are discussed below.

7.6.1 Detection of Leakage in the Distribution Pipes

The water leakage detection methods applied in the underground water mains are (i) direct observations method, (ii) sounding rod method, (iii) plotting hydraulic gradient line method, and (iv) using waste detecting meters method.

7.6.1.1 Direct Observation Method

The presence of wet soft stain on the unpaved ground, or a lavishly grass emerging in a lawn, or the sudden appearance of a water spring at an unusual place provides us an indication of leakage in the underground water pipe somewhere in the surrounding area. These indications can be easily detected at places of the pipes being laid below the loamy or the clayey soils. In case of sandy soils such observations are difficult to be traced out. However, if initial indications of such a type at some places are available or exist the same may be confirmed by using sounding rod method.

7.6.1.2 Sounding Rod Method

The presence of moistness or muddiness at a particular spot is an initial indication of leakage near that point. Therefore, for initial assessment with a sharp pointed metal is driven into the ground along the pipeline and the ear is placed on the top of the inserted rod thus hearing the sound of the escaping water, if leakage is there. In order to confirm this sound in a better way the sound can be magnified by a stethoscope type instrument called an aquaphone or a sonoscape. Such a method is more effective during the night hours as the general atmosphere is calm and still.

7.6.1.3 Plotting the Hydraulic Gradient Line

The pressure at various points in the pipeline is determined and loss of head at different points along the length of the pipe line is calculated. The difference of total head and head loss is the available pressure at that particular point and is represented by hydraulic gradient line. The appearance of any sudden deviation or any drastic change in the slope of the hydraulic gradient line thus indicates the location of a leakage in the pipeline.

7.6.1.4 Using Waste Detection Meters

The meter measures any abnormal high flow passing through a water main during the period of low consumption, such as during night or early hours of morning. However, no unusual consumption occurs during night hours therefore, if such an unnatural excess flow from a portion of the pipe is observed by using this meter it indicates leakage of water in that section of pipe. Therefore, the test is recommended to be conducted at night time.

The method is carried out by closing down the supply of all the area except that particular section selected for the testing of leakage therefore isolating that particular suspected locality. Therefore, after fitting the meter at the head of the main supply pipe the water is then allowed to flow into the area after recording in the meter. Thereafter, the laterals or the branches of the area are shut off gradually and the flow values at each shut-off are recorded each time and in each drop. It is an indication of inconsistent drop in the recorded flow during shut-off point to the leakage in the particular pipe. Thereafter, proper repair and maintenance may be carried out.

Dencon's wastewater detection meters are the widely used for measuring the flow in the pipeline because of their higher sensitivity and accuracy. The meter containing brass disc balanced by a counterweight is forced down by water passing through it and creates movement in it. This movement of the disc can be transferred through a rod to a pencil point attached with graph paper mounted on a drum and moves vertically up and down on the graph paper and moving horizontally with time. The quantity of the water flowing through the meter is automatically gets recorded and traced on the graph paper which can be removed for analysis after the test. This method proves to be quite scientific and is widely used for detecting leakage in the distribution pipes by using such waste detection meters. Although it is a costlier method as it requires high charges of crew operation and the cost on installing additional valves and cutouts for inserting meters in the pipelines.

7.7 DESIGN OF A DISTRIBUTION SYSTEM

The various steps involved in the design of a distribution system are described below.

7.7.1 Survey and Preparation of Contoured Plans and Maps

The strip of land between the treatment plant (water works) and the distribution area is surveyed to obtain rates to repair the main pipeline alignment that will carry treated water into the distribution area. The distribution area (city or town) is also thoroughly surveyed and the area's contoured plan is prepared to identify the distribution areas, distribution or service reservoir locations, pumping stations, etc. Further comprehensive maps of the distribution area are prepared to show the locations of highways, avenues, lanes, shopping areas, industrial areas, parks, gardens, etc. Also planned are the cross sections of highways, avenues, lanes, etc., showing the locations of the existing underground service lines such as electric lines, telephone lines, gas lines, sewage lines, existing water supply lines, if any, etc.

7.7.2 Tentative Layout

The entire distribution area (town or city) is divided into separate distribution areas, and the same are shown on the detailed distribution area map. For each district the population density (i.e., average number of people per hectare) is also marked. The design scheme to be followed is decided and tentative alignment is designated for all mains, sub-mains and branches, as well as for delivery or service reservoir locations, valves, hydrants and other appurtenances.

7.7.3 Discharge in Pipelines

The optimal discharge to be carried by each pipeline is determined based on population density, form of distribution zone (i.e., residential, industrial, etc.) and demand for fire. The size of the distribution pipes is so defined, that at all points the minimum residual pressure is preserved. The pipes are designed for a discharge range between 2.25 and 3 times the average supply volume. For population over 50,000 the distribution mains are designed for a discharge of 2.25 times the average supply rate, whereas the distribution mains are designed for a discharge of 3 times the average supply rate for population under 50,000.

7.7.4 Design of Water Distribution Network

For designing a proper distribution network various factors such as the layout of the road network in the city along with comparison of levels of the different zones of the city needs to be properly analyzed and assessed as the arrangement of the distribution pipes should be decided accordingly. The various important water distribution network such as the dead-end system, the grid-iron system, the circle or ring system, and the radial system have already been discussed. These systems can be adapted in individual or a combination of the two or more, depending on the requirement for a particular town or a city.

7.7.4.1 Computation of Available Residual Pressure Heads

(i) **Elevation head:** This is the sum of flow potential energy described by the elevation in one cross section. This corresponds to Z in the cross section

Pressure head is a quantity of the flow potential energy in one cross section described by the water pressure.

(ii) **Piezometric head:** This is the sum of one cross section of the height and the head of strain.

(iii) **Velocity head:** This is a quantity of kinetic flow energy described by the water velocity in one cross section.

(iv) **Darcy Weisbach equation:** Fundamental equation describing the head loss due to friction for flow in pipes is the Darcy friction for flow in pipes is the Darcy Weisbach equation in Eq. 7.2.

$$h_f = f\frac{L}{D}\left(\frac{V^2}{2g}\right) = f\frac{L8Q^2}{\pi^2 gD^5} \tag{7.2}$$

where, f = Darcy friction factor (dimensionless)
L = Pipe length (L) †
D = Pipe diameter (L)

The friction factor depends on the Reynolds number (Re or NR) and the relative roughness (ε/D), where, ε is the absolute roughness. It therefore depends on the pipe diameter and pipe material.

(v) Pipe flow formula: The factors that control the design of supply conduits are flow resistance, available pressure or head, permissible flow velocities, scour, transport of sediments, water quality and relative cost. These are numerous formulae available for use in calculating flow velocity as shown in Eq. 7.3-Eq. 7.4. Hazen-Williams formula has been adopted in the loop method of designing distribution system.

$$V = 1.318\, C\, R^{0.63}\, S^{0.54} \text{ (English units)} \tag{7.3}$$

$$V = 0.85\, C\, R^{0.63}\, S^{0.54} \text{ (SI Units)} \tag{7.4}$$

where,

Q = Discharge in cubic metre per hour
d = Diameter of pipe in mm
v = Velocity in mps
r = Hydraulic radius in m
S = Slope of hydraulic gradient line
C = Hazen-William coefficient. Hazen-Williams' formula has been used, co-efficient C as shown in Table 7.5.

Table 7.5: Hazen-Williams' co-efficient values for different pipe materials

Pipe material	Recommended C values	
	New pipes	Design purposes
Unlined metallic pipes, cast iron, ductile iron	130	100
Mild steel	140	100
Galvanised iron above 50 mm diameter	120	100
Galvanised iron 50 mm diameter and below used for house service connections	120	55
Centrifugally lined metallic pipes, cast iron Ductile iron and mild steel pipes lined with cement mortar		
Up to 1200 mm diameter	140	140
Above 1200 mm diameter	140	145
Asbestos cement	150	140
PVC, GRP and other plastic pipes	150	145

(*Source*: Central Public Health and Environment Engineering Organization (CPHEEO) 1999; Manual on Water Supply and Treatment, New Delhi)

The loss of head (h_f) through any pipe discharging at the rate of Q can be expressed as in Eq. 7.5

$$h_f = rQ^n \tag{7.5}$$

where r is a proportionality factor; and n is an exponent.

According to Hazen-Williams formula the loss of head h_f, due to friction in a pipe of length L, diameter D, and roughness coefficient C_H, when carrying a discharge Q is given as (Eq. 7.6-Eq. 7.9)

$$h_f = \frac{10.68 L Q^{1.852}}{C_H^{\ 1.852} D^{4.87}} \tag{7.6}$$

Thus by comparison, we obtain

$$r = \frac{10.68 L}{C_H^{\ 1.852} D^{4.87}}; \text{ and } n = 1.852 \tag{7.7}$$

Similarly, according to Darcy-Weisbach formula the loss of head (h_f) is given as;

$$h_f = \frac{f\,LQ^2}{2g(\pi/4)^2 D^5} = \frac{f\,LQ^2}{12.10 D^5} \tag{7.8}$$

Thus, by comparison, we obtain

$$r = \frac{f L}{12.10 D^5} = \text{and } n = 2 \tag{7.9}$$

Where the pipe lengths are large, small losses can be overlooked. However, if the minor losses are high, consideration may be given to them in terms of the head loss due to friction in equal lengths of the pipe. In order to evaluate the pressures and flows available in any section of the system, it is also important to examine the pipe networks of a given distribution system and to recommend ways to build on it if found to be insufficient.

7.7.5 Calculation and Fixing the Sizes of Pipes of a Simple Distribution System

The size of the pipe is decided after finalizing the layout plan of the distribution pipes and the locations of other appurtenances such as valves, bends and hydrants. The size of such a pipe should be such as that the minimum allowable pressure at the time of maximum demand of water should be ensured at the end points of the pipes. The distribution system should be designed on such a terminal pressure value in all the pipes that remains above the designed minimum allowable pressure and also not to exceed the maximum allowable pressure chosen for the design.

The size of the pipes in a water distribution network is computed based on trial and error procedure where size is assumed and terminal pressure is computed and then compared with designed minimum and maximum pressure value. The procedure is repeated till the required size matches with designed values. However, no direct method is available to compute the size of the pipes in a distribution system. The head loss through each pipe is evaluated by computing the terminal pressure in all the pipes. Darcy-Weisbach or Hazen-William formulas are the most commonly used equations for determining the head loss in a pipe when the discharge of the pipe is already provided or known and thereby equating the inflow and outflow at each point of the pipe network.

The total head loss in a pipe usually comprises head loss due to pipe friction with minor head losses due to installed valves, fittings, head loss in pipes at the bends and also with sudden contraction and expansion of the pipes. All these minor losses are considered in order to compute the total residual pressure to be maintained in the system. The frictional head loss (H_L) can be calculated by using the Hazen-Williams formula and is equal to the total head loss in the pipe is given by Eq. 7.10.

$$V = 0.85\ C_H\ R^{0.63}\ S^{0.54} \tag{7.10}$$

where,

$S = H_L/L$ = Slope = Head loss per unit length of the pipe or conduit (m/m)

V = velocity of the flow (m/s)

R = Hydraulic radius = $A/P = D/4$ (for circular conduits)

D = Diameter of the conduit (m)

A = Area of the conduit (m^2)

P = Wetted parameter of the pipe (m) = πD, when flowing at full

C_H = Hazen Williams smoothness constant of pipe

Therefore the equation can be written as in Eq. 7.11:

$$H_L = \frac{1}{0.094}\left(\frac{Q}{C_H}\right)^{1.85} \cdot \frac{L}{d^{4.87}} \tag{7.11}$$

where,

L = Length of the pipe (m)

d = Diameter of the pipe (m)

Q = Flow rate or pipe discharge (m^3/sec)

H_L = Total head loss in the pipe or conduit (m)

C_H = Hazen-Williams smoothness constant of pipe

Assuming the pipe diameter the head loss in each pipe can be computed by using this formula. The diameter of the pipe can be easily finalized by assessing the residual pressure at the terminal points and comparing it with the designed minimum and maximum allowable pressure values. However, instead of formula the Hazen-William's chart as shown in Fig 7.17. is recommended for various trial values of the pipe diameter in designing the suitable network system. This will save the time and reduce the efforts for such bulky calculations. The charts are available at different C_H values and are very helpful in deciding the size and length of the pipes to be used in the water distribution network. The chart as shown in Fig. 7.17 is based on C_H value of 100 used for cast iron pipes and such charts are easily available in the office and can be used easily. The method of using this chart: a straight edge is placed by putting a straight scale on any two known values, such as discharge and diameter, and the other two unknown values lying on the line of scale such as head loss per 1000 m, and velocity can be read out directly. The loss of head is obtained in terms of per 1000 m value, therefore the head loss value should always be multiplied by the length of the pipe in thousand metres and that will be the exact head loss value of the pipe of that particular diameter.

This calculated head loss value in different pipes for used in assessing the terminal pressure at the starting point of the reservoir or the point of pumping. The size of the pipe to be laid in a particular zone can be suitably increased or decreased. If the terminal pressure in any such zone is observed to be less than the minimum permissible or more than the maximum permissible values, and the number of iterations are carried out for the whole system till the required pressures are obtained.

The design velocity to be considered generally depends upon the diameter of the pipe to be used. However, it is advised that to achieve the optimum velocity in the range of 0.9 to 1.8 m/sec the pipe diameter should be in the range of 10 cm to 40 cm or so. Therefore, in order to obtain the optimum velocity at a given discharge condition and with assumed diameter, the pipe diameters should be decided cleverly from the monogram. The optimum velocity values at different pipe diameters are as (a) 0.9 m/s for 10 cm dia pipe, (b) 1.2 m/s for 15 cm dia pipe, (c) 1.5 m/s for 25 cm dia pipe, and (d) 1.8 m/s for 40 cm dia pipe.

However, it is strongly recommended that the distribution mains should be designed for carrying the flow value at maximum hourly draft of the maximum day which is about 2.7 times the average daily demand or for coincident draft with fire, whichever is more. The distribution pipes are designed on flow condition of maximum demand which is 3 times the average daily demand with a minimum size of 10 cm diameter. Also the service pipes laid in the street are designed for twice the average daily flow with a minimum size of 19 mm or $\left(\frac{3''}{4}\right)$ diameter for high pressure system. If the pressure is less the diameter of service pipe can be kept around 13 mm or $\left(\frac{1''}{2}\right)$. Even the distribution pipe of diameter 10 cm is not generally used, because such a pipe can provide enough water for domestic needs up to 1000 persons yet but not suits the minimum needs of single fire hose. Therefore, mostly the distribution pipes of minimum size of 15 to 20 cm are used and recommended.

These simple distribution networks such as dead end systems can be solved easily by calculating the terminal pressure at various points starting from the known pressure point values and then calculating the head losses in the different pipe sections. The

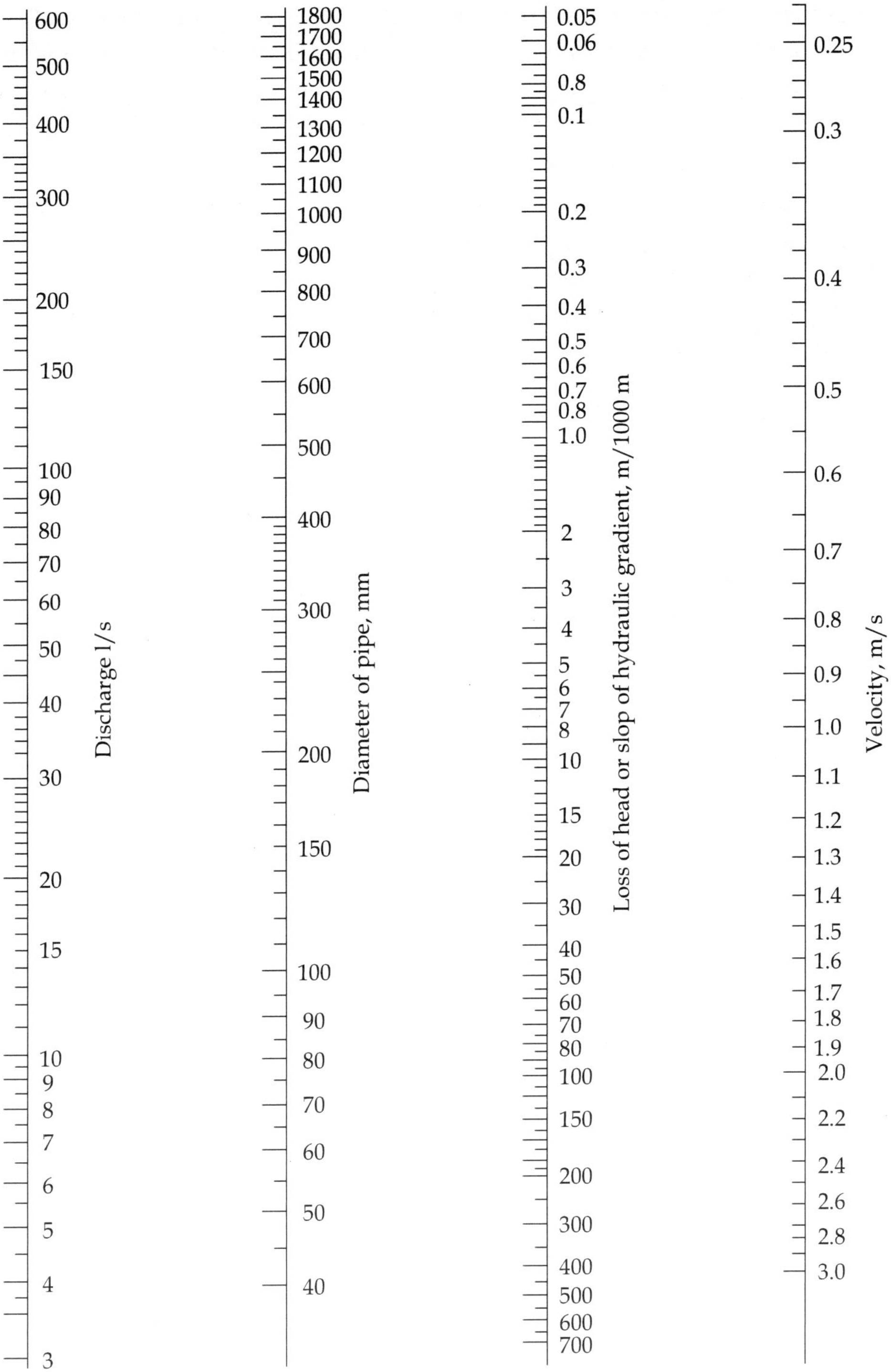

Fig. 7.17: Nomogram

examples pertaining to such a network has been solved and is discussed in the chapter. However, examples associated with complicated networks of water distribution in a city can be solved by using different types of methods such as equivalent pipe method, method of sections and Hardy-Cross method and the same has been explained with numerical examples of each method.

Example 7.2: A dead end type of a distribution pipe network is used to supply water to a certain zone of the city with different blocks having different populations, as shown in Fig. 7.18. The average water supply may be assumed to be 220 litres per capita per day. Design the sizes of the distribution pipes with the following assumptions:

The population in different blocks may be assumed as shown as *P* in Fig. 7.18.

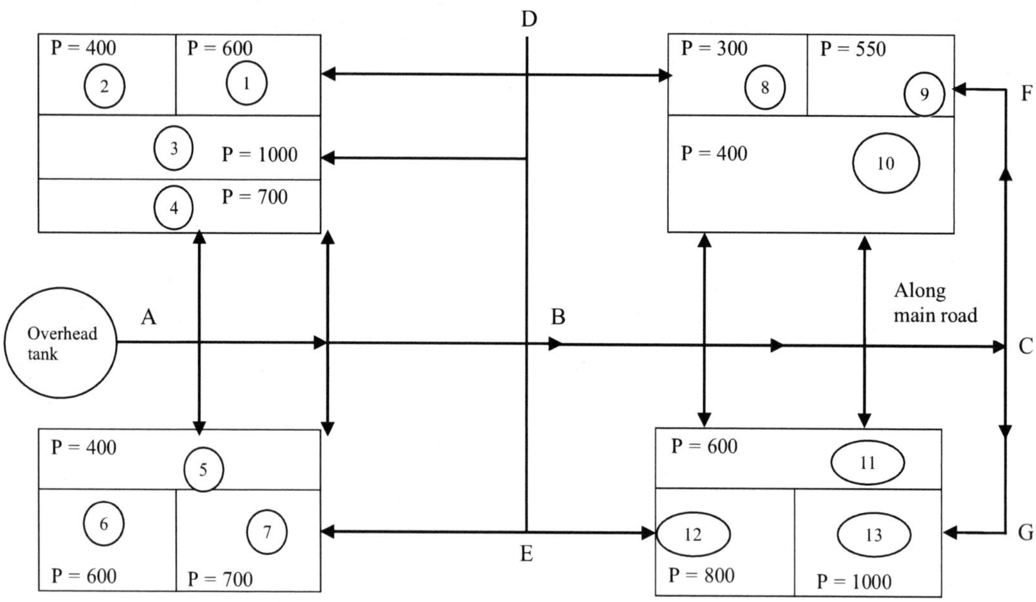

Fig. 7.18

Provided that:

- The R.L. of the bottom of the storage tank may be taken 240.00 m.
- The R.L. of points along main road may be taken as:

  A-225.00 m, B- 210.00 m, C-200.00 m
- The lengths *AB* and *BC* may be taken as 600 m and 450 m, respectively.
- The distribution system may be designed for a maximum demand equal to 2.7 times the average demand.
- The minimum pressure head to be maintained at any point in the distribution system may be assumed as 15 m.

Solution: The size of each pipe needs to be determined in the main pipeline *AC* consisting of two portions as *AB* and *BC*. We have to compute the population served at each point on the line *AC* with portions *AB* and *BC* by calculating backward from the point and going towards the overhead tank at *A*, as shown in Fig. 7.19.

All the calculations are summarized in Table 7.17. The maximum water demand has been considered to be as 2.7 times of the average daily demand and the total flow has been assessed based on the given population of each block with respect to the feeding pipes in column 4 has been calculated as shown in column 5. Thereafter, the diameter of the pipe has been assumed and as shown in column 6 with respect to flow value of column 5, the rate of loss of head per 1000 m is then read out from the Hazen's nomogram. The total head loss in a particular pipe has been calculated by multiplying the length of the pipe with head loss determined from the nomogram as shown in column 9 and is known as actual head loss. The hydraulic levels of different points are then worked out by starting from the last point *A*, after completing the calculations for both the pipes *AB* and *BC*. These hydraulic gradient levels are obtained after subtracting the assessed head loss values from the available head and are compared with the actual ground level values. Thereafter, the available pressure heads are worked out as shown in column (12).

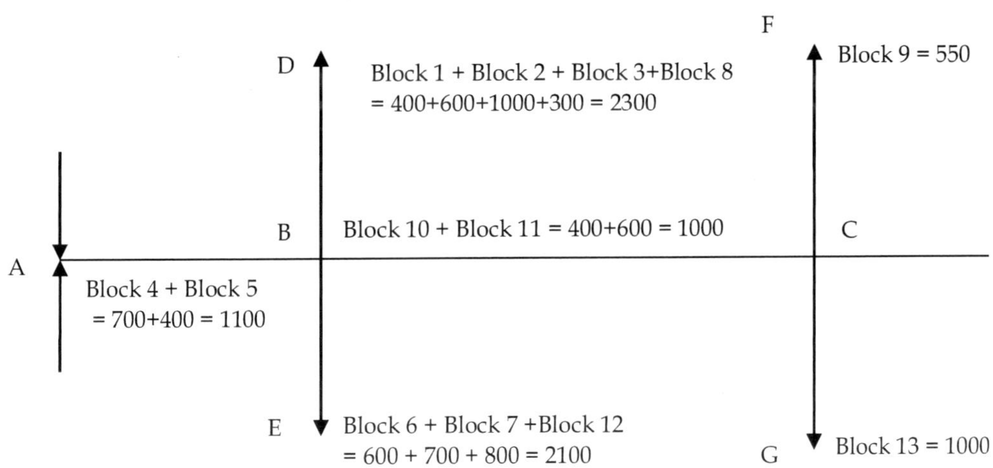

Fig. 7.19

In case for pipe *BC*, the available head at point *C* as calculated by assuming the pipe diameter of *BC* to be 120 mm comes out be 9.85 m which is less than the permissible head of 15 m as given in the problem. Therefore, the assumed diameter of the pipe is increased from 120 mm to 140 mm for the length *BC* and head loss is recalculated in the second trial for the entire network. In the second trial the net available head comes out to be 18.85 m which is greater than 15.0 m as provided in the problem. Therefore, the pipe diameter of 140 mm for pipe *BC* with length 450 m and diameter of 200 mm for pipe *AB* with length 600 m is acceptable and the same can be explained from the Table 7.6.

Table 7.6: Pressure distribution and fixing the size of pipe diameter

First trial

Pipe section	Population served			Max. demand Q_{max} = $\frac{2.7 \times 220 \times P}{24 \times 60 \times 60} \cdot \frac{L}{s}$ = 0.006875 P	Pipe dia assumed (mm)	Head Loss			Hydraulic level worked out from last point A (m)	Ground level (m)	Net head available at the point in Column 10 (m)
	Previous	Local	Total (*P*)			Rate/ 1000 m*	Length (m)	Total head loss = Rate × length			
1	2	3	4	5	6	7	8	9	10	11	12
BC	550 + 1000 =1550	1000	2550	17.53	120	35	450	15.75	C-209.85***	200	9.85 (less than 15.0) change the dia & recalculate
AB	2550 + 2300 + 2100 = 6950	1100	8050	55.34	200	24	600	14.4	B-225.60**	210	15.6 (O.K.)
									A-240.00	225	15.0 (O.K.)
Second trial is needed											
BC	550 +1000 =1550	1000	2550	17.53	140	15	450	6.75	218.85	200	18.85 (O.K.)
AB	2550 + 2300 + 2100 = 6950	1100	8050	55.34	200	24	600	14.4	B-225.60	210	15.6 (O.K.)
									A-240.00	225	15.0 (O.K.)

*Head loss values are calculated from nomogram with calculated Q_{max} and available pipe length values

**240.00 m – 14.4 m = 225.60 m (first trial)

***225.60 – 15.75 m = 209.85 m (first trial) and 225.60 m – 6.75 m = 218.85 m (second trial)

7.7.6 Analysis of Complex Networks

For solving out any complex pipe network system certain assumptions are needed to be fulfilled:

- There should be no discontinuity in pressure and the algebraic sum of the pressure drop around a closed loop must be zero.
- The law of continuity must be fulfilled as the flow entering a junction must be equal to the flow leaving the same junction.

Therefore, by using the methods of successive approximation and on the basis of these two above basic principles, the pipe networks are generally solved easily. The direct analytical solution is not possible, because of many non-linear equations are involved in the calculations and to solve them simultaneously is a tedious task. The various important methods used for solving these complex water distribution networks are given below.

7.7.6.1 Hardy-Cross method

The procedure suggested by Hardy and Cross requires that the flow in each pipe is assumed by the designer (in magnitude as well as direction) in such a way that the principle of continuity is satisfied at each junction (i.e., the inflow at any junction becomes equal to the outflow at that junction).

This approach is to assume a distribution of flow inside the network in such a way that at each junction the principle of consistency is satisfied. A correction to these assumed flows for each pipe loop in the network is then measured successively, until the correction is reduced to an appropriate magnitude. Now, if Q_a is the assumed flow in a pipe and Q be the actual flow in that pipe, then correction will be given by the Eq. 7.12:

$$\Delta = Q - Q_a \text{ therefore, } Q = Q_a + \Delta \tag{7.12}$$

The head loss in the pipe as H_L, can be determined by the formula in Eq. 7.13:

$$H_L = KQ^x \tag{7.13}$$

where, K is a constant depending upon the size of pipe and its conditions, the head loss can also be determined by Hazen-William formula as:

Now, putting the value of $Q = Q_a + \Delta$ in $H_L = KQ^x$, getting Eq. 7.14;

$$H_L = K(\Delta + Q_a)^x$$

$$= K(Q_a^x + x\,Q_a^{x-1}\Delta) \text{ (neglecting the terms of higher power of } \Delta\text{), therefore} \tag{7.14}$$

$$= K(Q_a^x + x\,Q_a^{x-1}\Delta)$$

However, in a closed loop network of pipeline, total head loss must be zero as indicated in Eq. 7.15;

$$\therefore \quad \Sigma K(Q_a^x + x\,Q_a^{x-1}\Delta)) = 0$$

$$\text{or,} \quad \Sigma K\,Q_a^x = -\Sigma K\,x\,Q_a^{x-1}\Delta) \tag{7.15}$$

However, the value of Δ is considered to be same in all the pipes of the selected loop network under consideration, and is a common parameter, therefore

$$\Sigma K\, Q_a^x = -\Delta \Sigma K\, x\, Q_a^{x-1}$$

$$\Delta = \frac{\Sigma K\, Q_a^x}{\Sigma x K Q_a^{x-1}}$$

The Δ in all the pipes of the selected loop of the network has the same sign or direction and denominator of the above derived equation is considered the absolute sum of individual pipes in the summation. (Eq. 7.16-Eq. 7.17). Therefore,

$$\Delta = \frac{\Sigma K\, Q_a^x}{\Sigma \left| x K Q_a^{x-1} \right|} \tag{7.16}$$

$$\boxed{\Delta = -\frac{\Sigma H_L}{x.\Sigma \left| \frac{H_L}{Q_a} \right|}} \tag{7.17}$$

The flow is assumed and the head losses in all the connected pipes of the loop is computed and added algebraically as ΣH_L on the numerator side. After assuming the diameters of the pipes, the magnitude and direction of flow in these pipes is also assumed. Considering the flow pattern the head loss in each pipe is computed with due regard to sign and is considered the head loss to be +ve in clockwise direction and –ve in anticlockwise direction. The actual sum is calculated from the values of $K Q_a^{x-1}$ and $\frac{H_L}{Q_a}$. The assumed flow values in each loop are corrected by putting the obtained Δ values of each loop using Eq. 7.17. With due regard to the sign the common pipe will need both corrections in two loops.

In a network consisting of one or more than one loop, the second correction needs to be applied to the already corrected flow of the first assumed flow values in a loop of the network. It has already been discussed that the common pipe will receive both the corrections in each iteration of the calculations. For better results and precision the procedure can be repeated.

The value of "x" in Hardy Cross method is assumed to be constant with a definite value of 1.85 when using Hazen-Williams formula, and is considered 2 when using Darcy-Weisbach formula. However, the analysis in a given distribution system carried out by dividing it into two or more loops, such that each pipe in the network should at least be included once in each loop. The entire procedure of using Hardy-Cross method to analyze a given flow network system is given below:

- A normal flow distribution is assumed internally and it should be ensured that the sum of the flows entering any junction must be equal to the sum of the flow leaving that junction.

- Using the required equations and diagram compute the head loss in each pipe by assuming that the flow in clockwise flows direction to be positive and will produce a positive head loss and vice versa.
- Considering the flow conditions and with due regard to sign, calculate the total head loss around each circuit by $\Sigma K Q_a^x$ as, x being the constant thus depending on the equation to be used.
- Also considering the flow conditions and with due regard to sign calculate in the same circuit the sum of $x\,K Q_a^{x-1}$.
- The flow in each pipe should be corrected by using Eqs. 10.17 and 10.18 with dual correction in pipes common to the two loops with due regard to sign.

Table 7.7: Hardy-Cross method for flow corrections (first correction)

Pipe	Assumed flow, Q_a (L/s)	K (given)	$H_L = K\,Q_a^2$	$\left\lvert\frac{H_L}{Q_a}\right\rvert$	After first correction Corrected Q as $Q_{a1} = Q_a + \Delta_1$ (L/s)
(1)	(2)	(3)	(4)	(5)	(6)
For loop ABDA (Loop I)					
AB	25	2	1250	50	(+)22.5
BD (common pipe)	25	1	625	25	(+)35*
DA	-25	2	(-) 1250	50	(-)27.5
		Σ	(+) 625	125	
$\Delta_1 = \dfrac{\Sigma H_L}{x.\Sigma\left\lvert\frac{H_L}{Q_a}\right\rvert}$ or $\Delta_1 = -\dfrac{625}{2\times 125} = (-)2.5$					
For loop BCDB (Loop II)					
BC	50	4	(+) 10000	200	37.5
DB (common pipe)	-25	1	(-) 625	25	(-) 35*
CD	-25	3	(-) 1875	75	(-) 37.5
		Σ	(+) 7500	300	
$\Delta_1 = -\dfrac{7500}{2\times 300} = (-)12.5$					
*Correction of flow in common pipe, BD = 25 + Δ_1 - Δ_1'					
*Therefore, flow in pipe, BD = 25 + (-2·5) - (-12·5) = 25 - 2·5 + 12·5 = 35 L/s					

Corrected flow values after first correction as shown in Tables 7.7 and 7.8.

Table 7.8: Corrected flow values after first correction

Pipe	First correction discharge values (L/s)
For loop ABDA	
AB	(+) 22.5
BD (common pipe)	(+) 35
DA	(-) 27.5
For loop BCDB (Loop II)	
BC	37.5
DB (common pipe)	(-) 35
CD	(-) 37.5

Now putting these corrected discharge values the pipe system in both the loops is re-analyzed for discharge values in the second correction as below in Table 7.9.

Table 7.9: Hardy-Cross method for second correction

Pipe	**Assumed flow, Q_{a1} (L/s)**	**K (given)**	**$H_L = K Q_{a1}^2$**	**$\left\|\frac{H_L}{Q_{a1}}\right\|$**	**After first correction Corrected Q as $Q_{a2} = Q_{a1} + \Delta_2$ (L/s)**
(1)	**(2)**	**(3)**	**(4)**	**(5)**	**(6)**
For loop ABDA (Loop I)					
AB	(+) 22.5	2	1012.5	50	(+) 19.8
BD (common pipe)	(+) 35	1	1225.0	25	(+) *32.6**
DA	(-) 27.5	2	(-) 1512.5	50	(-) 30.2
		Σ	(+) 725	125	
		$\Delta_2 = -\frac{725}{2\times 135} = (-)2.7$			
For loop BCDB (Loop II)					
BC	37.5	4	(+) 5625	150	37.2
DB (common pipe)	(-) 35	1	(-) 1225	35	*(-)32.6**
CD	(-) 37.5	3	(-) 4218.75	112.5	(-) 37.8
		Σ	(+) 181.25	297.5	
		$\Delta_2 = -\frac{181.25}{2\times 297.5} = (-)0.3$			

*Correction of flow in common pipe, BD = 35 + Δ_1 - Δ_1'
*Therefore, flow in pipe BD = 35 + (-2.7) - (-0.3) = 25 - 2.5 + 0.3 = 32.6 L/s

In order to check the accuracy in flow values and to balance the flow at each point, the corrected flow obtained in final correction after the second trial is plotted on the pipe network as shown in Fig. 7.20.

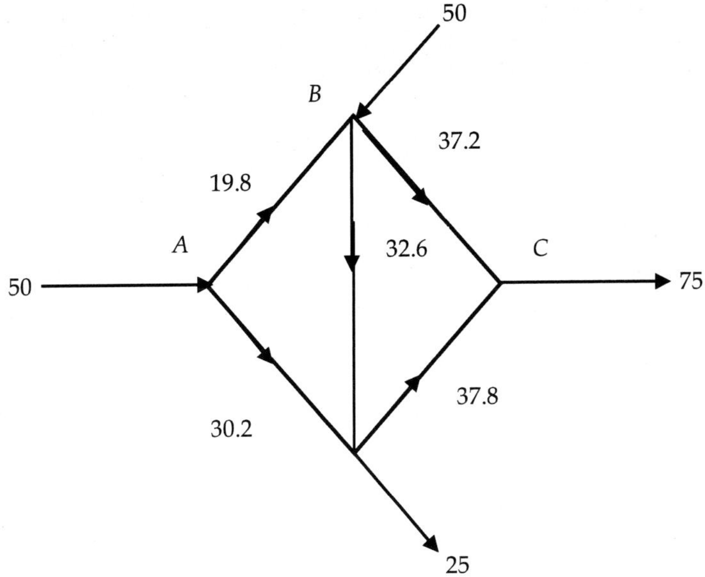

Fig. 7.20

Example 7.3: Calculate the head losses and the corrected flows in the various pipes of a distribution network shown in Fig. 7.21. The diameters and lengths of the pipes used are given against each pipe. Make use of Hardy-Cross method with William-Hazen's formula. Compute the corrected flows after two corrections?

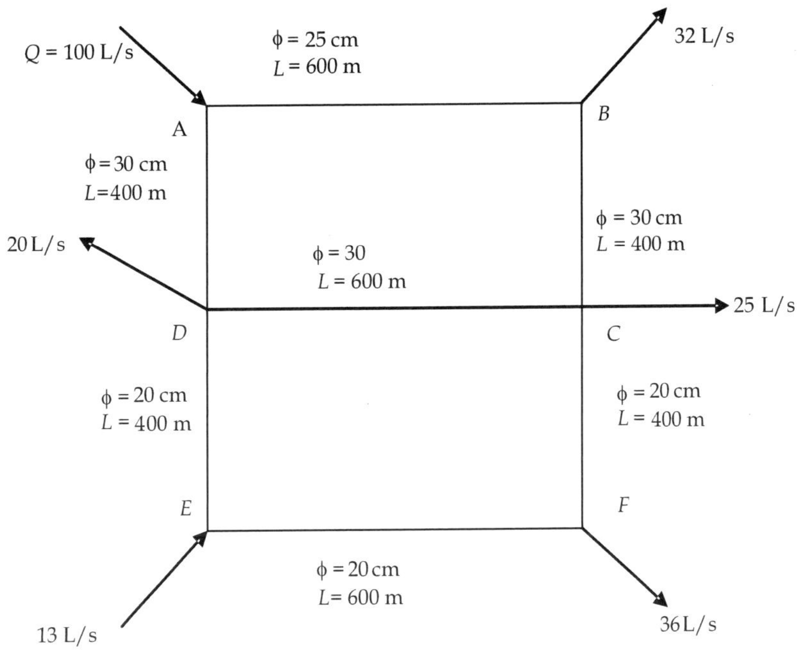

Fig. 7.21

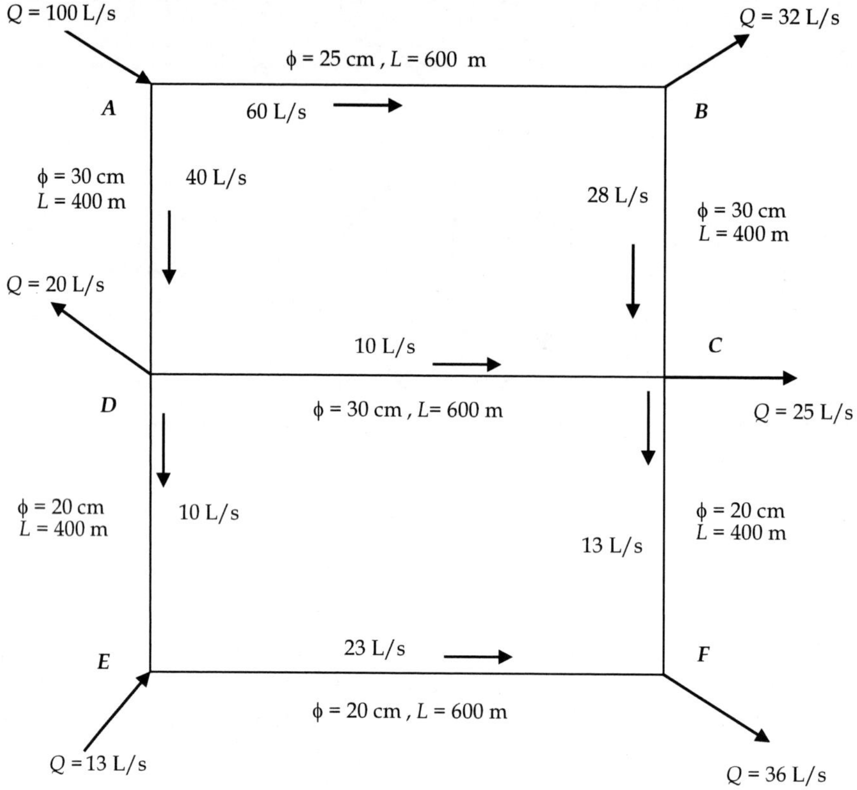

Fig. 7.22

Hazen-Williams equation required to evaluate the head loss (HL) in each can be represented and further by modifying as (Fig. 7.22)

$$H_L = \frac{1}{0.094}\left(\frac{Q}{C_H}\right)^{1.85} \frac{L}{d^{4.87}} \text{ where } C_H = 100 \text{ (assumed)}$$

$$H_L = \frac{1}{0.094}\left(\frac{Q}{100}\right)^{1.85} \cdot \frac{L}{d^{4.87}} = \frac{L}{470} \cdot \frac{Q_a^{1.85}}{d^{4.87}} = K.\, Q_a^{1.85}$$

where, $K = \frac{L}{470\, d^{4.87}}$

The K values can be computed for pipes of different diameters and lengths as shown in Table 7.10 and **Hardy-Cross method for flow corrections (first correction) is shown in Table 7.11, followed by Hardy-Cross method for flow corrections (second correction).**

Table 7.10: *K* values calculations for Hardy cross method

Pipe	Length (m)	Diameter (m)	$\frac{L}{470\,d^{4.87}}$
For Loop ABCD			
AB	600	25	1092
BC	400	30	300
CD	600	30	450*
DA	400	30	300
For Loop DCFE			
DC	600	30	450*
CF	400	20	2182
FE	600	20	3273
ED	400	20	2182
*Common pipe			

Table 7.11: Hardy-Cross method for flow corrections (first correction)

Pipe	Assumed flows		K value calculated as above	$H_L = K\,Q_a^{1.85}$	$\left\lvert\frac{H_L}{Q_{a1}}\right\rvert$	After first correction Corrected Q as $Q_{a1} = Q_a + \Delta_1$ (L/s)
	L/s	m³/s				
1	2	3	4	5	6	7
For Loop ABCD						
AB	60	0.06	1092	(+) 6.0	100	(+) 39.1
BC	28	0.028	300	(+) 0.4	14	(+) 7.1
CD (Common pipe)	(-) 10	0.01	450	(-) 0.09	9	(-) 36.95*
DA	(-) 40	0.04	300	(-) 0.78	20	(-) 60.9
			Σ	(+) 5.53	143	

$$\Delta_1 = -\frac{\Sigma H_L}{x.\Sigma\left|\frac{H_L}{Q_a}\right|} = -\frac{5.53}{1.83\times 143} = (-)0.0209\frac{\text{m}^3}{\text{s}} = (-)20.9\frac{\text{L}}{\text{s}}$$

Pipe	L/s	m³/s	K value	H_L	$\left\lvert\frac{H_L}{Q_{a1}}\right\rvert$	Corrected Q (L/s)
For Loop DCFE						
DC (Common pipe)	(+) 10	0.010	450	(+) 0.09	9	(+) 36.95*
CF	(+) 13	0.013	2182	(+) 0.71	54.62	(+) 19.05
FE	(-) 23	0.023	3273	(-) 3.05	132.61	(-) 16.95
ED	(-) 10	0.010	2182	(-) 0.44	44	(-) 3.95
				(-) 2.69	240.23	

$$\Delta_1' = -\frac{(-)2.69}{1.85\times 240.23} = (+)0.006\frac{\text{m}^3}{\text{s}} = 6.05\frac{\text{L}}{\text{s}}$$

*Correction of flow in common pipe, DC = 10 + Δ_1 - Δ_1'

*Therefore flow in pipe DC = –10 + (–20.9) – (6.05) = –10 – 20.9 – 6.05 = –36.95 L/s

Table 7.12: Second correction in the network using Hardy-Cross method

Pipe	Assumed flows		K value calculated as above	$H_L = K\ Q_a^{1.85}$	$\left\|\frac{H_L}{Q_{a2}}\right\|$	After first correction corrected Q_{a1} as Q_{a2} $= Q_{a1} + \Delta_1$ (L/s)
	L/s	m³/s				
1	2	3	4	5	6	7
For Loop ABCD						
AB	(+) 39.1	0.0391	1092	(+)2.71	69.31	(+) 38.93
BC	(+) 7.1	0.0071	300	(+) 0.03	4.23	(+) 6.93
CD (Common pipe)	(–) 36.95*	0.0369	450	(–) 1.01	27.24	(–) 35.59 *
DA	(–) 60.9	0.0609	300	(–) 1.69	27.75	(–) 60.73
			Σ	(+) 0.040	128.53	

$$\Delta_1 = -\frac{\Sigma H_L}{x.\Sigma\left|\frac{H_L}{Q_a}\right|} = -\frac{0.04}{1.85 \times 128.53} = (-)0.000175\frac{\text{m}^3}{\text{s}} = (-)0.175\frac{\text{L}}{\text{s}}$$

Pipe	L/s	m³/s	K value	H_L	$\left\|\frac{H_L}{Q_{a2}}\right\|$	Corrected (L/s)
For Loop DCFE						
DC (Common pipe)	(+) 36.95*	0.0369	450	(+) 1.01	27.24	(+) 35.59*
CF	(+) 19.05	0.019	2182	(+) 1.43	75.26	(+) 19.05
FE	(–) 16.95	0.0169	3273	(–) 1.72	101.78	(–) 16.95
ED	(–) 3.95	0.0039	2182	(–) 0.08	20.51	(–) 3.95
				(+) 0.64	224.79	

$$\Delta_1' = -\frac{0.64}{1.85 \times 224.79} = -0.00154\frac{\text{m}^3}{\text{s}} = (-)1.54\frac{\text{L}}{\text{s}}$$

*Correction of flow in common pipe, DC = –36.95 + Δ_1 – Δ_1'

*Therefore, flow in pipe, DC = –36.95 + (-0.175) – (–1.54) = –36.95 – 0.175 + 1.54 = -35.59 L/s

Corrected flow in all the pipes after second correction are as shown in Table 7.13 & Fig. 7.23.

Table 7.13: Corrected flow values after second correction

Pipe	AB	AD	DC	DE	EF	BC	CF
Corrected flow (L/s)	38.93	60.73	35.59	3.95	16.95	6.93	19.05

It should be understood that in this approach the head loss between any two points worked out via different routes is found to be the same then the assumed pipe diameters and the discharge will be treated as satisfied. However, the computed head loss value should be equal to or less than the available head loss or otherwise if it is greater, then the assumed diameter values should be changed. The computed head loss is again compared with available head loss and if it matches or below this value then the assumed pipe diameter is considered suitable value and the process is called as adjustment of diameters. Thereafter, using Hardy-Cross correction factor (Δ), the assumed discharge values should be corrected with the obtained Δ values for each pipe in the loop. A number of iterations are carried out till the value of head loss along two different routes are nearly equal, and the Δ values are negligible thus indication of flow balanced conditions in the network system.

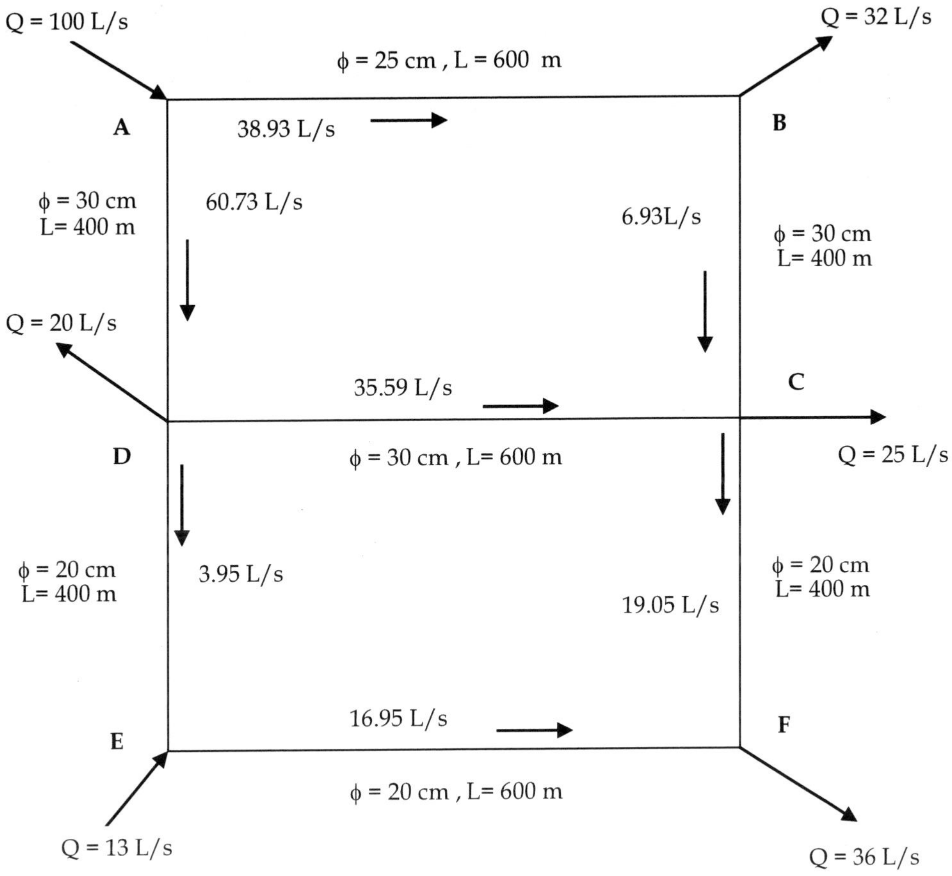

Fig. 7.23: Flow in each pipe after final corrections

7.7.6.2 Equivalent Pipe Method

This technique is often used as an aid in solving large pipe networks, through which it is useful to; first of all, substitute the different small loops with single equivalent pipes with the same head loss. In this method a complex network of pipes is replaced by an equivalent pipe system giving the same discharge with same loss of head as in the complex network, of pipes. The method of pipe equivalent is used by substituting a single equivalent pipe with the available circuit to be analysed in a network and is based on the following two principles of hydraulics:

(i) The head loss is additive under given flow conditions through the pipes connected in series.

(ii) The head loss through each pipe will cause equal head loss under the flow conditions through pipes connected in parallel.

For pipes in series the discharge is same in each pipe and the total head loss is equal to the amount of the head loss in each pipe. On the other side, the discharge through each pipe is different for the pipes in parallel, and is so distributed that the loss of head through each pipe is the same. The length and diameter of an equivalent pipe required to replace a system of pipes in series or in parallel may be determined as indicated below.

Pipes in series: According to Hazen-Williams formula the loss of head due to friction in a pipe of length L, diameter D, and roughness coefficient C_H when carrying discharge Q is given as in Eq. 7.18.

$$h_f = \frac{10.68 L Q^{1.852}}{C_H^{\ 1.852} D^{4.87}} \tag{7.18}$$

Thus the total head loss due to friction in pipes AB and BD which are in series as shown in Eq. 7.19 and Eq. 7.20

$$h_f = (h_f)_{AB} + (h_f)_{BD} \tag{7.19}$$

or,

$$h_f = \frac{10.68 L_1 Q_1^{\ 1.852}}{C_{H_1}^{\ 1.852} D^{4.87}} + \frac{10.68 L_2 Q_2^{\ 1.852}}{C_{H_2}^{\ 1.852} D^{4.87}} \tag{7.20}$$

where, C_{H_1} and C_{H_2} are the roughness coefficients for pipes AB and BD respectively.

If pipes AB and BD can be replaced by an equivalent pipe of length $L_{E,}$ diameter D_E, roughness coefficient C_H and carrying the same discharge Q_1 then the loss of head due to friction in the equivalent pipe is given by Eq. 7.21;

$$h_f = \frac{10.68\ L_g Q_1^{\ 1.852}}{C_H^{\ 1.852} D_E^{\ 4.87}} \tag{7.21}$$

Since the head loss due to friction is same in both cases in Eq. 7.20 and 7.21. we have

$$\frac{10.68\ L_E Q_1^{\ 1.852}}{C_H^{\ 1.852} D_E^{\ 4.87}} = \frac{10.68\ L_1 Q_1^{\ 1.852}}{C_{H_1}^{\ 1.852} D^{4.87}} + \frac{10.68\ L_2 Q_2^{\ 1.852}}{C_{H_2}^{\ 1.852} D^{4.87}}$$

$$\text{or,} \quad \frac{L_E}{C_H^{\ 1.852} D_E^{\ 4.87}} = \frac{L_1}{C_{H_1}^{\ 1.852} D^{4.87}} + \frac{L_2}{C_{H_2}^{\ 1.852} D^{4.87}}$$

Assuming that, $C_{H_1} = C_{H_2} = C_H$, we have,

$$\frac{L_E}{D_E^{\ 4.87}} = \frac{L_1}{D^{4.87}} + \frac{L_2}{D^{4.87}}$$

Solving for L_E, we get Eq. 7.22

$$L_E = D_E^{4.87}\left[\frac{L_1}{D^{4.87}} + \frac{L_2}{D^{4.87}}\right] \tag{7.22}$$

If there are n no. of pipes in series to be replaced by an equivalent pipe, then we have Eq. 7.23;

$$L_E = D_E^{4.87}\left[\sum_{i=1}^{i=n} \frac{L_i}{D_i^{4.87}}\right] \tag{7.23}$$

If instead of Hazen-Williams formula, Darcy-Weisbach equation is used, then Eq. 7.24 will be obtained:

$$L_E = D_E^{5}\left[\sum_{i=1}^{i=n} \frac{L_i}{D_i^{5}}\right] \tag{7.24}$$

From Eq. 7.31 or 7.32, the length L_E of equivalent pipe can be determined if its diameter D_E is given or specified, and its diameter D_E can be determined if its length L_E is given or specified. The length L_E of equivalent pipe may be taken equal to the sum of the lengths of all the pipes in series, i.e., $L_E = L_1 + L_2 + \ldots.. + L_n$, and its diameter D_E may be determined.

However, the diameter (D_E) determined for any given or specified length (L_E) will usually be a non-standard diameter. On the other hand if for a given or specified standard diameter D_E, the length L_E of equivalent pipe is determined, it will usually be different from the sum of the lengths of all the pipes in series.

Pipes in parallel: For pipes *ABD* and *ACD* which are in parallel as shown in Fig. 7.24, we get Eq. 7.25

$(h_f)_{ABD} = (h_f)_{ACD}$

or, $(h_f)_{AB} + (h_f)_{BD} = (h_f)_{AC} + (h_f)_{CD}$ (7.25)

Using Hazen-Williams formula, we obtain Eq. 7.26;

$$\frac{10.68\, L_1Q_1^{1.852}}{C_{H_1}^{1.852}D_1^{4.87}} + \frac{10.68\, L_2Q_2^{1.852}}{C_{H_2}^{1.852}D_2^{4.87}} = \frac{10.68\, L_3Q_3^{1.852}}{C_{H_3}^{1.852}D_3^{4.87}} + \frac{10.68\, L_4Q_4^{1.852}}{C_{H_4}^{1.852}D_4^{4.87}}$$

$$\text{or, } \frac{L_1Q_1^{1.852}}{C_{H_1}^{1.852}D_1^{4.87}} + \frac{L_2Q_2^{1.852}}{C_{H_2}^{1.852}D_2^{4.87}} = \frac{L_3Q_3^{1.852}}{C_{H_3}^{1.852}D_3^{4.87}} + \frac{L_4Q_4^{1.852}}{C_{H_4}^{1.852}D_4^{4.87}} \tag{7.26}$$

where, C_{H_1}, C_{H_2}, C_{H_3} and C_{H_4} are the roughness coefficients for pipes *AB*, *BD*, *AC* and *CD*, respectively.

Also assume, $C_{H_1} = C_{H_2} = C_{H_3} = C_{H_4}$, (Eq. 7.27)

we have;

$$\frac{L_1Q_1^{1.852}}{D^{4.87}} + \frac{L_2Q_2^{1.852}}{D_2^{4.87}} = \frac{L_3Q_2^{1.852}}{D_3^{4.87}} + \frac{L_4Q_2^{1.852}}{D_4^{4.87}}$$

$$\text{or, } \frac{Q_1}{Q_2} = \left[\frac{\dfrac{L_3}{D_3^{4.87}} + \dfrac{L_4}{D_4^{4.87}}}{\dfrac{L_1}{D^{4.87}} + \dfrac{L_2}{D_2^{4.87}}} \right]^{\frac{1}{1.852}} \tag{7.27}$$

$Q_1 = kQ_2$, we get Eq. 7.28;

$$\text{where } k = \left[\frac{\dfrac{L_3}{D_3^{4.87}} + \dfrac{L_4}{D_4^{4.87}}}{\dfrac{L_1}{D^{4.87}} + \dfrac{L_2}{D_2^{4.87}}} \right]^{\frac{1}{1.852}} \tag{7.28}$$

Further, if Q is the total discharge, then

$$Q = Q_1 + Q_2 = Q_2\,(k + 1)$$

Again if pipes ABD and ADC can be replaced by an equivalent pipe of length L_E, diameter D_E, roughness coefficient C_H, and carrying the total discharge Q, then the loss of head due to friction in the equivalent pipe is given by

Assuming, $C_{H_1} = C_{H_2} = C_H$, we have

$$\frac{L_E Q^{1.852}}{D_E^{4.87}} = \frac{L_1 Q_1^{1.852}}{D_1^{4.87}} + \frac{L_2 Q_1^{1.852}}{D_2^{4.87}}$$

Solving for L_E, we get Eq. 7.29

$$L_E = \frac{D_E^{4.87}}{Q^{1.852}} \left[\frac{L_1 Q_1^{1.852}}{D_1^{4.87}} + \frac{L_2 Q_1^{1.852}}{D_2^{4.87}} \right]$$

or,

$$L_E = \frac{D_E^{4.87}}{(k+1)^{1.852}} \left[\frac{L_3}{D_1^{4.87}} + \frac{L_4}{D_2^{4.87}} \right] \tag{7.29}$$

If instead of Hazen-Williams formula, Darcy-Weisbach formula is used then the following equation relating length L_E and diameter D_E of equivalent pipe is obtained as in Eq. 7.30.

$$L_E = \frac{D_E^{5}}{(k+1)^{2}} \left[\frac{L_3}{D_3^{5}} + \frac{L_4}{D_4^{5}} \right] \tag{7.30}$$

The method can be described in a better way with suitable example as:

Let us assume a pipe circuit $ABCD$ with different pipe diameters and lengths as shown in Fig. 7.24. For the required analysis the entire network of pipes can be carried out by splitting the circuit into two portions or routes as route ABC and route ADC and splitting the discharge accordingly.

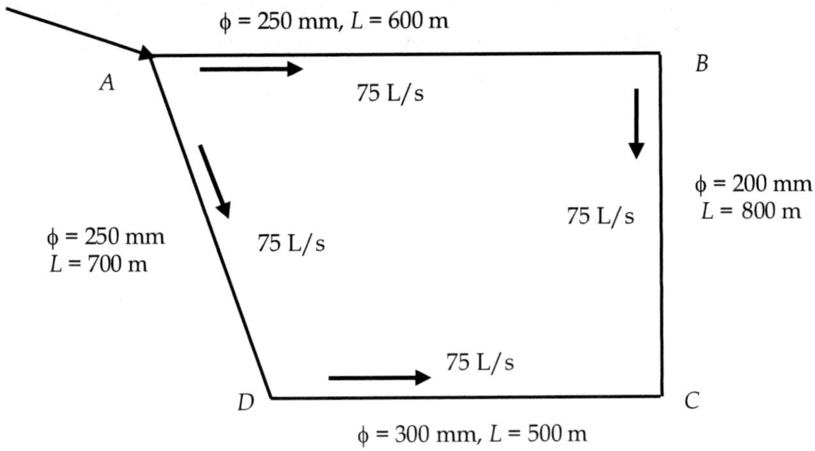

Fig. 7.24

The flow rate of 75 L/s is assumed in the route *ABC* and *ADC*, the head loss in both the circuits can be assessed by using William- Hazen's nomogramas:

$$\text{Head loss for circuit } ABC = \frac{14}{1000} \times 600 + \frac{40}{1000} \times 800 = 8.4 + 32 = 40.40$$

$$\text{Head loss for circuit } ADC = \frac{14}{1000} \times 700 + \frac{5.5}{1000} \times 500 = 9.8 + 2.75 = 12.55$$

It is assumed that the route *ADC* and *ABC* can be replaced by 250 mm diameter pipe in each route then the length of pipe in each route can be obtained by considering that the head loss equivalent of 14 m per 1000 m length in the equivalent 250 mm diameter pipe as:

Hence, replacing the entire route *ABC* with 250 mm diameter pipe, the length of that pipe will be

$$= \frac{40.40}{14} \times 1000 = 2886 \text{ m}$$

Similarly, replacing the entire route *ADC* with 250 mm diameter pipe, the length of that pipe will be

$$= \frac{12.55}{14} \times 1000 = 896 \text{ m}$$

When the flow values are properly divided the head loss in both the routes should be overall equal. Therefore, head loss with respect to the head loss value of 40.40 of route *ABC* with route *ADC* length 896 m and diameter 250 mm can be calculated as

$$= \frac{40.40}{896} \times 1000 = 45.09 \text{ m}$$

Therefore, with equivalent pipe diameter of 250 mm and head loss of 45.09 m the discharge assessed from the nomogram is 120 L/s. Therefore, the total discharge carried by the pipe route *ABC* and *ADC* is 75 + 120 = 195 L/s.

Therefore, the equivalent pipe with 250 mm diameter carrying a discharge of 195 L/s, the head loss as calculated from nomogram is 70 m per 1000 m of length. However, the head loss of 40.40 m should be ensured, therefore the length of pipe required is

$$= \frac{1000}{70} \times 40.40 = 577 \text{ m} \qquad \text{Ans.}$$

Therefore the total circuit *ABCD* can now be replaced with equivalent pipe of diameter 250 mm and length of 577 m by maintaining the same flow as required in the network.

EXERCISE

1. Calculate the head losses and the corrected flows in the various pipes of a distribution network shown in figure below. The diameters and lengths of the pipes used are given against each pipe. Make use of Hardy-Cross method with William Hazen's formula. Compute the corrected flows after two corrections?

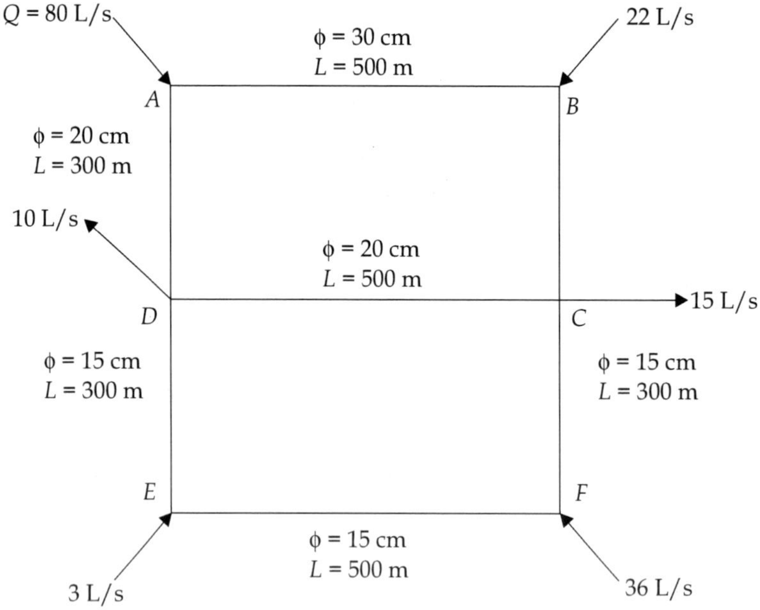

2. Illustrate with sketches the different types of layouts of pipe systems in distributing water with their merits and demerits?

REFERENCES

1. James, L.G. (1988). *Principles of Farm Irrigation System Design*. John Wiley & Sons, New York.
2. Modi, P.N. and Seth S.M. (1998). *Hydraulics and Fluid Mechanics*. Twelfth Edition, Standard Book House, New Delhi.
3. Central Public Health and Environment Engineering Organisation (CPHEEO) 1999; *Manual on Water Supply and Treatment*, New Delhi).

8

CHAPTER

Water Transmission Pipes, Pumps and Pumping Stations

8.1 WATER TRANSMISSION USING PIPES

The water supply system consisting of intake of water from various types of sources and thereafter transportation of this water to the water treatment plant in the city or town along with post-treatment distribution to the end users requires various sizes and types of pipe and piping networks. In case of surface sources such as rivers or streams the intake structures are constructed in order to pump this water to the water treatment plants. Thereafter, this water is transported in storage tanks and is distributed to the consumers in the city through pipe network system only. When water is transmitted from nearby places like river Yamuna in Delhi the length of pipes required is less that results in low cost and maintenance required for the water being transmitted. However, when water is transported from far distances the length of pipes makes the system more costly. The underground water sources such as wells and tube wells are generally economical as the facilities are generally established within the city itself that makes the transportation of water easy.

However in case of water to be transported in cities located on higher elevation, the transportation of water is carried out by using pumps and pressure conduits. The water to be taken from rivers requires efforts while constructing the intake structures and wells near the waterbody and then pumping and transmitting it to the desired places for treatment and distribution to consumers. When the source of water is available at higher elevations the transportation of water to the public proves to be easy at it does not require any pumping and the water can be conveyed through gravity flow to the required places. The example of such a system is available in Mumbai where the water is extracted and conveyed from lakes through gravity without any pumping system.

Types of pipes

(i) Gravity pipes

(ii) Pressure pipes

8.2 GRAVITY PIPES

The conduits in which the flow is to be conveyed by using the natural gravity system are known as gravity pipes. As shown through Fig. 8.1 the hydraulic gradient line coincides with water surface with respect to the available heads at various points and is almost parallel to the bed of the conduit. This is due to the fact that the flow takes place by gravitational force and all along the atmospheric pressure and therefore follows the natural slope available on the ground. The pipes cannot be laid on undulations on valleys and hills as the existing topography does not support the gravitational flow easily. These pipes or conduits are to be laid above the normal surface level on trestles when to be laid on valleys and depressions. However, when the pipes are to be taken through canals or tunnels the same may be laid below the normal supply levels. This is because when the conduits are to be laid by using the alternative routes in such types of areas requires zigzag patterns following the roads and highways requires enormous length of pipes and proves to be costly.

The different types of water conveyance conduits working under gravitational force are discussed in brief below:

(i) Canals: The canals, open channels used for irrigation purpose, are constructed artificially to transport the water by cutting the ground in deep and constructing banks on lower ground level. These canals are generally constructed on various parameters such as available slopes, nature of the ground and soil, velocity of the water to be conveyed, quality of water and various losses that take place during transportation of water through these conduits. When the quality of water matters such as for public water supplies these canals are provided linings.

(ii) Flumes: When the water is transported through open channels or canals over the trestles it is known as flumes. The system can be used easily to convey the water anywhere across the valleys, over the other drains or over any such obstructions on their way. They can be constructed in rectangular or circular cross sections and can be made of wood, metals, R.C.C. or masonry.

(iii) Aqueducts: Aqueducts are closed rectangular or circular or horse-shoe shaped closed channels made of R.C.C. or masonry and are generally constructed across the roads, railways and highways with gradual slopes. These aqueducts are covered but water is allowed to flow through mere action of gravity only and therefore are designed on $\frac{1}{2}$ or $\frac{3}{4}$th full capacity. These are constructed as closed system but still the water is prone to pollution and contamination through cracks and leaky joints in them. The circular section proves to be most efficient in carrying out water with maximum flow conditions as from hydraulic point of view they are designed with maximum hydraulic radius (R) that covers the maximum area per unit of wetted perimeter. Although the rectangular sections are hydraulically inefficient and require more construction material but due to higher stability are mostly used.

8.3 PRESSURE PIPES

The pressure conduits are closed pipes carrying water at a pressure higher than the atmospheric pressure with no air entering in the system. As shown in Fig. 8.1 the hydraulic gradient line is generally determined by installing the piezometers at different sections of the conduit. The bed of these pressure conduits does not depend on the hydraulic gradient line and therefore velocity is not governed by the hydraulic gradient line. The important feature of these pressure conduits is that they can carry water at any desired places even with lesser length requirements as they can freely be laid on undulations on the hilly areas, beneath the valleys, sags and can be effectively used even above the hydraulic gradient lines also. These pressure pipes can also be used for shorter routes as compared to flumes and canals and therefore prove to be economical one. However, during rising of water the negative pressure develops and therefore a limiting value of negative pressure development up to 8.3 m is generally permitted. The main benefit of using these pressure pipes is that being a closed system the water is not exposed to atmosphere and therefore rarely gets polluted. Therefore, in the water supply system the pressure conduits are generally preferred while in conveyance of sewage or irrigation water the gravitational flow system is preferred.

8.4 HYDRAULIC FLOW AND DESIGN OF PIPES

The parameters such as pressure, flow rate, velocity and hydraulic gradient line are important parameters that generally govern the water supply through pressure conduits. From economical consideration the hydraulic gradient line and pressure line should be close to each other and should also be straight when pipes are buried in depth. This may be due to the fact that the higher the depth of pipe the higher will be the pressure and therefore this makes necessary that the pipe material should be strong enough to resist the pressure of the conduit. Due to high strength requirement more material is needed thus making the overall facility to be costlier. Therefore, considering the economy and hydraulics these pipes should be optimized in terms of size, shape and material.

From the flow hydraulics it is generally recommend that the flow velocities should neither be too large as it may result in high pressure loss nor it should be too small so as to require large size diameter pipes. The velocity should also be non-silting to prevent the frequent clogging in the pipe and non-scouring so that the pipe material may not be damaged or disintegrated. The normal velocity through pressure pipes is recommended in the range of 0.9 m/sec to 1.5 m/sec. The maximum velocity is restricted in the range of 3-6 m/sec as it depends on the pipe material and size of the available pipes in market. The design of the pipelines in water distribution system is based on the principle that available pressure head in between the source and the city consumers is lost in overcoming the frictional resistance offered by the flow inside that pipe. This head loss is represented by H_L and is generally computed by using various

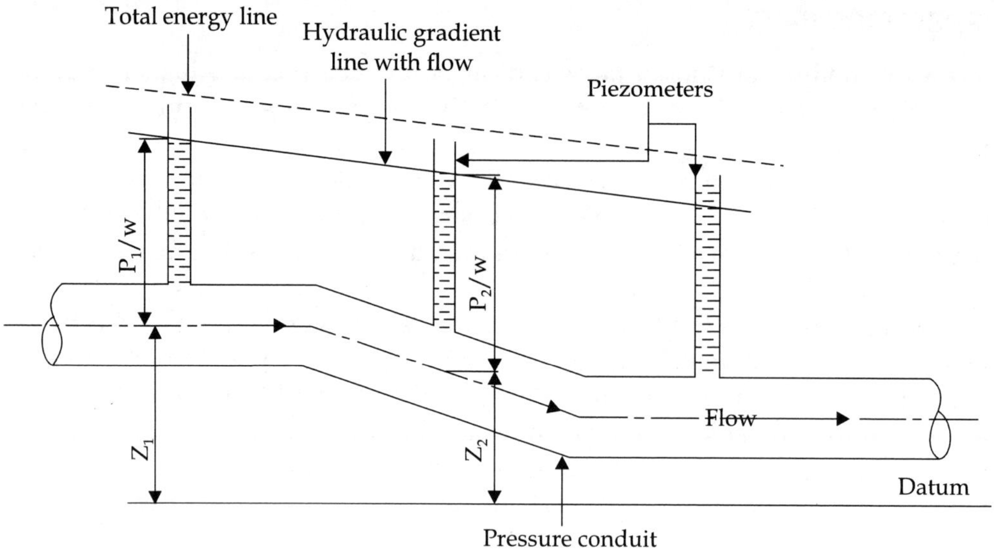

Fig. 8.1: Hydraulic gradient line and flow representation in pressure conduits

mathematical equations given by various researchers. The minor losses such as loss due to sudden contraction and expansion of the pipes, bends, valves and other fittings are represented as H_{Lm} and are also considered while designing any water distribution system. In case of pipes conveying water at greater distances such minor losses may be neglected. The frictional head loss in a pipe can be calculated by using the following equations:

(i) Darcy-Weisbach formula
(ii) Manning's formula
(iii) Hazen-Williams formula
(iv) Modified Hazen-Williams formula

The Hazen-Williams formula is mostly used in assessing the head loss in pressure type pipes as compared to others. The above-mentioned formulae are explained in details below.

Darcy-Weisbach formula

It is almost more than 100 years old and can be explained as in Eq. 8.1.

$$H_L = \frac{f' L \cdot V^2}{2gd} \tag{8.1}$$

where,

H_L = Head loss (m)
V = Velocity through pipe (m/sec)
g = Acceleration due to gravity
f' = Friction factor depends on Reynolds number (R_e) and relative roughness (δ) of pipes. The δ can be calculated as $\frac{2e}{d}$,

where,

e = Absolute roughness of inside surface of the pipe having diameter d. Therefore, f values are given as:

= 0.02 (for new and smooth pipes)

= 0.075 (for old rough pipes)

d = Diameter of pipe (m)

The f' value can be calculated as per Eqs. 8.2-8.3.

$$f' = 0.04\left[1+\frac{1}{35d}\right] \text{ for old pipes} \tag{8.2}$$

$$f' = 0.02\left[1+\frac{1}{35d}\right] \text{ for new pipes} \tag{8.3}$$

where, d = Diameter of the pipe (m)

The exact value of f' depends on R_e and δ and can be determined by using the formula as shown in Eq. 8.4:

$$f' = \alpha + \frac{b}{R_e^m} \tag{8.4}$$

where, α, b and m are constants that depend on δ values.

(i) For laminar flow conditions with R_e values = 2000 , the f' can be computed as per Eq. 8.5;

$$f' = \frac{64}{R_e} \tag{8.5}$$

(ii) When the Reynolds number value (R_e) is equal to or greater than 4000 the turbulent flow conditions exists and f' value can be calculated by using the Nikuradse equation as per Eqs. 8.6-8.7:

$$\left[\frac{1}{\sqrt{f'}} = 2\log_{10} R_e\sqrt{f'} - 0.8\right] \text{ for new and smooth pipes} \tag{8.6}$$

$$\left[\frac{1}{\sqrt{f'}} = 2\log_{10} R_e\frac{2}{2.e} - 0.8\right] \text{ for old and rough pipes} \tag{8.7}$$

For smooth pipes and the Reynolds number (R_e) varying between 20,000 and 20,00,000 the f' can be calculated by using the formula given by Slitcher (Eq. 8.8).

$$f' = 0.005 + \frac{0.396}{R_e^{0.3}} \tag{8.8}$$

The Nikuradse has given another equation for Reynolds number varying between 20,000 to 32,40,000 as per Eq. 8.9,

$$f' = 0.0032 + \frac{0.221}{R_e^{0.237}} \tag{8.9}$$

The above-mentioned equations can be utilized for the turbulent flow conditions in pipes. However, Cole-Brook and White suggested equation under critical flow conditions with Reynolds number, (R_e) lying in between 2000 and 4000 value and recommended to make use of R_e and in Eq. 8.10 as:

$$\frac{1}{\sqrt{f'}} = 1.74 - 2\log_{10}\left[\frac{2e}{d} + \frac{18.7}{R_e\sqrt{f'}}\right] \tag{8.10}$$

The 2*e* values for different types of pipes is summarized in Table 8.1. Dary-Weisbach equation is not generally very much preferred by the engineers working in area of water supply systems. This is because of the fact that the equation is too old and the uncertain values of $\frac{2e}{d}$ and f' obtained while solving the water transportation problems of water supply and its distribution. A trial and error method is required for solving the *e*, *d* and f' values makes it more complicated.

Table 8.1: Absolute roughness values of pipes made with different materials

S. No.	Pipe material	2 e values (mm)
1.	Glass, copper, lead, asbestos	0.0015
2.	Steel or wrought iron	0.0458
3.	Asphalted cast iron	0.122
4.	Galvanized iron	0.152
5.	Cast iron	0.30
6.	Concrete	0.30-3.0
7.	Riveted steel	0.9-9.0

Manning's Formula

The formula is mostly used for solving problems of gravity conduits but is also applicable for pressure pipes with turbulent flow conditions and provides better results as the roughness coefficient (n) value can be estimated precisely. The Manning's equation is described in Eq. 8.11;

$$Q = \frac{1}{n} A.R^{\frac{2}{3}}\sqrt{S_0} \tag{8.11}$$

where,

Q = Flow rate or discharge (m^3/sec)

V = Flow through velocity (m/sec)

n = Roughness coefficient

R = Hydraulic radius or hydraulic mean depth

$$= \frac{\text{Area}}{\text{Wetted perimeter}} = \frac{\frac{\pi}{4}d^2}{\pi d} = \frac{d}{4} \text{ (for circular closed conduits)}$$

d = Diameter of pipe (m)

S_0 = Slope of energy line = $\frac{H_L}{L}$ = Head loss per unit length of pipe

The above equation therefore can be written as Eq. 8.12.

$$V = \frac{1}{n} \cdot R^{\frac{2}{3}} \sqrt{\frac{H_L}{L}}$$

Therefore,

$$H_L = \frac{n^2 \cdot V^2 \cdot L}{R^{\frac{4}{3}}} \tag{8.12}$$

Hazen-Williams Formula is illustrated in Eq. 8.13;

$$V = 0.85\, C\, R^{0.63}\, S^{0.54} \tag{8.13}$$

where,

V = Flow through velocity (m/sec)

C_H = Hazen-Williams coefficient also known as coefficient of hydraulic capacity

R = Hydraulic radius or hydraulic mean depth

$$= \frac{\text{Area}}{\text{Wetted perimeter}} = \frac{\frac{\pi}{4} d^2}{\pi d} = \frac{d}{4} \text{ (for circular closed conduits)}$$

d = Diameter of pipe (m)

S = Slope of energy line = $\frac{H_L}{L}$ = Head loss per unit length of pipe

The Hazen-Williams coefficient (C) values vary with the type of pipe and its material and is summarized in Table 8.2 and Table 8.3 shows recommended CR values in modified Hazen-Williams formula at 20°C.

Table 8.2: Hazen-Williams coefficient values for different pipe materials

Pipe material	Recommended C values	
	New pipes	Design purposes
Unlined metallic pipes, cast iron, ductile iron	130	100
Mild steel	140	100
Galvanized iron above 50 mm diameter	120	100
Galvanized iron 50 mm diameter and less used for house service connections	120	55
Centrifugally lined metallic pipes, cast iron, ductile iron and mild steel pipes lined with cement mortar		
Up to 1200 mm diameter	140	140
Above 1200 mm diameter	140	145
Asbestos cement	150	140
PVC,GRP and other plastic pipes	150	145

(*Source*: Central Public Health and Environment Engineering Organization (CPHEEO) 1999; *Manual on Water Supply and Treatment*, New Delhi)

Modified Hazen-Williams Formula

The Hazen-Williams formula suffers the following limitations:

(i) The Hazen-Williams coefficient (C_H) is has the units $L^{-0.37}T^{-1}$ and is not dimensionless, any change in units can lead to the change in overall values is a major drawback of the equation.

(ii) The pipe equations should be used in all types of pipes and their sizes, but the present equation having numerical constant value of 0.85 is calculated based on the hydraulic radius of 0.3 m and slope of 1 in 1000. This makes the equation applicable on very narrow range. However, when the equation is used in normal practice it can lead to ± 30% error in velocity and around ± 55% error in head loss values thus makes it practically not feasible.

(iii) The head loss in pressure pipes is caused due to friction that, in turn, depends on the pipe roughness and Reynolds number values. But in the present Hazen-Williams equation the pipe roughness and Reynolds number is not considered other than velocity, diameter of the pipe and slope value. An error of around ± 45% has been depicted in the literature when the flow estimation using Hazen-Williams equation is compared by Darcy-Weisbach equation. It has been reported that in Hazen-Williams the flow in pipes at higher C_H values with large diameter pipes the errors are less as compared to lower C_H values with low diameter pipes.

Considering the above-mentioned facts in view and limitations, the Hazen Williams formula has been modified by using the Darcy-Weisbach equation and Cole-Brook-White equation. The modified Hazen-Williams equation thus eliminates the limitations and can be written as in Eq. 8.14.

$$V = \frac{3.83 C_R [d^{0.6675} (g \cdot S)^{0.5525}]}{\vartheta^{0.105}} \tag{8.14}$$

or

$$V = 143.534\ C_R \cdot R^{0.6575}\ S^{0.5525}$$

where,

C_R = Roughness coefficient (a dimensionless constant)

d = Pipe diameter (m)

g = Acceleration due to gravity (m/sec^2)

S = Slope of energy line $= \frac{H_L}{L}$ = head loss per unit length of pipe

ν = Kinematic viscosity of the water (m^2/sec) at 20°C the value is 10^{-6}

R = Hydraulic radius or hydraulic mean depth

$$= \frac{\text{Area}}{\text{Wetted perimeter}} = \frac{\frac{\pi}{4} d^2}{\pi d} = \frac{d}{4} \text{ (for circular closed conduits)}$$

Substituting the value of S as $\frac{H_L}{L}$ and $R = \frac{d}{4}$ with $V = \frac{Q}{A}$

The equation can be written as Eq. 8.15.

$$H_L = \frac{L.\left(\frac{Q}{C_R}\right)^{1.81}}{994.62.d^{4.81}} \quad (8.15)$$

Q = Discharge or flow rate in pipe (m^3/sec) with pipe diameter d (m)

Table 8.3: Recommended CR values in modified Hazen-Williams formula at 20°C

S. No.	Pipe Material	Diameter (mm)		Velocity (m/sec)		C_R value new	CR value for design period 30 years
		From	To	From	To		
1.	R.C.C.	100	2000	0.3	1.8	1.0	1.0
2.	A.C.	100	600	0.3	6.0	1.0	1.0
3.	HDPE/PVC	20	100	0.3	1.8	1.0	1.0
4.	C.I. (for water with +ve Langelier's index)	100	1000	0.3	1.8	1.0	0.85
5.	C.I. (for water with -ve Langelier's index)	100	1000	0.3	1.8	1.0	0.53
6.	Steel (for water with -ve Langelier's index)	100	2000	0.3	2.1	1.0	0.73
7.	SGSW	100	600	0.3	2.1	1.0	1.0
8.	G.I. (for water with +ve Langelier's index)	15	100	0.3	1.5	0.87	0.74

Example 8.1: Estimate the size of pipes and gradient required for a service reservoir for water supply to a town having 50,000 population and large city with industrial area having population around 7 lakh. Assume required data suitably?

Solution: Data given:

(i) Town with population = 50,000

(ii) City with population = 7,00,000

Assume the per capita demand of water required for

(i) Town with population of 50,000 = 135 L/capita/d

(ii) City with population of 7,00,000 = 280 L/capita/d

The average quantity of water required for

(a) Town with population of 50,000 = $135 \times 50{,}000$ L/d = 6.75×10^6 L/d

$= 6.75 \times 10^3 \text{m}^3/\text{d}$

(b) City with population of 7,00,000 = $280 \times 7{,}00{,}000$ L/d = 1.96×10^8 L/d

$= 1.96 \times 10^5 \text{ m}^3/\text{d}$

Assuming that the daily water demand is 1.8 times the average water quantity and therefore the maximum quantity of water required can be calculated as:

(a) Town with population of 50,000 $= 6.75 \times 10^3 \text{ m}^3/\text{d} = \dfrac{1.8 \times 6.75 \times 10^3}{24 \times 60 \times 60}$

$$= 0.14 \text{ m}^3/\text{sec}$$

(b) City with population of 7,00,000 $= 1.96 \times 10^5 \text{ m}^3/\text{d} = \dfrac{1.8 \times 1.96 \times 10^5}{24 \times 60 \times 60}$

$$= 4.08 \text{ m}^3/\text{sec}$$

Assume the flow velocity to be 1.25 m/sec in circular pipe system, the area of the pipe can be calculated as:

$$Q = A \times V$$

(a) Area of pipe for a town

$Q = 0.14 \text{ m}^3/\text{s}$ and $V = 1.25$ m/sec

Area $(A) = \dfrac{0.14}{1.25} = 0.112 \text{ m}^2$

The diameter of the pipe can be calculated as:

$$\frac{\pi}{4} d^2 = 0.112$$

$$d = \sqrt{\frac{4 \times 0.112}{\pi}} = 37.7 \text{ cm} = 38 \text{ cm diameter pipe}$$

(b) Area of pipe for a city

$Q = 4.08 \text{ m}^3/\text{s}$ and $V = 1.25$ m/sec

Area $(A) = \dfrac{4.08}{1.25} = 3.264 \text{ m}^2$

The diameter of the pipe can be calculated as:

$$\frac{\pi}{4} d^2 = 3.264$$

$$d = \sqrt{\frac{4 \times 3.264}{\pi}} = 2.0 \text{ m} = 200 \text{ cm diameter pipe}$$

Hydraulic gradient by Hazen-Williams formula

The equation can be written as:

$$V = 0.85\, C_H \,.\, R^{0.63} \,.\, S^{0.54}$$

where, the value of $C_H = 110$

Therefore,

$$V = 0.85 \times 110 \cdot \left(\frac{\pi}{4}\right)^{0.63} \cdot S^{0.54} = 93.5 \cdot \left(\frac{d}{4}\right)^{0.63} \cdot S^{0.54}$$

(a) For a town the slope is as

$$V = 93.5 \cdot \left(\frac{d}{4}\right)^{0.63} \cdot S^{0.54}$$

or $$1.25 = 93.5\cdot\left(\frac{0.38}{4}\right)^{0.63}\cdot S^{0.54}$$

or $$1.25 = 93.5\cdot\left(\frac{0.38}{4}\right)^{0.63}\cdot S^{0.54}$$

$$\frac{1.25}{21.22} = S^{0.54}$$

$$S^{0.54} = 0.058$$

$$S = (0.058)^{\frac{1}{0.54}}$$

Therefore, $$S = (0.058)^{\frac{1}{0.54}} = 0.0051$$

Therefore, $S = \frac{1}{196}$ indicates that we require hydraulic gradient with 1 m fall at every 196 m length

(b) For a city

$$V = 93.5\cdot\left(\frac{2}{4}\right)^{0.63}\cdot S^{0.54}$$

or $$1.25 = 93.5\cdot\left(\frac{2}{4}\right)^{0.63}\cdot S^{0.54}$$

or $$1.25 = 93.5\cdot\left(\frac{2}{4}\right)^{0.63}\cdot S^{0.54}$$

$$\frac{1.25}{60.4} = S^{0.54}$$

$$S^{0.54} = 0.02$$

$$S = (0.02)^{\frac{1}{0.54}}$$

Therefore, $$S = (0.02)^{\frac{1}{0.54}} = 0.0071$$

Therefore, $S = \frac{1}{1390}$ indicates that we require hydraulic gradient with 1 m fall at every 1390 m length.

Example 8.2: Determine the hydraulic gradient of a 1.8 m concrete pipe carrying a discharge of 2.89 m/sec having 10°C temperature by using (a) Darcy-Weisbach formula, (ii) Mannings's formula, (c) Hazen-Williams formula, and (d) modified Hazen-Williams formula, Assume any other data suitably?

Solution: Given that: $Q = 2.89$ m/sec

$$d = 1.8 \text{ m}$$

Area of pipe $A = \frac{\pi}{4}d^2 = \frac{d}{4}(1.8)^2 = 2.54 \text{ m}^2$

Therefore, velocity of flow $V = \frac{Q}{A} = \frac{2.89}{2.54} = 1.14$ m/sec

(a) From Dary-Weisbach equation it is stated that:

$$H_L = \frac{f' L \cdot V^2}{2gd}$$

where,

f' = Friction factor depends on Reynolds number, $R_e = \frac{V \cdot d}{v}$

where ν = Kinematic viscosity of water at 10°C = 1.31×10^{-6} m²/sec

$$R_e = \frac{V \cdot d}{v} = \frac{1.14 \times 1.8}{1.31 \times 10^{-6}} = 1.56 \times 10^6 = 15{,}66{,}412$$

For smooth pipes and the Reynolds number (R_e) varying from 20,000 to 20,00,000 the f' can be calculated by using the formula given by Slitcher as:

$$f' = 0.005 + \frac{0.396}{R_e^{0.3}}$$

$$f' = 0.005 + \frac{0.396}{(1566412)^{0.3}} = 0.01$$

Therefore, $$H_L = \frac{0.01 \times Lx \cdot (1.14)^2}{2 \times 9.8 \times 1.8}$$

or $$S = \frac{H_L}{L} = \frac{0.01 \times 1.3}{35.28}$$

Therefore, $S = 0.003 = \frac{1}{2713}$

It indicates that 1 m fall in every 2713 m length of pipe.

(b) Manning's formula

$$H_L = \frac{n^2 L \cdot V^2}{R^{\frac{4}{3}}}$$

Given that

$n = 0.013$

$$R = \frac{d}{4} = \frac{1.8}{4} = 0.45 \text{ m}$$

Therefore, putting the values in Manning's equation, we get

$$H_L = \frac{(0.013)^2 L \cdot (1.14)^2}{(0.45)^{\frac{4}{3}}}$$

$$\frac{H_L}{L} = \frac{0.00169 \times 1.3}{0.345} = 0.00063 = S$$

Therefore,

$$S = \frac{1}{1570}$$

The hydraulic gradient is 1 m at every 1570 m length of pipe

(c) Hazen-Williams formula:

$$V = 0.85\ C_H \cdot R^{0.63} \cdot S^{0.54}$$

where $C_H = 130$ (from Table 8.2)

Therefore, we have

$$1.14 = 0.85 \times 130 \cdot \left(\frac{1.8}{4}\right)^{0.63} \cdot S^{0.54}$$

or

$$1.14 = 0.85 \times 130 \times 0.604 \times S^{0.54}$$

or

$$\frac{1.14}{66.8} = S^{0.54}$$

Therefore,

$$S = \left(\frac{1.14}{66.8}\right)^{\frac{1}{0.54}} = 0.0005 = \frac{1}{1865}$$

The hydraulic gradient obtained is 1 m of fall at every 1865 m length of pipe

(d) Modified Hazen-Williams equation

The given equation is:

$$V = 143.534\ C_R \cdot R^{0.6675} \cdot S^{0.5525}$$

where,

$V = 1.14$ m/sec

$C_R = 1$ (for concrete pipe in Table 3.2)

$R = \frac{d}{4} = \frac{1.8}{4} = 0.45$ m

Therefore, putting the values in above equation, we get

$$1.14 = 143.534 \times 1 \times (0.45)^{0.6675} \cdot S^{0.5525}$$

or

$$S^{0.5225} = \frac{1.14}{84.23}$$

$$S = \left(\frac{1.14}{84.23}\right)^{\frac{1}{0.5525}} = 0.00041 = \frac{1}{2410}$$

The hydraulic gradient obtained is 1 m fall at every 2410 m length of pipe.

The value obtained using modified Hazen-Williams formula is $\frac{1}{2410}$ while the value obtained in Darcy-Weisbach formula is $\frac{1}{2713}$ is quite close to each other.

8.5 FLOW HYDRAULICS

Flow in pipe systems: In a flow system when two pipes are laid in parallel, the head loss through each pipe will be equal as shown in Fig. 8.2.

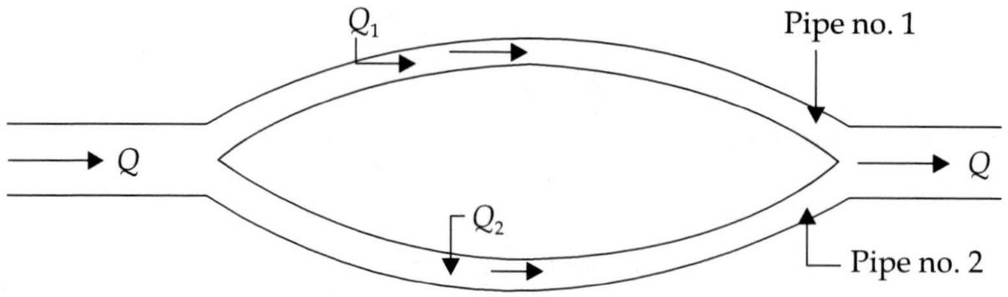

Fig. 8.2: Flow in pipe systems

Considering the fact that the characteristics of the pipe are known, the flow distribution in the pipe system would be such that by applying the equation of continuity such that

$$Q = Q_1 + Q_2$$

and calculating the head loss simultaneously. The requirement of laying the pipes in parallel is due to the following reasons:

- When the demand of water increases the flow requirement increases, therefore in order to increase the capacity of the water line additional pipes are laid in parallel with the previously laid pipes.
- For requirement of operation and maintenance sometimes more than two pipes are laid in parallel such that if one pipeline is under repair the other standby pipe line can be utilized.
- Whenever the flow values are too high and one single large size diameter pipe is not available in the market, two or three pipes of lesser diameter are generally laid in parallel.

The water supply pipes are sometimes need to be connected in series and the head loss in individual pipes is calculated and the summation of head loss in all the connected pipes in series is calculated. The other minor losses that contribute to the total head loss include the losses due to bends, expansions, contractions, valves, gates, etc. in the pipes in slopes, trestles and on undulations. However, in case of lesser pipe lengths the minor losses are sometimes even higher than the frictional losses in the pipes and therefore needs to be considered in the design aspect. These minor losses are generally assessed and are represented in terms of $K\frac{V^2}{2g}$ and are summarized in Table 8.4.

Table 8.4: Head loss values in various types of minor losses

S. No.	Losses type and causes	*K* values
1.	Losses in entrance	
	Bell mouth entry	0.04
	Square edged entry	0.5
2.	Loss at exit	1.0
3.	Loss in sudden contraction	0-0.5 (varying with dia)
4.	Loss in sudden enlargement	0.17-1.0 (varying with angles)
5.	Losses in fittings and valves	
	Standard 90° elbow	0.9
	Standard tee	1.8
	Return bend	2.2
	Gate valve (half open)	5.6
	Gate valve (wide open)	0.2
	Swing check valve (wide open)	2.5
	Glove valve (wide open)	10.0
	Angle valve	5.0
	Meter venturi	0.3
	Orifice	1.0

The head loss in pipe system connected either in parallel or in series or the combination of both types of system is generally expressed by using the Eq. 8.16:

$$H_L = KQ^x \tag{8.16}$$

The K value depends on diameter, length and roughness and arrangements of the pipes in a system. In the equation the "x" is a factor of which the value in Hazen-Williams formula is taken as 1.85 and in Darcy-Weisbach and Manning's formulae this value is 2.

Example 8.3: The lowest level of an impounding reservoir, situated in hills and used as a source of supply, is higher than the level of the service reservoir by 175 m and the distance between them is 60 km. The storage reservoir and the service reservoir were originally connected by a single pipe line designed to carry a maximum daily demand of 35 MLD. However, it was later found necessary to increase the supply to 40 MLD. It was, therefore, decided to lay another pipe of the same diameter alongside the first one in a certain required length and cross-connected. Calculate the diameter of the pipes and the length of the second pipe required. Make use of William Hazen's equation. Also derive an expression expressing H, in terms of land (d) from the original William-Hazen's formula. Assume $C_H = 120$.

Solution:

$$\text{Initial discharge } Q_1 = 35 \text{ MLD} = \frac{\frac{35 \times 10^6}{10^3} \text{m}^3}{d} = \frac{35000}{24 \times 60 \times 60} = 0.405 \text{ m}^3/\text{sec}$$

It is assumed that the total head available between storage reservoir and service reservoir with a distance of 175 m and the same is assumed to be dissipated in overcoming the frictional resistance offered by the pipe material during the water flow.

Now using the Hazen-Williams equation:

$$V = 0.85\, C_H\, R^{0.63} \cdot S^{0.54}$$

where $C_H = 120$ (given)

$$S = \frac{H_L}{L} = \frac{175}{60000} = 0.003$$

and

$$V = \frac{Q}{A} = \frac{0.405}{\frac{\pi}{4} d^2} = \frac{0.515}{d^2}$$

Therefore, we have

$$V = \frac{0.515}{d^2} = 0.85 \times 120 \times \left(\frac{d}{4}\right)^{0.63} (0.003)^{0.54}$$

$$\frac{0.515}{d^2} = \frac{102}{2.39} \times (d)^{0.63} (0.003)^{0.54}$$

$$\frac{0.515}{d^2} = \frac{102}{2.39} \times (d)^{0.63} \times 0.043$$

$$\frac{1.231}{4.3} = (d)^{2.63}$$

$$(d)^{2.63} = 0.28$$

$$d = (0.28)^{\frac{1}{2.63}} = 0.616 \text{ m}$$

The diameter of the pipe is 0.616 m.

The present discharge is required to be increased to 40 MLD $= \frac{40{,}000}{24 \times 60 \times 60} = 0.463$ m^3/sec

It is assumed that pipe of length L km is to be laid in parallel along the length of the old pipe and the discharge in both the pipes is distributed equally. Therefore, the discharge flowing through the pipe with length (L-60) km would be 0.463 m^3/sec and discharge in both the parallel pipes would be half of main flow, i.e., 0.231m^3/sec with both pipes of equal diameter. (Fig. 8.3)

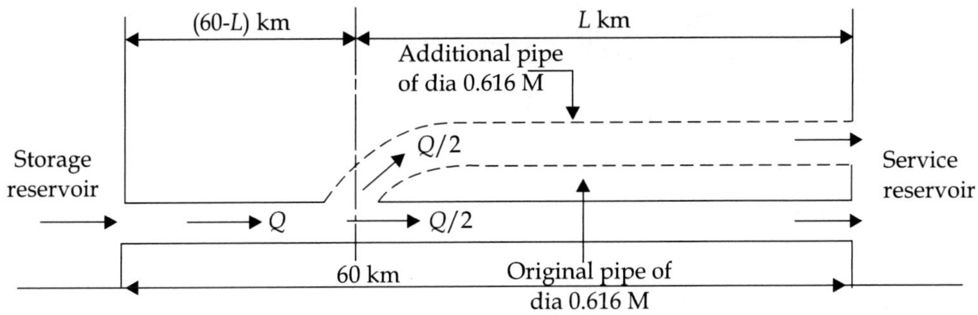

Fig. 8.3: Cross section of pipe

The total head loss in both the pipes is summarized below as:

175 = Head loss in pipe having length (60-L) km with discharge of 0.463 m^3/sec + head loss in length L km with discharge of 0.231 m^3/sec.

Therefore, in order to assess the length (L) of the parallel pipes the head loss can be represented with Hazen-Williams equation in terms of H_L and Q:

$$V = 0.85\ C_H\ R^{0.63}\ .\ S^{0.54}$$

$$V = \frac{Q}{\frac{\pi}{4}d^2} = 0.85\ C_H \cdot \left(\frac{d}{4}\right)^{0.63} \cdot S^{0.54}$$

$$Q = \frac{0.85\ C_H}{2.4}\frac{\pi}{4}\cdot(d)^{2.63}\cdot S^{0.54} = 0.278\cdot C_H \cdot d^{2.63}\cdot\left(\frac{H_L}{L}\right)^{0.54}$$

Therefore,

$$H_L^{0.54} = \frac{Q\ L^{0.54}}{0.278 C_H d^{2.63}}$$

$$H_L = \frac{1}{0.094}\cdot\left(\frac{Q}{C_H}\right)^{1.85}\frac{L}{d^{4.87}}$$

Putting the values of Q, C_H, d and H_L in the above equation, the required L value can be calculated as:

The C_H value common to all = 120

Therefore,

$$H_L = \frac{1}{670}\cdot(Q)^{1.85}\frac{L}{d^{4.87}} = \frac{1}{670}\cdot(Q)^{1.85}\frac{L}{d^{4.87}}$$

Substituting the values in the above equation, we get

$$175 = \frac{1}{670}\cdot(0.463)^{1.85}\frac{(60-L)\times 1000}{(0.616)^{4.87}} + \frac{1}{670}\cdot(0.231)^{1.85}\frac{L\times 1000}{(0.616)^{4.87}}$$

$$175 = \frac{1}{670}\times 0.24\times\frac{(60-L)\times 1000}{0.094} + \frac{1}{670}\times 0.066\frac{L\times 1000}{0.094}$$

$$175 = \frac{1000}{670 \times 0.094} \cdot [0.24 \times (60 - L) + 0.066 L]$$

$$\frac{175 \times 670 \times 0.094}{1000} = (0.24 \times (60 - L) + 0.066\ L)$$

$$11.02 = 14.4 - 0.24\ L + 0.066\ L$$

$$-3.38 = -0.174\ L$$

$$L = 19.42 \text{ km}$$

Therefore, an additional length of 19.42 km of diameter 0.616 m is required.

8.6 PRESSURE CONDUITS AND FORCES

Pipes should be designed so as to withstand enough pressure and therefore, should be structurally designed so that the pipes can withstand the various forces likely to come on them. The various forces that generally act on the pressure pipes are:

- The force acts in terms of internal pressure of water includes water hammer pressure and therefore the pipe should be made of materials strong in tension and can resist it.
- The force that is caused by the pressure due to external loads in the form of backfill, traffic loads, etc., and is resisted by using pipe material strong in compression.
- The forces caused in the pipe due to longitudinal temperature stresses when pipes are laid above the ground and therefore it is to be resisted by providing such type of pipes with expansion joints.
- The forces caused due to longitudinal stresses developed due to unbalanced pressure at bends, or at points of changes of cross section. Such a force can be controlled by anchoring the pipe in massive blocks of concrete or stone masonry.
- The forces developed in pipes due to flexural stresses produced and supported over trestles, etc.

The above mentioned various forces generated during flow of water in pipes are discussed below in detail.

8.6.1 Internal Pressure of Water

The internal pressure in a pipe is caused by the flowing water with available water head to which it is subjected and also the additional transient pressure that develops during the sudden stop of flowing water. When a closed pipe is running with water under full flow conditions, the overall water pressure in the pipe (p/w) is equated to the vertical ordinate between the hydraulic gradient (HG) line and the centre-line of the pipe. However, when the water comes at rest, the available pressure in the pipe is equal to the total static head of water at that time (the reservoir water level ordinate and the centre of the pipe) as shown in Fig. 8.4.

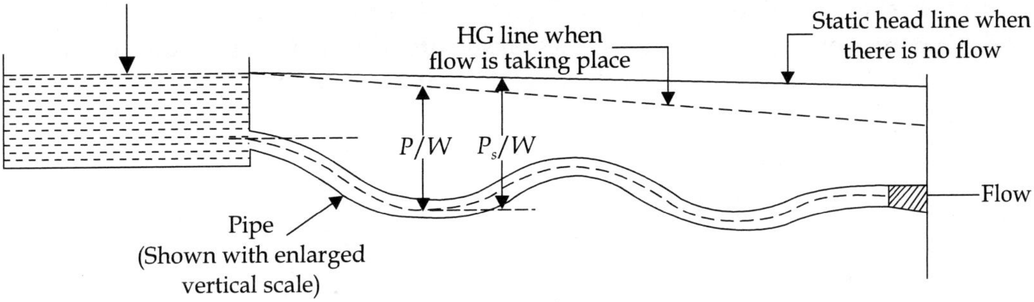

Fig. 8.4: Internal pressure of water with hydraulic gradient line

Also an increase in transient pressure is caused for a short period of time, when gates, valves, pumps, etc., are closed or opened. This pressure is in addition to the above-mentioned static water head pressure. The additional pressure created by such a temporary or transient phenomenon is known as water hammer pressure and is denoted by p_h.

The maximum internal pressure likely to come under worst circumstances is usually taken equal to the sum of full static pressure and the water hammer pressure. Hence, if the static pressure is denoted by p_s, and the water hammer pressure is represented by P_h, then the total internal maximum pressure that can occur in the pipe is $p = p_s + p_h$. The internal pressure creates transverse stresses or circumferential tension called Hoop tension, in the pipe walls. The magnitude of Hoop tension developed in this manner is approximately given by Eq. 8.17.

$$\sigma = \frac{pd}{2 \cdot t} \quad (8.17)$$

where,

σ = Circumferential tensile stress produced. It has to be counteracted by providing Hoop reinforcement in R.C.C. pipes and should not exceed the safe allowable tensile stress of the pipe material, otherwise bursting of pipe may occur.

p = Total internal pressure including the static head water pressure under full and water hammer pressure.

d = Pipe diameter (m)

t = Pipe thickness (m)

8.6.2 Water Hammer Pressure

When flowing water in a pipe is stopped suddenly by closing of valves, gates, etc. and therefore the behind velocity of the water in that pipe gets retarded. Due to which the conversion of kinetic energy into elastic energy takes place and the momentum of the stopped water as slug gets dissipated. It produces a series of positive and negative pressure waves that travel back and forth in the pipe itself. This continues till these waves are damped out by friction. This phenomenon is known as *water hammer*. It is generally considered that the water is an incompressible or inelastic fluid. In this case of water hammer pressure the water shows elastic properties in the pipe during such actions. The same can be explained below and is depicted through Fig. 8.5.

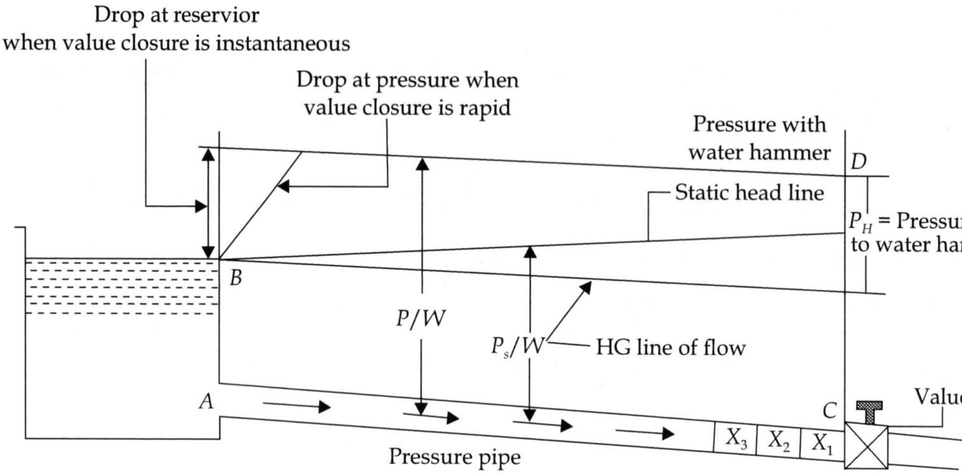

Fig. 8.5: Development of water hammer pressure in pipe

Let us assume that when a valve is closed an element of water as ΔX exits at the upstream side of the valve comes to rest whereas the water at the upstream side of this ΔX is in continuous motion with full velocity (V). It, however, strikes at an element ΔX_1 and this element gets compressed by the running water. This increases the pressure on the element ΔX_1. This can be well understood such that a part of the pressure is utilized in compressing the water and the remaining part of this pressure is utilized in stretching or extension of the wall surroundings of the pipe at element ΔX_1. In the next step this forward motion of water in the pipe is stopped at the point of element ΔX_2 which exists at upstream of ΔX_1. However, the forward movement of water at upstream of the element ΔX_2 compresses it. This process continues and repeats on such successive elements for a short duration. During this short period of time the pressure wave travels back to the reservoir and the moving water in the pipe stops completely. This creates an excessive pressure at the end of pipe in reservoir end A. It is not possible in a reservoir that pressure at point B cannot be greater than pressure at point A. Therefore, the pressure at point A suddenly drops to a normal value with depth AB.

But as represented by CD in Fig. 8.5 the pressure in almost whole of the pipe is more than the pressure at point A. Therefore, the water of the pipe starts flowing back into the reservoir, and a wave of pressure unloading (or reflected wave) travels along the pipe from A to C. At the instant this pressure unloading wave reaches C, the entire water is under the static head (equal to depth AB). The water is still moving back into the reservoir and this reverse velocity must now be stopped instantaneously. This will produce a drop in pressure at C, which will be below the static head and its magnitude will be p_h which is same as it was an instant before. A wave of negative pressure called, waves of rarefaction, therefore, gets set up, which travels towards the reservoir from C to A. As soon as this negative wave reaches the point A, the velocity of the water in the entire pipe is zero, but the pressure everywhere in the pipe is negative $-p_h$ except at A, where it has to be equal to the depth AB. This pressure difference starts the flow from the reservoir into the pipe, and the wave of pressure unloading, known as wave of reflected negative pressure, travels from A to C. As this wave reaches to point C, the entire water in the pipe is moving with its original velocity towards the valve located at C and the pressure in the pipe is brought to its initial value. This completes one full cycle of waves of positive and negative pressures generated due to the instantaneous valve closure. The process goes on repeating until damped out by friction, as shown in Fig. 8.6. The time required for a round trip of the pressure wave from C to A, and back from A to C, is represented by the term $\frac{2S}{U_p}$

where, S is the distance of the reservoir from the valve, and U_p is the velocity or speediness of the pressure wave.

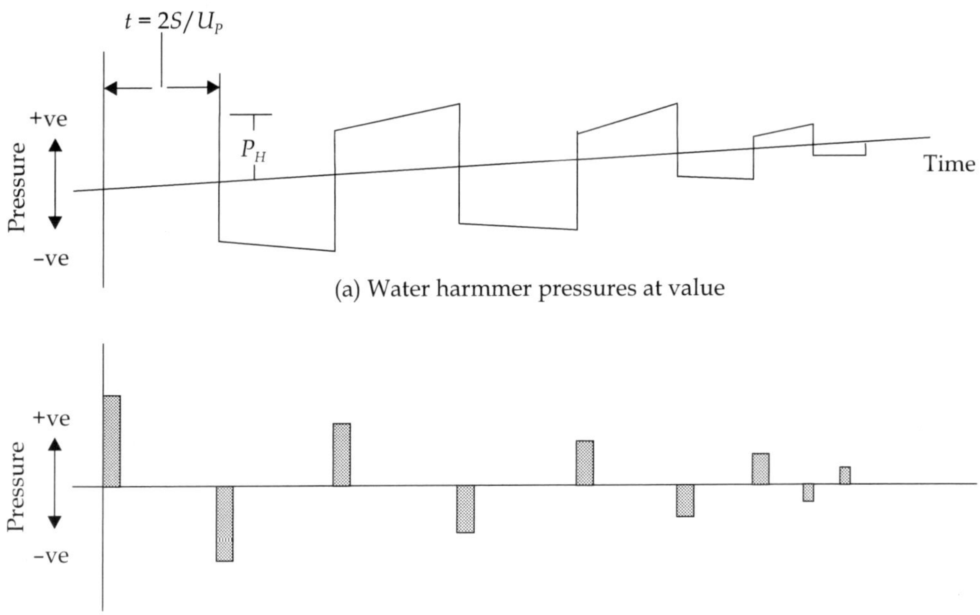

Fig. 8.6: Water hammer pressure developed due to closure of a valve

In case a valve is suddenly opened, expansion of the water upstream of the valve occurs, and a wave of negative pressure travels to the reservoir. The wave is reflected back up to the valve, again producing a wave of positive pressure, which finally gets reflected.

8.6.3 Estimating Water Hammer

It is evident that the additional pressure is caused by opening and closing of values or gates during flow of water in a pipe. This additional pressure depends on the approach by which these values are closed or opened as the more rapidly the valve is closed there is more rapid change in momentum, and therefore higher will be the developed additional pressure. This additional pressure, also known as water hammer pressure, may be as high as it can cause bursting of the pipe shell that may be due to the increased circumferential tension in the pipe. Therefore this water hammer pressure needs to be accounted during the design of pipes required for public water supplies.

The maximum water hammer pressure is developed when the valve is suddenly or immediately closed. The maximum pressure developed can be described as Eq. 8.18.

$$P_{h(\max)} = \rho_w \cdot U_p \cdot V \tag{8.18}$$

$P_{h(\max)}$ = Maximum water hammer pressure developed (N/m^2)

ρ_w = 1000 kg/m^3

U_p = Velocity of the pressure wave generated

V = Velocity of water in the pipe (reduced V to 0)

The velocity of pressure wave generated is known as U_p and is given by Eq. 8.19:

$$U_p = \sqrt{\frac{E_w}{\rho_w}}, \frac{1}{\sqrt{1 + \frac{E_w}{E_p} \cdot \frac{d}{t}}} \tag{8.19}$$

where, the term $\sqrt{E/\rho}$ corresponds to velocity of the pressure wave in any medium and is assumed to be same as the velocity of sound in that medium. Due to elasticity in the walls of pipe carrying water the velocity of the pressure wave developed due to water hammer is less and is denoted by the term $\sqrt{\frac{E_w}{\rho_w}}$.

However, during the circumferential stretching that takes place in the pipe and the longitudinal extension of the pipe is opposed, the velocity of the pressure wave generated in terms of U_p as Eq. 8.20;

$$U_p = \sqrt{\frac{E_w}{\rho_w}}, \frac{1}{\sqrt{1 + \frac{E_w}{E_p} \cdot \frac{d}{t}}} \tag{8.20}$$

where,

E_w = Modulus of elasticity of water

E_p = Modulus of elasticity of pipe material

d = Diameter of the pipe

t = Thickness of the pipe shell

ρ_w = Density of water (γ_w/g)

Putting the value in Eq. 6.15, we get

$$P_{h(\max)} = \rho_w \cdot \sqrt{\frac{E_w}{\rho_w}}, \frac{1}{\sqrt{1+\frac{E_w}{E_p}\cdot\frac{d}{t}}} \cdot V$$

where,

$\sqrt{\frac{E_w}{\rho_w}}$ = Sound velocity in water = 1433 m/sec

The equation in terms of S.I. units can be written as Eq. 8.21:

$$P_{h(\max)} = \left(\frac{9.81}{9.81}\right) \cdot (1433) \cdot \frac{V}{\sqrt{1+\frac{E_w}{E_p}\cdot\frac{d}{t}}} \cdot \text{kN/m}^2$$

$$P_{h(\max)} = \frac{1433 \cdot V}{\sqrt{1+\frac{E_w}{E_p}\cdot\frac{d}{t}}} \cdot \frac{\text{kN}}{\text{m}^2}$$

$$P_{h(\max)} = \frac{1433 \cdot V}{\sqrt{1+k\cdot\frac{d}{t}}} \cdot \frac{\text{kN}}{\text{m}^2} \tag{8.21}$$

where,

$$k = \frac{E_w}{E_p} = \frac{\text{Modulus of elasticity of water}}{\text{Modulus of elasticity of pipe material}} \tag{8.22}$$

The k value as specified in literature is 0.01 for steel pipes, 0.02 for cast iron pipes and 0.1 for cement-concrete pipes as shown in Eq. 8.22.

E value can be taken as:

E_w for water = 21×10^5 kN/m^2

E_p for steel = 21×10^7 kN/m^2

E_p for cast iron = 10.5×10^7 kN/m^2

E_p for cement concrete = 21×10^6 kN/m^2

When the valve is closed slowly, small pressure waves in series are transmitted up to the pipe and these waves reach the reservoir and return back to the point of valve. The distance travelled by these waves from the valve to the reservoir can be denoted by S. The time taken by the wave while travelling to the reservoir and returning back to the valve is known as critical time (T_c) and is represented as in Eq. 8.23;

$$T_c = \frac{2S}{U_p} \tag{8.23}$$

If the actual time of closure (T) is less than or equal to the critical time (T_c), thus resulting in maximum water hammer pressure in the pipe. However, when the critical time (T_c) is less than the actual closure time (T), the positive waves are superimposed by the negative pressure waves and therefore full pressure never develops. This developed pressure is calculated as in Eq. 8.24;

$$P_h = P_{h(\max)} \cdot \left(\frac{T_c}{T}\right) \tag{8.24}$$

Therefore, the water hammer pressure can be reduced significantly closing valves the valves slowly. Moderate velocities are kept in smaller pipelines and therefore water hammer pressure can be kept moderate. In larger pipe systems when operated at relatively high velocities, surge tanks, pressure relief valves, etc., are provided that help in maintaining the water hammer pressure under control. Alternatively, if such arrangements are not made the system becomes uneconomical and costly as pipes are required to endure such a high water hammer pressure and are generally made very strong and therefore requires very costly material.

The water hammer pressure can be calculated by using the above-mentioned equations from (8.18) to (8.24). However, in municipal water work systems under normal range of water hammer pressure encountered the values of water hammer pressure for cast iron and for steel pipes may be adopted as provided in Table 8.5. However, for centrifugally cast iron pipes, somewhat higher allowances may be made for larger sizes. Table 8.5 shows normal values of water hammer pressure to be adopted for steel and cast iron.

Table 8.5: Normal values of water hammer pressure to be adopted for steel and cast iron pipes

S. No.	Pipe diameter (cm)	Water hammer pressure (kN/m²)
1.	7.5 to 25	840
2.	30	770
3.	40	700
4.	50	630
5.	60	600
6.	75	560
7.	90	530
8.	105-150	490

Example 8.4: The water is flowing from a reservoir into a steel pipe with 1.2 m diameter, thickness of 2.5 cm and velocity of 1.8 m/sec. The valve is situated at a distance of 2.25 km from the reservoir, estimate the water hammer pressure when the valve is closed in the following times:

(i) The closure time is 2.5 sec

(ii) The closure time is 6.5 sec

Solution: Given that:

$$\frac{E_w}{E_p} = \frac{\text{Modulus of elasticity of water}}{\text{Modulus of elasticity of pipe material}} = \frac{21 \times 10^5 \text{ kN/m}^2}{21 \times 10^7 \text{ kN/m}^2} = 0.01$$

$d = 1.2$ m

$t = 2.5$ cm $= 0.025$ m

and $\sqrt{\frac{E_w}{\rho_w}} = 1433$ m/sec

Using equation

$$U_p = \sqrt{\frac{E_w}{\rho_w}} \cdot \frac{1}{\sqrt{1 + \frac{E_w}{E_p} \cdot \frac{d}{t}}}$$

Putting the value in the above equation the value of U_p can be calculated as:

$$U_p = 1433 \cdot \frac{1}{\sqrt{1 + 0.01 \cdot \frac{1.2}{0.025}}} = 1433 \cdot \frac{1}{\sqrt{1 + 0.48}} = 1178 \text{ m/sec}$$

The critical time (T_c) can be calculated as :

$$T_c = \frac{2S}{U_p} = \frac{2 \times 2250}{1178} = \frac{4500}{1178} = 3.8 \text{ sec}$$

The maximum water hammer pressure $= P_{h(max)}$

$$P_{h(max)} = \rho_w \cdot U_p \cdot V = \left(\frac{9.81}{9.81}\right) \times 1178 \times 1.8 \frac{\text{kN}}{\text{m}^2} = 2120 \text{ kN/m}^2$$

Case (i) when closure time is 2.5 sec

$T = 2.5$ sec and $T_c = 3.8$ sec

Here, we have $T < T_c$

Therefore, the full water hammer pressure developed will be = 2120 kN/m^2

Case (ii) when closure time is 6.5 sec

$T = 6.5$ sec and $T_c = 3.8$ sec

Here, we have $T > T_c$

Therefore, the full water hammer pressure developed will be $P_h = P_{h(\max)} \cdot \left(\frac{T_c}{T}\right)$

$$= 2120 \times \left(\frac{3.8}{6.5}\right) = 1239 \text{ kN/m}^2$$ **Ans.**

Example 8.5: Water is to be transmitted from a reservoir to a city by means of a steel pipe with 1.2 m diameter across a ground with undulations. The water level in the reservoir is 425 m whereas the lowest level of the pipe in its L-section is 300 m. Determine the thickness of the pipe required with respect to the maximum pressure, if the allowance for water hammer is made for 900kN/m^2 over static head. The steel pipe may be taken to be riveted with efficiency of joint as 75%. The allowable working stress in steel may be taken as 122600 kN/m^2.

Solution: Given that:

Pipe diameter = 1.2 m

Water level in reservoir = 425 m

Lowest water level = 300 m

Allowable water hammer pressure = 900 kN/m^2

Efficiency of the joint = 75%

Working stress in steel = 1,22,600 kN/m^2

Now the static head of water P = (425 – 300) = 125 m

Because 1 m of water head = 9.81 kN/m^2

Therefore, $P = 125 \times 9.81 = 1226$ kN/m^2

Allowable water hammer pressure, $P_h = 900$ kN/m^2

Total internal pressure, $p = P + P_h = 1226 + 900 = 2126$ kN/m^2

The pipe thickness equation is given as:

$$t = \frac{1}{\eta} \cdot \frac{p \cdot d}{2 \cdot f} = \frac{1}{0.75} \times \frac{2126 \times 1.2}{2 \times 122600} = 0.01387 \text{ m} = 1.39 \text{ m}$$

Add 3 mm extra proviso for corrosion.

The steel pipe thickness required = 1.34 + 0.3 = 1.69 cm

Therefore use 17 mm thickness pipe

Example 8.6: In water supply system the water is to be supplied to a town using a cast iron gravity main pipe with diameter of 0.6 m and 40 km long, across a rolling ground from the headworks to the service reservoir in the town. The pipe off-take at the headworks is at RL + 510 m and the maximum water level of the reservoir is at RL +375 m. The lowest ground level in the L-section of the main is at RL + 290 m at a point 20 km from the off-take. However, it is desired to increase the supply to the town by 25%.

(i) Calculate the distance of duplicate pipe of the same size to be laid.

(ii) Determine the thickness of this pipe at a point 20 km away (where the RL of the centre of the pipe is 290 m).

The water hammer pressure may be neglected as additional arrangements have been made for dissipating the same.

Solution: Given that:

Pipe diameter = 0.6 m

Pipe length = 40 km

R.L. of pipe off-take = 510 m

Water level maximum = 375 m

Lowest ground level = 290 m

Off-take distance = 20 km

Increase in supply = 25%

Head available between the source and service reservoir = 510 – 375 = 135 m

The pipe should be designed in such a manner that the entire available head is lost in overcoming the frictional and minor losses.

Make use of Hazen-Williams equation and calculate the required velocity as:

$$V = 0.85\ C_H \cdot R^{0.63} \cdot S^{0.54}$$

where,

$C_H = 120$ (from Table 8.1, for cast iron pipe)

$$R = \frac{d}{4} = \frac{0.6}{4} = 0.15 \text{ m}$$

$H_L = 135$ m

$L = 40$ km = 40,000 m

Putting the values in above equation as:

$$V = 0.85 \times 120 \times (0.15)0.63 \times \left(\frac{135}{40000}\right)^{0.54}$$

$$= 0.85 \times 120 \times 0.30 \times 0.046 = 1.41 \text{ m/sec}$$

$$Q = A \cdot V = \left(\frac{\pi}{4} d^2\right) \cdot V = \frac{\pi}{4} \times (0.6)^2 \times 1.41 = 0.4 \text{ m}^3/\text{sec}$$

Additional discharge of 25% will be

$$Q' = 1.25\ Q = 1.25 \times 0.4 = 0.5 \text{ m}^3/\text{sec}$$

A pipe of length L km is to be laid in parallel along the length of previous pipe and is shown through Fig. 8.7 below:

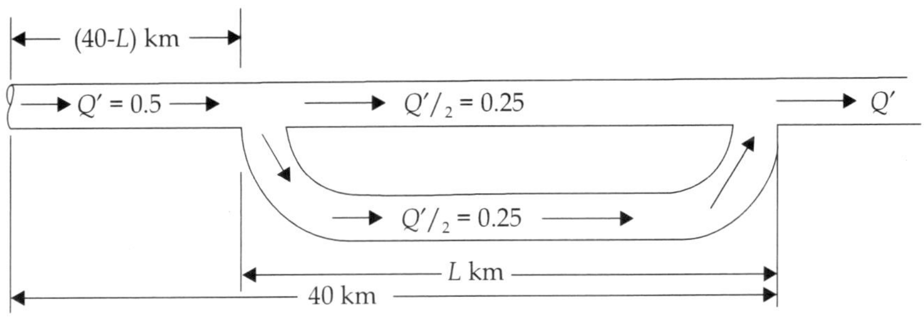

Fig. 8.7: Equivalent pipe length and diameter

The available head is equal to the head loss during water travelling to the destination from the reservoir and the same can be equated as below:

$$H_L = \frac{1}{0.094} \cdot \left(\frac{Q}{C_H}\right)^{1.85} \frac{L}{d^{4.87}}$$

Putting the values of Q, C_H, d and H_L in the above equation, the required L value can be calculated and depicted in Fig. 8.8.

The C_H value will be same = 120

Therefore,

$$H_L = \frac{1}{0.094} \cdot \left(\frac{Q}{120}\right)^{1.85} \frac{L}{d^{4.87}} = \frac{1}{670} \cdot (Q)^{1.85} \frac{L}{d^{4.87}}$$

Substituting the values in the above equation, we get

$$135 = \frac{1}{670} \times (0.5)^{1.85} \frac{(40 - L) \times 1000}{(0.6)^{4.87}} + \frac{1}{670} \times (0.25)^{1.85} \frac{L \times 1000}{(0.6)^{4.87}}$$

$$135 = \frac{1}{670} \times 0.277 \times \frac{(40 - L) \times 1000}{0.08} + \frac{1}{670} \times 0.077 \times \frac{L \times 1000}{0.08}$$

$$135 = \frac{1000}{670 + 0.08} \cdot [0.277 \times (40 - L) + 0.077\ L]$$

$$\frac{135 \times 670 \times 0.08}{1000} = [0.277 \times (40 - L) + 0.077\ L]$$

$7.236 = 11.08 - 0.277\ L + 0.077\ L$

$-3.844 = -0.2\ L$

$L = 19.22$ km

Therefore, the additional flow can be compensated by dividing the total flow equally by placing an additional pipe of length 19.22 km parallel.

Now the thickness of the pipe can be calculated as:

The maximum static head available = 510 – 290 = 220 m of water head

= 220 × 9.81 kN/m² = 2158 kN/m²

Therefore, the total internal pressure = 2158 kN/m²

Now using equation,

$$t = \frac{1}{\eta} \cdot \frac{p \cdot d}{2 \cdot f}$$

where,

f = Allowable tensile stress in cast iron pipe = 75,000 kN/m²

d = 0.6 m

η = Assume efficiency of the pipe joint (0.75)

Putting the values in above equation as:

$$t = \frac{1}{\eta} \cdot \frac{p \cdot d}{2 \cdot f} = \frac{1}{0.75} \times \frac{2158 \times 0.6}{2 \times 75000} = 0.01151 \text{ m} = 1.15 \text{ m}$$

Now taking an extra provision of 3 mm for corrosion, the pipe shell thickness is as :

= 1.15 + 0.3 = 1.45 cm

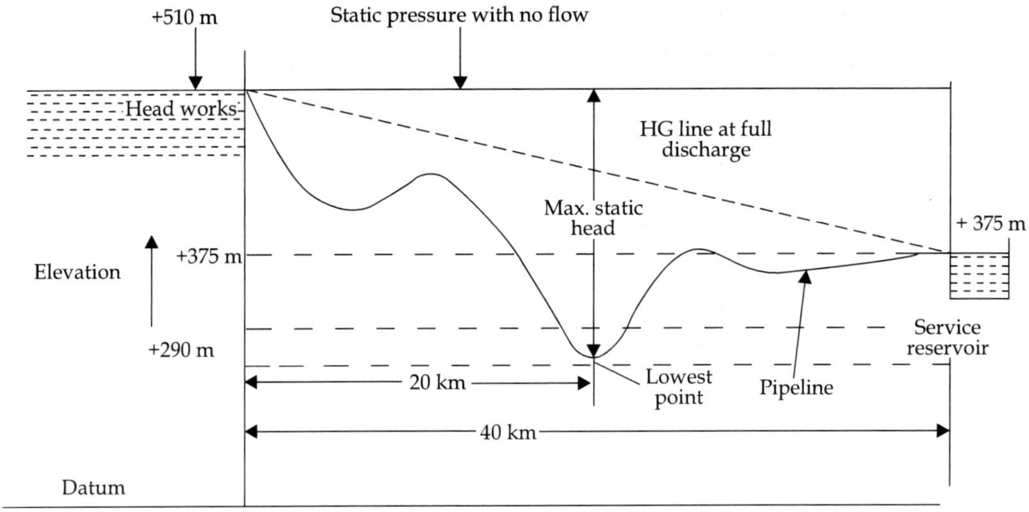

Fig. 8.8: Representation of L-section of the pipe in solved example

8.6.4 Pressures Due to External Loads

In many construction works and practices the pipes are often buried under the ground and placed in trenches which are backfilled on the roads or in streets. These pipes are often loaded with weight of the backfill, the superimposed traffic loads if any and the self-weight of the pipe is always, transferred to the pipe. This transferred load produces compressive stresses in the pipe material specially when the pipe is empty and there is no internal pressure to counteract these loads. This material may fail in compression, if it exceeds the allowable compressive stress of the pipe material. While designing these pipes such aspects are considered and the pipes should also be checked for this possible failure. In case of pipes used for water supply systems this is not an issue as such pipes are generally laid narrower and not much depth. However, the pipes commonly used for transporting water are made of steel, reinforced concrete and the cast iron pipes are quite good in compression. Moreover, the stresses that develop due to such types of external loadings can be estimated by using the empirical formulae based on the experiments carried out by Marston, etc. at lowa State College in U.S.A. and are discussed here. These formulae are expressed under different types of loading conditions along with different types of backfill materials as given below.

(i) For pipes to be placed in highway culverts resting on or projecting above the undisturbed ground in cohesion-less soil (sand) and covered with backfills can be represented by using the equation with external load (W) per unit length of pipe is given by Eq. 8.25 and Fig. 8.9.

$$W \cdot C_p \cdot \gamma \cdot D^2 \tag{8.25}$$

C_p = Pipe coefficient the value depends on pipe type and characteristics of foundation & backfill (values in Table 8.6)

γ = Unit weight of backfill material

D = External pipe diameter = (Pipe internal diameter + 2 × pipe thickness)

Table 8.6: Values of pipe coefficient (C_p) under different conditions

Loading and placing condition of pipe	**Cover depth / Pipe dia external** $= \frac{H}{D}$	C_p **values**	
		Rigid pipe	**Flexible pipe**
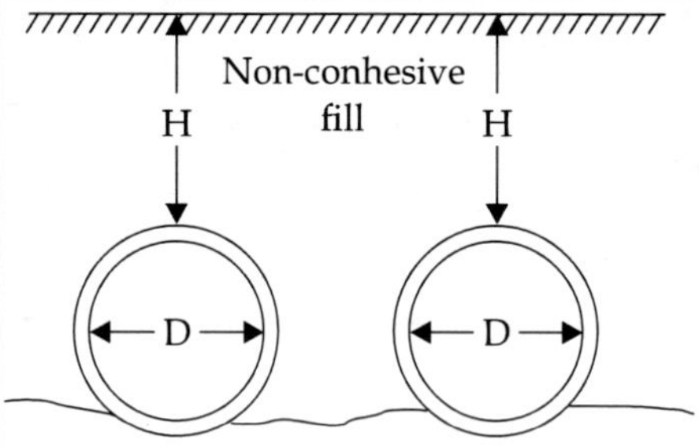 **Fig. 8.9**	1.0	1.2	1.1
	2.0	2.8	2.6
	3.0	4.7	4.0
	4.0	6.7	5.4
	6.0	11.0	8.2
	8.0	16.0	11.0

(ii) For flexible pipes such as steel pipes and are buried in narrow trenches and thoroughly compacted side fills, such as shown in Fig. 8.10, the external load per unit length of pipe is given by Eq. 8.26;

$$W = C_P \cdot \gamma \cdot B \cdot D \tag{8.26}$$

where,

C = A coefficient characteristic of the fill material and depends on H/B ratio (values in Table 8.7)

(iii) For pipes such as concrete, cast iron, vitrified clay, etc., considered as rigid and are buried in narrow trenches and thoroughly compacted with cohesion-less fills, the external load per unit length of the pipe is given by Eq. 8.27.

$$W = C_P \cdot \gamma \cdot B^2 \tag{8.27}$$

where, C, γ and B have the same meaning as given in the previous equation.

(iv) The amount of superimposed load such as traffic load, etc., transferred to the pipe can be evaluated by using Boussinesq's equation. Assuming the fill surface to be horizontal, the Eq. 8.28:

$$p_t = \frac{3 \cdot H^3 \cdot P}{2\pi \cdot Z^5} \tag{8.28}$$

where,

p_t = Unit pressure developed at any point in the fill at a depth (H) below the surface due to traffic load

P = Superimposed load such as traffic load, etc.

Z = The slant height of the considered point from the load P

H = Distance of the top of pipe below the surface of the fill

Table 8.7: C values of pipes buried under narrow trenches with different material conditions

Material type and pipe condition	**Fill material**	**Dry sand & gravel**	**Saturated topsoil**	**Clay**	**Saturated clay**
	Unit weight (kN/m^3)	**16**	**16**	**19.2**	**20.8**
	H/B ratio	**C values**			
	1.0	0.84	0.86	0.88	0.92
	2.0	1.45	1.50	1.55	1.62
	3.0	1.90	2.00	2.10	2.26
	4.0	2.22	2.33	2.49	2.65
	5.0	2.45	2.60	2.80	3.03
	6.0	2.60	2.78	3.04	3.33
	7.0	2.75	2.95	3.23	3.57
	8.0	2.80	3.03	3.37	3.76
	9.0	2.88	3.11	3.48	3.92
	10.0	2.92	3.17	3.56	4.04
	12.0	2.97	3.24	3.68	4.22
	14.0	3.00	3.28	3.75	4.34

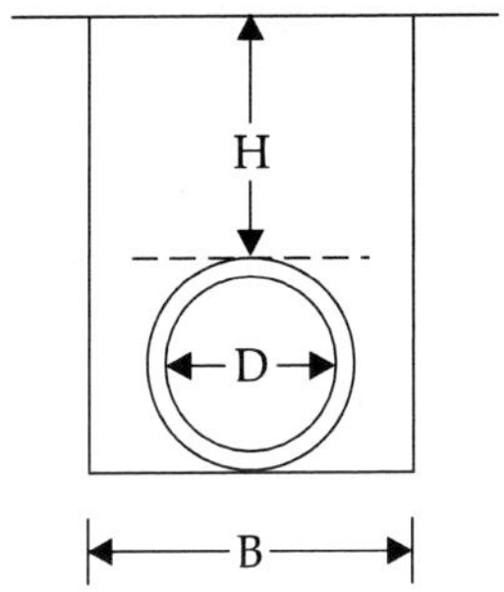

Fig. 8.10

Also the total traffic load (W′) developed per unit length of pipe is determined by integrating Eq. (8.26) as per the availability of the projected area of the pipe. This projected area of the pipe can be subdivided into small cells and then integrating the same. This is then computing the load on each sub-division and finally adding these values to obtain the total. However, the impact value from moving loads may be taken as double in case of unsurfaced roads and the same is computed. This is because the paving of roads reduces this effect of impact significantly. Also, the effect of superimposed loads affects with respect to the depth and decreases rapidly as the depth of cover increases, this is because the slant height (Z^5) increases as compared to vertical distance of the top of pipe below as H^3.

The total load per unit length of the pipe from the backfill and from traffic can be found by adding W with W'. The compressive stress produced during such conditions in a pipe is checked in an empty pipe and is given by (Eq. 8.29):

$$\text{Stress produced} = \frac{W + W'}{t} \frac{\text{kN}}{\text{m}} \tag{8.29}$$

where,

t = Thickness of pipe (m)

Also sometimes there is possibility of creation of vacuum in the pipe due to water hammer, which will increase the compressive stress. The pipe thickness should be checked for that possibility also, to avoid collapse in such conditions.

Example 8.7: A pipes 1.2 m in diameter, buried in a trench 1.5 m wide, is backfilled with dry sand. The top of the pipe is 2.75 m below the surface of the fill. The pipe passes at right angles under a one-lane road which carries a vehicle whose loading (including impact) consists of two concentrated 10 kN loads located at 2 m apart transverse to the roadway. Find the maximum vertical force exerted on a unit length of the pipe, if

(a) using steel pipe, and

(b) using a cast iron pipe.

The thickness of the pipe may be neglected for calculating the external diameter of the pipe. (c) Calculate the stress produced in a steel and cast iron pipe. The thickness of both pipes is 1.25 cm?

Solution: First calculate the forces caused by the backfill in both the given cases:

Given that:

$d = 1.2$ m

$B = 1.5$ m

For dry sand $\gamma = 16$ kN/m^3

$H = 2.75$ m

Case (i)

The steel pipe is considered a flexible pipe

Therefore, using equation 6.24, we have

$$W = C_P \cdot \gamma \cdot B \cdot D$$

The value of C_p can be calculated by using the H/B ratio from Table 3.6 which can be read out as:

$\frac{H}{B} = \frac{2.75}{1.5} = 1.83$, therefore the $C_p = 1.32$ (approx.)

Now putting the value in the above equation as:

$$W = C_P \cdot \gamma \cdot B \cdot D = 1.32 \times 16 \times 1.5 \times 1.2 = 38.3 \text{ kN/m}$$

Case (ii)

For cast iron pipe as it is considered a rigid pipe the equation is:

$$W = C_P \cdot \gamma \cdot B^2 = 1.32 \times 16 \times (1.5)^2 = 47.5 \text{ kN/m}$$

Similarly, the forces caused by traffic load can be calculated as:

Pressure developed due to traffic loads $= p_t = \frac{3 \cdot H^3 \cdot P}{2\pi \cdot Z^5}$

In the above equation the value of slant height Z can be calculated as:

$$Z = \sqrt{\left(\frac{\text{Spaceing between the wheels}}{2}\right)^2 + H^2} = \sqrt{\left(\frac{2}{2}\right)^2 + (2.75)^2} = 2.92 \text{ m}$$

Putting the value of Z, P and H in equation, we get

$$p_t = \frac{3 \cdot H^3 \cdot P}{2\pi \cdot Z^5} = \frac{3 \times (2.75)^3 \times 10}{2 \times \pi \times \cdot (2.92)^5} = 0.467 \text{ kN/m}^2$$

Now the total force due to traffic load/m length of pipe (W')

$$W' = p_t x \cdot \text{Number of wheels} \times \text{projected area}$$

Projected area = Diameter × 1 metre length = 1.2 × 1 = 1.2

Putting the value in the above equation, we get

$$W' = 0.647 \times 2 \times 1.2 = 1.12 \frac{\text{kN}}{\text{m}} \text{ length of pipe}$$

Total external forces including the backfill and traffic per metre length of pipe

Case (i) for steel

$W + W' = 38.3$ kN/m + 1.12 kN/m = 39.42 kN/m

Case (ii) for cast iron

$W + W' = 47.5$ kN/m + 1.12 kN/m = 48.62 kN/m

Stress produced

(i) Stress in steel pipe $= \dfrac{W + W'}{t} = \dfrac{39.42}{0.0125} = 3153 \text{ kN/m}^2$

(ii) Stress in cast iron pipe $= \dfrac{W + W'}{t} = \dfrac{\text{kN}}{\text{m}} = \dfrac{48.62}{0.0125} = 3890 \text{ kN/m}^2$

8.6.5 Temperature Stresses

Sometimes the pipes are laid above the ground and are exposed to atmosphere and are subjected to temperature changes with respect to the change in climatic conditions. Also during the diurnal variations in temperature these pipes expand during daytime and contract at night. However longitudinal stresses develops in the pipe material when the prevention against this expansion or contraction is done by fixation or friction over the supports. The magnitude of these stresses can be calculated by using Eq. 8.30 as below:

Elaongation = $L \cdot \alpha \cdot T$

Strain = Elongation per unit length $= \dfrac{L \cdot \alpha \cdot T}{L} = \alpha . T$

Stress = E · Strain = $E_p \cdot \alpha \cdot T$

or,

$$f = E_p \cdot \alpha \cdot T \tag{8.30}$$

where,

E_p = Modulus of elasticity of the pipe material (discussed earlier)

α = Thermal co-efficient of expansion of pipe material (values in Table 3.8)

T = Temperature difference (°C)

Special type of expansion joints should be provided at definite intervals (say 20 m or so) to counteract the temperature stresses in pipes exposed to atmosphere or are placed above the ground. Thermal coefficient of expansion (α) values for different pipe materials are shown in Table 8.8.

Table 8.8: Thermal coefficient of expansion (α) values for different pipe materials

S. No.	Material	α values in m/m/°C (20-100°C)
1	Steel	12.6×10^{-6}
2	Cast iron	11.0×10^{-6}
3	Concrete	9.8×10^{-6}
4	Copper	16.5×10^{-6}

8.6.6 Flexural Stresses

Sometimes in order to overcome the obstacles in transporting water crossing the railway track or crossing a waterbody the pressure pipes are carried supported between arrangements or structures such as trestles, piers, bridge hangers like beams, etc. Also sometimes the rainwater, soil erosion, etc. may wash off the ground from below the pipes resting on the ground, this exposes the pipes and their supports. During such exposure the bending stresses are produced in the pipe, as under such instances the pipe acts like a beam with loads resulting from the self weight of the pipe along with weight of the water in the pipe along with other superimposed loads that might be acting on the pipe. These stresses in the pipe can be assessed by utilizing the usual methods of structural analysis applied to the beams. However, under normal conditions, the stresses are often neglected when pipes have smaller spans, and are therefore considered for long spans subjected to huge superimposed loads.

Pressure types: The various important terms mostly used in India for defining the various pressure systems by the manufacturers in defining the pressure pipes.

Working pressure: The pressure is defined as the actual maximum pressure that includes abnormal conditions such as water hammer pressure to which the pipes may be subjected during their operation.

Design pressure: It may be defined as the maximum pressure for which the pipe has been designed. This is equal to the product of the working pressure and suitable factor of safety to cover abnormal increase in pressure due to unforeseen circumstances.

Test pressure: It may be defined as the maximum pressure which the pipe can withstand without any leakage when tested for hydrostatic pressure in accordance with the standard methods of testing.

8.7 PIPES AND REQUIREMENTS

Pipes bring raw water from the source via the delivery system to the treatment plants. Water is still under pressure and the pipe material will also withstand stress due to internal strain, vaccum pressure, when the pipes are empty, water hammer when the valves are closed and temperature stress.

8.7.1 Requirements of Pipe Material

- It should be capable of withstanding internal and external pressures .It should have easy joint facility.
- It should be available in all sizes, should be easy to transport.
- It should be durable.
- To alter its quality, it should not react with water.
- Pipes expense should be less.
- Loss of frictional head should be minimum.
- The affected units could be quickly replaced.

8.7.2 Various Types of Pressure Pipes

Depending upon the construction material, the pressure pipes are of the following types:

- Cast iron pipes
- Steel pipes
- Reinforced cement concrete pipes
- Hume steel pipes
- Vitrified clay pipes
- Asbestos cement pipes
- Miscellaneous types of pipes

The selection of a particular type of material for a pipe depends mainly upon the relative economy, the pressures likely to come and the working pressures, maximum permissible sizes and capacities, availability of materials and labour for their construction, etc. The various types of pipes and their relative utilities are discussed below and summarized in Table 8.10.

Cast iron pipes: Cast iron pipes are widely used for city water supplies. They are sufficiently resistant to corrosion and may last as long as 100 years or so. They are generally manufactured in lengths of about 3.5 m, but may be manufactured up to 6 m or so, on special orders. Cast iron pipes can be manufactured by two methods. One is the ordinary sand moulding and the other is the centrifugal process. Pipes cast in horizontal position by ordinary sand moulding are called "McWane" pipes, while those cast vertically are called "pit cast pipes". Horizontal cast pipes are 100 per cent stronger in tension and 50 per cent stronger in rupture than vertically cast iron pipes.

Centrifugally cast iron pipes are made either in sand or metallic moulds. The former are called sand spun pipes and the latter are called delavaud pipes. The spun iron pipes are denser and tougher than the pipes moulded by ordinary methods. In spun iron casting in metallic moulds, the molten metal is applied to the interior of a water cooled cylindrical metal mould which is rotated rapidly. Centrifugal force holds the metal against the mould and forms a homogeneous pipe with a cylindrical bore. After the pipe cools down to a temperature of about 800°C, it is taken out from the machine to a furnace where it is raised to about 925°C, and then slowly cooled to about 650°C so as to anneal it. It is cooled after hydrostatic testing.

Indian standards: Centrifugally cast iron pipes (called spun iron pipes) are generally available in India in three types (i.e., Class LA. Class A, and Class B) as per the provisions of I.S. 1536-1976 as given in Table 8.9:

Table 8.9: Types of cast iron (spun) pipes as per IS:1536-1976

S. No.	Category of pipes in order of increasing thickness for same diameter	Test pressure (kN/m^2)	Working pressure (kN/m^2)
1.	Class LA	3500	1200
2.	Class A	3500	1800
3.	Class B	3500	2000
Test pressures in vertically cast iron pipes as per IS: 1537-1976			
S. No.	**Diameter (mm)**	**Test pressure using socket and spigot joints (kN/m^2)**	
		Class A	**Class B**
1.	Up to 300	2000	2500
2.	300-600	2000	2500
3.	600-1000	1500	2000
4.	1000-1500	1000	1500

The nominal internal diameters available are 80, 100, 125, 150, 200, 250, 300, 350, 400, 450, 500, 600, 700 and 750 mm. Their respective thicknesses for class LA are 7.2, 7.5, 7.9, 8.3, 9.2, 10.0, 10.8, 11.7, 12.5, 13.3, 14.2, 15.8, 17.5, and 18.3 mm, respectively. For class A, the respective thicknesses are 7.9, 8.3, 8.7, 9.2, 10.1, 110, 119, 12.8, 13.8, 14.7, 15.6, 17.4, 19.3 and 20.2 mm. And for Class B, the respective thicknesses are 8.6, 9.0, 9.5, 10.0, 11.0, 13.0, 14.0, 15.0, 16.0, 17.0, 19.0, 21.0 and 22.0 mm.

Vertically cast iron pipes are generally available in India in two types (Class A and Class B) as per the provisions of IS 1537- 1967. The thicknesses of Class B pipes are more than those of Class A for the same diameters. These pipes are available in internal nominal diameters of 80, 100, 125, 150, 200, 250, 300, 350, 400, 450, 500, 600, 700, 750, 800, 900, 1000, 1100, 1200, and 1500 mm. Their respective thicknesses for Class A pipes are 7.9, 8.3, 8.7, 9.2, 10.1, 110, 11.9, 12.8, 13.8, 14.7, 15.6, 17.4, 19.3, 20.2, 21.1, 22.9, 24.8, 26.6, 28.4 and 33.9 mm. For Class B, the respective thicknesses are: 8.6, 9.0, 9.5, 10.0, 11.0, 12.0, 13.0, 14.0, 15.0, 16.0, 17.0, 19.0, 210, 22.0, 23.0, 25.0, 27.0, 29.0, 31.0 and 37.0 mm. The test pressures are as given in Table 8.9.

As is evident from the above data, the spun iron pipes are thinner than the vertically cast iron pipes, and hence are lighter and can be manufactured in longer lengths. In order to protect these pipes from corrosion, they are generally counted with bituminous linings. Larger sizes may be provided with cement mortar (1:2) lining, whereas smaller sizes, such as those to be used in small service lanes for water supply distribution system, are generally coated with zinc and are known as galvanised iron pipes. The advantages of cast iron pipes are:

- Moderate in cost
- Easy to join
- Strong and durable
- Corrosion resistant
- Long life up to 100 years or so
- Service connections can be easily made.

The disadvantages of cast iron pipes are:

- Water carrying capacity decreases with time, as the value of friction factor increases due to tuberculation in certain waters.
- They cannot be used for high pressures. Generally, not used for pressures above 700 kN/m.
- When large they are very heavy and uneconomical.
- They are likely to break during transportation or while making connections joints.

The following types of joints are generally used in cast iron pipes:

- Socket and spigot joint
- Flanged joint
- Mechanical joint called dresser coupling
- Flexible joint
- Expansion joint.

These joints are described below.

(i) Socket and spigot joint. Cast iron pipes are joined by a "socket and spigot joint" (sometimes called bell and spigot joint) since olden days, perhaps a century or so. This joint is till today used on a large scale.

The cast iron pipes, to be joined by the "socket and spigot joint", are made in such a way that one of their ends is larger than the other end. The larger end is called "socket" or "bell", while the normal end is called "spigot". The spigot is fitted into the socket. A few strands of jute are wrapped around the spigot before inserting it into the socket (or bell) and then, more jute is packed into the joint. The remaining space the spigot is finally filled with molten lead, which gets solidified and tightly caulked into the joint after cooling, and thus making a water-tight joint. The quantity of lead required per joint varies from 3.5 to 4 kg for 15 cm diameter 1.2 m diameter pipe

8.7.3 Different Types of Pipes

Table 8.10: Advantages and disadvantages of different types of pipes

S. No.	Type of pipe	Advantages	Disadvantages
1.	Cast iron pipes	Cost is moderate.	Breakages of pipes are large.
		The pipes are easy to join. The pipes are not subjected to corrosion.	The carrying capacity of these pipes decreases with the increase in life of pipes.
		The pipes are strong and durable. Service connections can be easily made.	The pipes are not used for pressure greater than 0.7 N/mm^2. The pipes are heavier and uneconomical beyond1200 mm diameter.
2.	Steel pipes	No. of joinings are less because these are available in long lengths. The pipes are cheap in first cost The pipes are durable and strong enough to resist high internal water pressure. The pipes are flexible to some extent and they can therefore laid on curves. Transportation is easy because of lightweight.	Maintenance cost is high. The pipes are likely to be rusted by acidic or alkaline water. The pipes require more time for repairs during breakdown and hence not suitable for distribution pipes. The pipes may deform in shape under combined action of external forces.
3.	Prestressed concrete pipes	The inside surface of pipes can be made smooth.	The pipes are heavy and difficult to transport.
		Maintenance cost is low. The pipes are durable with life period 75 years.	Repairs of these pipes are difficult. The pipes are likely to crack during transport and handling operations.
		No danger of rusting. These pipes do not collapse or fail under normal traffic.	There pipes are affected by alkalies and acids.
4.	RCC pipes	There are pipes are most durable with usual life of about 75 years.	Transportation is difficult repair work is difficult.
		The pipes can cast at site work and thus there is reduction in transport charges.	Initial cost is high. These pipes are affected by acids, alkalis and salty waters.
		Maintenance cost is less. Inner surface of pipe can made smooth. No danger of rusting.	

5.	AC pipes	The inner surface of pipes are very smooth.	The pipes are brittle and therefore handling is difficult.
		The joining of pipe is very good and flexible.	The pipes are not durable.
			The pipes cannot be laid in exposed places.
		The pipes are anticorrosive and cheap.	The pipes can be used only for very low pressures.
		Light in weight and transport is easy.	
		The pipes are suitable for distribution pipes of small size.	
6.	Galvanized iron pipes	The pipes are cheap.	The pipes are affected by acidic or alkaline waters.
		Light in weight and easy to handle.	The useful life of pipes is short about 7 to 10 years.
		The pipes are easy to join.	
7.	PVC pipes	Pipes are cheap.	The co-efficient of expansion for plastic is high.
		The pipes are durable. The pipes are flexible.	
			It is difficult to obtain the plastic pipes of uniform composition.
		The pipes are free from corrosion.	The pipes are less resistance to heat.
		The pipes are good electric insulators	
		The pipes are light in weight and is easy to mould any shape.	Some types of plastic impart taste to the water.

8.7.4 Maintenance of Pipes

Hygienic quality and adequate flow in the pipelines are to be maintained, preventive maintenance of pipes includes the following:

- Detection of leaks inside the user's premises at defective joints ferrule ties, pipes and fittings.
- Corrosion identification in tubing, cracks and repair of those portions.
- Water wastage of 15 to 25 per cent by pipe joints should be reduced by taking effective preventive steps.
- Pipes washing by flushing and disinfecting of pipes.
- Pollution defense.
- Records of pipe laid lengths, pipe length repaired or replaced, cost incurred, fire hydrant no, service connections no, and other related data disconnected from the delivery network should be held for ready reference.

8.7.5 Pipe Corrosion: Causes and Prevention

Corrosion denotes the loss of pipe material due to the action of water (internal corrosion of the pipe) and the action of waterlogged soil above the pipe surface (external corrosion of the pipe) due to the effects of corrosion. The various factors contributing to the pipe corrosion are as follows:

- Acidity: Due to the presence of carbonic acid or other acids, water with low pH value can cause corrosion.
- Alkalinity: The water with ample alkalinity to calcium bicarbonate is anti-corrosive in nature.
- Biological action: iron-bacteria growth, and sulphur bacteria that develop aerobic and anaerobic corrosion, respectively.
- Chlorination: Free chlorine or chloramine makes the water corrosive.
- Electrical currents: Corrosion canals are often formed by uniting dissimilar metals or by electrical device earthing to water pipes.
- Mineral and organic constituents: The presence of high total solids in water speed up the corrosion process.
- Oxygen: Oxygen is found in both the corrosive and non-corrosive waters. The aeration, in fact, is employed in some cases for preventing corrosion.

Effects of pipe corrosion

- Corrosion of the pipe may result in tuberculation (forming small projections on the inner surface of the pipe) which decreases water carrying capacity.
- Corrosion of the pipe contributes to pipeline disintegration, and it needs significant repairs.
- Corrosion of the tubing gives the flowing water colour, taste, and odour.
- The pipe corrosion imparts colour, taste and odour to the flowing water which significantly affects the connections to the pipes.
- Corrosion of the pipe will make the water hazardous to drinking and other uses.

Prevention of pipe corrosion

Pipe corrosion is not possible to completely eliminate but we can minimize by the following methods.

- Cathodic protection: By connecting the pipeline with DCs negative edge. Generator or magnesium-like anode elements, such that the whole pipe functions as a cathode. This therapy is most successful in cathodic conditions. It is costly and requires various practical problems.
- Proper pipe material: More resistance is found in iron or steel alloys with chromium, copper or nickel.
- Protective linings: Asphalt, bitumen, cement mortar, paints, resins, tar, zinc, etc., coat minimize corrosion.
- Water treatment: Proper PH treatment and modification, calcium carbonate control, removal of dissolved oxygen and carbon dioxide, addition of silicate sodium, etc.

8.8 PUMPS FOR LIFTING WATER

8.8.1 Introduction

In general, for public water supplies pumps are required at different stages in various units in order to lift and transmit water from one unit to the other unit. The requirement of pump at various stages is as:

- The surface water in low lying areas or from tube well or groundwater cannot flow through gravity by its own and therefore needs to be lifted up and is taken to the mains.
- When sufficient slope is not available for gravitational flow at the treatment plants to transmit the water from one unit to other unit, pumping of water is required and therefore is must.
- Sometimes the water after treatment is required to be distributed directly or water is pumped into the reservoirs located at enormous heights.
- In order to reach the water at greater heights or in multistoried buildings when enough pressure is not available or during fire breaks or such emergency the water needs to be boosted up in the intermediate points and therefore pumping is required.

When the source of water is located at high hilly or mountainous areas the water can be transmitted to the consumers by using gravitational flow systems. However, such types of locations are seldom available in the country therefore lifting of water is required at one or more points. Initially, at such instances the water is required to be lifted at the initial stage at the source itself from where it is transmitted to the water treatment plant from the sump well through the intake conduits. Lift is required after the treated water is taken to the distribution systems. This treated water may either be distributed directly or may be stored in elevated tanks or underground reservoirs for further distribution to the consumers or end users.

In case of groundwater sources such as wells or tube wells the first lift of water is required and is stored either in collection basin. The water is then taken to the water treatment plant from the collection basin and thereafter water is again lifted after treatment into the distribution system. However, direct lifting of water to distribution mains is carried out rarely. The lifting of water at intermediate points or boosting required is based on the economy and construction and design of distribution system.

Pumps are mainly incorporated in industries for the distribution of liquids. Pumps convert mechanical energy into hydraulic energy. For the operation of pumps, electrical energy is needed. Pumps have two main purposes. Transfer of liquid from one place to another place (e.g., water from an underground into a water storage tank) and circulation of liquid throughout the system (e.g., cooling water or lubricants through machines and equipment).

8.8.2 Components of a Pumping System

The pumping system used for water lifting and transmission consisting of various components:

- Pump casing and impellers
- Prime movers: electric motors, diesel engines or air systems
- Piping used to carry the fluid
- Valves, to control the flow in the system
- Other fittings, controls and instrumentation
- End-use equipment, which have different requirements (e.g., pressure, flow) and therefore determine the pumping system components and configuration. Examples include heat exchangers, tanks and hydraulic machines.

8.8.3 Types of Pumps and Their Suitability in Water Supply

Based on the mechanical principle of water lifting pumps are classified as the following (Table 8.11).

Table 8.11: Types of pumps and their suitability in water supply

S. No	Type of pump	Examples	Suitability
1.	Velocity pumps	Centrifugal pumps, deep well, turbine pumps, jet pumps	Widely used in water supply schemes which contain sand, silt, etc.
	Displacement pumps	Reciprocating pumps. Rotary, chain, gear wheel, pump and wind mills.	Is suitable for moderate heads and small discharges suitable for fire safety, individual houses water supply.
2.	Buoyancy pumps	Airlifting pumps	Are used to pump water from deep wells to a lift of around 60 m containing mud, silt, rubble, etc.
3.	Impulse pumps	Hydraulic ram	Used for small water supply projects to leave the water for around 30 m or so in height.

8.8.4 Classification of Pumps

The general classification of pumps is done on the basis of its mechanical configuration and their working principles (Fig. 8.11) may be classified as:

1. Rotodynamic pumps / kinetic pumps
2. Displacement pumps / positive displacement pumps

8.8.4.1 Rotodynamic Pumps

Rotodynamic pump is also called kinetic pump as it imparts velocity and pressure to liquid as it conveys to impeller and convert that velocity into additional pressure.

Kinetic pumps are subdivided into two major groups and they are centrifugal pumps and positive displacement pumps. Also in case of rotodynamic pressure pumps, tangential force is imparted during pumping action which usually accelerates the fluid by spinning the impeller. Some systems which involve rotodynamic pumps can require positive priming displacement pumps. Usually, they are used for low-to-high rate of discharge. With this type of pump the pressure differential scale is in a scale from low to high.

Classification of Rotodynamic Pumps

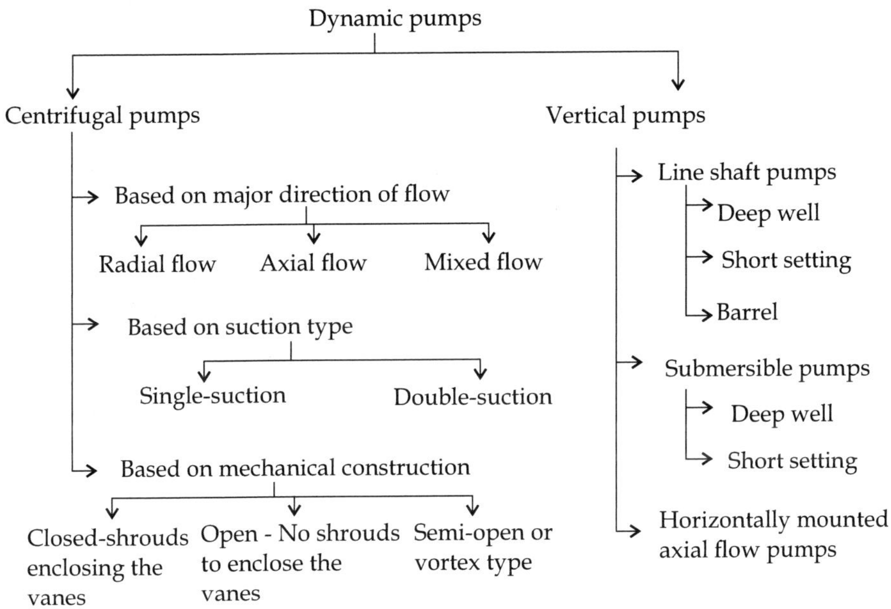

Fig. 8.11: Classification of rotodynamic pumps

8.8.4.1.1 Centrifugal Pumps

A centrifugal pump is a rotating machine in which flow and pressure is generated rotodynamically. The impeller and the volute or casing are the two main components of a pump by the virtue of which energy changes occur. The casing of a pump collects the liquid being expelled by impeller and converts it into pressure energy. Centrifugal pumps are subdivided into several categories, as shown in above Fig. 8.11.

To increase the friction of a fluid they use a revolving impeller. Commonly, centrifugal pumps are used to transfer liquids via a piping network. The fluid joins the pump impeller around or near the spinning axis and is accelerated by the impeller, flowing radially outwards through a diffuser or volute chamber (casing) from which it emerges into the downstream piping network as seen in Fig. 8.12. Centrifugal pumps are used for large discharge through smaller heads. These types of pumps are used for supply of water and handling of milk in dairy plants.

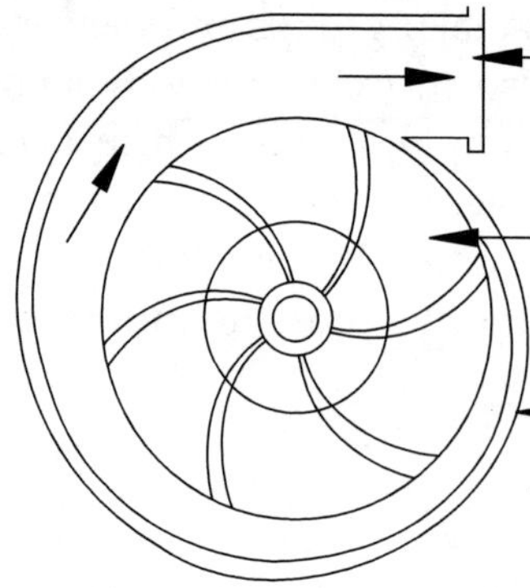

Fig. 8.12: Centrifugal pump

Radial flow: In radial flow system of a pump the impeller discharges the fluid to the shaft axis at right angles. In such a centrifugal pump, the centrifugal force fully creates the strain. The radial style pumps are used for high-head and low-discharge applications. The ratio of impeller outside diameter (D_2) to eye diameter (D_1) in radial flow pumps is 2 or more, and the impeller is small in diameter.

Mixed flow: In these pumps the direction of flow is partly axial, and partly radial. Therefore, the resulting flow is diagonal. The mixed flow style pumps are used for the application of medium head and high discharge. For this centrifugal system, under which the energy is produced partly through centrifugal force and partly through raising the impeller's vanes into the air. In mixed flow pumps ratio of impeller outside diameter (D_2) to eye diameter (D_1) is less than 1.5 and the impeller has wider width.

Axial flow: In such type of pumps the movement into the impeller is low head and very high discharge parallel to the shaft axis. The axial flow style pumps are used for medium head and high discharge applications. In centrifugal system pumps the energy is produced by the propelling or rising operation of the impeller's vanes on the air. The flow through impeller is parallel to shaft axis low head and very high discharge. In axial flow type pumps the axial flow pumps ratio of impeller outside diameter (D_2) to eye diameter (D_1) is one and impeller has no width.

Impeller dimensions of axial, radial and mixed flow pumps

Specific speed (Ns) of the centrifugal pump identifies the approximate acceptable ratio of the impeller eye diameter (D_1) to the impeller outside diameter (D_2) in designing a good impeller as shown for different types of pumps in Table 8.12.

Table 8.12: Pump types and specific speed

Pump type	Specific speed (Ns) rpm	(D_1/D_2) Ratio
Radial flow	500 to 5000	> 1.5
Mixed flow	5000 to 10,000	< 1.5
Axial flow	10,000 to 15,000	= 1

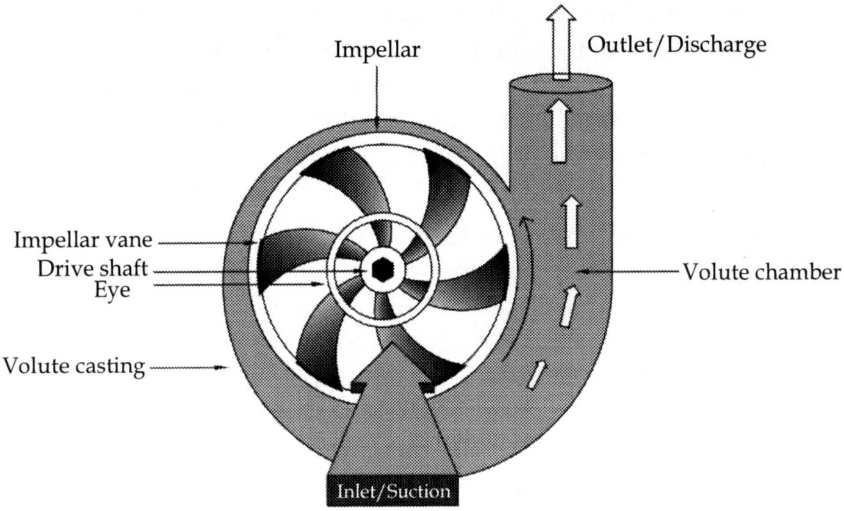

Fig. 8.13: Centrifugal with volute chamber

The **centrifugal pump** consisting of a radial or a mixed flow impeller may either be in closed casing or open. In an open impeller the vanes are attached to a hub. The "open impeller centrifugal pump" has less efficiency as compared to a "closed impeller centrifugal pump" in case of water lifting. However, the "open impeller" pumps are less likely to be clogged by debris, floating material, etc., therefore such pumps are mostly used for pumping raw waters or sewage containing floating materials, solids and other impurities. The casing of such radial flow centrifugal pump may either be of a volute type or of a turbine type.

In a **volute type** of pump, the required pressure head is created as the impeller is allowed to reduce the velocity of water as it enters the discharge pipe when water is discharged into a gradually expanding spiral casing, being designed to produce an equal velocity of flow around the circumference of the impeller. A centrifugal pump functions as a reversed turbine except that special device to improve the output are made in this unit. Centrifugal pumps are all of the form of external flow, where the centrifugal movement induced by the moving vanes raises the radial velocity of water. The pump will be full of water when it begins and, thus, it should not be left empty. External power is used to turn the vanes which gives a centrifugal head to the water collected in the pump.

At a high velocity and pressure this water is pushed to leave the rotating vanes at the outer perimeter. This creates a partial vacuum in the middle of the pump and subsequently the water from the suction pipe flows into it. It is the high pressure from the water that leaves the vanes, used to overcome the pump's delivery head. In the pumps made in the earlier years, the kinetic energy of the water leaving the vanes was not utilized and there used to be considerable loss of energy in the formation of eddies in the surrounding circular chamber.

This flow was rectified in later models by translating the kinetic energy into pressure energy by letting the water exit the vanes and flow through a passage is passed slowly at a distance. The change in the pressure head generated in this way is used to change the pump's transmission head, thus increasing the pump's output. Generally, centrifugal pumps are made of the radial flow type only. But there are also axial flow or propeller pumps which are particularly adopted for low heads.

A pump is described with respect to the number of impellers installed in it. The pump installed with one impeller is known as a single-stage pump. The pump installed with two or more impellers and placed in such a way that the discharge from one impeller enters the eye of the next impeller, are known as double-stage or multistage pump. Such types of pumps are advantageous for lifting water at greater heights. In the case of a borehole pump, such as a deep well turbine pump, used for pumping groundwater, several impellers are installed on a vertical shaft, which is suspended and rotated by the prime mover motor placed at the ground surface. The pump bowels are thus kept below the well water level. The type of pump is installed in a well casing of limited size and entire assembly has to be of small diameter.

Submersible pump is another type of a deep well pump, and is different from the usual deep well turbine pump as the driving motor in such pumps is placed below the pump bowels within the well water level. Therefore, in these pumps the driving shaft is completely eliminated thus make the water to rise at the surface through a riser pipe in the suspended assembly. In such case the bearing friction is reduced by eliminating the driving shaft from the ground surface to the pump. It helps the lift pipe for the delivery of water to the surface without any obstruction. This eliminates the well water to come in contact with the electrical parts or the motor bearings, by enclosing it in oil filled case with a mercury seal where the shaft passes through at the top. The recent advancements allow the windings of these pumps to remain surrounded by the water which acts as coolant and bearing lubricant.

Priming and operation of centrifugal pumps

Before starting the centrifugal pump it has to be primed with water, that is filling water in the pump casing to remove the air trapped in the pump through an exhaust valve. This trapped air in the pump casing generally hinders its operation and ultimately reduces its drawdown efficiency. Priming is not required if the pumps are installed below the normal water level of the area. Priming is required when the pumps are placed on horizontal shafts. Recent advancements have made provisions and the pumps are provided with vacuum pumps to remove the air from the casing without the use of water from outside.

The installation of centrifugal pump with horizontal shaft is shown in Fig. 8.14. To make the pump efficient and effective, the suction and the delivery pipe should be appropriately arranged and it is mandatory that the pump should be started with utmost care after priming only. Also it is suggested that the diameter of the pump casing at the inlet end and at the discharge end should be kept smaller than that the pipe used for delivery of water and connected to it. This proves to be economical as well as provides better efficiency. In case of horizontal reducer to be used at the inlet end, at the same time an eccentric reducer must be used. This prevents air accumulation in the pump. To prevent the backwater flow after stoppage of pump a foot valve should also be fitted at the end of the suction pipe. This acts as self-priming process and therefore priming is not required every time. For preventing the water backflow during power failure, the delivery pipe connected to the pump is often installed with a check valve. and a gate valve. The gate valve is also provided at the delivery pipe that opens slowly only after the pump has picked up its full speed and the discharge pipe is filled with water. During this stage full back pressure exists on the pump and the motor picks up the load gradually at this stage. However, the pump started with empty discharge pipe along with the discharge valve open could develop an overloading and burning of the motor. To prevent the entry of debris during pumping of raw water a strainer or a screen is generally provided at the entrance of the suction pipe.

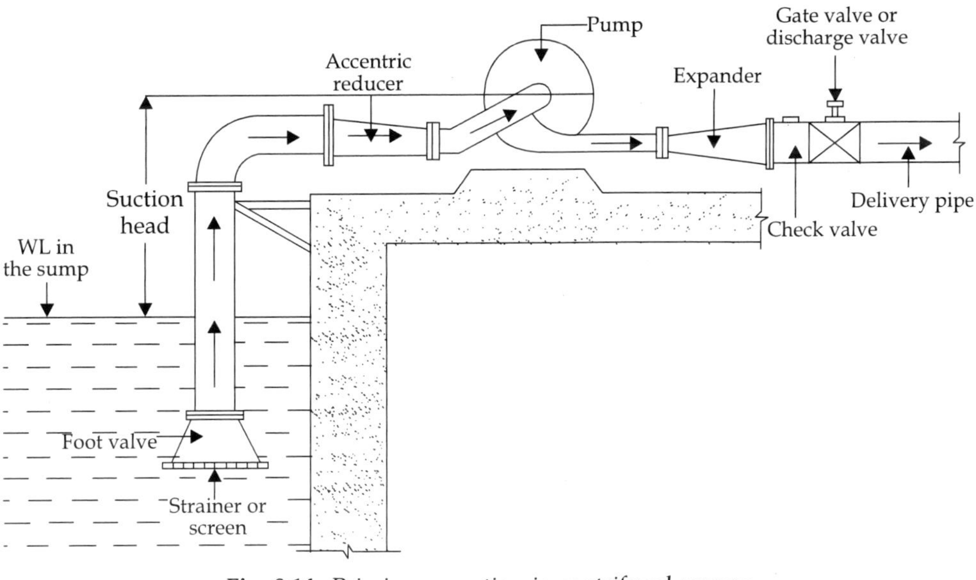

Fig. 8.14: Priming operation in centrifugal pumps

Characteristics of centrifugal pumps

The operating characteristics of a centrifugal pump and axial flow pump depend on its design and speed of operation. The typical characteristic curves chart indicating the discharge against head, efficiency and horsepower is shown through Fig. 8.15. The curves are obtained and plotted at constant speed, with variations of head, efficiency, and brake horsepower, against the corresponding rate of discharge. The obtained curves helps in selecting a particular pump for a particular use. From the plotted curves it can be inferred that at a particular speed of operation:

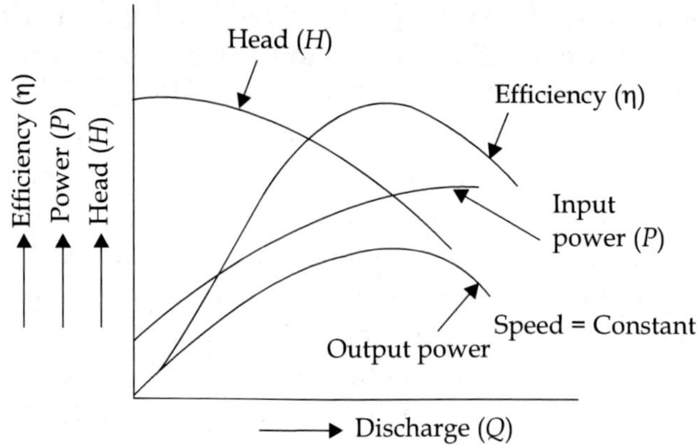

Fig. 8.15: Typical characteristic curves of a centrifugal pump for a given speed

- An increase in discharge thus decreases the head produced. The value of maximum discharge is limited under minimum head conditions as beyond which the pump will stop working. However at zero discharge condition with discharge valve being closed the maximum head is developed and this head is known as shut-off head. The same should be restricted to last long, otherwise, the pressure will abnormally rise above the design head.
- At a particular discharge or at a particular use the maximum efficiency can be obtained. The discharge at which the efficiency of the pump is maximum is known as the normal discharge. This discharge is also known as the rated discharge or the rated capacity of the pump at a given speed and is used in rating the pumps. However, it indicates that during the requirement of variable discharge condition for a pump working at a particular speed the pump will not be able to function at its maximum efficiency thus resulting in reduced efficiencies. Therefore, to obtain the maximum efficiency under variable discharge conditions the best option is to vary the speed with belts and pulleys or by using a variable speed motor. However, the best alternative for regulating the flow under variable discharge conditions is to provide several pumps of smaller capacities, in parallel. This would be helpful that a variable number may be operated at required capacity depending on the flow requirements and all the pumps will be operated comfortably at the maximum efficiency. This would be helpful in managing the water demand variations in water supply schemes throughout the day. During such situations any method may be adopted or sometimes all the pumps are operated at full capacity. The excess water during the period of less demand is stored in the "service reservoir" and can be used during high demand periods.
- While increase in discharge and being maximum at the shut-off head the brake horsepower required from the motor as prime mover to drive the pump, increases. Also this prime mover will not be overloaded when the discharge valve is closed under this condition.

Note:

- Between specific speeds of 1000 to 4000, the radial flow centrifugal pumps perform well.
- Between specific speed of 4000 to 7000, mixed flow centrifugal pumps are most efficient.
- At specific speed greater than 7000, axial flow pumps have higher efficiencies.
- High specific speed is associated with high discharge at low heads, hence axial flow pumps are particularly suitable to storm water pumping.

Advantages and disadvantages of centrifugal pumps

The rotodynamic pumps, especially those of centrifugal type, are widely used for pumping, and have almost replaced the reciprocating types of pumps.

The various advantages possessed by these pumps are:

- Initial cost as well as the maintenance cost are comparatively low.
- Size is compact and can be installed in limited space.
- Mechanism is simple, and thus, less skilled labour is required for their operation and repairs.
- Can be operated with high speed electric motors, or gas engines and steam turbines.
- Discharge obtained is steady and is non-pulsating
- Can be used for pumping water containing silt, sand, etc.
- Are quite durable and safe against high pressures.

The limitations or the various drawbacks of such pumps are:

(i) Such pumps require priming.

(ii) Such a pump should neither be started with discharge valve open, so as to avoid overloading, nor should the discharge valve be kept closed for longer time after starting the pump, so as to avoid too high pressures.

(iii) The discharge pipe has to be provided with check valve, so as to avoid back flow when the pump suddenly stops due to power failure, etc., with the discharge valve open.

(iv) The discharge from such a pump varies with the head of water. Hence, when uniform discharge is required at varying heads the pump has to be rotated at variable speeds which is neither, practical nor economical.

(v) For high heads, the efficiency is low (say about 50 to 80 per cent).

(vi) Their ordinary suction lift is limited (say up to about 6 m or so) and the pump has to be placed under water where higher suction lifts are involved, such as in turbine pumps or in submergible pumps; thereby, necessitating special types of centrifugal pumps with their increased costs.

8.8.4.1.2 Propeller Pump

A propeller pump is a device of the high-flow, low-lift impeller type with a linear flow path. The propeller pump may be mounted vertically, horizontally or bent, and usually has the motor above the sea level with the impeller under the sea as shown in Fig. 8.16. These pumps operate by drawing water from an outer enclosure and form a discharge outlet from an impeller head bladed by a propeller.

Fig. 8.16: Propeller pump

8.8.4.1.3 Turbine Pump

Turbine pumps are centrifugal motors, which incorporate friction and flow with a rotary fluid transfer system. They usually employ blade geometry, which brings pressure from inlet to outlet by allowing fluid movement across the vanes. Turbine pumps use kinetic energy and an impeller and transfer air. The centrifugal force forces the liquid within close proximity to the impeller or propeller vanes to the building surface. The impeller's cyclical motion creates friction inside the rotating vessel. Even the form of turbine pumps relates to the rate of suction and discharge (Fig. 8.17). In a turbine type or diffuser type of pump, the impeller is surrounded by stationary guidevanes which reduce the velocity of water before it enters the casing, thus converting the velocity head into pressure head in the casing itself. The casing surrounding the guidevanes is generally circular and concentric with the impeller. The velocity of the water is thus more completely converted into pressure head. This results in higher efficiency than a volute pump.

8.8.4.1.4 Axial Split-Case Pumps

Axial split-case pumps have a shell broken up around the shaft 's middle axis. The impellers can be easily revealed by cutting the upper half of the casing for repair and

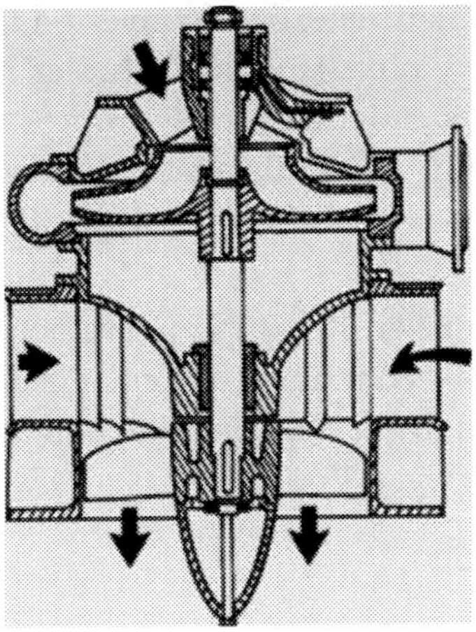

Fig. 8.17: Turbine pump

testing. This is sometimes called the split horizontal or the single-case horizontal pump. For higher pressures, the axial-split pumps can be single or multi-stage. The pumps are usually mounted with shafts in the horizontal position, but vertically mounted pumps for reduced floor space are also available.

The following methods are used to convert the kinetic energy of the water leaving the vanes into pressure energy.

(i) **The volute chamber:** The volute chamber is a circular casing circling the wheel also known as the impeller. The water that exits the vanes is guided to circumferentially travel through the volute chamber. The area of the volute chamber increases gradually and thus accompanied by corresponding pressure increase, the velocity is decreased as depicted in Fig. 8.18.

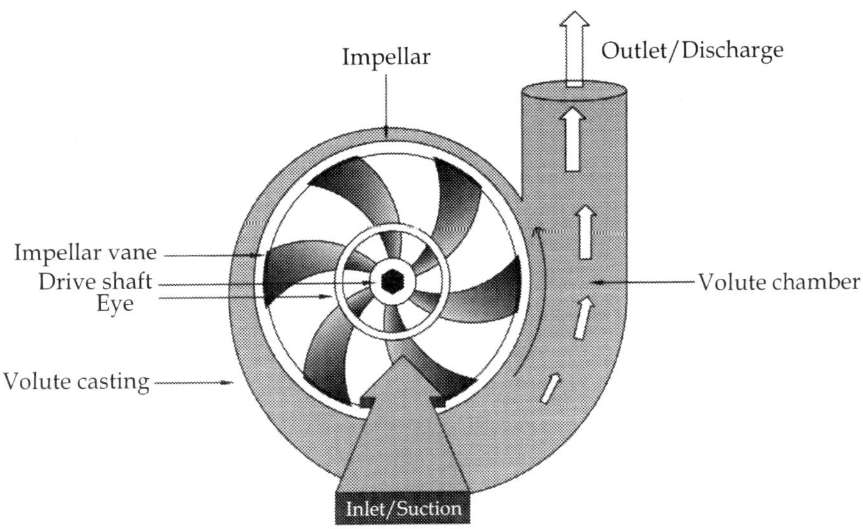

Fig. 8.18:

A considerable part of kinetic energy is converted into pressure energy as the water reaches the delivery pipe. However, studies have shown that the pump efficiency in this system is only marginally improved. This means that eddies are not entirely eliminated, and some energy depletion happens in the eddies due to the continually rising volume of water flowing through the volute chamber.

(ii) Vortex or whirlpool chamber: It is an improvement on the ordinary chamber of the volute. In this case, by combining a circular chamber and a spiral chamber, the impeller is encircled by a chamber. In this system the performance is considerably improved. (Fig. 8.19).

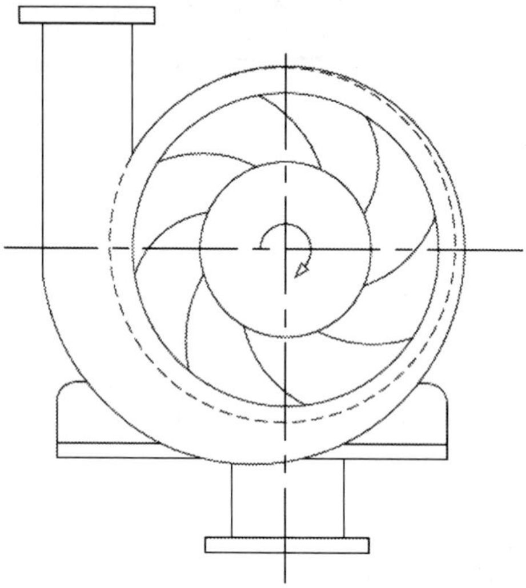

Fig. 8.19: Centrifugal pump with whirlpool chamber

(iii) Guide blades: There are fixed vanes that collect the water leaving the vanes in motion. The guidevanes provide a slowly increasing flow area which reduces kinetic energy and increases pressure energy. A pump containing these reception guidevanes is called a turbine engine. At their receiving tips the guidevanes angles are such that the water reaches the guides without shock. The guides are fixed over a circular ring. The ring with the guide blades is called a diffuser. The pump is also provided with a volute chamber which directs the water to the delivery pipe (Fig. 8.20).

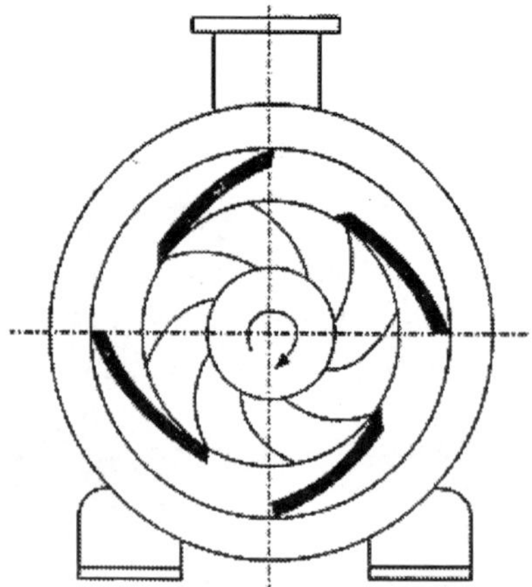

Fig. 8.20: Centrifugal pump with guide blades

Single suction and double suction impellers

Impellers may be single suction impellers or double suction impellers.

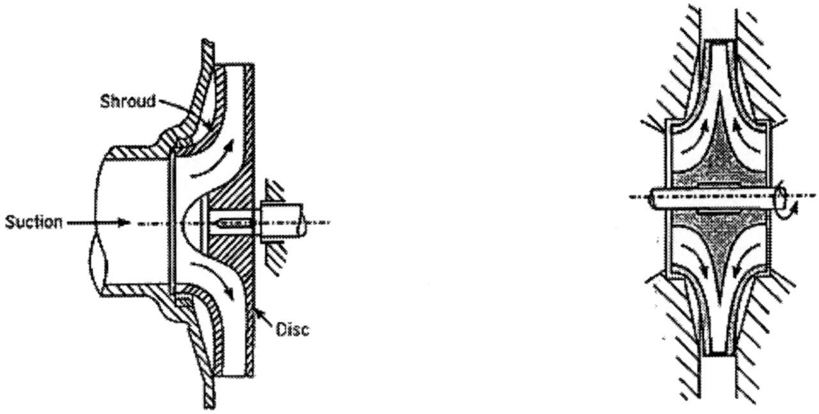

Fig. 8.21: Single suction and double suction impellers

Water exits just at one side of the casing in a single suction impeller. But it is on both sides of the casing, water reaches the impeller with a double suction impeller. This sort of impeller can handle a great flow rate. Whereas in the case of a single suction impeller a lateral force is transferred to the impeller due to the entrance of water only on one side of the casing. But in a double suction impeller, the thrusts act on both sides of the impeller and are balanced.

(i) **The suction pipe:** It is a pipe which connects the pump and the pump inlet. On the lower end of this pipe is provided a non-return valve. If the pump is not running, this valve does not allow the water to drain out of the pipe. This assists in priming by holding the water in the suction pipe. In general, a strainer is also installed at the lower end of the pipe which only lets clear water into the pipe. This pipe's upper end is attached to the impeller's inlet and this portion is called pump's head. This pipe is called the suction pipe since the air pressure in this pipe is below (Fig. 8.21).

The negative pressure at the pump inlet is kept under limits to prevent cavitation. The height of the suction pipe above the sump water level is kept as low as possible by this reason. Also, the diameter of the suction pipe is made wide from this aspect. Bends should be stopped in the suction pipe to minimize pressure drops. Due to larger diameter provided, the velocity in the suction pipe is lowered and accordingly, the loss of head due to friction in the pipe is minimized.

(ii) **The delivery pipe:** This is a pipe connecting the pump outlet to the delivery end or delivery reservoir. A delivery valve is provided with strainer near the outlet of the pump, in order to regulate the flow into the delivery pipe. (Fig. 8.22).

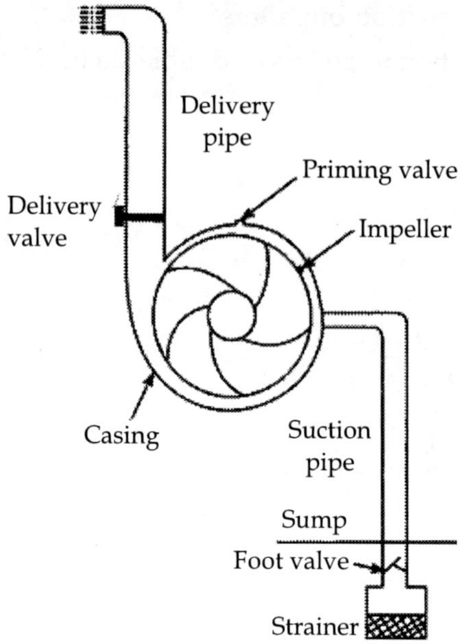

Fig. 8.22: Centrifugal pump

8.8.4.1.5 Vertical Pumps

Vertical pumps were originally developed for well pumping. The bore size of the well limits the outside diameter of the pump and so controls the overall pump design. Vertical pumps can be subdivided into three major categories:

- Lines-shaft pumps
- Submersible pumps
- Horizontally mounted axial-flow

Line-shaft pumps: The driver is mounted on the discharge head for these type of motors. The line-shafting extend through the column to the bowl assembly and transmits torque to the pump rotor.

Submersible pumps: Submersible pumps are close-coupled pumps powered by a submersible motor designed to be submerged in a wet tube. The engine is mounted beneath the bowl and is directly connected to the rotor shaft of the pump.

8.8.4.2 Displacement Pumps

The displacement pumps or more strictly speaking positive displacement pumps, work on the principle of mechanically inducing vacuum in a chamber, and thus, sucking in a certain volume of water, which is then mechanically displaced and forced out of the chamber. The two types of displacement pumps are :

(a) reciprocating pumps ; and

(b) rotary pumps

8.8.4.2.1 Reciprocating Pumps

The simplest type of a reciprocating pump, which is still widely used in Indian villages and towns and sometimes in cities too, is a hand operated well pump (commonly called hand pump). In such a pump (Fig. 8.23), a piston or a plunger reciprocates in a closed vertical cylinder. This is moved up and down by hand. On the upstroke, a vacuum gets created in the cylinder, below the piston, thereby opening the check valve, *V*, at the base of the pump, thus withdrawing water into the space below the piston. Simultaneously, the water above the piston is forced out of the spout. On the down stroke, the check valve, *Vi*, gets closed and the piston valve, *V*, gets opened, thus permitting water to enter the cylinder above the piston. On the next upstroke, the piston valve, *V*, is closed, and the water above the piston comes out of the spout, and at the same time, the check valve, *Vi*, opens, permitting water to enter the space above the piston. The cycle is thus continued, and continuous supplies of water can be obtained.

These pumps can be used in any depth. For wells, where water table is within 6 m or so, the cylinder is placed above the ground. The pump, however, requires a foot valve at the lower end of the suction pipe, so as to avoid priming. Where the water table (or the static water level in the dug well, from which the water is lifted by such a hand pump) is more than 6 m below the ground surface, the cylinder is attached to a drop pipe and placed in the well (Fig. 8.23).

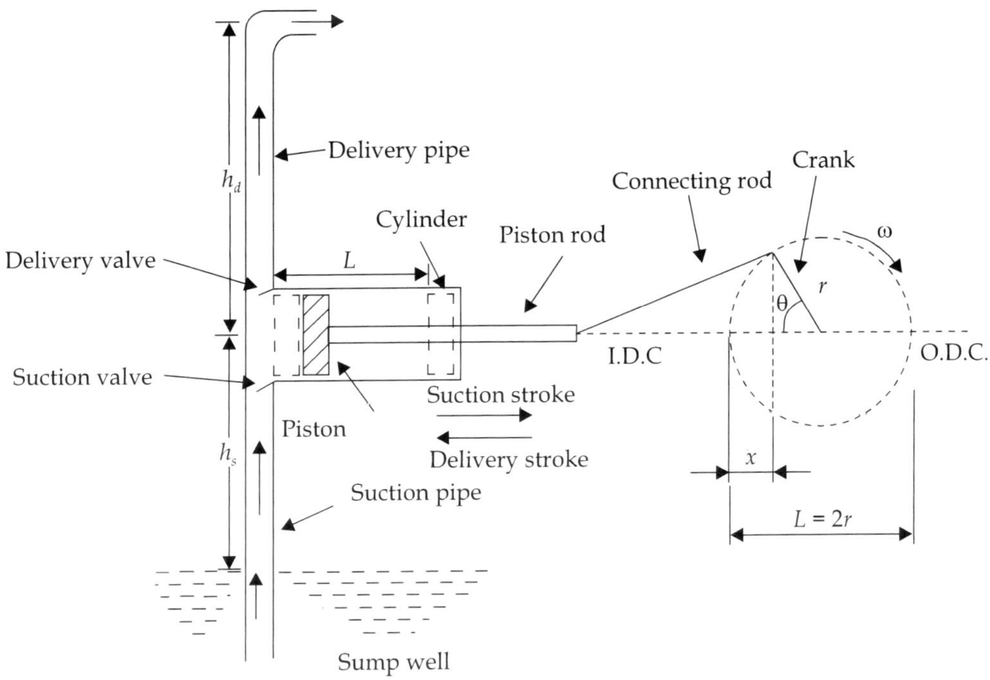

Fig. 8.23: Reciprocating pump

A simple hand pump type of a reciprocating pump described above, is a single acting pump, in which the discharge occurs only on alternate piston strokes. However, double acting pumps, wash deliver water with each stroke, are also sometimes used, and provide more uniform flow.

Advantages and disadvantages of reciprocating pumps

Large reciprocating pumps operated by power were widely used for pumping water during the nineteenth century even in the developed countries like U.S.A. However, nowadays, except that of small hand pumps, they are almost outdated, and rotodynamic pumps especially of centrifugal type are replacing them. Their advantages are given below.

- Since a definite volume of water is discharged with each stroke, the discharge rate of such a pump depends only on the speed with which the piston is moved and is independent of the head. It, therefore, gives a constant discharge under variable heads. Such pumps are very useful for very high and variable head with low discharges.
- They are durable and flexible
- High efficiency is possible, provided the valves and the packings are in good condition. In fact, the efficiency depends upon the slippage (i.e., the amount of water leaking between the piston and the cylinder walls during the stroke). Hence, if the walls and packings are in good condition, slippage may be less than 10 per cent of the total output.

The disadvantages of the reciprocating pumps are:

- The initial cost of installing such a pump is four times that of a centrifugal pump.
- Reciprocating pumps occupy large space (6 to 8 times the space required by the horizontal centrifugal pumps and needs skilled supervision during operations.
- They are unsuitable for pumping water containing sediments, as in such a case, there may be a sudden stoppage of the pump leading to serious damage to the pump.
- Maintenance charges are high, because parts like valves require constant attention produce pulsating.

8.8.4.2.2 Rotary Pumps

In rotary type of displacement pumps, the rotary motion is used in place of a reciprocating motion as is used in reciprocating type of displacement pumps. The rotary motion is obtained by using two cams or two gears which mesh together and rotate in opposite directions. The rotating elements fit the casing closely. The water enters through the suction pipe and is trapped between the cams (or the teeth of gears) and the casing. It is then forced out though the discharge pipe, as the cams or the gears rotate. A definite quantity of water depending on the size and the shape of the gears (or the cams) is thus raised with each revolution.

Advantages of rotary pumps

- Do not require any priming and are self-primed. They are, therefore, often used to prime large centrifugal or reciprocating pumps.
- The flow from such a pump is nearly free from pulsations.
- Since they have no valves, they are simple to construct and easy to repair as compared to the reciprocating pumps.
- They are often used for fire protection systems for buildings and for small domestic water systems.
- Their efficiency is high at low to moderate heads and for smaller discharges of the order of 2000 litres per minute, although they have been used for pressures up to 70 kg/cm^2 (700 m head of water) and discharges over 1,35,000 litres per minute.

Disadvantages of rotary pumps

- The initial cost is high.
- They are not durable, as they require frequent replacement of their cams or gears.
- Water containing sediments are injurious to such pumps, as the wear and abrasion destroy the seal between the cams and the casing.

8.8.4.3 Miscellaneous Types of Pumps

Sometimes the special types of pumps such as airlift pump, jet pump, the hydraulic ram, etc., are used for some specific needs. These are discussed below.

8.8.4.3.1 Air Lift Pumps

Air lift pumps are commonly used for lifting water from deep wells having depth around 60-80 m. In such type of pumps the compressed air is forced into the well using a small air pipe. The air is then released through a diffuser into the eduction pipe at the bottom of the well. The air rises up in the form of bubbles and forms a mixture with air and water in the eduction pipe. Therefore, by hydrostatic pressure the water is forced upward in the pipe by this resulting mixture of air and water inside the eduction pipe as it is lighter than the water outside the pipe and therefore, water finally comes out through the outlet at the top. To remove air from water a separator is often placed at the outlet. This reclaimed air is cooler than the atmospheric air and can be compressed more cheaply. These air lift pumps can be utilized to lift water as high as 150 m but in this case the obtained efficiency is as low as 25 to 50 per cent.

The parameter known as *percentage submergence* is generally used to assess the effectiveness of an air lift pump and is given by Eq. 8.31.

$$\text{Percentage submergence} = \frac{\text{Depth of submergence } (D_s) \times 100}{\text{Depth of submergence } (D_s) + \text{ Effective pump lift } (H_e)}$$

$$= \frac{D_s \times 100}{D_s + H_e} \tag{8.31}$$

where,

$D_s + H_e$ represents the effective length of eduction pipe.

At a submergence value of about 70%, the air lift pump operates at best efficiency. The minimum submergence value of 25% is required in order to operate the pump even at low efficiency.

Advantages of air pumps

- In spite of low efficiency, an air lift pump can deliver large amounts of water from small diameter wells.
- The air lift is not harmed by sedimented water and therefore, such pumps are particularly suitable for use in crooked or damaged wells.
- Air lift pumps are generally cheaper, reliable and simple in operation.
- There are no moving parts of the pump, which may be in contact with the water. Hence, such pumps can be easily used for highly acidic or alkaline waters.
- Pumping from a number of wells can be done by installing a common compressor unit.
- The yield of a well, using such a pump, can be increased by using more amount of compressed air.

Disadvantages of air lift pumps

- To obtain sufficient value of "percentage submergence" the depth of submergence has to be increased, and thus the well has sometimes to be made deeper than required This increases the cost.
- An air lift pump is generally not adopted for raising water much above the ground level, and if this is necessary, a second pump has to be installed.
- Their efficiency is generally less.
- The flow obtained is not continuous but is intermittent. This is due to the fact that small air bubbles have a tendency to unite, and large bubbles thus formed, rise more rapidly than water.
- These are less flexible in fulfilling variable demands.

8.8.4.3.2 Jet Pumps

Sometimes for small construction works such as for dewatering trenches, etc., portable pumps, used for pumping water from small wells, are known as jet pumps. A typical example of such a pump and its arrangement is shown through Fig. 8.24. From the figure it can be observed that the compressed air or steam or water is made to enter the pipe A, which is nozzled at its discharging end, thus forcing out through the nozzle. The jet of compressed air (or water or steam) thus comes out of the nozzle at a high velocity, and discharges into the throat of another pipe (B) as shown in Fig. 8.24. This high velocity jet creates a suction which draws the water up the pipe B and finally discharges through the discharge end.

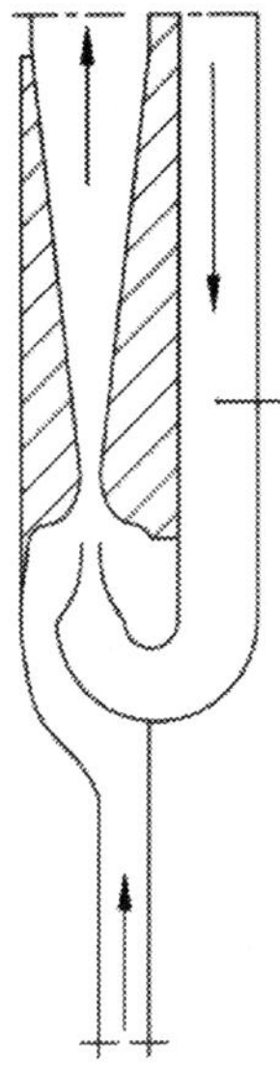

Fig. 8.24: Jet pump

The jet pumps possess lower efficiency of around 25 per cent or so but such type of pumps are easy to handle and portable as they are compact and light in weight. The sedimented waters can be easily withdrawn or transmitted by using these pumps.

8.8.4.3.3 Hydraulic Ram

In a hydraulic ram pumping arrangement is made such that no outside power is required to withdraw water. It utilizes the basic principle of water hammer pressure that develops in a moving mass of water when it is suddenly stopped. However, for a hydraulic ram to function properly a large amount of water must be available at moderate head because it helps to lift a small amount of water to higher head comfortably. The simple layout of a hydraulic ram is depicted through Fig. 8.25.

The system consists of a supply pipe for conveying water from the source to the valve box of the ram. This valve box comprises two automatic valves as V_1 and V_2. Valve V_1 is used for opening downward as the waste valve and valve V_2 for opening upward as the delivery valve. A delivery pipe is attached in the base of air chamber above the delivery valve, V_2.

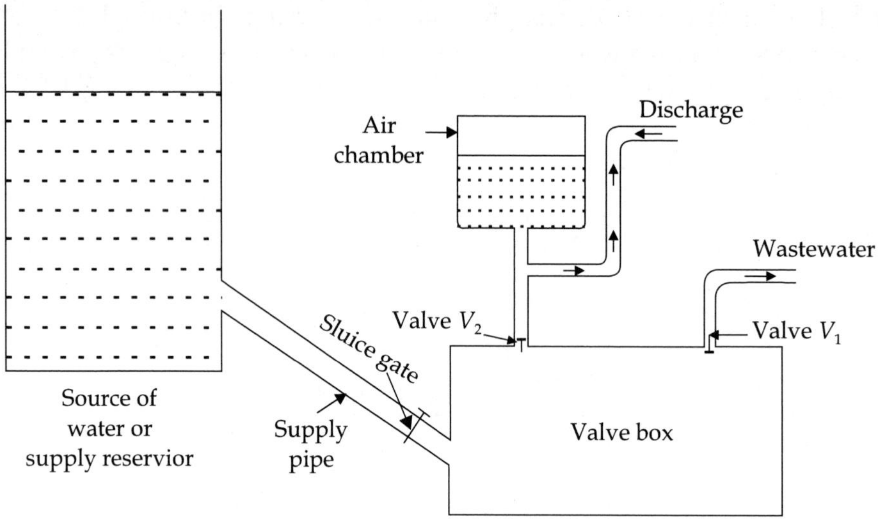

Fig. 8.25: Layout of hydraulic ram

During the starting of the ram the waste valve remains open due to its self weight as the water when flows from the source to the valve box, it goes out of the waste valve. During this process, the velocity of the water increases more and more this ultimately results in increase in the net force that exerts upward on the valve till it becomes high enough to overcome the weight of the valve *V*1 thus forcing it upward and closing it automatically. This results in the development of a wave of positive water hammer pressure with closing of valve *V*1. This developed positive water hammer pressure opens the delivery valve *V*2 (upward) and thus allowing some boosted water to pass through the delivery pipe. The delivery valve automatically closes due to removal of pressure when the wave of negative or suction pressure returns from the supply reservoir. During this instant of time the waste valve opens due to self-weight. However, in the next new cycle the flow again starts from the waste valve which accelerates gradually and thus closing the waste valve suddenly and automatically. Overall, the withdrawal of water continues with 50 to 200 cycles per minute.

The air chamber helps in reducing the fluctuations of flow through the pipe and it all depends on designing a ram properly. For small water supply systems and projects the hydraulic ram is quite suitable and is installed at lower levels than the source. It can be utilized to lift water even up to 30 m or so.

Advantages of hydraulic ram

- It is durable as it works on simple principle and after starting no due attention is required.
- It proves to be economical as it is cheap and does not require any fuel.
- It can also be used when power is not available.

Disadvantages of hydraulic ram

- It should not be used at place of water scarcity as significant amount of water is wasted through the waste valve.
- Depending upon the supply head, the lift and other factors, this ratio of the wasted water to the pumped water may vary from 6: 1 to 2:1.
- It produces considerable noise while working.

8.8.5 Factors Affecting the Selection of a Particular Type of Pump

The various factors which must be thoroughly considered while selecting a particular type of a pump for a particular project are:

- Capacity of pumps
- Importance of water supply scheme
- Initial cost of pumping arrangement
- Maintenance cost
- Space requirements for locating the pump
- Number of units required
- Total lift of water required
- Quantity of water to be pumped

Recent advancements in water lifting equipment has led to the fact that these reciprocating pumps are outdated. The use of rotodynamic pumps is very common under all the ordinary conditions of pumping as they provide satisfactory and economic services. However, under extreme conditions the pumps other than centrifugal ones may be used. The selection of different types of pumps as per the usage can be done on the basis of the following considerations: The rotary pumps may prove to be equally satisfactory as the centrifugal pumps for very small discharges and are economically better as they are less costly if the sediment-free water is to be pumped.

However, during the variable heads when constant discharge is needed, the centrifugal pump may result in operational problems. Whereas, the discharge through a reciprocating pump depends only on the speed of the pump. The reciprocating pumps are found to be useful when water is to be pumped against very high but variable heads with a higher suction lift. These reciprocating pumps can be used to lift and transmit the sediment-free water and no financial constraints exit for the installation of these pumps.

However, for the pumping of waste water and water containing solids the centrifugal pumps are very useful and suitable as good as pumping of treated waters. The horizontal shaft centrifugal pump and the vertical spindle turbine pump and the submersible pump are generally compared and choice is made accordingly. The horizontal pump can be used under a wide range of pumping conditions and is economical too. While the pumps with vertical spindle and submersible pumps may be preferred for handling large quantities of water under low heads, and for wells and tueb wells involving higher suction lift.

However when water is to be pumped concurrently from a number of wells the air lift pumps are economical and therefore selected accordingly with a common compressor unit that can be used to feed all the pumps. The efficiency in the above process is generally low. The hydraulic ram and jet pumps may also be used accordingly.

8.8.5.1 Head, Power and Efficiency of Pumps

The total head (H) against which a pump has to work consists of :

(i) Suction lift (H_s)

(ii) Delivery head (H_d)

(iii) Head lost due to friction , entrance and exit in rising main (H_L) is given by Eq. 8.32;

$$\therefore \quad H = H_S + H_d + H_L \tag{8.32}$$

The work done by the pump in lifting Q m^3/sec of water by head H is given by Eq. 8.33;

$$H = \gamma_w \cdot Q \cdot H \text{ kN. m/sec} \tag{8.33}$$

where,

$$\gamma_w = \text{unit wt. of water } \frac{\text{kN}}{\text{m}^3} \text{ in} = 981 \text{ kN/m}^3$$

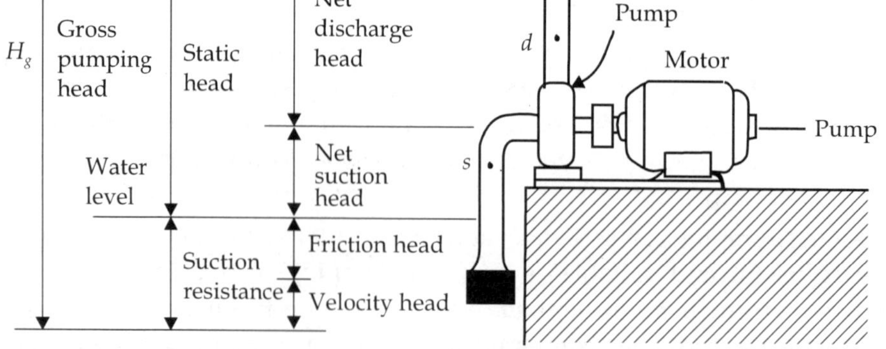

Fig. 8.26: Various heads during water pumping

Q = Discharge to be pumped in m^3/sec

The water horsepower of the pump is given as Eq. 8.34

$$\text{WHP} = \frac{\gamma_w \cdot Q \cdot H}{0.735} \tag{8.34}$$

($\because$ 1 HP = 735 W = 0.735 kW)

If η is the efficiency of the pump set, then the brake horsepower of the pump (BHP) is given by Eq. 8.35

$$\text{BHP} = \frac{\text{W.H.P.}}{\eta} = \frac{\gamma_w \cdot Q \cdot H}{\eta \times 0.735} \tag{8.35}$$

8.8.5.2 Economical Diameter of the Pumping Mains

In a water supply system or transmitting a certain fixed discharge, the diameter of a pipe can be reduced by increasing the flow velocity through the pipe. This increased velocity will increase frictional head loss and therefore is costly as the cost of pumping increases. However, at higher flow velocity the diameter and the cost of pipe can be reduced. This increases the required horsepower of the pump and ultimately increases the cost of pumping. The total annual expenses can be minimized by optimizing the optimum conditions while choosing the pipe diameter, cost of pumping and optimizing the discharge. The diameter thus obtained under such optimum conditions is known as the economical diameter of the pipe.

However, when the pipe diameter is less than the economical diameter it may result in lower cost of pipe but it will result in high head loss. This increase in pumping cost is much higher than the resultant saving in the cost of pipe. On the other hand, when the pipe diameter is more than the economical diameter, it will result in low pumping cost but the increase in size of the pipe diameter will result in the increase in the cost of pipe and will be much higher than the resultant saving obtained in the cost of pumping. Lea has given an empirical formula for calculating the economical diameter of the pipe as Eq. 8.36.

$$D = 0.97 \text{ to } 1.22\sqrt{Q} \tag{8.36}$$

where,

D = Economical diameter(m)

Q = Discharge to be pumped (m^3/sec)

The optimum flow velocity may vary between 1.35 and 0.8 m/sec. For optimization of cost of pipe and pumping cost a rigorous analysis is required and should be worked out at different assumed flow velocities in between 0.8 to 1.8 m/sec. A graph plotted between the yearly cost and the size of the pipe indicating the economical size at the least cost, and thus selected accordingly.

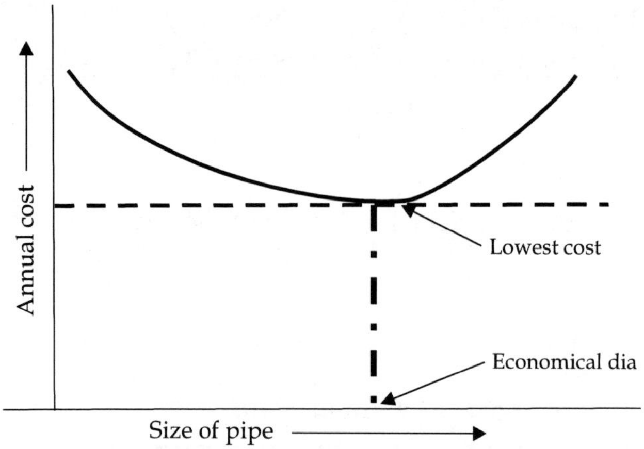

Fig. 8.27: Economical diameter of pipe

Example 8.8: An electric motor with centrifugal pump lifts water through a total height of 75 m from the storage reservoir to the discharge end. The efficiency of the pump is assumed to be 80% and the efficiency of motor to be 85%. The lift is to be carried out through 350 m length using 12 cm diameter pipe with a pumping rate of 1700 L per minute. If the power costs around 75 paise per kWh, what is the cost of power for pumping 5 million litres of water, assume $4f = 0.025$?

Solution: Given that

$H_s + H_d$ = Total lift including suction and delivery = 75 m

$L = 350$ m

$$Q = 1700 \text{ L/minute} = \frac{1.7}{60} \text{ m}^3/\text{sec} = 0.28 \text{ m}^3/\text{sec}$$

Pipe diameter = 12 cm = 0.12 m

$$\text{Head loss in pipe friction} = H_f = \frac{f \cdot LV^2}{2gd} = \frac{f \cdot LQ^2}{2g\left(\frac{\pi}{4}\right)^2 d^5}$$

where, $f = 4f = 0.025$

$$\therefore\ H_f = \frac{0.025 \times 350 \times (0.028)^2}{2 \times 9.81 \times (0.785)^2 (0.12)^2} = \frac{0.025 \times 350 \times 0.00078}{19.62 \times 0.616 \times 0.0000248} = 22.77 \text{ m}$$

Total head against which pump has to work = 75 + 22.77 = 97.77 m

$$\text{Water power required} = \gamma_w QH = (9.81 \times 0.028 \times 97.77) \text{ kN}\cdot\frac{\text{m}}{\text{s}} \text{ or watts}$$

$$= 26.85 \text{ watts} = \frac{26.85}{0.735} \text{ HP} = 36.5 \text{ HP}$$

($\because$ 1 HP = 735 W = 0.735 kW)

The water horsepower of the motor required is:

$$= \frac{36\cdot5}{\eta_p\cdot\eta_m} = \frac{36.5}{0.80\times0.85} = 53.73 \text{ HP}$$

Therefore, the power required = 53.73 × 0.735 kW = 39.49 kW

$$\text{Time taken to lift 5 million litres of water} = \frac{\text{Vol. of water to be pumped}}{\text{Pumping rate}}$$

$$= \frac{5\times10^6}{1700}\text{ minutes} = \frac{5\times10^6}{1700\times60}\text{ hr} = 49 \text{ hours}$$

Therefore, the energy required = 36.5 × 49.0 kWh = 1788.5 kWh

Cost of the electrical energy @ 75 paise per unit

= Rs. 1788.5 × 0.75 = Rs. 1341 **Ans.**

8.9 PUMPING STATION AND ITS LOCATION

The establishment of a pumping station should be selected on the basis of location from which water is to be received and location at which water is to be supplied. However, before selection a suitable site of establishment of pumping station the following points should be considered:

- The source of water in a water body should be away from the point of contamination.
- The selected site should be above the point of highest flood level of a river and the future growth and expansion can be done easily and comfortably.
- During water intake from a river the setup of pumping station should be such that one can draw sufficient water during the dry weather flow at the peak demand of the day.
- Also the possibilities of fire hazard should be considered while selecting the location of pumping station.
- Also when coal is used as the source of energy generation the site for establishing a pumping station should be selected in close proximity of a railway station so that the cost of coal transportation would be minimum and can be made available in very less time for the power generation.
- The pumping station should be neat and clean and the interior as well as exterior of the station should be appealing and vibrant. The pumping stations should be properly managed with enough lighting and provided with a healthy ventilation system. There should be enormous space available for individual motor and pump units for running, operation and their maintenance.

Pumps setup in pumping station

There are three ways of setting, considering the type of pump and the inlet chamber.

- Pumps of vertical shaft sunk in the water of the pump well.

- Pumps with vertical or horizontal shafts set in a dry chamber located beside the pump well.
- Pump of horizontal shaft is located above the water level.

The pumps can be installed either below the water level in a sump well or above it. When pumps are located below water level the priming is not required, however, the pumps should be isolated in dry area or a pit and is connected with common suction pipe provided with sluice valves. This is helpful during maintenance and repair of the pumps such that these values can be closed and repaired easily. This eliminates the priming requirement and positive suction head develops thus it proves to be useful for the centrifugal pumps that generally do not have self-priming system. The electric motors be placed at an elevation such that the place should be safe and dry too and is connected to the pumps by means of vertical shafts. When the driving motor and main pump are above the water and the first stage under the water level thus by applying a first stage submerged part, the pump gets water from the inflow. There is no need for priming until we proceed. The water distribution speed is regulated by the upper component, hence the pump essentially has a double stage consisting of a low-pressure (submerged stage) and a higher-pressure stage. This solution partly eliminates the above-mentioned shortcomings, by having several advantages (power take-off shaft with lower capacity, main part of the pump is easy to maintain).

When the pumps are placed above the water level in the sump well such that the possible suction lift of the pumps is not be affected. The suction of the pump depends on the atmospheric pressure, frictional losses, velocity and temperature of the water in the suction pipe. It is generally limited to a value of 5 metres or so. The suction line should be installed as short as possible as this would be helpful in keeping the losses to a minimum value. Also to avoid the air pockets the line should be provided with continuous upward slope from the foot valve to the pump. Therefore, the foot valves are provided at the end of suction pipe and is allowed to be submerged in water sufficiently. This helps in preventing the air to enter the pipe and therefore avoids the priming problems.

The driving engines are also placed on the same heights as of the pumps and are coupled horizontally. The alignment of horizontal shafts can be carried out without any difficulty. The benefit of this approach is the comparatively less space needed (there is no suction pipe and foot valve) and the simple start and operation (the air cannot reach the suction) pipe). Its drawbacks are the difficulties in fitting the bearings in the vertical shaft, lack of output (due to many guide bearings), increased corrosion and difficulty in testing, maintenance and repair (the piston should be drained first), especially in the case of a significant gap in the level between the motor and the piston. This solution is applied mostly to pumps of high pressure. There are many pumps of this type working well with a suction head of 5-7 m, nevertheless, suction head higher than 3-4 m should be avoided.

8.9.1 Characters of Pump Setting

A permanent pump station should be used if:

- The pumping delivery is extremely high.
- The annual utilization of the pump station is high (close to 600 hours).
- High operational safety is needed.

A mobile pump station should be used if:

- The use of the pump is occasional, and the utilization is less than 200 hr annually.
- The location of pumping is not permanent.

There is an in-between solution, a temporary pump station, where the structural parts (tubes, mountings) are built, and the driving motor and pump are mobile.

8.9.2 Pipe System in a Pump House

A suction screen should be used on the suction pipe, with as little resistance as possible. For suction pipes, particularly if the pump is not an inflow system, reduction of the losses is very necessary to obtain the lowest suction head possible and complete sealing. The suction pipe resistance can be minimized with a broad cross section and small suction tube, and sharp curves can be avoided. The inlet speed of 0.8-1.0 m/sec and 1-2 m/sec in the suction pipe minimizes the lack of entry. The suction pipe should be attached to the pump inlet with the shortest reduction section possible in order to avoid speed reduction below 0.8 m/sec and to prevent air splitting and corrosion. With the application of the shortest possible vertically or slightly (1-2%) declining suction pipe, the formation of air locks can be prevented.

The flanged joints are provided in the cast iron pipes at the bends, turns, etc. The flow through velocity in these pipes at average flow ranges from 0.6 to 1.2 m/sec.

If air segregation happens at any section of the pipeline it can be overcome by correctly attaching it to a deaerator with deaeration of the actual portion. Reducing sections installed in a horizontal suction pipe would be asymmetrical to the higher, horizontal brace. Hydrostatic checks should be used to verify the suction pipe after contact, to avoid leakage. Sub-surface suction pipes and syphons will also be made of welded steel. A velocity of 2 m/sec is most economical in the case of transmission pipes at the pump house. Connection to a bigger cross section external pipeline should take place outside the pump house. Velocity above 2 m/sec is only allowable for acute water volume fluctuations and for as brief a period as possible.

In case of marked water quantity fluctuation, two delivery pipes should be applied which result in a safer operation besides ensuring an even velocity. In this case all pumps will be hooked to a common collection main at both ends of which pipes are connected to the supply. With the help of a gate valve in the middle of the collecting main, the two delivery pipes and the pump units can be operated separately. An appropriate gate valve or other cut-off mechanism should be inserted into the transmission pipe of each pump, preferably in the vicinity of the pump outlet or in the case of an inflow system before the pump inlet, to shut the pump off. A flap valve or regulation flap should be placed between the pump and the gate valve on the delivery side to avoid reflux of water in the distribution pipe, in the event of a sudden pump stop, particularly if multiple pumps are operating on a common delivery pipe.

Pipes can be made of flange joined cast iron or welded steel in the pump room. To dewater the common delivery pipe, a gate valve operated branch pipe would be inserted into the deepest section of the common delivery pipe, depending on the depth, size and quantity of water to be drained.

8.9.3 Point to be Observed in Selecting a Pump

- Capacity and performance: The pump should have the necessary capacity and optimum performance.
- Lift: Suction head up to pump level from water level.
- Head: It is also called delivery head. Generally, the total head (suction and delivery head) should meet all possible situations with respect to the head.
- Reliability: The failure rate and types of troubles can be provided by a reputed fabrication or similar pump already in use.
- Initial cost: The cost and the cost of installing should be minimal.
- Power: The criteria for power should be lower than for operation
- Maintenance: The cost of maintenance should be low. It is important to determine the availability and the cost of spares.

8.9.4 Horsepower of Pump

The horsepower (H.P.) of a pump can be determined by calculating the work done by a pump in raising the water up to H height is given by Eq. 8.37 and break horsepower is given by Eq. 8.38.

Let the pump raise W'' kg of water to height H'' m

Then work done by pump $= W \times H$ kg-m

$= \text{WQH m kg/sec}$

where, W = Density of water in kg/m^3

Q = Water discharge by pump in m^3/sec

$$\text{Water horsepower} = \frac{\text{Discharge} \times \text{Total head}}{75}$$

$$\text{Water horsepower} = \frac{\text{WQH}}{75} \tag{8.37}$$

$$\text{Break horsepower} = \frac{\text{WHP}}{\eta} \tag{8.38}$$

8.9.5 Valves

Then various types of valves used in pipes used for pumping water at various positions in enumerated below as:

(i) Sluice or gate valves are provided in such a manner that one at suction side and another at delivery side.

(ii) A check valve should be provided on the delivery side in between the pump and the gate valve.

(iii) At delivery side a pressure relief valve and air valve should always be provided.

When shut-off the piping as well as to control the volume of water flowing is required in such case gate valves, ring valves or flap / butterfly valves are provided and are run manually. If the gate valve is larger than 800 mm and easy and regular operation is needed or automatic or remote control is necessary, hydraulic or electrical control may be used to operate the system. A jam-proof wedge gate valve or a parallel slide valve, cup valve, or flap valve can be used in remote operation. A butterfly valve or flap valve can be used with manual, electrical or hydraulic operation, if the space is limited. Pipelines with high pressure and cross section of 400 mm or more ring or cup valves can be inserted to ensure smooth operation and protection shut-off.

If the pump fails, a flap valve or regulating flap may prevent efflux. Automatic butterfly valves or a ring valve creates less pressure loss or resistance than the flap valve. A trap fixed at the end of the tube can be added for intake pipes of ponds and reservoirs, with a pressure of 0-2 m water column, with automatic cut-off. Sluices that pass between the well walls or moving sluice valves are used to remove arbitrary pressure or to break off well intake pipes. They are worked at water pressure of up to 3-4 m with a spindle drive. At higher water pressure the best solution is given by diameter rack drive or sliding sluices, or sluice valves. The sluices are generally controlled manually. Electric or hydraulic drive is occasionally used (e.g., at very quick cut-off, or open, in case of automatic operation).

8.9.6 Sizing of Pump Units and the Standby Capacity

The pumping stations are not always operated at full load capacity at water works. The efficiency of pumps in a pumping station varies with change in load. Therefore, the pumping stations are designed in such a manner that some pumps may be operated at full capacity at all times. Therefore, always two to four pumps should be installed. The sizes of various pumps are designed and fixed on the basis of local conditions such as the demand, available storage and the desirability of the off-peak pumping. The same can be explained with an example such that if four pumps are installed then two pumps with largest capacity may be used for supplying the maximum demand. On the other hand, the smaller capacity pump can be utilized to handle the normal demand and the other pump may be utilized for the minimum requirement.

There should always be a provision of standby capacity that can be utilized during emergency such as to take care during repairs, breakdowns, etc. There should be 100% standby capacity. However, the standby capacity of 33.5% to 50% is considered sufficient and therefore should be provided at pumping stations against the peak demand.

Example 8.9: Design a raw water pumping station for obtaining water from a perennial river with the following data:

(a) Population to be served = 4 lakh

(b) Daily demand for water = 150 litres per person per day.

(c) Summer requirements = 150% of average.

(d) Normal water level in river = R.L. 200.00 m

(e) Low water level in river = R.L. 198 m

(f) The water is to be pumped in two shifts of 10 hr each daily, and lifted to the inlet of water treatment plant situated nearby at R.L. 217.00 m.

(g) The losses due to friction in the C.I. rising main of 0.9 m dia, are low due to its short length, and can be assumed to be 2 metres in summer season and 1.2 m on an average.

(h) The efficiency of the pumps and the driving motors can be taken as 75% and 90% respectively.

Solution: Average daily water demand = $(4 \times 10^5)(150)$ L/d = 60×10^6 L/d

$$= \frac{60{,}000}{24 \times 60 \times 60}\,\text{m}^3/\text{sec} = 0.694\ \text{m}^3/\text{sec}$$

Maximum daily demand in summer = 1.5 times the average demand

$$= 1.5 \times 0.694\ \text{m}^3/\text{sec} = 1.04\ \text{m}^3/\text{sec}$$

The pumps have to work for 20 hours a day, the capacity of the pumps should be designed for the following daily demand:

For summer day $= 1.04 \times \frac{24}{20} = 1.24\ \text{m}^3/\text{sec}$

For an average day $= 0.694 \times \frac{24}{20} = 0.832\ \text{m}^3/\text{sec}$

Height through which water is to be lifted is given as:

For an average day = 217.00 – 200.00 = 17.00 m

For a summer day= 217.00 – 198.00 = 19.00 m

Therefore, the total lift against which water is to be lifted is given as:

For an average day = 17.00 + 1.2 = 18.2 m

For a summer day = 217.00 – 198 = 19.00 m

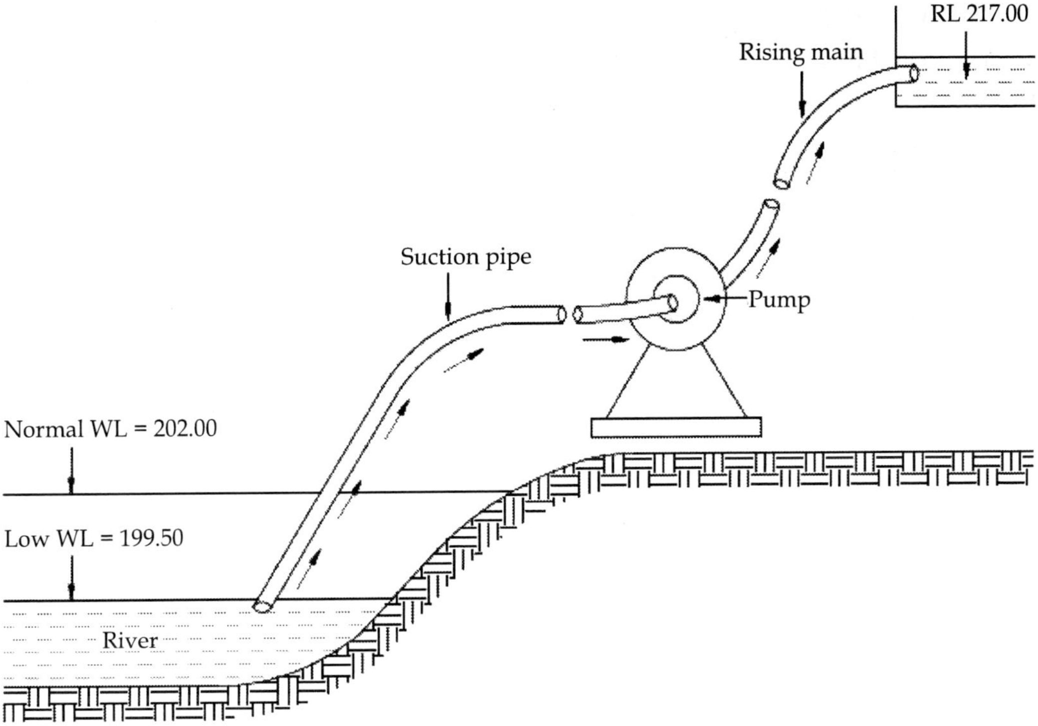

Fig. 8.28: Water intake and distribution

∴ Water horsepower required is given as:

$$\text{For an average day} = \frac{\gamma_w \cdot Q \cdot H}{0.735} = \frac{9.81 \times 0.832 \times 18.2}{0.735} = 202.10 \text{ H.P.}$$

$$\text{For a summer day} = \frac{9.81 \times 1.24 \times 19}{0.735} = 314.45 \text{ H.P.}$$

∴ Brake horsepower required is given by:

$$\text{For an average day} = \frac{202.10}{0.75 \times 0.90} = 299.40 \text{ H.P.}$$

$$\text{For a summer day} = \frac{314.45}{0.75 \times 0.90} = 465.85 \text{ H.P.}$$ **Ans**

(i) Therefore, provide 4 centrifugal pumps, each with a capacity 0.31 m^3/sec, with driving unit of 120 BHP motors.

(ii) Provide 2 centrifugal pumps, each with capacity of 0.16 m^3/sec, driven by say 60 BHP motors.

Hence, during average days, 2 units of 120 BHP will work giving a standby capacity of 100%, and with increasing demand the 60 BHP units can be switched on, and 33% standby capacity will always be available at maximum speed.

8.9.7 Rising Main

General guidelines

The through main should be laid with an even gradient with minimum sharp curves and twists, as far as possible. This should be laid in a trench with a minimum earth cover of 0.6 m, which should be tightly held by clamps to concrete blocks if carried aboveground. This may be required, even if under the ground, where pressure is high to anchor the pipe at bends, and again this is normally achieved via concrete blocks. The pipe should be protected in the trench around its barrel and not on its joints. It requires further drilling at each joint, so to have the bottom of the joint not lying on the floor of the trench. In the rocky trench, the pipe should be laid on prepared bed of sand.

To prevent the washwater from flowing back to the pump, a wash-out valve is inserted into the rising main immediately outside the pump house with a stop valve or non-return valve immediately downstream of it. A non-return valve or reflux valve is safer than a stop valve, as it cannot be left closed unintentionally when the pump begins again. Unless the rising main is a long one, the inclusion of an air valve would seldom be required and this can only be done for a short rising main under exceptional conditions. This is also a bad idea to cut connections directly from the rising main, and the rising main will usually carry the water directly from the pump to the high-level storage tank without being tapped or having water drawn off in any way.

In other cases, it might be required, e.g., where the rising main passes near an isolated house which is a long distance from the storage tank and which will require any costly gravity pipe back from the tank to supply it. Care should be taken in such situations that the link feeds a storage tank with a ball valve and does not discharge freely at the end. The diameter of the rising main should keep the water velocity fairly low, and the head lost in friction, as a result. A maximum speed of 1.5 m/sec should not be surpassed and, the installation of a larger diameter pipe should be considered, balancing this against the lower friction head and the smaller pipe pumping systems.

Example 8.10: A pump is to deliver water from an underground tank against a static head of 40 m. The suction pipe is 50 m long and is of 25 cm diameter with Darcy-Weisbach friction factor, $f = 0.02$. The delivery pipe is of 20 cm diameter, 1600 m long and has $f = 0.022$. The pump characteristics can be expressed as:

$$H_p = 100 - 6000\ Q^2$$

where,

H_p = Pump head in m

Q = Discharge in m^3/sec

Calculate the head and discharge of the pump? (U.P.S.C. Engg. Services, 1991)

Solution:

(i) Static head (H_1) = 40 m

(ii) Head loss in suction pipe (H_2)

$$H_2 = \frac{f \cdot L \cdot Q^2}{2 \cdot g \cdot \frac{\pi^2}{16} \cdot d^5}$$

$$= \frac{0.02 \times 50 \times Q^2 \times 16}{2 \times 9.81 \times \pi^2 \times (0.25)^5} = 84.68\, Q^2$$

(iii) Head loss in delivery pipe (H_3)

$$= \frac{0.02 \times 1600 \times Q^2 \times 16}{2 \times 9.81 \times \pi^2 \times (0.2)^5} = 90.98\, Q^2$$

Total head loss, $H_p = H_1 + H_2 + H_3 = 40 + 84.68\, Q^2 + 90.98\, Q^2$

$$H_p = 40 + 175.66\, Q^2$$

Or, Using given equation of pump

But, $$H_P = 1000 - 6000\, Q^2$$

$$100 - 6000\, Q^2 = 40 + 175.66\, Q^2$$

$$100 - 6000\, Q^2 = 6175.66\, Q^2$$

$$Q = \sqrt{\frac{60}{6175.66}} = 0.0986 \text{ m}^3/\text{sec}$$

$\therefore$ $$H_p = 100 - 6000 \times (0.0986)^2 = 41.7 \text{ m}$$ **Ans.**

Example 8.11: For water supply of a town, water is pumped from a river 3 km away into a reservoir. The maximum difference of levels of water in river and the reservoir is 20 m. The population of the town is 40,000 and per capita water demand is 110 litre per day. If the pumps are to operate for a total of 8 hours and the efficiency of pumps is 80%, determine the horsepower of the pumps. Assume friction factor as 0.03, velocity of flow at 2 m/s and maximum daily demand as 1.5 times the average daily demand.

Solution: Population of towns = 40,000

Per capita demand = 110 L/d

$\therefore$ Average water required $= \dfrac{40{,}000 \times 110}{1000}$

$= 4400 \text{ m}^3/\text{day}$

Maximum daily water required $= 1.5 \times 4400 \text{ m}^3/\text{day} = 6600 \text{ m}^3/\text{day}$

The above quantity has to be pumped from the river in 8 hours to the reservoir.

$\therefore$ Water lifted per second $= \dfrac{6600}{8 \times 60 \times 60} \text{ m}^3/\text{s}$

$= 0.229 \text{ m}^3/\text{s}$

Velocity through the pipe lifting the above discharge = 2 m/s (given)

$\therefore$ Dia. of pipe(d) is obtained as: Velocity $\times$ Area = Discharge or $$2\cdot\left(\frac{\pi}{4}\cdot d^2\right) = 0.229$$

$\therefore$ $$d = \sqrt{\frac{0.229\times 4}{2\pi}} = 0.38 \text{ m}$$

H_L = Head loss through the pipe of 3000 m long and of diameter 0.38 m

$$= \frac{f\cdot L\cdot V^2}{2\cdot g\cdot d} = \frac{0.03\times 3000\times 2^2}{2\times 9.81\times 0.381} = 48.1 \text{ m} = 48.1$$

Total lift (H)= Actual diff. of W.L. in reservoir and river + H_L

$$= 20 + 48.1 = 68.1 \text{ m}$$

$$\text{H.P. of pump} = \frac{\gamma_w QH}{\eta\, 0.735} = \frac{9.81\times 0.229\times 68.1}{0.735\times 0.8}$$

$$= 260.17 \text{ H.P.}$$ **Ans**

Example 8.12: For a population of 2 lakh having per capita daily demand of water of 150 litres , determine the water horse power to raise the water from a river of RL 120 m to treatment plant of RL 140 m through a rising main 1 m in diameter. Assume 2 m as the total head loss due to friction, etc. and pumping efficiency of 70%.

(ESE-Mains Civil Engineering : 2010)

Solution: Population = 2×10^5

Per capita daily demand = 50 litres

Discharge required for the pump will be

$$Q = \frac{2\times 10^5\times 150\times 10^{-3}}{24\times 60\times 60}$$

$$= 0.3472 \text{ m}^3/\text{s}$$

Delivery head, H_d = 140 – 120 = 20 m

Loss of head due to friction, h_f = 2 m

Total head against which the pump has to work will be

$H = H_d + h_f = 20 + 2 = 22$ m

Water horsepower of the pump is given by;

$$\frac{\gamma_w QH}{0.745} = \frac{9.81\times 0.3472\times 22}{0.745} = 100.486 = 100.5 \text{ HP}$$

If is the efficiency of pump, then brake horsepower is given as:

$$\text{BHP} = \frac{\text{WHP}}{\eta} = \frac{100.5}{0.70} = 14.357 \text{ HP}$$

EXERCISE

1. The velocity of water flowing from a reservoir into a1 metre diameter steel pipe is 2 m/sec. If a valve is situated in the pipeline at a point 2 km from the reservoir, evaluate the water hammer pressure developed by the closure of this valve, if (i) closure time is 2 sec, and (ii) closure time is 6 sec. Take thickness of pipe shell as 2.5 cm.
2. Estimate the gradient in a 2 m diameter smooth concrete pipe carrying a discharge of 3 m^3/sec at 10°C temperature by using (a) Darcy-Weisbach formula, (b) Mannings formula, (c) Hazen-Williams formula, and (d) modified Hazen-Williams formula. Assume suitable data if not given.
3. A pipe 1 metre in diameter buried in a trench 1.4 m wide is backfilled with dry sand. The top of the pipe is 2.5 m below the surface of the fill. The pipe passes at right angles under a one lane road which carries a vehicle whose loading consists of two concentrated 8 kN loads located at 1.8 m apart transverse to the roadway. Find the maximum vertical force exerted on a unit length of the pipe, if

   (a) The pipe is made of steel

   (b) The pipe is a cast iron pipe

   The thickness of the pipe may be neglected for calculating the external dia of the pipe.

   (c) The stress produced in steel pipe if its thickness is 1 cm.
4. A centrifugal pump driven by an electric motor lifts water through a height of 65 metres from the reservoir to the discharge end. The pump efficiency is 77 per cent and the motor efficiency is 85 per cent. The lift is through 350 metres length of 15 cm diameter pipe and the pumping rate is 2500 L per minute. If $F = 0.025$ and power costs Rs. 3.25 per kilowatt hour what is the cost of power for pumping five million litres of water?
5. What is meant by water hammer and how it is produced in pipes conveying water under pressure along with suitable equation? Design a rapid sand filter unit for 5 MLD supply with all its principal components including under-drainage system and washwater troughs.
6. A 0.6 m dia cast iron pipe is subjected to internal pressure of 420 kN/m^2. Assume safe working stress of 60,000 kN/m^2 in tension. Calculate the thickness of shell necessary to resist the internal pressure.
7. Define the typical operating characteristics of centrifugal pumps.
8. What is meant by water hammer and how it is produced in pipes conveying water under pressure?
9. Describe various types of pumps. What are the advantages and disadvantages of each in brief?

10. Water has to be supplied to a town with one lakh population at the rate of 150 L/c/d from a river 2000 m away. The difference in elevation between the lowest water level is the sump and the reservoir is 36 m. If the demand has to be supplied in 8 hours determine the size of the main and the BHP of the pumps required. Assume the maximum demand as 1.5 times the average demand? Assume f = 0.0075, velocity in the pipe 2.4 m/sec and efficiency of the pump 80%.
11. Estimate the roughly sizes of the supply conduits leading to an aqueduct service reservoir serving (a) a relatively small town of 15,000 population, and (b) a relatively large city with industrial establishments having a population of 5 lakh people. Also find the hydraulic gradient at which the pipelines are proposed to be laid? Assume any suitable data according to Indian conditions?
12. The lowest level of an impounding reservoir situated in hills and being used as a source of supply is higher than the level of the service reservoir by 150 metres and the distance between them is 50 km. The storage reservoir and the service reservoir were originally connected by a single pipeline designed to carry a maximum daily demand of 24 million litres per day. However, it was later found necessary to increase the supply to 35 million litres per day. It was therefore decided to lay another pipe of the same diameter alongside the first one in a certain required length and cross connected. Calculate the dia of the pipes and the length of the second pipe required to be laid. Make use of William-Hazen's equation. Also derive an expression expressing H_L in terms of L and d from the original William-Hazen's formula. Assume CH = 130 and any other data suitably.
13. For water supply of town, water is pumped from a river 3 km away into a reservoir. The maximum difference of levels of water in river and the reservoir is 20 m. The population of the town is 50,000 and per capita demand is 120 L/d. If the pumps are to operate for a total of 8 hours and the efficiency of pumps is 80%, determine the horsepower of the pumps. Assume friction factor is 0.03, the velocity of the flow as 2 m/s, and maximum daily demand as 1.5 times the average daily demand.
14. Mention the method to calculate the most economical size of a rising main. List the appurtenances necessary to be installed on a rising main between pumps and overhead tank. Indicate their relative position on line diagram.
15. What design criteria is commonly employed for finding out the sizes of the gravity mains used for conveying water from high level impounding reservoir to towns situated miles away?
16. Describe the common stresses produced in pipelines used for conveying water?
17. What is meant by water hammer and how it is produced in pipes conveying water under pressure?

18. A reflux valve on 75 cm dia steel main, 4.5 km long is closed in 1.5 minutes when the rate of flow was 70 MLD. What extra pressure developed due to closing the valve assume $E_{steel} = 2.1 \times 10^6$ kg/cm^2 and pipe thickness to be 12 mm.
19. Explain the corrosion phenomenon in brief and also suggest its remedial measures?
20. A 0.6 m dia cast iron pipe is subjected to internal pressure of 420 kN/m^2. Assume safe working stress of 60,000 kN/m^2 in tension calculate the thickness of shell necessary to resist the internal pressure.
21. If the above pipe is buried in cohesive soil weighing 20 kN/m^3 with a cover of 1.5 m calculate the compressive force produced if the pipe is under vehicle with two concentrated loads of 10 kN 1.75 m apart.
22. How do you find the most economical size of the rising mains?
23. Describe the different types of pumps along with examples required to lift water from a source to the city in water supply scheme.
24. Define priming and its importance in pump operation.
25. Define the typical operating characteristics of centrifugal pumps.
26. Define specific speed of a centrifugal pump, how it is calculated.
27. Describe the working of hand operated reciprocated pumps.
28. What do you understand by the term head, power and efficiency of pumps?

9

CHAPTER

Water Main Appurtenances

In a water supply and sanitation system the water pipes are buried under the ground and have to travel ups and downs as the water is to be supplied to multistoried buildings at great heights. Pipes in such water supply system require frequent cleansing and timely maintenance. Also the negative and positive pressure caused by excessive air in empty pipes requires to be released in order to save the pipelines from bursting. To isolate and drain the pipelines and connected pipes for repairs and maintenance various components such as valves, fittings, plugs, gates, joints, manholes and caps are provided at required places in the pipe line. These fixed devices along the water distribution system are known as **appurtenances**. They are provided to control the rate of flow of water, to release or admit air into pipeline, detecting leakages, to meet the demand during emergency and ultimately to improve the efficiency of the distribution system. Some of the important fixtures used in the water distribution system are:

- Valves
- Fire hydrants
- Water meters

9.1 VALVES

Valves are required for controlling the flow of water, regulating heat, releasing or accepting air, preventing water flow in the opposite direction. The following are the various types of valves according to their function:

- Sluice valves or gate valves
- Check valves or reflex valves
- Air valves
- Drain valves or blow off valves or scour valves
- Pressure relief valves

9.1.1 Sluice Valves

The most popular type of valves used to control and regulate the flow of water through the pipelines are sluice valves or gate valves. A sluice valve consists of a parallel sided or wedge-shaped disk or circular gate and has a nut that connects with the thread of an operating spindle as shown in Fig. 9.1. The valves comprise two types of gate raising spindles such that one of the rising spindle is connected to the gate and does not rotate with the hand wheel. The other type of non-rising spindle such that it rotates in a screwed gate connection. The rotating direction for the valve to open is normally anticlockwise.

The gate is separated from the waterway in the completely open state and thus this valve is often known as a full-way valve. Sluice valves with threaded, flanged, or bell-and-spigot ends are available. In order to divide the pipe line into different sections these values are provided at a regular interval of every 3 to 5 km distance when water is to be transported from source to the city using large pipelines. Often, sluice valves are difficult to operate as they require a great force to open or close against a high unbalanced pressure. The valves are generally provided at point of lowest pressure in the pipe line known as summit so as to use the conduits of economical and cheap material that is suitable and can easily be operated at low pressure. However, for large diameter conduits the small diameter valves proves to be economical and effective.

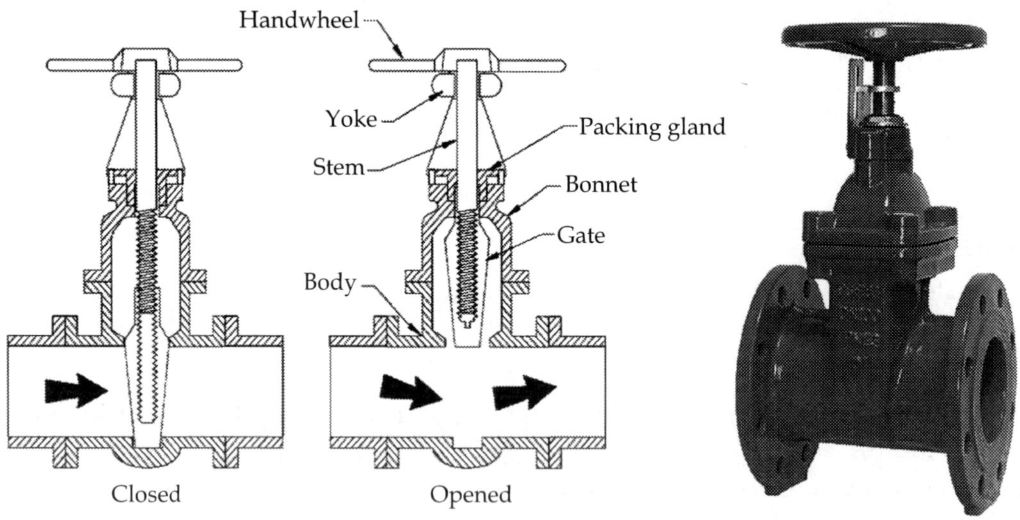

Fig. 9.1: Sluice valves (Hard Hat Engineer, 2020)

Butterfly valves

A butterfly valve consists of a round, cross-sectional elliptical tube, housed in a cylindrical valve body that essentially forms a continuation of the pipe itself as shown in Fig. 9.2.

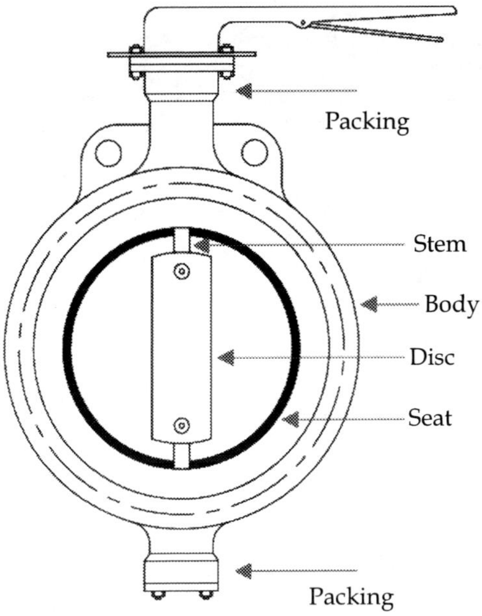

Fig. 9.2: Butterfly valve (Louzada, 2017)

The rotation of the valve spindle either places the disk in a closed position or in a completely open position where it is parallel to the pipe axis. Butterfly valves are used particularly in large size conduits to control and stop the flow. For larger sizes, they are often cheaper than sluice valves, and take up less room. Butterfly valves have the advantage of being easy to operate due to lack of sliding components, compact size, reduced chamber or valve house and improved closing and retarding characteristics. Butterfly valves, however, require a significantly higher head loss than sluice valves and even in a fully open position they give a relatively high flow resistance due to the thickness of the disc obstructing the flow. Also, as the gate can erode rapidly, butterfly valves are not ideal for operation in partially open positions and need high torques to open them under high pressure. Figure 9.2 shows the section of butterfly valve.

Globe valve

A globe valve consists of a disc attached axially to a vertical spindle, and the hand is raised by the incoming water pressure and thus the valve is opened as shown in Fig. 9.3.

In globe valves the flow changes direction twice by 90° resulting in high loss of head. These valves are commonly used as water taps and in small diameter (less than 100 mm) pipes.

Check valve or reflux valve

These are also known as non-return valves or reflux valves as it prevents the backflow of water. A check valve allows water to flow in one direction only, and it automatically prevents the flow in the opposite direction. Such type of valves are generally installed

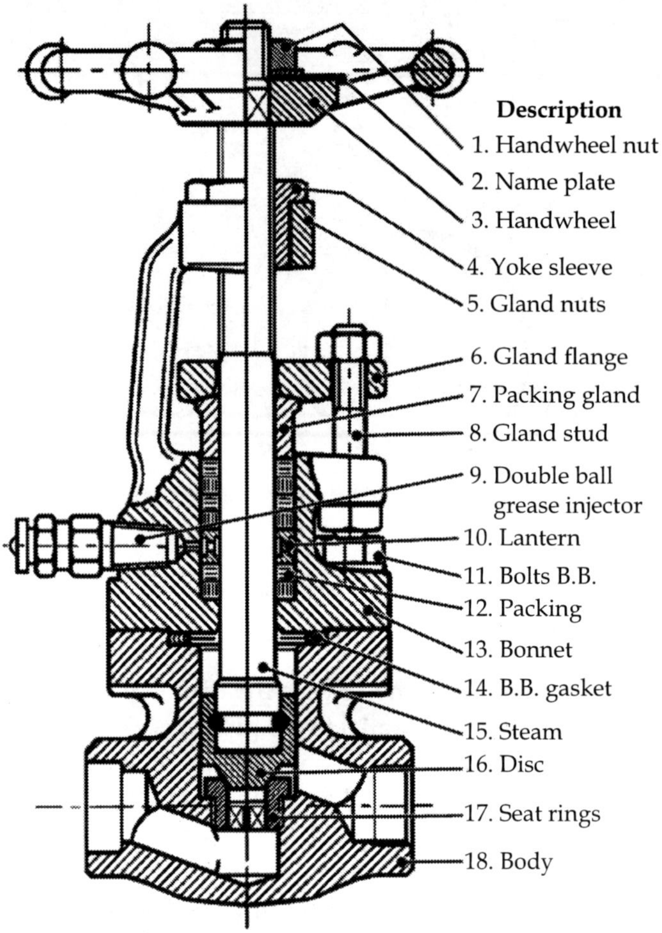

Fig. 9.3: Globe valve (globe valves)

on the delivery side of a pumping set. This prevents the backflow of pumped or stored water when the pump is stopped. Such type of valves are swing check or ball devices and the test valve for swing form is commonly used as shown in Fig. 9.4.

The swing style valve consists of a gunmetal door (or disc) hinged at the top edge of it. The door fits comfortably under its self-weight against an annular gunmetal bench. As water flows in the arrow direction, the flowing water pressure lifts the door and leaves the door open. The open door poses no barrier to the flow of water. When the water flow in this direction ceases, the door swings down and fits over its seat again, so the valve is closed and the water flow in the opposite direction is stopped. Invariably, the reflux valve is installed in water from flowing back to the pump if the pump fails or stops and thus saves the pumping equipment from potential damage.

Another simplest type of check valves is flap shutter hinged at the outlet of the pipe which is governed by the pressure of flowing water and is generally kept open in the direction of flow of water itself. It, however, closes by its self-weight when the flow in permeable direction stops and the valve cannot open out in the opposite direction.

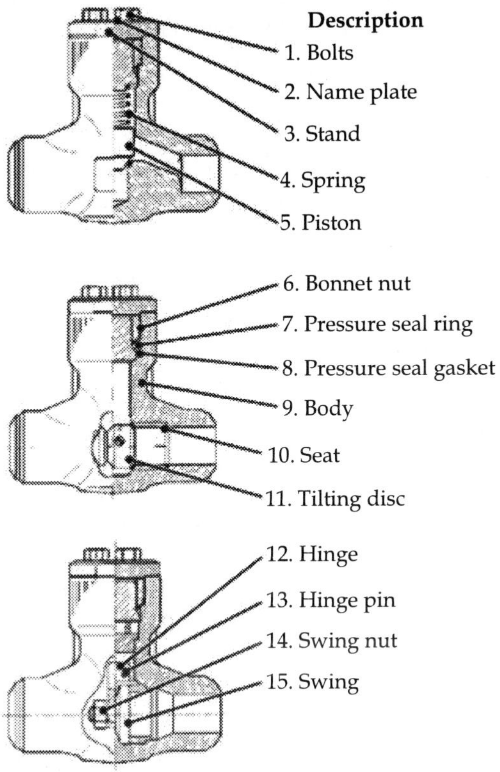

Fig. 9.4: Check valve (check valves)

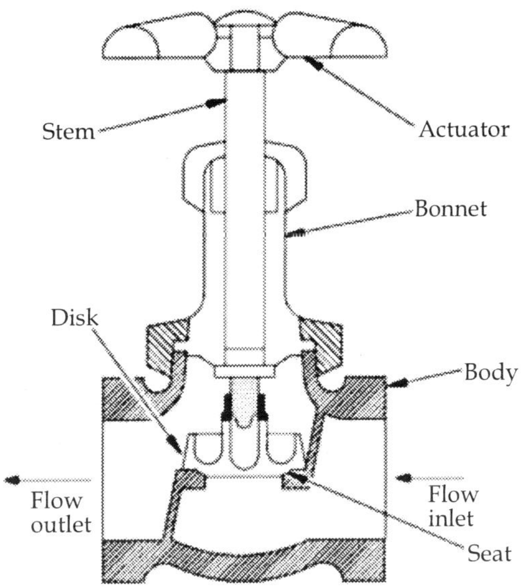

Fig. 9.5: Typical check valve

Such valves are provided at the pump discharges in order to minimize the water hammer pressure on the pump. These valves are also used to isolate the polluted water and potable water system. In order to prevent the draining of the suction water during the stoppage of the pump the check valve is installed at the end of suction pipe and is known as foot valve as illustrated in Fig. 9.5.

9.1.2 Air Valves

In order to release the inside air and negative pressure in a pipeline the air valves are provided on both sides of the sluice valves and along the point of summit of the pipe line. Such valves are also provided at the downstream. However, for ensuring the safety of pipes against collapse these valves are provided at a point of summit near the HGL or above the HGL line.

These are automatic valves and are of two types, namely:

- Air inlet valves
- Air relief valves

Air inlet valves

These valves are installed below the ordinary sluice valves in order to protect the pipe against the developed negative pressure during the sudden stoppage of water under steady state flow conditions. These valves open automatically and allow air to enter the pipeline, so that negative pressure in the pipelines can be circumvented.

Air relief valves

Often at the top of the pipes gas is trapped and due to air lock blocks the water flow. In these situations it is important to remove the accumulated air from the pipelines. The vacuum pressure created by sudden closure of sluice valves in pipelines at the downstream side can be diminished by opening the air relief valve thus prevent the damage of conduits under such abnormal conditions. It is achieved by means of air relief valves automatically as section is shown in Figs. 9.6 and 9.7.

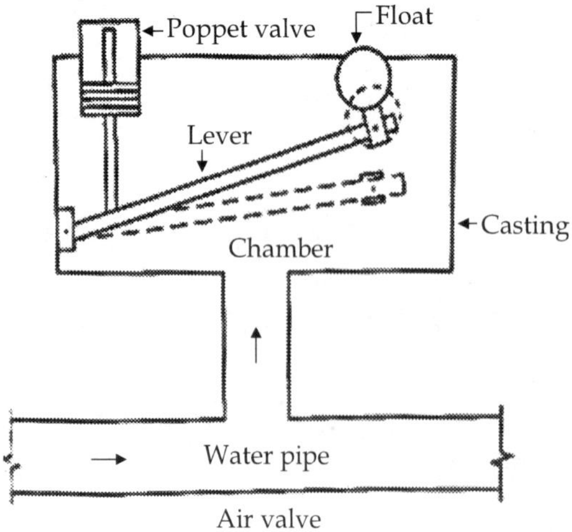

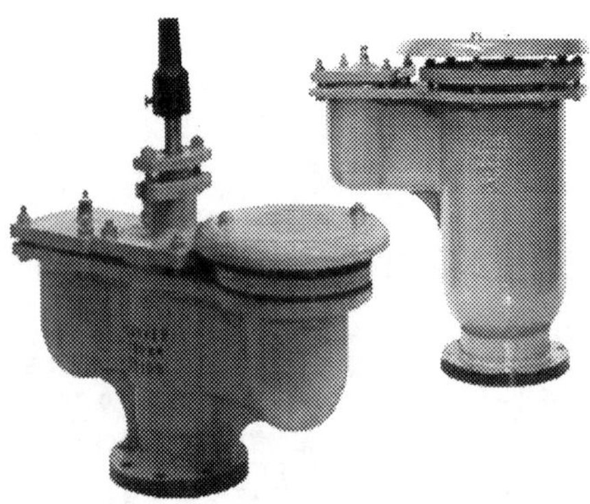

Fig. 9.6: Air relief valve (air valves in piping systems)

This valve consists of a chamber in which one or two floats are mounted and attached to the pipeline. Within the pipeline water fills the float chamber while there is flow under pressure, which makes the float close the outlet. Yet where the air in the tank accumulates, air reaches the chamber, causing the float to descend, thereby opening the exit. The collected air is discharged through the outlet.

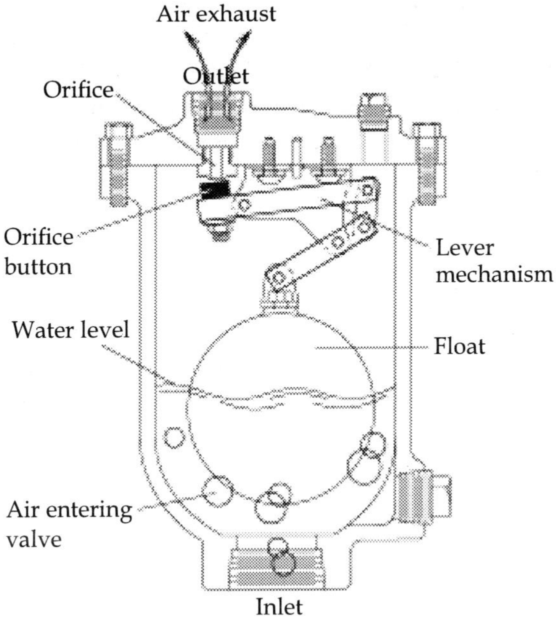

Fig. 9.7: Typical air relief valve

9.1.3 Drain Valves or Blow-off Valves

These are also called wash out valves they are provided at all the dead ends and depression of pipelines to drain out the wastewater. These are ordinary valves operated by hand.

9.1.4 Scour Valves

The scour valves or blow-off valves or drain valves are needed for the pipe being fully cleaned or drained to remove the accumulated sand or silt in the pipe and for inspection, repairs and maintenance. They are ordinary sluice valves found in the pipeline at the dead ends and depressions, or low points. We discharge or drain into a sump from which water can be pumped to waste. To avoid the risk of contamination of water in the tube, there should be no direct contact between the valve and the sewer or drain, and two scour valves are mounted in series as well. The outlet into the channel will be higher above the high level of water. When the outlet needs to be under high water, however, a test valve has to be installed to avoid backflowing.

9.1.5 Pressure Relief Valves

The pressure relief valves, also known as overflow towers, are installed to maintain the pressure below a specified value in a pipeline and thus protect it from the possible risk of bursting due to excessive pressure. If the pressure in a pipe reaches a specified value, the valve automatically opens and allows some amount of water to flow out of the pipe, thus reducing the pressure in the pipe. A pressure relief valve consists of a disk charged to the spring. The load on the disk can be changed by releasing or compressing a handle to the spring. The valve is placed on an opening at the top of the pipeline. As long as the pressure in the pipe is less than the design value, the disk remains securely attached to the opening and the valve remains closed. Therefore such valves are generally used in order to eliminate the water hammer pressure from a conduit when water flow is stopped suddenly.

As the pressure in the pipe reaches the design value, the disk is forced to rise and some amount of water is allowed to flow out of the pipe, thus decreasing the pressure in the pipe. With the pressure reduction in the pipe the disk will be pulled back to suit against the opening and the valve will be closed again. The pressure-relief valves are positioned at any point in the pipeline where maximum pressure is likely. Thus, these valves are also mounted at appropriate intervals in the pipeline at low points, where the pressure is high. In addition, on the upstream side of a sluice valve, a pressure relief valve is normally given so that the pipe on the upstream side of the valve is released from the water hammer pressure resulting from the sudden closing of the sluice valve. Generally speaking, the pressure relief valves are most effective on small pipelines for which the escape of a fairly limited volume of water would ease water hammer pressures. The pressure relief valves, however, are not sensitive enough to rapid water hammer pressure fluctuations. Various other devices such as air vessels, surge tanks, etc., are used.

9.2 MANHOLES

Manholes are provided along the pipeline at appropriate intervals. They are beneficial during building and used for inspection and repair later on. They are normally located in wide pipelines between 300 and 600 m apart. Cross section of rectangular manhole is shows in Fig. 9.8.

Their most important positions are at the summits and the main valves downstream. These are commonly supplied for steel and concrete pipelines, and are less common in cast iron and cement pipelines for asbestos.

9.3 FIRE HYDRANTS

A fire hydrant is an outlet installed in a pipeline to pump water mainly for firefighting (or fire extinguishing) purposes. Sometimes, however, these can also be used to remove water for other purposes, such as street sprinkling, street flushing, etc.

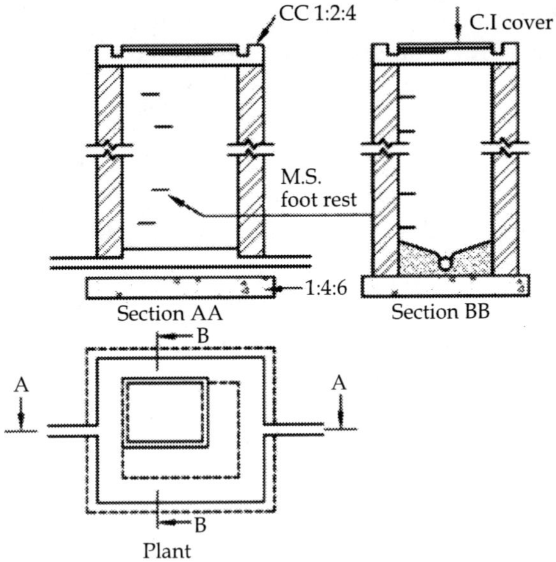

Fig. 9.8: Rectangular manhole for 0.9 m × 0.8 m clear in plan and depth less than 0.9 m (*Source*: CPHEEO,1993)

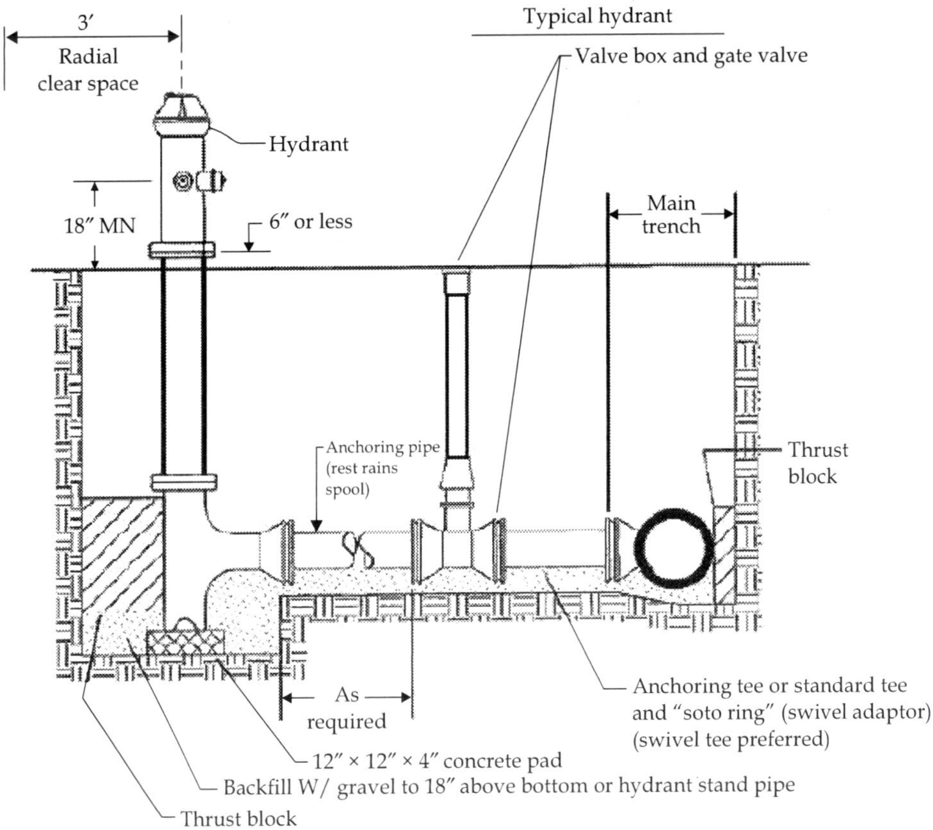

Fig. 9.9: Fire hydrants

Water is collected from a nearby fire hydrant via a fire hose when a fire breaks out. Typically, large amount of high pressure water is needed for firefighting in order to make it enter the place of fire occurrence. Therefore, if water is available from a fire hydrant at the appropriate pressure, it can be used directly for firefighting by means of a fire hose connected to the fire hydrant outlet. However, when water is needed at a much higher pressure, the same is produced by adding a fire engine or pump to the fire hydrant outlet. Figure 9.9 shows a Mueller Super Centurion 250 Fire Hydrant which has three ports (or nozzles): one 4-1/2″ pumper nozzle used to supply water to a pumper truck, and two 2-1/2″ hose nozzles when the hydrant is used directly.

The fire engine or pump draws water from the fire hydrant enhances the pressure and the high pressure water from the fire engine outlet, or the pump is used for firefighting by a fire hose connected to the fire engine outlet or pump outlet. A nozzle is installed at the end of the fire hose to produce a powerful stream of water. The number and location of fire hydrants in a distribution system depend*s on different factors, such as* chances of fire occurrence, firefighting water requirement, building need, area population, etc. Generally, fire hydrants are located at all significant junctions of the road and at intervals not exceeding 300 m.

9.4 WATER METERS

These are the instruments that are mounted on the pipes to calculate the amount of water that flows through the pipe at a given point. The readings obtained from meters allow the amount of water supplied to be figured out and hence the customers can be charged accordingly. Typically, the water meters are built to supply water to factories, hotels, major institutions, etc. Metering avoids the wasting of filtered water.

The water meters may be classified into the following two categories:

- Inferential type of meters or velocity meters
- Displacement type meters

9.4.1 Inferential Type of Meters or Velocity Meters

The inferential style meters or velocity meters operate on the theory that the discharge or rate of water flow through a pipe of a given cross-sectional area is proportional to the flow velocity, higher the flow velocity more would be the discharge. The commonly used inferential type meters or velocity meters are:

(a) Rotary or turbine meters
(b) Venturi meters

Rotary or Turbine Meters

A rotary meter used for limited domestic flow measurement consists of a limited wheel with a set of radial blades or vanes and mounted on a shaft. It is wrapped in a box fitted with inlet and outlet. The water which flows through the meter causes the wheel to rotate. The number of revolutions per unit time the wheel will make will depend on the velocity of the water flow, i.e. the higher the flow velocity will be the rotation velocity and vice versa. The wheel rotation actuates a clock-like mechanism through a gear system, and the amount of water flowing through the meter is registered through a dial system. The meter is set such that the total amount of water flowing through it for a given period of time is read directly. These meters may be used to measure small domestic supplies as well as large commercial and industrial supplies. Further, these meters can be used even if water contains suspended particles or sediments. The precision of these meters is, however, comparatively less at low flows.

Venturi Meters

The basic theory on which a venturi meter operates is that a pressure difference is created by the cross-sectional area of the pipe, and the pressure difference calculation helps to calculate the discharge through the pipe. A venturi meter consists of:

- an inlet segment followed by a convergent cone;
- a cylindrical throat; and
- a gradually divergent cone.

Inlet portion of a venture meter's is of the same diameter as the pipe, followed by a convergent cone. The converging cone is a small pipe that tapers from the pipe's original size to that of the venturi meter throat. The venturi meter throat is a short parallel sided tube with less of its cross-sectional area than the drain. The venturi meter's divergent cone is a increasingly divergent pipe with its cross-sectional area rising from that of the throat to the original pipe level. Pressure taps are provided at the inlet and throat portion of the venturi meter to allow the calculation of the difference in pressure between those parts. The discharge or the rate of flow of water through a venturi meter is given by Eq. 9.1:

$$Q = KC\sqrt{h} \tag{9.1}$$

where,

Q = Discharge through venturi meter

K = Constant of venturi meter

C = Coefficient of discharge of venturi meter

h = Difference of pressure head between the inlet and the throat sections of venturi meter.

A venturi meter is preferably used for measuring the high flows in large pipes, especially for the pipes carrying raw water from the source to the water treatment plant. It is, however, not suitable for measuring small flows.

9.4.2 Displacement Type Meters

A displacement style meter tests the rate of water flow by recording the number of times it fills and empties a container of known volume. Different displacement type meters are available, which include reciprocating type, rotary type, oscillating type and nutating-disk meters, depending on the type of motion of the moving component inside the meter.

Nutating-Disc Meter

A nutating-disk meter consists of a hard rubber disc mounted within a chamber that is supplied with inlet and outlet. The water that enters the chamber causes the disk to oscillate with a circular motion around its centre. The oscillations imparted by one complete filling and emptying are reported by the meter in terms of the volume of water flowing through it through a gear system.

9.5 WATER SUPPLY PLUMBING SYSTEMS IN BUILDING AND HOUSES

It is necessary to know the following terms relating to plumbing, principles and the common practices used in the house plumbing.

- Water main: A water supply pipe vests in the administrative boundaries for the public use.
- Ferrule: It is gun-metal or bronze that is screwed into the hole in the CI pipe mains. Connection pipe departs from the ferrule. Through changing the opening of the ferrule, the strain in the domestic supply and fair distribution between the house link is affected. Usually, the opening of the ferrule in the contact pipe is equal in area to the area of flow.
- Saddle: It is used in place of ferrule for mains of AC or PVC pipes.
- Communication pipes: It is a house attachment pipe that takes off from the ferrule. The water-supply authority owns and operates it. The contact pipe stops at the customer premises boundary.
- Service pipe: After the stop cock it is the part of the house relation. This is owned and held by the owner. There are no pumps mounted on the pipe.
- Water meter: It is built for flow analysis. It is an integrated meter which records the total flow up to the measurement time. In general, rotary water meters of 12.5 mm to 18.75 mm are mounted either at the beginning of the service pipe or at the middle. A masonary pit is built around it. It has sealing facility by the water supply authority.
- Residual pressure: It is generally measured at the ferrule and should be about 7 m head of water.
- Goose nech: This is the short bent pipe which allows for minor length shifts due to expansion which movement of pipes due to settling of the soil.

9.5.1 Plumbing Systems in Water Supplies

The following are the requirements of plumbing systems in water supplies

- Plumbing of water lines should be such that the flowback from cistern and sinks is not permitted. All joints should be completely watertight, and no leakage or spill should be allowed under taps or cocks.
- Pipelines should not be carried under walls or foundations. It should not be near sewers or the disposal of waste water. Cross-connections should be no possibility.

- If pipe lines are connected to electrical cables adequate insulation procedures should be followed.
- Plumbing lines should be such that simple inspection and repair of fixtures and joints should be made possible.
- Number of joints should be less and the number of bends and tees should be less.
- This will provide sufficient material-efficient and corrosion-protected discharge at installations, air lock, negative pressure and noise due to tube flow and flushing.

9.5.2 House Water Connections

House water connection is shown in Fig. 9.10.

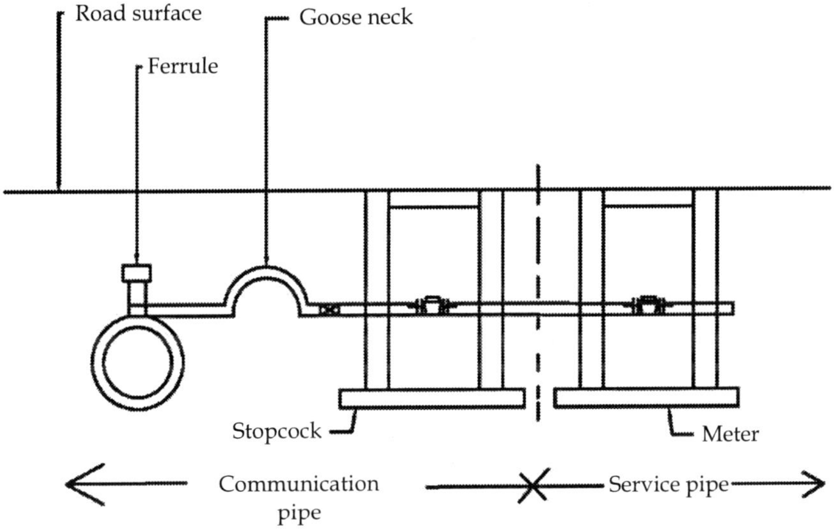

Fig. 9.10: House water connection

(i) **Stop Cocks:** It is a valve that is mounted at the end of the contact pipe and it is under the jurisdiction of water supply regulations. The purpose of stop cock is to stop water supply. Temperature disconnections occur at the stopcock while permanent disconnections occur at the ferrule. Figure 9.11 shows the section of stop cock.

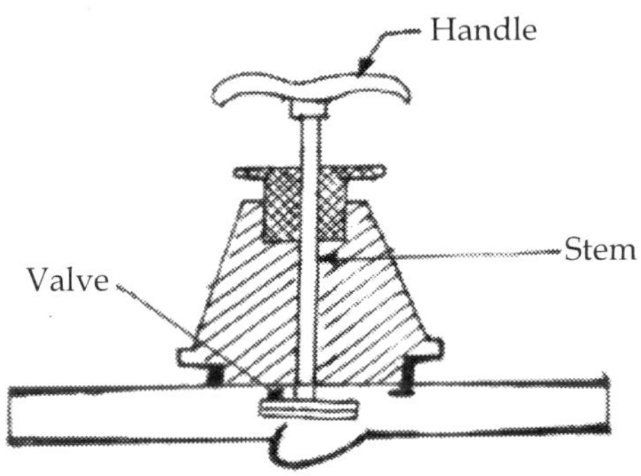

Fig. 9.11: Stopcock

(ii) Water taps or bib cocks: Water taps that are located at the end of the water pipes and from which water is collected by customers. It is works with a handle, with the water coming out of the gate. The bibcocks can also be of push type and are automatic in action. Figure 9.12 shows the section of bib cock.

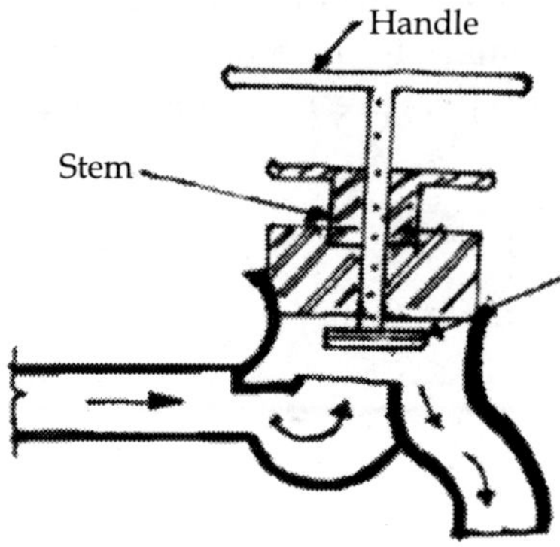

Fig. 9.12: Bib cock

The cocks of the bib should be close to water. The leaky bib cocks trigger water waste.

9.5.3 Pipe Fittings

In addition to the pipes, valves, tapes, various types of pipe fittings such as unions, caps, plugs, flanges, nipples, crosses, tees, elbows, bends, etc., are used during laying of distribution pipes as shown in Fig. 9.13.

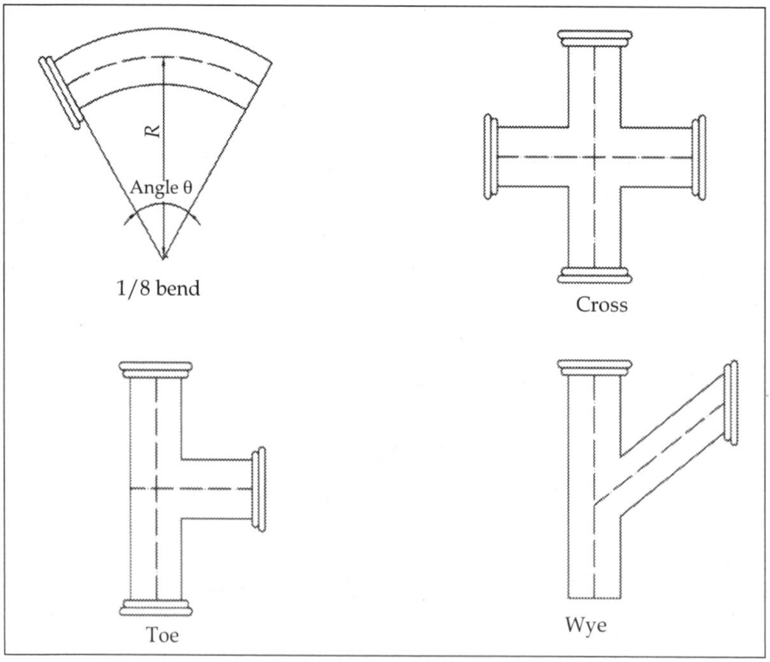

Fig. 9.13: Pipe fittings

9.6 STORAGE OF WATER IN BUILDINGS

In the buildings, the storage of water is required for the following purposes:

- For supplying water to the consumers during non-supply hours.
- For reducing the maximum rate of demand on the water mains.
- For storage of water during interruption to damage repair, etc., of the water main, when the available head is insufficient to supply the water in each storey in multistorey buildings.

Water storage in buildings is made of cast iron, wrought iron, galvanized stainless steel plates or water tanks from R.C.C. Storage tanks can be stored on the roofs or on the floor, and should be watertight. In such a location the storage tank should be mounted so that the water discharge can be readily seen. To clean up the tank, the tank should be fitted with overflow pipe and drain pipe close to the bottom. The storage tanks are fitted with drain pipes for drawing the water.

9.6.1 Estimating Storage Capacity

The quantity of water to be stored depends on the following factors:

- Rate of supply of water from water works
- Type of building such as residential, public or industrial
- Whether water supply is continuous or intermittent
- Frequency replenishment of overhead tanks, during the 24 hours As per IS 2065-1963 the storage capacities are given in the Tables 9.1 and 9.2.

Table 9.1: Flushing storage capacity

S. No.	Classification of Buildings	Storage Capacities
1.	For tenements having common conveniences	900 L net per w.c. seat
2.	For residential premises other than tenements having common conveniences	270 L net for w.c. seat and 180 L for each additional w.c. seat in the same flat
3.	For factories and workshops	900 L per w.c. seat and 180L per urinal seat
4.	For cinemas, public assembly halls, etc.	900 L per w.c. seat and 350L per urinal seat

Table 9.2: Domestic storage capacities

S. No.	Floor	Storage	Remarks
1.	For premises occupied tenements with common conveniences		
1a.	Ground floor	Nil	Providing no downtake fitting installed
1b.	1st , 2nd, 3rd, 4th and upper floors	500 L per tenement	
2.	For premises occupied as flats of block		
2a.	Ground floor	Nil	Providing no downtake fitting installed
2b.	1st , 2nd, 3rd, 4th and upper floors	8000 L per tenement	

9.6.2 Overhead Storage, Under Ground Storage Tanks

When water is to be distributed at very high pressure elevated tanks may be constructed with steel or R.C.C. R.C.C elevated tanks are very popular because of following reasons:

- Long life
- Little maintenance
- Decent appearance

Recently pre-stressed R.C.C. tanks complement each other, as they are more economical than regular R.C.C tanks. Inlet, outlet, drain pipe, overflow pipe, water level indicator, manhole, ladder, ventilating pipe, lighting conductor, etc., are installed for all overhead tanks. Approximately 60 to 100 cm long balconies are built around the tank for inspection and maintenance. These tanks can store large quantity of water as shown in Fig. 9.14.

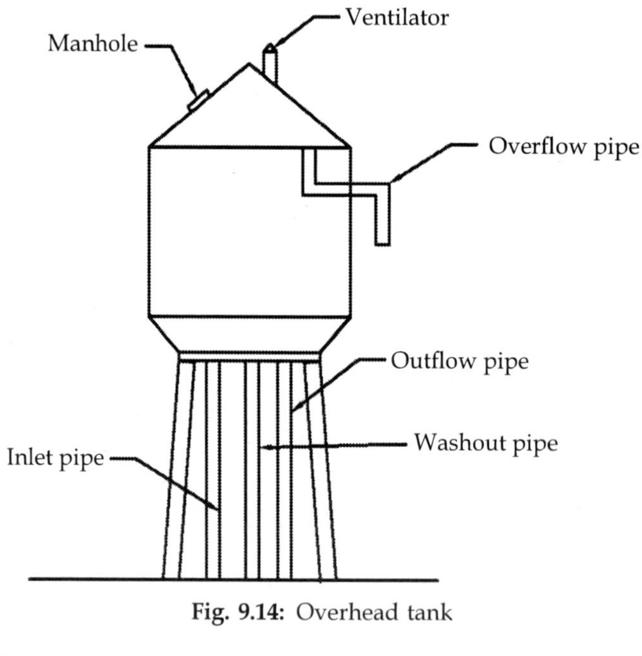

Fig. 9.14: Overhead tank

Underground storage reservoir

Such reservoirs are used for clear water storage and distribution. Such reservoirs are constructed on high natural grounds and are usually made of concrete blocks, bricks, simple or reinforced cement. The lateral walls are built to take up the water pressure when the reservoir is full and the earth pressure when empty. The groundwater level location is often taken into account when constructing such reservoirs. Such reservoir floors can be built with R.C.C slab or square blocks of stone resting on columns. Bitumen compounds are used at all building joints to get water tightness. Around 60 cm thick earth layer is built on the top of the roof and green lawns are preserved from heat and cold. For aeration of water and inspection, ventilation pipes and stairs are provided respectively as shown in Fig. 9.15.

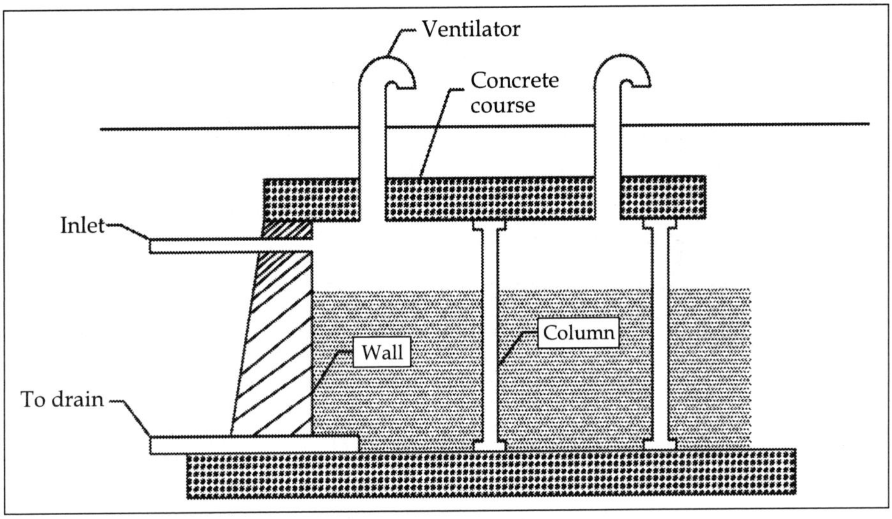

Fig. 9.15: Underground reservoir

Types of tanks

(a) **R.C.C. tanks:** R.C.C. tanks are popular because of following reasons:

- They have long life
- Very little maintenance
- Decent appearance

(b) **G.I. tanks:** G.I. tanks are generally in rectangular or square in shape. Nowadays G.I. tanks are not preferred because of following reasons:

- Life of the tank is short
- Corrosion of metal
- Maintenance cost may be more

(c) **HDPE tanks:** HDPE tanks are very popular for storing small quantity of water and hence useful for residential purpose. The following are the advantages of HDPE tanks:

- Handling is easy because of light weight
- Cheap in cost
- Maintenance cost is low
- Cleaning of tanks are easy

General requirement of domestic water storage

- The filtered water is stored until it is distributed.
- To absorb the hourly fluctuations in the demand for water and thus allow the treatment units and pumps to operate at the average constant rate. This raises the expenses of care service & repair as well as increase their performance.

- To meet the demands for water during fires.
- In case of pump failure, water will be supplied to fix the storage reservoir.

9.7 WATER PIPING SYSTEM IN BUILDING

The following are the requirements of piping system in buildings.

- Waterline plumbing should be such as not to allow backflow from cisterns and sinks.
- All joints must be completely watertight and no leakage or spill should be allowed at taps or cocks.
- Do not hold pipelines under walls or foundations.
- It should not be near sewers or the disposal of waste water. Cross-connections should not be necessary. If pipelines are connected to electrical cables, adequate insulation procedures should be observed.
- Plumbing lines should be such that the inspection and repair of the fixtures and joints should be simple.
- The number of joints will be lower and the number of bends and tees fewer.

Piping system using direct supply: When the residual pressure at the ferrule is greater than 7 m and there is continuous supply in the mains, water can be supplied directly from the service pipe for various fixtures for a single storey house.

Piping system using over head tanks: If the supply is intermittent and the residual pressure is low then water is pumped into overhead tanks and then supplied by gravity to the distribution pipes at the necessary pressure.

Piping system using underground and overhead tank supply: If the supply is intermittent and the residual pressure is low, then a ground level storage tank and an overhead storage tank are installed to supply water. Water from the overhead tank is drawn by pushing pipes down and then into the mounting pipes.

Pumped systems: If the residual pressure is less than 7 m at the ferrule and there is continuous supply in the mains, water can be supplied by pumping from the service pipes.

EXERCISE

1. Distinguish between expansion joints and flanged joints.
2. Write short notes on any three of the following:
   (a) Metering in distribution system
   (b) Level measurement
   (c) Pressure measurement devices and instruments
   (d) Management information system (MIS)
   (e) Leakage control

3. What do you understand by the term water auditing? How water auditing is helpful in water saving and economical point of view for industries and commercial usage?
4. What are the different methods of monitoring of leakage and its control? Define the various causes of failure of pipelines in water treatment system.
5. What are the basic objectives for operation and maintenance of water treatment plant? Discuss the checklist required in connection with the operation and maintenance of water treatment plant.
6. Distinguish between RCC pipes and cast iron pipes? Why copper pipes are used to convey hot water.
7. What are the different types of air relief valves used in a water distribution system for public water supply.
8. Describe the importance of drain valves or blow-off valves used in water distribution system.
9. Write short notes on:
   (a) Sluice valves
   (b) Butterfly valves
   (c) Globe valve
   (d) Reflux valve
   (e) Scour valves
   (f) Pressure relief valves
10. Describe manholes and fire hydrants used in water supply and sanitation system?
11. Enumerate the importance and significance of different types of water meters used in municipal water supply?
12. Describe various water fittings used in water supply systems?

REFERENCES

1. *Hard Hat Engineer.* (2020). Retrieved from Types of Gate Valve and Parts - A Complete Guide for Engineer: https://hardhatengineer.com/gate-valve-types-parts/
2. *Air Valves in Piping Systems.* (n.d.). Retrieved from https://www.valvemagazine.com/magazine/sections/back-to-basics/9170-air-valves-in-piping-systems.html
3. *Check Valves.* (n.d.). Retrieved from Flow Control: https://www.flowcontrol.no/valves/check-valves/
4. *Globe Valves.* (n.d.). Retrieved from Flow Control: https://www.flowcontrol.no/valves/globe-valves/

5. Louzada, R. (2017). A Repairable System Subjected to Hierarchical Competing Risks: Modeling and Applications. *IEEE Access*.
6. Modeling a Fire Hydrant by Engineered software knowledge base. http://kb.eng software.com/eskb/pipe-flo/modeling-piping-system-devices/modeling-a-fire-hydrant

Index

D

E

F

G

H

I